17th Annual International Astrophysics Conference (AIAC 2018)

Journal of Physics: Conference Series
Volume 1100

Santa Fe, New Mexico, USA
5 - 9 March 2018

Editor:

G.P. Zank

ISBN: 978-1-5108-7436-7
ISSN: 1742-6588

Printed from e-media with permission by:

Curran Associates, Inc.
57 Morehouse Lane
Red Hook, NY 12571

Some format issues inherent in the e-media version may also appear in this print version.

Institute of Physics
Dirac House, Temple Back
Bristol BS1 6BE UK

Phone: 44 1 17 929 7481
Fax: 44 1 17 920 0979

techtracking@iop.org

Additional copies of this publication are available from:

Curran Associates, Inc.
57 Morehouse Lane
Red Hook, NY 12571 USA
Phone: 845-758-0400
Fax: 845-758-2633
Email: curran@proceedings.com
Web: www.proceedings.com

Table of contents

Volume 1100

The 17th Annual International Astrophysics Conference

5–9 March 2018, Santa Fe, New Mexico, USA

Accepted papers received: 7 September 2018
Published online: 24 October 2018

Preface

Papers

IOP Conf. Series: Journal of Physics: Conf. Series **1100** (2018) 011001 doi:10.1088/1742-6596/1100/1/011001

Preface

G P Zank
Editor, *Proceedings of the 17th Annual International Astrophysics Conference,*
Center for Space Plasma and Aeronomic Research (CSPAR) and Department of Space Physics, University of Alabama in Huntsville, Huntsville AL 35805, USA

E-mail: garyp.zank@gmail.com

The 17th Annual International Astrophysics Conference was held at the La Posada de Santa Fe hotel, Santa Fe, New Mexico, USA from March 5 to 9, 2018. The meeting entitled, "Dissipative and Heating Processes in Collisionless Plasma: the Solar Corona, the Solar Wind, and the Interstellar Medium," addressed the theme of the transfer of energy from large-scales to small-scales, whether in the solar corona, the solar wind, or the interstellar medium. The topic is of critical importance in our understanding of the energetics and dynamics of collisionless plasma. The transfer of energy can take the form of turbulence in which the energy input at the largest scales is eventually transferred to very small or dissipative scales by some form of cascade mechanism, or from the generation of large compressible structures such as shock waves that dissipate energy on small scales associated with steepened structures, or from the energization of particles via a coupling of large- and small-scale processes, of which diffusive shock acceleration represents the best established example. The 17th AIAC explored these topics via a combination of invited 25 minute talks and a few 40 minute plenary talks. Of particular relevance to the meeting was the just-launched Parker Solar Probe and Solar Orbiter missions.

The 17th AIAC also celebrate Prof. Len Fisk's 75th birthday and his over 50 years engagement in space physics. Len has been one of our most important leaders intellectually, academically, administratively, and politically, and this represented an opportunity to celebrate Len's accomplishments with some selected plenary presentations. Len has written an engaging, entertaining, and informative overview of his life in science, with a good deal of useful advice to young and old alike.

Finally, we would like to thank Adele Corona and ICNS for her continued excellent organization of the AIAC meetings and her help in providing the logistical support for this volume of papers.

IOP Conf. Series: Journal of Physics: Conf. Series **1100** (2018) 011002 doi:10.1088/1742-6596/1100/1/011002

Peer review statement

All papers published in this volume of *Journal of Physics: Conference Series* have been peer reviewed through processes administered by the proceedings Editors. Reviews were conducted by expert referees to the professional and scientific standards expected of a proceedings journal published by IOP Publishing.

Evolution of Power Anisotropy in Magnetic Field Fluctuations at Different Solar Activity Levels

L Adhikari[1], G P Zank[1,2], L L Zhao[1], D Telloni [3], P Hunana[1,2], and D Shiota[4,5]

[1]Center for Space Plasma and Aeronomic Research (CSPAR), University of Alabama in Huntsville, Huntsville,
AL 35899, USA
[2]Department of Space Science, University of Alabama in Huntsville, Huntsville, AL 35899, USA
[3]INAF - Astrophysical Observatory of Torino, Via Osservatorio 20, 10025 Pino Torinese, Italy
[4]Institute for Space-Earth Environmental Research, Nagoya University, Nagoya, Aichi 464-8601, Japan
[5]National Institute of Information and Communications Technology (NICT), Koganei, Tokyo, 184-8795, Japan

E-mail: la0004@uah.edu

Abstract. Adhikari et al [33] recently developed a theoretical model to study anisotropy in magnetic field fluctuations using a dimensional analysis relating the power spectrum of the energy-containing and inertial ranges. We use the Adhikari et al and Zank et al [31] models to study the evolution of the power anisotropy in magnetic field fluctuations in the energy-containing and inertial ranges at different levels of solar activity. We obtain initial conditions at 1 au for the times 2003, 2009, and 2015 from Zhao et al [37] by assuming an 80:20 ratio between the turbulence energies, and a 2:1 ratio between the correlation lengths. The years 2003 and 2015 correspond to solar maxima and the year 2009 to a solar minimum. We find that the anisotropy in magnetic field fluctuations evolves differently during the solar minimum than during the solar maximum throughout the heliosphere.

1. Introduction

Anisotropy, an intrinsic property of the solar wind, arises due to a presence of a large-scale magnetic field $\mathbf{B_0}$ [1 - 3]. Anisotropy in the presence of $\mathbf{B_0}$ is indicated by unequal variances in the velocity and magnetic field fluctuations along the direction parallel and perpendicular to $\mathbf{B_0}$. Anisotropy is an important local property of the solar wind turbulence. Understanding the anisotropy of solar wind fluctuations may assist us in understanding solar wind turbulence and the propagation and acceleration of cosmic

IOP Conf. Series: Journal of Physics: Conf. Series **1100** (2018) 012001 doi:10.1088/1742-6596/1100/1/012001

rays [4 - 6]. Several theoretical and numerical methods [7 - 12] have been utilized to study anisotropy in a magnetized plasma such as the solar wind or a laboratory plasma. Some studies [13 - 18] investigate the anisotropy by calculating the energy spectrum of magnetohydrodynamic (MHD) turbulence. Others consider the variance of magnetic field fluctuations in directions parallel and perpendicular to the large-scale magnetic field [19 - 21].

The availability of single- and multi-spacecraft measurements has significantly increased our understanding of anisotropy in solar wind fluctuations [22 - 27]. The anisotropy of solar wind fluctuations in the fast and slow solar wind [28], and solar maximum and solar minimum [29] is found to be different. Moreover, the anisotropy may also depend on helio-latitude. Fluctuations in the fast wind tend to be more isotropic, while fluctuations in the slow wind are more anisotropic. Zank and Matthaeus [30] (and other studies [4, 28]) suggested that the solar wind at 1 au contains a majority of 2D fluctuations with a minority slab component. These studies put forward an 80:20 ratio between the 2D and slab turbulence, but this ratio can vary due to the dependency of anisotropy on the fast and slow solar wind, solar maximum and minimum, and higher and lower latitude.

Zank et al [31] (see also [32, 33, 34]) developed 2D and slab turbulence transport equations using a nearly incompressible (NI) phenomenology based on the inhomogeneous MHD model of Hunana and Zank [35]. Zank et al [31], and Adhikari et al [33] investigated the 2D and slab turbulence intensities using an 80:20 energy ratio between the 2D and slab turbulence at 1 au. Later, Adhikari et al [34] solved the 2D and slab NI turbulence transport equations throughout the heliosphere using 80:20, 70:30, 60:40, and 55:45 initial energy ratios between the 2D and slab turbulence at 1 au. Adhikari et al [34] found that the turbulence energies change a little, and the trends are similar. Adhikari et al [36] using the NI turbulence model of Zank et al [31] presented the evolution of anisotropy in magnetic field fluctuations in the energy-containing and inertial ranges throughout the heliosphere. Adhikari et al [36] found that an anisotropy in magnetic field fluctuations exists within $\sim$ 20 au, but the fluctuations in the outer heliosphere revert to an almost isotropic state. In this paper, we study the anisotropy in magnetic field fluctuations for different solar activity levels using the boundary conditions that Zhao et al [37] derived observationally. Table 2 of Zhao et al shows the values of the turbulence quantities at 1 au for the years 2003, 2009, and 2015. We note that the years 2003 and 2015 correspond to solar maxima and year 2009 to solar minimum. The 2003 solar maximum is stronger than the 2015 solar maximum. We present a theoretical model in Section 2, and discuss our results in Section 3. Finally, we summarize our results and present conclusions in Section 4.

2. Theory

In this section, we present a theoretical model for the anisotropy in magnetic field fluctuations in the inertial range. The details of the derivation of the model are described in Adhikari et al [36]. The variances of the 2D and slab magnetic field fluctuations are given by [36, 38]

$$\langle B_{2D}^2 \rangle = 2\pi \int g^{2D}(\mathbf{k}_\perp) d\mathbf{k}_\perp \quad \text{and} \quad \langle b_{sl}^2 \rangle = 4\pi \int g^{slab}(k_{||}) dk_{||}, \tag{1}$$

where $\mathbf{k}_\perp$ and $k_{||}$ are the perpendicular and parallel wave vectors. g is a function of the perpendicular and parallel wave number. We assume a power spectrum $g = Ck^{-1}$ and $g = Dk^{-5/3}$ in the energy-containing and inertial range, respectively, and C and D are constants. Suppose $g^{2D} = C^{2D}k_\perp^{-1}$ and $g^{slab} = C^{slab}k_{||}^{-1}$ for the 2D and slab turbulence, respectively, and then integrate Equation (1) from k_{inj} to k_b to obtain [36]

$$C_{ER}^{2D} \equiv \frac{\langle B_{2D}^2\rangle ER}{\log\left(\frac{(l_b^{2D})^{-1}}{k_{inj}}\right)} \quad \text{and} \quad C_{ER}^{slab} \equiv \frac{\langle b_{slab}^2\rangle ER}{\log\left(\frac{(l_b^{slab})^{-1}}{k_{inj}}\right)}, \tag{2}$$

where ER indicates energy-containing range, k_{inj} is a large-scale injection wave number, $k_b(\sim l_b^{-1})$ is a wave number that separates the energy-containing and inertial range, and l_b^{2D} and l_b^{slab} are the correlation lengths of the variances of the 2D and slab magnetic field fluctuations. From Equation (2), we obtain the ratio between the C_{ER}^{2D} and C_{ER}^{slab} as [36],

$$\frac{C_{ER}^{2D}}{C_{ER}^{slab}} = \frac{\langle B_{2D}^2\rangle ER}{\langle b_{slab}^2\rangle ER}\frac{\log\left(\frac{(l_b^{slab})^{-1}}{k_{inj}}\right)}{\log\left(\frac{(l_b^{2D})^{-1}}{k_{inj}}\right)}. \tag{3}$$

Equation (3) describes the ratio of the 2D and slab variances of magnetic field fluctuations in the energy-containing range. Now, we calculate the ratio of the variances of the 2D and slab magnetic field fluctuations in the inertial range. Since the wave number k_b separates the energy-containing and inertial range, we can write $Ck^{-1}|_{k_b} = Dk^{-5/3}|_{k_b}$, which yields $D = Ck_b^{2/3}$. Therefore, the power spectrum for the 2D and slab turbulence in the inertial range can be written as $g^{2D} = C_{ER}^{2D}(k_b^{2D})^{2/3}k_\perp^{-5/3}$ and $g^{sl} = C_{ER}^{sl}(k_b^{sl})^{2/3}k_{||}^{-5/3}$, respectively. We use g^{2D} in Equation (1) and integrate the equation from k_1^{2D} to k_2^{2D}, so that we consider the anisotropy in a wave number range $[k_1^{2D}, k_2^{2D}]$. Here, k_1^{2D} and k_2^{2D} are arbitrary wave numbers, and the $k_b^{2D} < k_1^{2D} < k_2^{2D}$ inequality is satisfied. The variance of the 2D magnetic field fluctuations is [36]

$$\begin{aligned}\langle B_{2D}^2\rangle IR &\equiv C_{ER}^{2D}(k_b^{2D})^{2/3}\int_{k_1^{2D}}^{k_2^{2D}} k_\perp^{-5/3}dk_\perp, \\ &= \frac{3}{2}C_{ER}^{2D}(k_b^{2D})^{2/3}((k_1^{2D})^{-2/3} - (k_2^{2D})^{-2/3}),\end{aligned} \tag{4}$$

where IR denotes the inertial range. Similarly, we obtain the variance of the slab magnetic field fluctuations as [36]

$$\begin{aligned}\langle b_{slab}^2\rangle IR &\equiv C_{ER}^{sl}(k_b^{slab})^{2/3}\int_{k_1^{slab}}^{k_2^{slab}} k_{||}^{-5/3}dk_{||}, \\ &= \frac{3}{2}C_{ER}^{slab}(k_b^{slab})^{2/3}((k_1^{slab})^{-2/3} - (k_2^{slab})^{-2/3}).\end{aligned} \tag{5}$$

Here, $g^{slab}(k_{||}) = C_{ER}^{slab}(k_b^{slab})^{2/3}k_{||}^{-5/3}$ is used in Equation (1), and the integral is integrated from k_1^{slab} and k_2^{slab}, where k_1^{slab} and k_2^{slab} are arbitrary wave numbers such that $k_b^{slab} < k_1^{slab} < k_2^{slab}$. Now, dividing Equation (4) by Equation (5) yields

$$\frac{\langle B_{2D}^2 \rangle IR}{\langle b_{slab}^2 \rangle IR} = \frac{C_{ER}^{2D}}{C_{ER}^{slab}} \left(\frac{k_b^{2D}}{k_b^{slab}} \right)^{2/3} \frac{((k_1^{2D})^{-2/3} - (k_2^{2D})^{-2/3})}{((k_1^{slab})^{-2/3} - (k_2^{slab})^{-2/3})},$$

and by assuming $k_1^{2D} = k_1^{slab}$ and $k_2^{2D} = k_2^{slab}$, we obtain

$$\frac{\langle B_{2D}^2 \rangle IR}{\langle b_{slab}^2 \rangle IR} = \frac{C_{ER}^{2D}}{C_{ER}^{slab}} \left(\frac{k_b^{2D}}{k_b^{slab}} \right)^{2/3}. \quad (6)$$

Equations (3) and (6) yield the ratio of the variances of the 2D and slab magnetic field fluctuations in the inertial range [36],

$$\frac{\langle B_{2D}^2 \rangle IR}{\langle b_{slab}^2 \rangle IR} = \frac{\langle B_{2D}^2 \rangle ER}{\langle b_{slab}^2 \rangle ER} \left(\frac{l_b^{slab}}{l_b^{2D}} \right)^{2/3} \frac{\log\left(\frac{(l_b^{slab})^{-1}}{k_{inj}} \right)}{\log\left(\frac{(l_b^{2D})^{-1}}{k_{inj}} \right)}, \quad (7)$$

We use Equation (7) to study the anisotropy of magnetic field fluctuations in the inertial range. In Equation (7), the turbulence parameters $\langle B_{2D}^2 \rangle ER$, $\langle b_{slab}^2 \rangle ER$, l_b^{slab}, and l_b^{2D} on the right hand side are associated with the energy-containing range, and are obtained from the NI turbulence transport model equations of Zank et al [31]. These large-scale turbulence quantities are calculated as [31]

$$\left\langle \frac{B^{2D,*2}}{\mu_0 \rho} \right\rangle = \frac{\langle z^{\infty,*+2} \rangle + \langle z^{\infty,*-2} \rangle - 2E_D^{\infty,*}}{4}, \text{ and}$$
$$l_b^{\infty,*} = \frac{L_+^{\infty,*} + L_-^{\infty,*} - L_D^{\infty,*}}{\langle z^{\infty,*+2} \rangle + \langle z^{\infty,*-2} \rangle - 2E_D^{\infty,*}}, \quad (8)$$

where μ_0 is the magnetic permeability of free space and ρ is the solar wind density. In Equation (8), "∞" indicates 2D turbulence, and "*" indicates slab turbulence. The parameters $\langle z^{\pm 2} \rangle$ and E_D are the energies in forward and backward propagating modes, and the residual energy, and $L^{\pm}$ and L_D are the corresponding correlation functions [31]. In Equation (7), $k_{inj} \sim 1.07 \times 10^{-9}$ km^{-1}, which corresponds to one solar rotation ($\sim$ 27 days). Moreover, the stream-shear sources of 2D turbulence for the forward and backward propagating modes, and the residual energy are [31, 33, 39]

$$S_{\langle z^{\pm 2} \rangle} = \frac{C_{sh}^{\pm} r_0 |\Delta U| V_{A0}^2}{r^2}; \quad S_{E_D} = \frac{C_{sh}^{E_D} r_0 |\Delta U| V_{A0}^2}{r^2}, \quad (9)$$

where r_0=1 au, V_{A0} is the Alfvén velocity at 1 au, and $C_{sh}^{\pm}$ and $C_{sh}^{E_D}$ are stream-shear driving constants for the forward and backward propagating modes, and the residual energy. Similarly, the stream-shear and pickup ion sources of slab turbulence are [31, 33, 39]

$$
\begin{aligned}
S_{\langle z^{\pm *2}\rangle} &= \frac{C^{*\pm}_{sh} r_0 |\Delta U| V_{A0}^2}{r^2}; \quad S^*_{E_D} = \frac{C^{*E_D}_{sh} r_0 |\Delta U| V_{A0}^2}{r^2}; \\
S_{\rm PUI} &= \frac{f_D n_H^\infty U V_{A0}}{n^0_{\rm sw} \tau^0_{\rm ion}} \exp(-L/r),
\end{aligned} \tag{10}
$$

where L (=8 au) is the ionization cavity length scale, f_D is the fraction of pickup ion energy transferred into excited waves, and $C^{*\pm}_{sh}$ and $C^{*E_D}_{sh}$ are stream-shear driving constants of the slab turbulence. The parameter n_H^∞ (=0.1 cm^{-3}) is the number density of interstellar neutrals entering the heliosphere, τ^0_{ion}(= 10^6 s) is the neutral ionization time at 1 AU, n^0_{sw} is the solar wind density at 1 AU, and U is the solar wind speed. Strengths of the stream-shear sources of 2D and slab turbulence, and the pickup ion source of slab turbulence are shown in Table 3.

3. Results

In this section, we show the ratio of the variances of the 2D and slab magnetic field fluctuations $\langle B^2_{2D}\rangle/\langle b^2_{slab}\rangle$ in the energy-containing and inertial ranges throughout the heliosphere from 1 to 75 au. First, we solve the NI turbulence transport equations of Zank et al [31] using the boundary conditions and solar wind parameter values shown in Tables 1 and 2, respectively, for three cases; i) no source of turbulence, ii) stream-shear sources of 2D and slab turbulence, and iii) stream-shear sources of 2D and slab turbulence, and a pickup ion source of turbulence. Table 1 shows the boundary conditions for the three years 2003, 2009, and 2015, from which we calculate the ratio $\langle B^2_{2D}\rangle/\langle b^2_{slab}\rangle$ in the energy-containing and inertial ranges for these three years. The boundary conditions are obtained from Zhao et al [37], and we assume an 80:20 energy ratio between the 2D and slab turbulence, and that $\lambda^* = 2\lambda^\infty$ for the correlation lengths, where λ^∞ is the 2D correlation length and λ^* is the slab correlation length.

The ratios of the variances of the 2D and slab magnetic field fluctuations with no source of turbulence, stream-shear sources of 2D and slab turbulence, and stream-shear sources of 2D and slab turbulence and a pickup ion source of slab turbulence are shown in Figures 1, 2, and 3, respectively. In the figures, the blue, red, and green curves correspond to the boundary conditions for the years 2003, 2009, and 2015, respectively. Furthermore, the solid curves denote the $\langle B^2_{2D}\rangle/\langle b^2_{slab}\rangle$ ratio in the inertial range, and the dashed curves the $\langle B^2_{2D}\rangle/\langle b^2_{slab}\rangle$ in the energy-containing range.

Figure 1 shows the evolution of anisotropy in magnetic field fluctuations in the energy-containing and inertial ranges when an interplanetary source of turbulence is not included in the turbulence transport model equations. The energy-containing range results (dashed curves) show that the $\langle B^2_{2D}\rangle/\langle b^2_{slab}\rangle$ ratio during solar minimum (i.e., 2009) increases monotonically with distance. However, the $\langle B^2_{2D}\rangle/\langle b^2_{slab}\rangle$ ratio corresponding to the solar maxima (i.e., 2003 and 2015) decreases slightly between $\sim$ 1.2−2 au and then increases with heliocentric distance. The results indicate that the $\langle B^2_{2D}\rangle/\langle b^2_{slab}\rangle$ ratio for the 2009 solar minimum is larger than the two $\langle B^2_{2D}\rangle/\langle b^2_{slab}\rangle$ ratios corresponding to the 2003 and 2015 solar maxima. Furthermore, the 2003 solar maximum $\langle B^2_{2D}\rangle/\langle b^2_{slab}\rangle$ ratio

Parameters	2003	2009	2015
$\langle z^{\infty+2}\rangle$ (km^{2}s^{-2})	474.98	176.81	303.32
$\langle z^{\infty-2}\rangle$ (km^{2}s^{-2})	1219.34	290.46	503.85
E_D^∞ (km^{2}s^{-2})	-346.23	-90.89	-160.94
L_∞^+ (km^{3}s^{-2})	1.93 $\times 10^8$	3.94 $\times 10^7$	9.2 $\times 10^7$
L_∞^- (km^{3}s^{-2})	6.95 $\times 10^8$	9.36 $\times 10^7$	2.6 $\times 10^8$
L_D^∞ (km^{3}s^{-2})	-4.30 $\times 10^8$	-7.18 $\times 10^7$	-1.5 $\times 10^8$
$\langle z^{*+2}\rangle$ (km^{2}s^{-2})	118.75	44.2	75.83
$\langle z^{*-2}\rangle$ (km^{2}s^{-2})	304.83	72.62	125.96
E_D^* (km^{2}s^{-2})	-86.56	-22.72	-40.23
L_*^+ (km^{3}s^{-2})	9.64 $\times 10^7$	1.96 $\times 10^7$	4.6 $\times 10^7$
L_*^- (km^{3}s^{-2})	3.48 $\times 10^8$	4.68 $\times 10^7$	1.3 $\times 10^8$
L_D^* (km^{3}s^{-2})	-2.15 $\times 10^8$	-3.59 $\times 10^7$	-7.5 $\times 10^7$

Table 1. Boundary conditions for the years 2003, 2009, and 2015 at 1 au. These values are obtained from Zhao et al [37] by assuming an 80:20 energy ratio between 2D and slab turbulence, and a 2:1 ratio between the correlation lengths for slab and 2D turbulence.

Parameters	2003	2009	2015
n (cm^{-3})	5.33	6.02	6.96
U (kms^{-1})	493.31	359.48	412.25
V_A (kms^{-1})	88.10	43.21	66.86
B (nT)	8.14	4.41	7.42
$\Delta U V_A^2$ (km^{3}s^{-3})	1.93 $\times 10^6$	2.29 $\times 10^5$	8.25 $\times 10^5$

Table 2. Solar wind parameter values at 1 au (Zhao et al [37]).

Parameters	case (i)	case (ii)	case (iii)
C_{sh}^+	0	0.9	0.9
C_{sh}^-	0	0.9	0.9
$C_{sh}^{E_D}$	0	-0.5	-0.5
C_{sh}^{*+}	0	0.3	0.3
C_{sh}^{*-}	0	0.3	0.3
$C_{sh}^{*E_D}$	0	-0.17	-0.17
f_+^D	0	0	0.25
f_-^D	0	0	0.24

Table 3. Strengths of the stream-shear sources of 2D and slab turbulence, and a pickup ion source of slab turbulence (Adhikari et al [33]).

is smaller than that of the 2015 solar maximum. Note that the 2003 solar maximum is stronger than the 2015 solar maximum. The energy-containing range anisotropy is larger during solar minimum than solar maximum. The energy-containing range anisotropy is quite sizable throughout the heliosphere, being between $\sim$ 4.5 and $\sim$ 5.5 for all solar conditions (maximum and minimum). The inertial range results exhibit similar

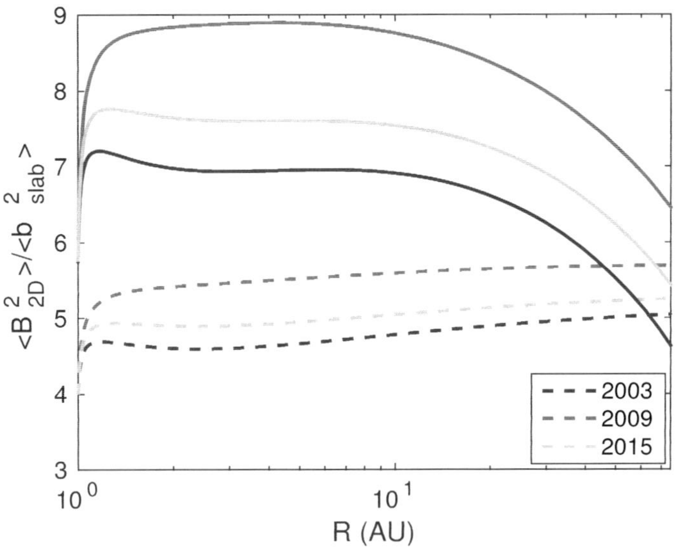

Figure 1. Evolution of the power anisotropy in magnetic field fluctuations as a function of heliocentric distance with no interplanetary source of turbulence. The solid curves correspond to the inertial range and the dashed curves to the energy-containing range.

characteristics for solar minimum and solar maximum. However, the $\langle B_{2D}^2\rangle/\langle b_{slab}^2\rangle$ ratio for the inertial range decreases beyond 10 au.

Stream-shear and pickup ion sources of turbulence are always present, and drive solar wind turbulence throughout the heliosphere [39]. Furthermore, these sources of turbulence depend on heliospheric location in that the stream-shear source of turbulence is effective within ~ 5 au, and the pickup ion source of turbulence beyond ~ 5 au. Figure 2 shows the evolution of $\langle B_{2D}^2\rangle/\langle b_{slab}^2\rangle$ with heliocentric distance when only stream-shear sources of 2D and slab turbulence are included in the NI turbulence transport model. Equations (9) and (10) are the stream-shear and pickup ion sources of 2D and slab turbulence. Zhao et al [37] calculated a form of stream-shear source of turbulence ($\Delta U V_A^2$) observationally at 1 au from 1995 through 2017. Zhao et al [37] found that the stream-shear source of turbulence is stronger during solar maximum than solar minimum. The bottom row of Table 2 shows the values of $\Delta U V_A^2$ for the years 2003, 2009, and 2015, which clearly makes the 2D and slab stream-shear sources of turbulence solar cycle dependent. In this case, the dashed blue, red, and green curves are very similar, unlike the dashed curves in Figure 1. Similarly, all the solid curves (blue, red, and green) are also almost identical. The results show that the anisotropy of magnetic field fluctuations in the inertial range is larger than in the energy-containing range. The anisotropy is ~ 5.5 for the inertial range and slightly less than ~ 4 for the energy-containing range throughout the heliosphere if only a shear-driving source is included.

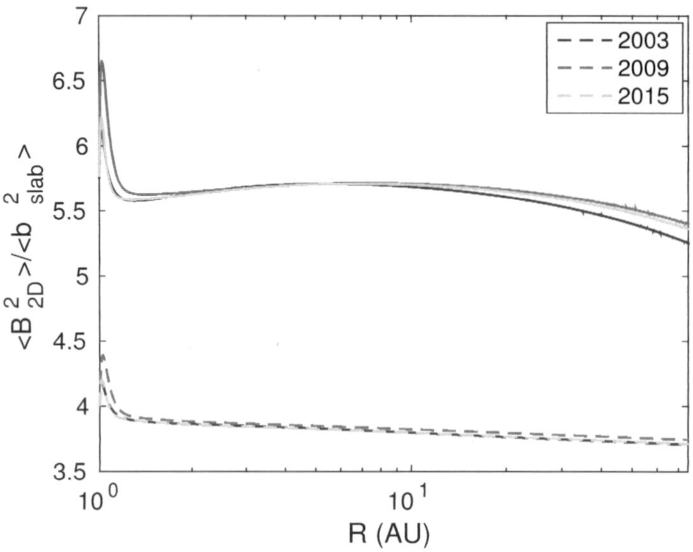

Figure 2. Evolution of the power anisotropy in magnetic field fluctuations as a function of heliocentric distance for a stream-shear source of turbulence. The solid and dashed curves are as in Figure 1.

When a pickup ion source of slab turbulence is also included in the NI turbulence transport model equations, the evolution of the anisotropy in magnetic field fluctuations in the energy-containing and inertial ranges with distance is shown in Figure 3. Equation (10) is a pickup ion source of slab turbulence for our turbulence transport model. The values of the solar wind speed, the solar wind density, and the Alfvén speed in Equation (10) for the years 2003, 2009, and 2015 are shown in Table 3. It makes a pickup ion source of turbulence solar cycle dependent, but a more detailed study is required in order to understand the dependence of pickup ion source of turbulence on solar cycle. We find that the $\langle B_{2D}^2\rangle/\langle b_{slab}^2\rangle$ ratio for the energy-containing and inertial ranges decreases with the inclusion of a pickup ion source of turbulence. However, it is interesting to note that the solid and dashed red curves associated with solar minimum decrease at a smaller heliocentric distance than the solid and dashed blue and green curves associated with solar maximum. Figure 3 indicates that the ratio of the variances of the 2D and slab magnetic field fluctuations for both the energy-containing and inertial ranges beyond ~ 3 au follow the ordering $(\langle B_{2D}^2\rangle/\langle b_{slab}^2\rangle)_{2003} > (\langle B_{2D}^2\rangle/\langle b_{slab}^2\rangle)_{2015} > (\langle B_{2D}^2\rangle/\langle b_{slab}^2\rangle)_{2009}$, and within ~ 3 au the $(\langle B_{2D}^2\rangle/\langle b_{slab}^2\rangle)_{2009}$ ratio is larger than the $(\langle B_{2D}^2\rangle/\langle b_{slab}^2\rangle)_{2003}$ and $(\langle B_{2D}^2\rangle/\langle b_{slab}^2\rangle)_{2015}$ ratios. It clearly shows that the ratio of the variances of 2D and slab magnetic field fluctuations associated with a solar minimum is different from the ratio of the variances associated with a solar maximum. Prior to the onset of pickup ion driving beyond ~ 5 au, the energy-containing and inertial ranges anisotropies are the same as in Figure 2. With the dominance of pickup ion driving turbulence beyond $10 - 20$ au, the anisotropies for both energy-containing and inertial ranges approach ~ 1.

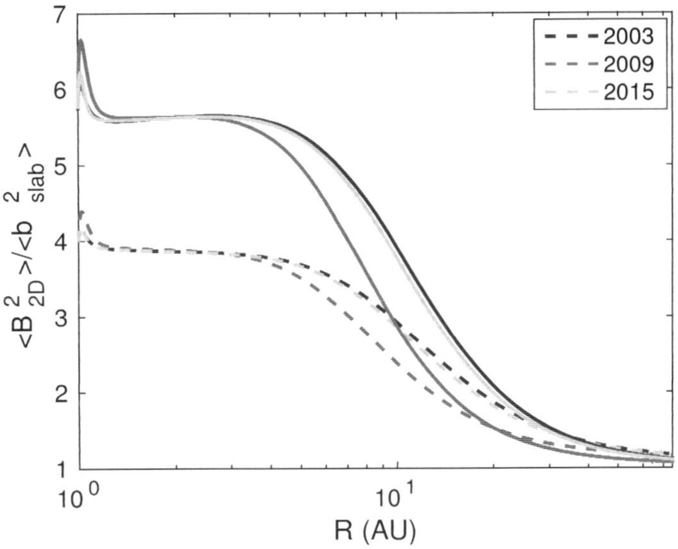

Figure 3. Evolution of the power anisotropy in magnetic field fluctuations as a function of heliocentric distance for both a stream-shear and pickup ion sources of turbulence. The solid and dashed curves are as in Figure 1.

4. Discussion and Conclusions

In this paper, we investigated the evolution of anisotropy in magnetic field fluctuations at different levels of solar activity as a function of heliocentric distance. For this we used the Zank et al [31] NI turbulence transport model equations, the Adhikari et al [33] model for anisotropy of magnetic field fluctuations, and the boundary conditions from the Zhao et al [37] observational analysis. We calculated the ratio of the variances of the magnetic field fluctuations $\langle B_{2D}^2\rangle/\langle b_{slab}^2\rangle$ for three cases; i) no source of turbulence, ii) stream-shear sources of 2D and slab turbulence, and iii) stream-shear sources of 2D and slab turbulence, and a pickup ion source of slab turbulence. We find that the ratios of the variances of the 2D and slab magnetic field fluctuations in the energy-containing and inertial ranges during the 2009 solar minimum are different from the ratios of the variances in the energy-containing and inertial ranges during the 2003 and 2015 solar maxima.

In the absence of a source of turbulence in the turbulence transport model, we find that the $\langle B_{2D}^2\rangle/\langle b_{slab}^2\rangle$ ratio associated with the 2009 solar minimum in the energy-containing range increases monotonically with heliocentric distance, while $\langle B_{2D}^2\rangle/\langle b_{slab}^2\rangle$ associated with the 2003 and 2015 solar maxima decreases slightly between $\sim 1.2 - 2$ au, and then increases with heliocentric distance. The ratio of the variances during solar minimum is larger than the ratio of the variances during solar maximum. Furthermore, the $\langle B_{2D}^2\rangle/\langle b_{slab}^2\rangle$ ratio in the inertial range during 2009 solar minimum is larger than the ratios of variances during the 2003 and 2015 solar maxima, in which the $\langle B_{2D}^2\rangle/\langle b_{slab}^2\rangle$ ratio in the inertial range decreases beyond $\sim$ 10 au. The anisotropies for both energy-containing and inertial ranges remain large throughout the heliosphere and the anisotropy in the inertial range is larger than the anisotropy in the energy-containing range. In this study, we assumed an 80:20 ratio between 2D and slab turbulence energies, and a 2:1 ratio between slab and 2D correlation lengths. Depending on the assumptions related to the turbulence energies and the correlation lengths, the evolution of anisotropies can be different as suggested by Verdini and Grappin et al [40], who find that the evolution of turbulent spectrum depends strongly on its initial anisotropy.

With the inclusion of 2D and slab stream-shear sources of turbulence in the NI turbulence transport model, we find that $\langle B_{2D}^2\rangle/\langle b_{slab}^2\rangle$ in the energy-containing and inertial ranges is less than the $\langle B_{2D}^2\rangle/\langle b_{slab}^2\rangle$ ratio when a source of turbulence is not included in the turbulence transport model. It indicates that a stream-shear source of turbulence reduces the anisotropy in magnetic field fluctuations. In this study, a stream-shear source of 2D turbulence is three times larger than a stream-shear source of slab turbulence, which is indicated by the chosen stream-shear driving constants shown in the Table 3. These anisotropies in magnetic field fluctuations results can be different for different 2D and slab stream-shear driving constants. Adhikari et al [33] calculated the anisotropy in magnetic field fluctuations in the energy-containing and inertial ranges for different stream-shear sources of slab turbulence with the same stream-shear source of 2D turbulence, and found that the anisotropy in magnetic field fluctuations decreases with an increase in the stream-shear source of slab turbulence. The $\langle B_{2D}^2\rangle/\langle b_{slab}^2\rangle$ ratio in the energy-containing range during 2009 solar minimum is larger than the $\langle B_{2D}^2\rangle/\langle b_{slab}^2\rangle$ ratio during the 2003 and 2015 solar maxima from 1 to 75 au. The $\langle B_{2D}^2\rangle/\langle b_{slab}^2\rangle$ ratio in the inertial range for the 2009 solar minimum is larger than that of the 2003 and 2015 solar maxima from 1 to $\sim$ 2 au, after which the three ratios are approximately similar between $\sim 2-10$ au. Eventually, the ratio during solar minimum becomes larger than the ratios during solar maxima beyond $\sim$ 10 au. We also find that the ratio of the variances in the inertial range is larger than the ratio of the variances in the energy-containing range. The energy-containing and inertial ranges anisotropies are $\sim$ 5.5 and $\sim$ 4, respectively, throughout the heliosphere.

The inclusion of both 2D and slab stream-shear sources of turbulence and a pickup ion source of slab turbulence in the turbulence transport model yields ratios of $\langle B_{2D}^2\rangle/\langle b_{slab}^2\rangle$ in the energy-containing range for solar maxima and minimum that are approximately similar from $\sim$ 1.2 to $\sim$ 4 au, after which the ratio corresponding to the 2009 solar minimum decreases beyond $\sim$ 4 au, while the ratios corresponding to the 2003 and 2015 solar maxima decrease beyond $\sim$ 5 au. The ratios of the 2D and slab magnetic field fluctuations in the energy-containing range from 1 au to $\sim 4-5$ au is $\sim$ 4, which eventually decreases to $\sim$ 1 at 75 au. The $\langle B_{2D}^2\rangle/\langle b_{slab}^2\rangle$ ratio in the inertial range during 2009 solar minimum increases sharply initially, then decreases rapidly, followed by a slight increase from $\sim$ 1.2 to $\sim$ 3 au, after which there is a gradual decrease with distance. Similarly, the $\langle B_{2D}^2\rangle/\langle b_{slab}^2\rangle$ ratio in the inertial range of the 2003 and 2015 solar maxima increases and decreases rapidly within $\sim$ 1.2 au, gradually increases until $\sim$ 4 au, and then decreases with increasing heliocentric distance. The $\langle B_{2D}^2\rangle/\langle b_{slab}^2\rangle$ ratio in the inertial range for the solar minimum is larger than the $\langle B_{2D}^2\rangle/\langle b_{slab}^2\rangle$ ratio for the solar maxima from 1 to $\sim$ 2 au, and then it reverses in the outer heliosphere beyond $\sim$ 2 au.

Acknowledgments
We acknowledge the partial support of the NSF EPSCoR RII-Track-1 Cooperative Agreement OIA-1655280, NSF-DOE award 1707247, NASA grants NNX08AJ33G, Subaward 37102-2, NNX14AC08G, NNX14AJ53G, A99132BT, RR185-447/4944336 and NNX12AB30G. G.P.Z. acknowledges the support of the International Space Science Institute (ISSI), both through the award of the 2017 Johannes Geiss Fellowship and in the framework of an International Team Project 504 entitled "Current Sheets, Turbulence, Structures and Particle Acceleration in the Heliosphere".

References
[1] Montgomery, D. & Turner, L. 1981, *Physics of Fluids*, 24, 825
[2] Matthaeus, W. H., Goldstein, M. L., & Roberts, D. A. 1990, *J. Geophys. Res.*, 95, 20673
[3] Narita, Y., Sahraoui, F., Goldstein, M. L., & Glassmeier, K.-H. 2010, Journal of *Geophysical Research (Space Physics)*, 115, A04101
[4] Bieber, J. W., Wanner, W., & Matthaeus, W. H. 1996, *J. Geophys. Res.*, 101, 2511
[5] Zank, G. P., Matthaeus, W. H., Bieber, J. W., & Moraal, H. 1998, *J. Geophys. Res.*, 103, 2085
[6] Shalchi, A., Bieber, J. W., Matthaeus, W. H., & Schlickeiser, R. 2006, *ApJ*, 642, 230
[7] Montgomery, D., & Turner, L. 1981, *PhFl*, 24, 825
[8] Shebalin, J. V., Matthaeus, W. H., & Montgomery, D. 1983, *JPlPh*, 29, 525
[9] Grappin, R. 1986, *PhFl*, 29, 2433
[10] Grappin, R., Velli, M., & Mangeney, A. 1993, *PhRvL*, 70, 2190
[11] Goldreich, P., & Sridhar, S. 1995, *ApJ*, 438, 763
[12] Ghosh, S., Matthaeus, W. H., Roberts, D. A., & Goldstein, M. L. 1998, *JGR*, 103, 23705
[13] Horbury, T. S., Forman, M., & Oughton, S. 2008, *PhRvL*, 101, 175005
[14] Podesta, J. J. 2009, *ApJ*, 698, 986
[15] Narita, Y., Glassmeier, K.-H., Sahraoui, F., & Goldstein, M. L. 2010a, *PhRvL*, 104, 171101
[16] Wicks, R. T., Horbury, T. S., Chen, C. H. K., & Schekochihin, A. A. 2010, *MNRAS*, 407, L31
[17] Bruno, R., & Telloni, D. 2015, *ApJL*, 811, L17
[18] Zank, G. P., Adhikari, L., Hunana, P., et al. 2017, *ApJ*, 835, 147
[19] Robinson, D. C., & Rusbridge, M. G. 1971, *PhFl*, 14, 2499
[20] Zweben, S. J., Menyuk, C. R., & Taylor, R. J. 1979, *PhRvL*, 42, 1270
[21] Belcher, J. W., & Davis, L., Jr. 1971, *JGR*, 76, 3534
[22] Montgomery, D. 1982, *PhyS*, 1982, 83
[23] Matthaeus, W. H., Goldstein, M. L., & Roberts, D. A. 1990, *JGR*, 95, 20673
[24] Milano, L. J., Dasso, S., Matthaeus, W. H., & Smith, C. W. 2004, *PhRvL*, 93, 155005
[25] Ruiz, M. E., Dasso, S., Matthaeus, W. H., Marsch, E., & Weygand, J. M. 2011, *JGRA*, 116, A10102
[26] Matthaeus, W. H., Dasso, S., Weygand, J. M., et al. 2005, *PhRvL*, 95, 231101
[27] Osman, K. T., & Horbury, T. S. 2007, *ApJL*, 654, L103
[28] Dasso, S., Milano, L. J., Matthaeus, W. H., & Smith, C. W. 2005, *ApJ*, 635, L181
[29] Smith, C. W. 2003, *in AIP Conf. Ser. 679, Solar Wind Ten, ed.* M. Velli et al. (Melville, NY: AIP), 413
[30] Zank, G. P., & Matthaeus, W. H. 1992, *JGR*, 97, 17189
[31] Zank, G. P., Adhikari, L., Hunana, P., et al. 2017, *ApJ*, 835, 147
[32] Zank, G. P., Adhikari, L., Hunana, P., et al. 2017, *Journal of Physics Conference Series*, 900, 012023
[33] Adhikari, L., Zank, G. P., Hunana, P., et al. 2017, *ApJ*, 841, 85
[34] Adhikari, L., Zank, G. P., Hunana, P., et al. 2017, *Journal of Physics Conference Series*, 900, 012001
[35] Hunana, P., & Zank, G. P. 2010, *ApJ*, 718, 148
[36] Adhikari, L., Zank, G. P., Telloni, D., et al. 2017, *ApJ*, 851, 117
[37] Zhao, L.-L., Adhikari, L., Zank, G. P., Hu, Q., & Feng, X. S. 2018, *ApJ*, 856, 94
[38] Zank, G. P. 2014, *in Transport Processes in Space Physics and Astrophysics, ed. G. P. Zank (Berlin: Springer)*
[39] Zank, G. P., Matthaeus, W. H., & Smith, C. W. 1996, *JGR*, 101, 17093

[40] Verdini, A., & Grappin, R. 2016, *ApJ*, 831, 179

A Model for Solar Wind Structure and Dynamics Based on Alfvén Wave Turbulence

M Asgari-Targhi, A A van Ballegooijen and M Shoda

Harvard-Smithsonian Center for Astrophysics, 60 Garden Street, Cambridge, MA 02138, USA

E-mail: masgari-targhi@cfa.harvard.edu

Abstract. We present a 3D reduced magnetohydrodynamic (RMHD) model of reflection-driven Alfvén wave turbulence in an open magnetic field at the center of a coronal hole. The non-linear interactions between outward (dominant) and inward (minority) propagating waves generate turbulence. The RMHD equations describing the turbulence include the effects of solar wind outflow velocity on the dissipation of waves. The Alfvén wave turbulence results in the transfer of energy to small perpendicular scales ("direct cascade") or to larger scales ("inverse cascade"). We first show the result of our calculation for a model where there is smooth variation of plasma parameters such as Alfvén speed and density with height along the flux tube. However, this model does not produce the required energy to heat the plasma and accelerate the fast solar wind. Therefore, we introduce our second model, where we include additional density fluctuation along the open field. These density variations simulate the effects of compressive MHD waves in the solar wind. We find that these variations in density enhance the turbulence dissipation rates, and thereby increase the heating rate and the acceleration of the solar wind.

1. Introduction

The fast solar wind at 1 AU has velocities in the range of 500-800 km s^{-1}, and originates from coronal holes on the Sun. It has long been known that Alfvén waves are present in the solar wind [1, 2], and *in situ* measurements have demonstrated that turbulent fluctuations in the magnetic field and plasma velocity are ubiquitous [3–7]. The turbulent waves are characterized by a broad spectrum of wavenumbers and frequencies. Transverse magneto-hydrodynamic (MHD) waves such as Alfvén waves likely originate in the solar photosphere, where they may be produced by interactions of the open magnetic field with subsurface convective flows. As the waves travel outward, they reflect due to spatial variations in Alfvén speed, which generates inward propagating waves [8, 9]. The nonlinear interactions between the inward and outward propagating waves can distort the wave fronts and produce turbulence [10–12]. Even though the outward propagating Alfvén waves are modified by turbulent interactions, they are capable of transporting energy and momentum over large distances before giving up their energy. Therefore, Alfvén wave turbulence (AWT) is believed to play an important role in the heating and acceleration of the fast wind [13–22]. Remote-sensing observations of the Sun further support the idea that Alfvén waves are present in the solar atmosphere [23–27].

Alfvén waves can affect the solar wind in two ways: they can heat the coronal plasma and provide a wave-pressure force [28, 29]. Turbulent cascade of the waves is considered to be responsible for transferring wave energy to small spatial scales, where the waves are dissipated and the energy is converted into heat of the plasma [11]. The nature of these collisionless dissipation processes is not well understood, but likely involves wave-particle interactions on the scale of the ion gyroradius or smaller [30]. Since the dissipation scale is so much smaller than the energy injection scale of the turbulence, it is generally assumed that the turbulent dissipation rate is not strongly dependent on the details of the dissipation process. Therefore, to estimate the plasma heating rate it is sufficient to simulate only the largest scales of the turbulence and to establish the energy cascade rates at those scales. The wave pressure force is caused by the spatial variation of the wave energy density along the open field lines [29], and plays an important role in driving the fast solar wind to the observed high velocity. Therefore, models of the acceleration of the fast solar wind by AWT should take into account the wave pressure forces.

Matthaeus and collaborators were the first to develop RMHD models of AWT in an open flux tube [11, 31, 32]. They demonstrated that a turbulent state can develop in the corona despite the fact that the waves can escape into the heliosphere. In a recent paper, Zank and co-authors introduced results of a model describing the transport and evolution of turbulence in open magnetic field regions [33]. In their model, a quasi-2D turbulence is generated in closed loops of the magnetic carpet, and interchange reconnection between the carpet loops and the open magnetic field causes the turbulence to be injected into the solar wind, where it is advected upward by subsonic, sub-Alfvénic flow. The turbulence has an Alfvénic component referred to as minority slab component and is evolved through the coupling between the majority quasi-2D and minority slab component. More recently, the effects of the wind outflow on the waves have been included in RMHD models [34–36]. In contrast to the model by Zank and collaborators [33], the RMHD picture provides only an Alfvén wave perspective.

MHD simulations of turbulent flows are difficult because they require high spatial resolution. To reach the required resolution, many authors focus on "homogeneous" turbulence in a box (often with periodic boundary conditions). This simulation domain may represent a small part of a much larger turbulent system such as the solar wind far from the Sun. This approach is not valid for the acceleration region of the solar wind (say, $1 - 20\ R_\odot$) because the wavelengths $\lambda_\parallel$ of the waves are expected to be comparable to the solar radius, so the large-scale gradients in the magnetic field and plasma density are expected to have important effects on the waves. The AWT model predicts that the waves in this region are highly anisotropic in the sense that the magnetic and velocity fluctuations are nearly perpendicular to the mean magnetic field. Also, the amplitude of the inward waves is likely to be much smaller than that of the outward waves, so the turbulence is highly imbalanced. It is not yet fully understood how imbalanced turbulence behaves in an inhomogeneous plasma.

In this paper we use a 3D RMHD model [35, 36] to study how reflection-driven AWT can heat and accelerate the fast solar wind emerging from an open magnetic field based at a coronal hole. The modeling includes the effects of the wind velocity on the waves. Also, the simulations take into account the inhomogeneity and non-local nature of the turbulence. We first show that the nonlinear interactions between the dominant $\mathbf{z}_+$ waves and minority $\mathbf{z}_-$ waves is responsible for the transfer of wave energy either to smaller perpendicular scales ("direct cascade") or to larger scales ("inverse cascade"). We then introduce density fluctuations in the open field and show that in a model with density variation, the wave reflection is increased significantly, thereby enhancing wave dissipations rates and turbulence and as a result increasing the heating rates.

2. Reduced MHD Model

We consider a thin magnetic flux tube inside a polar coronal hole at solar minimum. The tube extends radially outward from the Sun to a distance $r = 20\ R_\odot$. Alfvén waves are launched inside this tube by imposing incompressible "footpoint" motions at the coronal base ($r \approx 1.003\ R_\odot$). The assumed motions have a velocity amplitude $v_{\rm rms} \approx 40$ km s^{-1}, consistent with observed

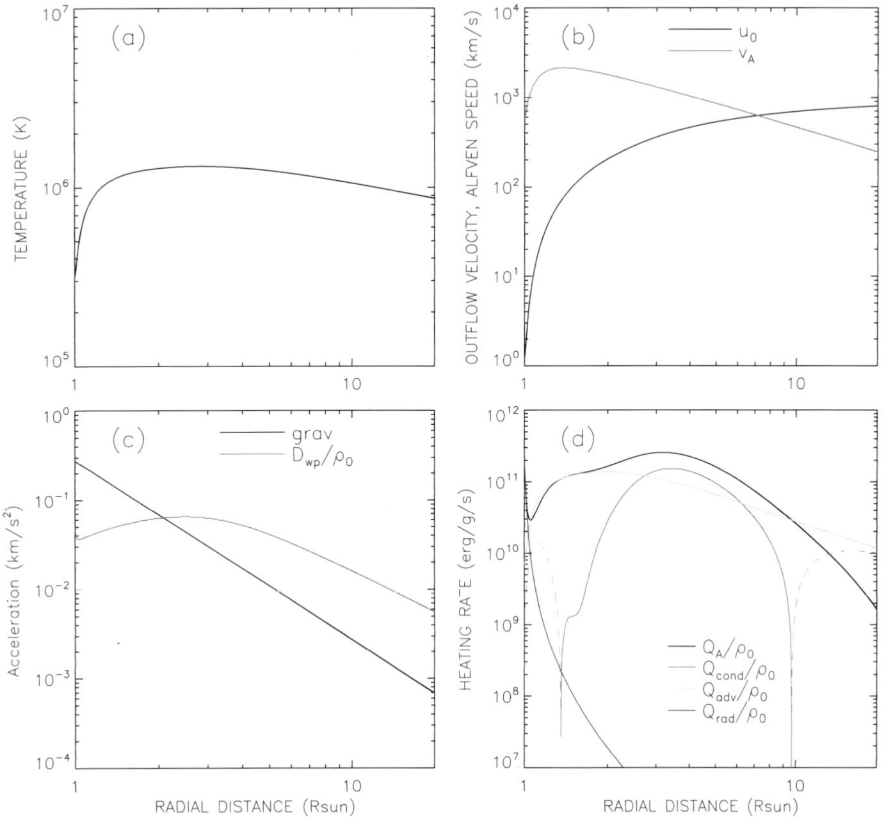

Figure 1. Properties of the background atmosphere: (a) temperature, (b) outflow velocity (black curve) and Alfvén speed (red curve), (c) wave pressure acceleration (red curve) and gravity (black curve), (d) heating rate Q_A needed to maintain the assume temperature, as well as the contributions from conduction (red), radiation (blue) and wind expansion (green).

non-thermal velocities in coronal holes [37], and a transverse length scale comparable to that of the solar granulation ($\sim$ 1 Mm). The waves are modeled using the RMHD approximation [38, 39], and are described in terms of Elsasser variables:

$$\mathbf{z}_\pm \equiv \mathbf{v}_1 \mp \mathbf{B}_1/\sqrt{4\pi\rho_0}, \tag{1}$$

where $\mathbf{B}_1(\mathbf{r}, t)$ and $\mathbf{v}_1(\mathbf{r}, t)$ are the magnetic- and velocity perturbations of the waves, and $\rho_0(r)$ is the mean plasma density. Here $\mathbf{z}_+$ describes the "dominant" waves, which always propagate radially outward, and $\mathbf{z}_-$ describes the "minority" waves, which can propagate inward or outward depending on how they are produced [40]. The form of the RMHD equations used here include the effects of a radially expanding magnetic field, density variations along the flux tube, and the solar wind outflow [34, 35]:

$$\begin{aligned} \frac{\partial \omega_\pm}{\partial t} &= -(u_0 \pm v_A)\frac{\partial \omega_\pm}{\partial r} + \frac{1}{2}\left(\frac{dv_A}{dr} \pm \frac{u_0}{2H_\rho}\right)(\omega_+ - \omega_-) + \frac{u_0}{2H_B}(\omega_+ + \omega_-) \\ &\quad \cdots - \frac{1}{2}[\omega_+, f_-] - \frac{1}{2}[\omega_-, f_+] \pm \nabla_\perp^2 \left(\frac{1}{2}[f_+, f_-]\right), \end{aligned} \tag{2}$$

where $f_\pm(\mathbf{r},t)$ are the velocity stream functions of the Elsasser variables, $\omega_\pm \equiv -\nabla_\perp^2 f_\pm$ are the vorticities, $u_0(r)$ is the solar wind outflow velocity, $v_A(r)$ is the Alfvén speed, $H_B(r) \equiv B_0/(dB_0/dr)$ is the magnetic scale length, $H_\rho(r) \equiv \rho_0/(d\rho_0/dr)$ is the density scale length, and $B_0(r)$ is the magnetic field strength. Note that u_0 and the radial gradients of B_0 and ρ_0 cause linear coupling between the ω_+ and ω_- waves. The brackets $[\cdots,\cdots]$ involve derivatives w.r.t. the perpendicular coordinates x and y, and are the only nonlinear terms in the equations. In addition there are wave damping terms (not shown).

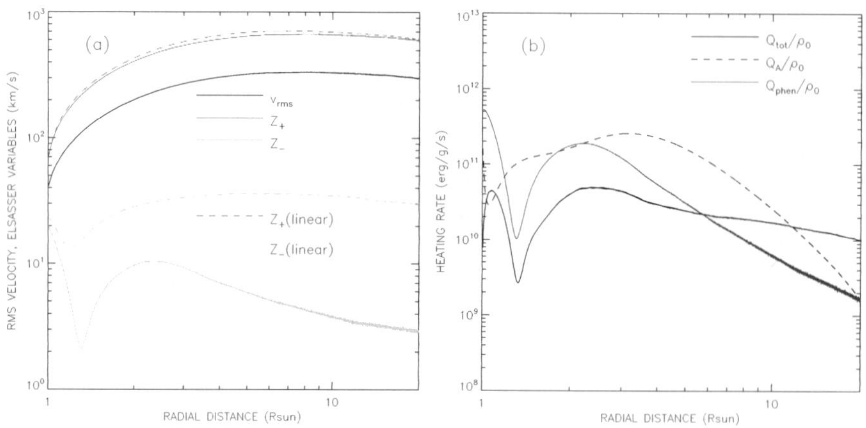

Figure 2. Radial dependence of wave-related quantities for a RMHD model of the solar wind. (a) Velocity amplitude of the waves (black curve), and Elsasser variables for dominant waves (red curve) and minority waves (green curve). The dashed red and green curves are for a model with the nonlinear terms switched off. (b) Wave energy dissipation rates per unit mass: rate derived from turbulence simulation (solid black curve), rate assumed in the setup of background atmosphere (dashed curve), and rate predicted by a phenomenological model [35].

In RMHD modeling, before the turbulence can be simulated we must first specify the quantities describing the background atmosphere. For the field strength $B_0(r)$ we use a potential field model. Figure 1 shows some properties of the background atmosphere. The temperature $T_0(r)$ is specified using a simple formula [35]. As shown in Figure 1(a), the temperature rises quickly at low heights and declines slowly at large height. The density $\rho_0(r)$ and outflow velocity $u_0(r)$ of the wind are determined by iteratively solving the mass, momentum and energy equations for the wind in a one-dimensional, time-steady approximation (for details see [35]). The heating rate $Q_A(r)$ is computed from the energy equation, which includes the effects of solar wind expansion and radiative and conductive losses [16]. The wave energy density $U_A(r)$ is computed by solving the wave action equation, where we use a boundary condition on the wave amplitude at the coronal base. The wave pressure force $D_{wp}(r)$ is computed from the gradient of $U_A(r)$. The momentum equation includes the wave pressure force.

Figure 1(b) shows the outflow velocity $u_0(r)$ of the wind (black curve) and Alfvén speed $v_A(r)$ (red curve). Note that the two curves cross at $r \approx 7\ R_\odot$, which is the Alfvén critical point. Panel (c) shows the acceleration associated with the wave pressure force (red curve) as well as the acceleration of gravity (black curve). Note that the wave pressure force is quite strong because $D_{wp}(r)$ exceeds gravity already at $r = 2.5\ R_\odot$. Finally, panel (d) shows the heating rate $Q_A(r)$ needed to maintain the temperature $T_0(r)$ of the background atmosphere (black curve). Some of this heat is used to compensate for the adiabatic cooling associated with the expansion of the solar wind; this contribution $Q_{adv}(r)$ is given by the green curve in Fig. 1(d). The blue and red curves give the radiative and conductive energy losses, respectively. Note that at low heights most of the energy is used to heat the expanding wind, while at 3 – 5 $R_\odot$ conduction is the dominant cooling mechanism. The dashed red curves show regions where thermal conduction causes heating of the plasma.

After the background atmosphere has been specified, the waves are simulated for a period of 30,000 s, which is sufficient for the turbulence to reach a statistically stationary state at all heights. Figure 2(a) shows the velocity amplitudes of the dominant waves $\mathbf{z}_+$ (solid red curve) and minority waves $\mathbf{z}_-$ (solid green curve), as well as the total wave amplitude (black). These values are averages over the tube's cross-section and over time. Note that the dominant waves are much stronger than the minority waves, so the turbulence is strongly imbalanced. Also, the minority waves have a minimum at $r \approx 1.4\ R_\odot$ where the Alfvén speed $v_{\rm A}(r)$ reaches its maximum. This indicates that the rate of production of minority waves depends strongly on the gradient of Alfvén speed, $dv_{\rm A}/dr$ in equation (2). The dashed red and green curves in Figure 2(a) are for a model in which the nonlinear terms in equation (2) are omitted, so there is no turbulence in this case. Comparison of the solid and dashed green curves shows that the nonlinear effects further suppress the minority waves.

Figure 2(b) shows the wave energy dissipation rate $Q_{\rm tot}(r)$ predicted by the RMHD model (solid black curve). This rate is again computed by averaging the dissipation rates over the flux tube cross-section and over time in the simulation. Note that $Q_{\rm tot}(r)$ is not necessarily the same as the heating rate $Q_{\rm A}(r)$ assumed in the setup of the background atmosphere. The latter is given by the dashed black curve in Figure 2(b), which is the same as the solid black curve in Figure 1(d). The goal is of course to construct a model in which $Q_{\rm tot} \approx Q_{\rm A}$, but we find that this does not happen in the present case. We find that in the height range from 1.2 to 10 $R_\odot$ the wave dissipation rate $Q_{\rm tot}$ is significantly smaller than $Q_{\rm A}$, indicating that the simulated wave heating is not sufficient to maintain the temperature of the background atmosphere. Note that $Q_{\rm tot}(r)$ has a dip at $r \approx 1.4\ R_\odot$ where the Alfvén speed has its maximum and the production of minority waves is suppressed; this is part of the reason for the discrepancy with $Q_{\rm A}$. The blue curve in Figure 2(b) shows results for a phenomenological turbulence model [35], but this model has only limited success in reproducing $Q_{\rm tot}$.

It is often stated that turbulence is generated by "counter-propagating" Alfvén waves. However, in equation (2) the nonlinear "bracket" terms do not contain radial derivatives and therefore do not depend on the direction of wave propagation. We find that in the present model with a smooth background atmosphere both the dominant (ω_+) and minority (ω_-) waves travel radially outward. This is demonstrated in Figures 3(a) and 3(b), where we plot $\omega_{\pm,k}(r,t)$ for a wave mode k with perpendicular wavenumbers $n_x = 8$ and $n_y = 4$ (see [36]). Note that both panels show patterns slanted to the right, indicating outward propagation. The reason that the minority waves travel outward is that they are produced by weak linear coupling with the dominant waves. This linear coupling is mainly due to the term with $dv_{\rm A}/dr$ in equation (2). The weakness of the coupling causes the minority waves to travel outward with their source (the dominant waves), consistent with earlier predictions [40]. In other RMHD simulations a combination of inward and outward propagation of the minority waves has been found [34]. It is likely that the behavior of the minority waves depends on the correlation time of the footpoint motions, which is only about 48 s in our model.

In most turbulence models for the solar wind it is assumed that the energy is transferred from larger to smaller spatial scales in a "direct" cascade. However, we find that in the acceleration region of the wind both direct and inverse cascades can occur. We computed the cascade rate of the dominant and minority waves in a RMHD model for the fast solar wind [36]. Figure 4 shows the cascade rate $\epsilon_+(a_\perp, r)$ for the dominant waves as a function of dimensionless perpendicular wavenumber $a_\perp$ and radial distance r (the wavenumber is scaled with respect to the flux tube width). In this model the waves are injected at the coronal base with wavenumbers $a_\perp$ in the

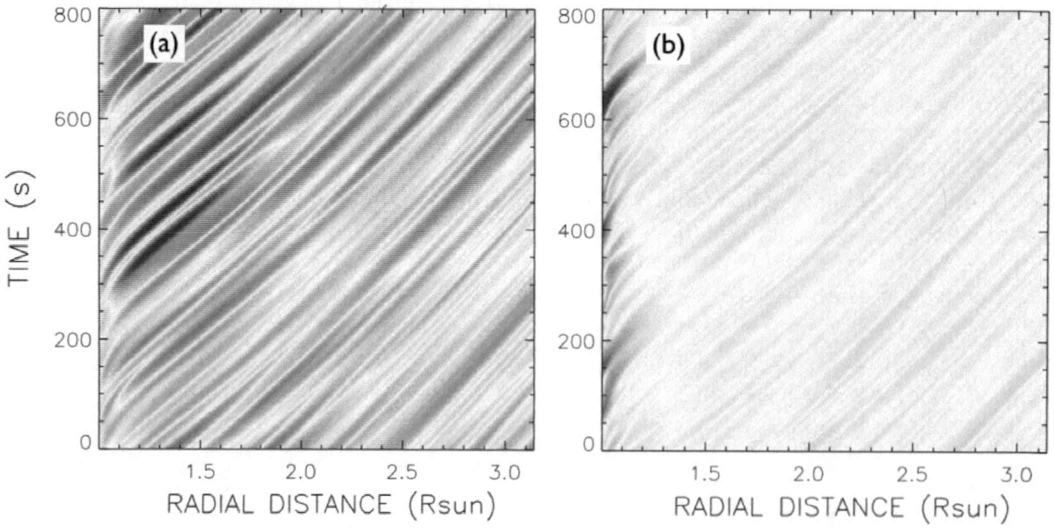

Figure 3. Vorticities as functions of radial distance r and time t for a wave mode k with perpendicular wavenumbers $n_x = 8$ and $n_y = 4$ in a RMHD model for the fast solar wind: (a) dominant wave $\omega_{+,k}$, (b) minority wave $\omega_{-,k}$. Note that both waves travel radially outward.

range 11 – 17, and the waves are dissipated at wavenumbers $a_\perp > 44$. The cascade rate ϵ_+ is computed by adding contributions from all mode triplets (a_1, a_2, a_3) that straddle a chosen wavenumber $a_\perp$ and transport energy from modes with $a_i < a_\perp$ to modes with $a_i > a_\perp$ (for details see [36]). Figure 4 shows that at low heights the dominant waves have a direct cascade (red region), but ϵ_+ changes sign at $r = 1.4\ R_\odot$ for wavenumbers in the range $15 < a_\perp < 44$, indicating an inverse cascade (blue triangular region). The inverse cascade continues up to $r = 2.5\ R_\odot$, but in a narrowing wavenumber range. The cascade rate ϵ_- of the minority waves is positive at all height (not shown). The negative cascade rates $\epsilon_+ < 0$ limit the amount of energy that can transported into the dissipation range ($a_\perp > 44$). Therefore, in the region between 1.4 and 2.5 $R_\odot$ the wave dissipation rate is reduced compared to what it would be if a direct cascade occurred at all heights and wavenumbers. This is part of the reason why the dissipation rates in our model are not sufficient to maintain the temperature of the background atmosphere.

The development of an inverse cascade is due to the change of sign of $dv_{\rm A}/dr$ at $r = 1.4$ $R_\odot$. As the dominant waves propagate outward they also cascade in wavenumber space, but the cascade is very slow because the amplitude of the minority waves is very small. Therefore, in the region between 1.4 and 2.5 $R_\odot$ the dominant waves are still in the process of gradually adjusting to the change in sign of $dv_{\rm A}/dr$. In contrast, the minority waves have a much shorter cascade time scale and can immediately respond to this change. This leads to a situation where the minority waves produce an inverse cascade of the dominant waves. This shows that the turbulent cascade in the acceleration region of the fast wind cannot be treated as a local process: the cascade of the dominant waves occurs over a distance comparable to the variations in the background atmosphere, and the cascade rate ϵ_+ is dependent on wavenumber.

3. Model with Density Variations

The difficulty with the solar wind model described in the previous section is that the energy produced by Alfvén wave turbulence is significantly less than the energy required to heat the

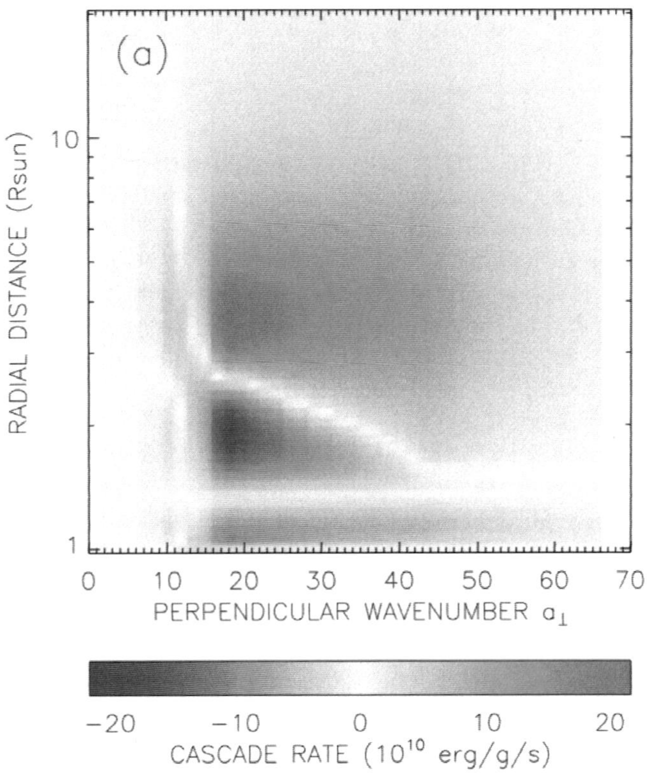

Figure 4. Energy cascade rate for the dominant waves in a model for the fast solar wind [36]. The rate $\epsilon_+(a_\perp, r)/\rho_0(r)$ is plotted as function of dimensionless perpendicular wavenumber $a_\perp$ and radial distance r from Sun center. Direct and inverse cascades are indicated by red and blue colors (see color bar).

plasma. The energy dissipation rate $Q_{tot}(r)$ generated as a result of RMHD turbulence is less than the heating rate $Q_A(r)$ necessary to heat the background atmosphere and accelerate the solar wind. Therefore, we need to raise the heating rate Q_{tot} without changing Q_A. In a system, where turbulence is generated by the nonlinear interactions between dominant and minority waves, the reflection of the waves creates the intermixing of the waves and thereby turbulence. Therefore, to enhance the rate of turbulence, we need to create more reflection of the dominant outward propagating waves. A way to achieve this is by introduction of density variations in the open field of our interest. The density fluctuations $\delta\rho(r, t)$ may be due to the sound waves coupling with Alfvén waves in the solar atmosphere. In the presence of sound waves, the Alfvén speed varies in time and space and Alfvén waves reflect as a result of these gradients in Alfvén speed. By imposing density variations in the open field, these gradients are increased and consequently the amplitude of the outward propagating waves are enlarged, creating more inward propagating waves by reflection and therefore increasing the turbulent propagation rate.

The existence of density fluctuations in the solar atmosphere and solar wind have been confirmed by observations. Intensity fluctuations corresponding to slow mode oscillating waves with periods of between 10 to 15 mins have been observed in polar plumes with the Extreme Ultraviolet Imaging Telescope (EIT) on the Solar and Heliospheric Observatory (SOHO) [41].

In coronal loops the intensity variations represent waves with periods of approximately 5 mins in the extreme ultraviolet [42, 43] and at visible wavelengths [44]. High-frequency intensity variations have been detected during the eclipses [45].

In solar wind, there is strong evidence of density fluctuations based on observations. In the past, a variety of radio observations were used to study the radial evolution of plasma at the distances of 4-6 $R_\odot$ [46] or natural radio sources and spacecraft signals were employed to measure electron density fluctuations and solar wind speed [47]. Recent observations of solar wind adopting radio interferometry have confirmed the fluctuations in the amplitudes of density ranging from $6\% - 15\%$ at heliocentric distances of 16-26 $R_\odot$ [48]. [49] used the Extreme-Ultraviolet images of comet Lovejoy from the Atmospheric Imaging Assembly (AIA) on the Solar Dynamics Observatory (SDO). They studied the striations of magnetic field structure in both closed and open magnetic field regions. These striations indicate the presence of large density variations (at least a factor of 6) associated with contrast in brightness between neighboring coronal flux tubes over scales of a few thousand kilometers.

Using observations from the Japanese Venus explorer Akatsuki [50] and applying radio occultation technique, [51] analyzed the frequency time series taken at heliocentric distances of 1.5-20.5 $R_\odot$. They detected quasi-periodic density fluctuations with periods of 100 to 2000 seconds at all distances. We will be drawing a comparison between the observations from [51] and our modeling results in this section.

We present a model of density fluctuations in the solar wind. In our simulations, we include static density variations in our prescribed background atmosphere, i.e., the variations are random in position along the flux tube and constant in time. Also, for simplicity the density is assumed to be constant over the cross-section of the flux tube. To implement the effect of density fluctuations, we first define a Fourier-filtered random function $\xi(r;\lambda)$

$$\xi(r;\lambda) = \int dk \int dl \, \mathrm{rand}\,(l)\, e^{-k^2\lambda^2} e^{2\pi i k(r-l)}, \tag{3}$$

where $\xi(r;\lambda)$ is a spatially randomly varying function with length scale larger than λ and rand (l) is a random function of l. We normalize $\xi(r;\lambda)$ as

$$\bar{\xi}(r;\lambda) = \frac{\xi(r;\lambda)}{\langle \xi(r;\lambda) \rangle_{\rm rms}}, \tag{4}$$

such that $\langle X \rangle_{\rm rms}$ denotes the root-mean-square value of X. Therefore the root-mean-square value of $\bar{\xi}$ is unity. The fractional density fluctuation $\delta\rho/\rho$ is given by

$$\delta\rho/\rho = n(r)\bar{\xi}(r;\lambda_\rho), \tag{5}$$

where $\lambda_\rho = 0.4 R_\odot$ and $n(r)$ represents an envelope of the fractional density fluctuation. A uniform density fluctuation with rms amplitude of 0.1 ($n(r) = 0.1$) was used in [35]. According to the radio wave observation by [51], the magnitude of density fluctuation is highly non-uniform and has a maximum around $r \approx 5 - 6 R_\odot$. In this study, we apply a nonuniform density variation based on this observation as

$$n(r) = \frac{1}{2}\beta^{-1/4}\frac{\delta B}{B}, \quad \beta = c_s^2/v_A^2. \tag{6}$$

where $\delta B_\perp(r)$ is the rms value of the magnetic fluctuation $\mathbf{B}_1$ and c_s is the sound speed. The resulting time-independent, one-dimensional model with spatial density fluctuations is used as the background atmosphere for the three dimensional RMHD model of Alfvén waves turbulence along the expanding flux tube. The methodology for solving the RMHD equations in this case is identical to the case of Smooth model discussed in the previous section. Figure 5(a) shows the dissipation per unit mass for the simulated waves without density fluctuation (dashed blue curve) and with density fluctuation (solid blue curve). As the Figure clearly indicates for a model with density variation , the AWT produces enough heating to heat the background atmosphere. At all heights (1-20 $R_\odot$) the density variation has increased the heating rate per unit mass along the open flux tube, and at many positions this increase is by an oder of magnitude. In Figure 5(b), we present the profile of $n(r)$ given by Eq. (6) (solid line) together with the observational values from [51] shown by circles.

We demonstrated in this section how density fluctuation can increase the level of turbulence and heat the open field structure. The density variation introduced in our model is in agreement with the observation of [51]. However, to create a full picture of the influence of density fluctuation on wave turbulence, heating and acceleration of the solar wind, the density variation should be included in the cross section of the flux tube as well as along the flux tube. To simulate these effects, a full 3D MHD modeling is required and not a RMHD modeling of waves.

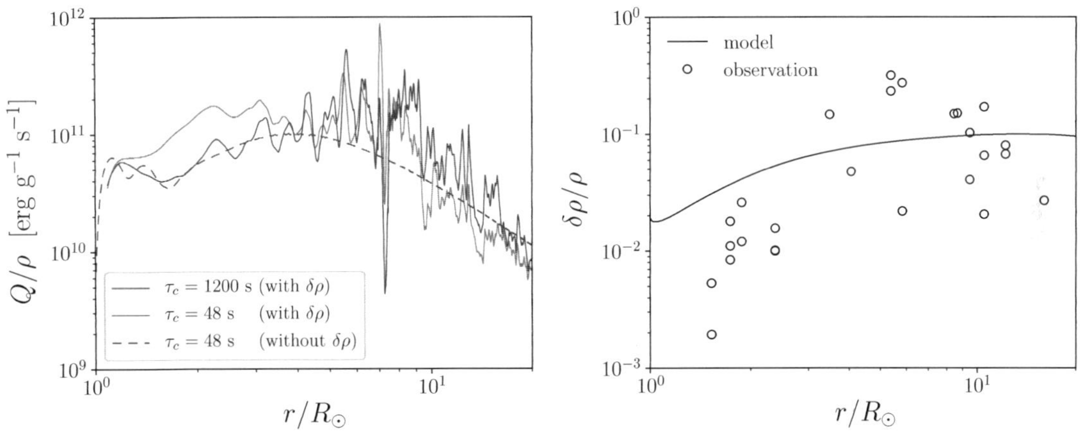

Figure 5. Figure on the left shows the heating rate for different correlation times. Figure on the right shows the amplitude of density fluctuations from our modeling (solid black line) and the observed values (circles) from [51].

Acknowledgments
This project was supported under contract NNM07AB07C from NASA to the Smithsonian Astrophysical Observatory (SAO) and contract SP02H1701R from Lockheed Martin Space and Astrophysics Laboratory (LMSAL) to SAO.

References

[1] Coleman Jr P J 1968 *Astrophys. J.* **153** 371

[2] Belcher J W and Davis L 1971 *J. Geophys. Res.* **76** 3534

[3] Tu C Y and Marsch E 1995 *Space Sci. Rev.* **73** 1

[4] Goldstein M L, Roberts D A and Matthaeus W H 1995 *Ann. Rev. Astron. & Astrophys.* **33** 283

[5] Bale S D, Kellogg P J, Mozer F S, Horbury T S and Reme H 2005 *Phys. Rev Lett.* **94** 215002

[6] Bruno R and Carbone V 2005 *Living Rev. Solar Phys.* **2** 4

[7] Borovsky J 2012 *J. Geophys. Res.* **117** A05104

[8] Heinemann M and Olbert S 1980 *J. Geophys. Res.* **85** 1311

[9] Velli M 1993 *Astron. & Astrophys.* **270** 304

[10] Zhou Y and Matthaeus W H 1990 *J. Geophys. Res.* **95** 10291

[11] Matthaeus W H, Zank G P, Oughton S, Mullan D J and Dmitruk P 1999 *Astrophys. J. Letters* **523** L93

[12] Chandran B D G, Quataert E, Howes G G, Hollweg J V and Dorland W 2009 *Astrophys. J.* **701** 652

[13] Suzuki T K and Inutsuka S I 2005 *Astrophys. J. Letters* **632** L49

[14] Suzuki T K and Inutsuka S I 2006 *J. Geophys. Res.* **111** A06101

[15] Cranmer S R and van Ballegooijen A A 2005 *Astrophys. J. Suppl.* **156** 265

[16] Cranmer S R, van Ballegooijen A A and Edgar R J 2007 *Astrophys. J. Suppl.* **171** 520

[17] Verdini A and Velli M 2007 *Astrophys. J.* **662** 669

[18] Verdini A, Velli M and Buchlin E 2009 *Astrophys. J. Letters* **700** L39

[19] Chandran B D G and Hollweg J V 2009 *Astrophys. J.* **707** 1659

[20] Chandran B D G, Dennis T J, Quataert E and Bale S D 2011 *Astrophys. J.* **743** 197

[21] van der Holst B, Sokolov I V, Meng X, Jin M, Manchester IV W B, Tóth G and Gombosi T I 2014 *Astrophys. J.* **782** 81

[22] Cranmer S R, Asgari-Targhi M, Miralles M P, Raymond J C, Strachan L, Tian H and Woolsey L N 2015 *Phil. Trans. R. Soc.* A **373** 20140148

[23] Pontieu B D *et al.* 2007 *Science* **318** 1574

[24] Tomczyk S, McIntosh S W, Keil S L, Judge P G, Schad T, Seeley D H and Edmondson J 2007 *Science* **317** 1192

[25] Tomczyk S and McIntosh S W 2009 *Astrophys. J.* **697** 1384

[26] Tian H *et al.* 2014 *Science* **346** 1255711

[27] Morton R J, Tomczyk S and Pinto R 2015 *Nat. Comm.* **6** 7813

[28] Belcher J W 1971 *Astrophys. J.* **168** 509

[29] Jacques S A 1977 *Astrophys. J.* **215** 942

[30] Chandran B D G, Li B, Rogers B N, Quataert E and Germaschewski K 2010 *Astrophys. J.* **720** 503

[31] Oughton S, Matthaeus W H, Dmitruk P, Milano L J, Zank G P and Mullan D J 2001 *Astrophys. J.* **551** 565

[32] Dmitruk P and Matthaeus W H 2003 *Astrophys. J.* **597** 1097

[33] Zank G P, Adhikari L, Hunana P, Tiwari S K, Moore R, Shiota D, Bruno R and Telloni D 2018 *Astrophys. J.* **854** 18

[34] Perez J C and Chandran B D G 2013 *Astrophys. J.* **776** 124

[35] van Ballegooijen A A and Asgari-Targhi M 2016 *Astrophys. J.* **821** 106

[36] van Ballegooijen A A and Asgari-Targhi M 2017 *Astrophys. J.* **835** 10

[37] Hahn M, Landi E and Savin D W 2012 *Astrophys. J.* **753** 36

[38] Strauss H R 1976 *Phys. Fluids* **19** 134

[39] Zank G P and Matthaeus W H 1992 *J. Plasma Phys.* **48** 85

[40] Velli M, Grappin R and Mangeney A 1989 *Phys. Rev. Letters* **63** 1807

[41] DeForest C E and Gurman J B 1998 *Astrophys. J. Letters.* **501**

[42] DeMoortel I, Ireland J and Walsh R W 2000 *Astron. & Astrophys.* **355** 23

[43] Marsh M A and Walsh R W 2006 *Astrophys. J.* **643** 540

[44] Minarovjech M *et al.* 2003 *Solar Phys.* **213** 269

[45] Pasachoff J M *et al.* 2002 *Solar Phys.* **207** 241

[46] Coles W A and Harmon J K 1989 *Astrophys. J.* **337** 1023

[47] Woo R 1996 *Astrophys. and Space Science* **243** 97

[48] Spangler S R 2002 *Astrophys. J.* **576** 997

[49] Raymond J C, McCauley P I, Cranmer S R and Downs C 2014 *Astrophys. J.* **788** 152

[50] Nakamura M *et al.* 2011 *Earth Planets Space* **63** 443

[51] Miyamoto M *et al.* 2014 *Astrophys. J.* **797** 51

IOP Conf. Series: Journal of Physics: Conf. Series **1100** (2018) 012003 doi:10.1088/1742-6596/1100/1/012003

Calculations in the theory of tearing instability

Stanislav Boldyrev[1,2] and Nuno F. Loureiro[3]

[1]Department of Physics, University of Wisconsin, Madison, WI 53706, USA

[2]Space Science Institute, Boulder, Colorado 80301, USA

[3]Plasma Science and Fusion Center, Massachusetts Institute of Technology, Cambridge MA 02139, USA

E-mail: boldyrev@wisc.edu

Abstract. Recent studies have suggested that the tearing instability may play a significant role in magnetic turbulence. In this work we review the theory of the magnetohydrodynamic tearing instability in the general case of an arbitrary tearing parameter, which is relevant for applications in turbulence. We discuss a detailed derivation of the results for the standard Harris profile and accompany it by the derivation of the results for a lesser known sine-shaped profile. We devote special attention to the exact solution of the inner equation, which is the central result in the theory of tearing instability. We also briefly discuss the influence of shear flows on tearing instability in magnetic structures. Our presentation is self-contained; we expect it to be accessible to researchers in plasma turbulence who are not experts in magnetic reconnection.

1. Introduction

Numerical simulations and analytic models have suggested that magnetic plasma turbulence tends to form anisotropic, sheet-like current structures at small scales [1–10]. Such structures are not necessarily associated with the dissipation scale of turbulence. Rather, a hierarchy of sheet-like turbulent eddies is formed throughout the whole inertial interval [11–15]. Recently, it has been realized that given large enough Reynolds number, such anisotropic structures may become unstable to the tearing mode at scales well above the Kolmogorov-like dissipation scale [16–21]. The Reynolds numbers for which such effects become significant are very large ($Re \gtrsim 10^6$), so their definitive study is beyond the capabilities of modern computers. Nevertheless, the rapid progress in in situ measurements of space plasma brings interest to small scales of magnetic turbulence, where such effects may be observed, e.g., [22]. It is, therefore, highly desirable to develop an understanding of the linear tearing theory in magnetic profiles such as those one might expect to find throughout the inertial range of turbulence, but not necessarily those associated with dissipative current sheets.

This brings attention to the two facets of the theory of tearing instability that are not traditionally covered in textbooks on magnetohydrodynamics or plasma physics. One is the theory of reconnection beyond the well-known Furth, Killeen & Rosenbluth regime of small tearing parameter [23]. This regime assumes limited anisotropy of a reconnecting magnetic profile, so it is not applicable to very anisotropic tearing modes relevant for our study. The other is the theory of tearing instability for the magnetic profiles that are different from the canonical tanh-like Harris profile [24]. Such a profile assumes that the reconnecting magnetic field is uniform in space except for the region where it reverses its direction. This is, arguably, not a general situation encountered in turbulence, where the magnetic fields strength varies in space on similar scales both inside and outside the reconnection region. Different magnetic profiles may lead to different scalings of the corresponding tearing growth rates, e.g., [18, 19, 25–27]. To the best of our knowledge, there are currently no texts methodically covering these aspects of tearing instability. Rather, various relevant analytical results are scattered over the literature, e.g., [28–32].

In this work we review the derivation of the standard Harris-profile tearing mode, and accompany it by a parallel derivation of the results for the sine-shaped profile. We devote special attention to the discussion of the exact solution of the inner equation, for which we use a method different from those previously adopted in the literature [28–32]. Although our work is mostly devoted to the tearing instability in magnetic profiles not accompanied by velocity fields, in the end of our presentation we discuss to what extent shear flows, typically present within turbulent eddies, can modify the results. The goal of our work is to give a self-contained presentation of the theory of tearing effects that are most relevant for applications to turbulence. We believe it will be useful for researchers in turbulence who are not necessarily experts in reconnection.

2. Equations for the tearing mode

We assume that the background uniform magnetic field is in the z direction, and the current sheet thickness, a, and length, l, are measured in the field-perpendicular plane. The current sheets are strongly anisotropic, $a \ll l$. We denote the reconnecting magnetic field, that is, the variation of the magnetic field across the current sheet, as B. Such structures can be created in turbulence if their life times are comparable to the Alfvénic time $\tau_A \sim l/V_A$, where V_A is the Alfvén speed associated with B. They are formed at all scales, the thinner the structure the more anisotropic it becomes. A theory describing a hierarchy of such magnetic fluctuations, or turbulent eddies, in MHD turbulence suggests that their anisotropy increases as their scale decreases, $a/l \propto a^{1/4}$ [11, 12, 33–36].

It has been proposed that at small enough scales the tearing instability of very anisotropic eddies can compete with their Alfvénic dynamics.[1] This means that below a certain critical scale the tearing time should become comparable to τ_A, so that the turbulence is mediated by tearing instability [16–18]. Such a picture has received some numerical and observational support [22, 25, 40]. The theory of tearing instability required to describe strongly anisotropic current sheets, goes beyond the simplified FKR theory and generally requires one to analyze structures that are different from the Harris-type current sheets.

It should be acknowledged that the first analysis of the tearing instability in structures formed by MHD turbulence dates back to 1990 [41].[2] That analysis was based on the Iroshnikov-Kraichnan model of MHD turbulence [42, 43] that treats turbulent fluctuations as essentially isotropic (that is, characterized by a single scale) weakly interacting Alfvén wave packets. The anisotropy of turbulent fluctuations has therefore not been quantified in [41]. Moreover, their model assumed the presence of a significant velocity shear in the current layer and adopted the tearing-mode growth rate calculated in the shear-modified FKR regime [44–46]. As a result, the approach of [41] was qualitatively different from that of the recent studies [16–18].

In our discussion we do not impose any limitations on the anisotropy of current structures, that is, we assume them to be anisotropic enough to accommodate the fastest growing tearing mode; this assumption is consistent with the model of MHD turbulence [12] adopted in [16, 18]. We do, however, make several important simplifications. First, we assume that the background magnetic field has only one component, $B_y(x)$. An optional uniform guide field in z-direction may also be present; it has no effect on the problem. Second, in most of the work we assume that the configuration is static, that is, there is no background flows. In section 8 we briefly

[1] These ideas stem from the observation that if current sheets were allowed to have arbitrarily large aspect ratios, they would be tearing unstable at rates that diverge when the Lundquist number tends to infinity [37–39].

[2] We were not aware of this important early work at the time when our previous studies [16, 18] were published.

discuss the effects of a shear flow, where, similarly to the magnetic field, the shearing velocity field is assumed to have only one component, $v_y(x)$. (We refer the reader to, e.g., [47–49] for a broader discussion of the possible effects of background flows and strong outflows). Finally, our analysis is limited to the MHD framework.

To obtain the equations governing the tearing mode, we follow the standard procedure and represent the magnetic field as $\mathbf{B}(x,y) = B_0 f(x)\hat{\mathbf{y}} + \mathbf{b}(x,y)$, where $f(x)$ describes the profile of the reconnecting field, see Fig. (1). Its typical scale, the thickness of the reconnection layer, is a. The weak perturbation field can be represented through the magnetic potential $\mathbf{b} = -\hat{z} \times \nabla\psi = (\partial\psi/\partial y, -\partial\psi/\partial x)$. We assume that the background velocity is zero. The incompressible velocity perturbation is represented through the stream function $\mathbf{v} = (\partial\phi/\partial y, -\partial\phi/\partial x)$. We will neglect the effects of viscosity, but will keep the magnetic diffusivity. The magnetic induction and velocity momentum equations take the form e.g., [2]:

$$\frac{\partial\psi}{\partial t} = \frac{\partial\phi}{\partial y} B_0 f + \eta\nabla^2\psi, \tag{1}$$

$$\frac{\partial}{\partial t}\nabla^2\phi = B_0 f \frac{\partial}{\partial y}\nabla^2\psi - B_0 f'' \frac{\partial\psi}{\partial y}. \tag{2}$$

We can use the Fourier transform in the y direction, and represent the fluctuating fields as

$$\psi = \tilde{\psi}(\xi)\exp(ik_0 y)\exp(\gamma t), \tag{3}$$

$$\phi = -i\tilde{\phi}(\xi)\exp(ik_0 y)\exp(\gamma t). \tag{4}$$

In these expressions and in what follows we will use the dimensionless variables

$$\xi = x/a, \tag{5}$$

$$\tilde{\eta} = \eta/\left(k_0 V_A a^2\right), \tag{6}$$

$$\lambda = \gamma/\left(k_0 V_A\right), \tag{7}$$

$$\epsilon = k_0 a, \tag{8}$$

where v_A is the Alfvén velocity associated with B_0. The anisotropy parameter ϵ can be arbitrary. In applications to turbulence, however, the most relevant cases are those corresponding to $\epsilon \ll 1$, as only very anisotropic turbulent eddies become significantly affected by the tearing instability, e.g., [16–21, 25]. We, therefore, assume $\epsilon \ll 1$ in our discussion, which is the most difficult case to analyze. The results we derive can be extended to $\epsilon \approx 1$ if necessary.

In what follows we will use the tilded variables and omit the tilde sign. The dimensionless equations take a simple form:

$$\lambda\psi = f\phi + \eta\left[\psi'' - \epsilon^2\psi\right], \tag{9}$$

$$\lambda\left[\phi'' - \epsilon^2\phi\right] = -f\left(\psi'' - \epsilon^2\psi\right) + f''\psi, \tag{10}$$

where we denote by primes the derivatives with respect to ξ. An additional simplification of these equations can be obtained from the following consideration. We assume that η and λ are small parameters (this assumption can be verified a posteriori, from the obtained solution). The range of scales where the terms including these parameters can be neglected will be called the *outer region.* The range of scales where they become significant will be called the *inner region.*

We will solve equations (9) and (10) in the outer and inner regions separately, and then asymptotically match the solutions. In the inner region, where as we will see, $\xi \ll 1$, we have $\partial^2/\partial\xi^2 \gg 1$. Due to the smallness of η and λ, the terms in the square brackets can be relevant

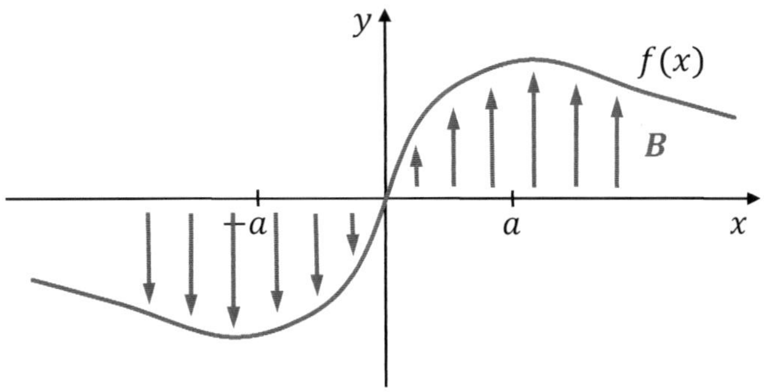

Figure 1. Sketch of a general profile of the background magnetic field.

only in the inner region, and, therefore, small ϵ^2 terms can always be neglected in the square brackets. The equations then take the form

$$\lambda\psi - f\phi = \eta\psi'', \tag{11}$$

$$-f\left(\psi'' - \epsilon^2\psi\right) + f''\psi = \lambda\phi''. \tag{12}$$

Those are the equations describing the tearing instability in the very anisotropic case $\epsilon \ll 1$, and they are the main equations we are going to discuss in this work. The right-hand-side terms in these equations are relevant only in the inner region. In the outer region, they can be neglected.

3. The outer equation

We start with the outer region. We need to solve the equations

$$\phi = \frac{\lambda}{f}\psi, \tag{13}$$

$$\psi'' = \left(\frac{f''}{f} + \epsilon^2\right)\psi, \tag{14}$$

subject to the boundary conditions $\psi, \phi \to 0$ at $\xi \to \pm\infty$ (or to the periodic boundary conditions if the magnetic profile $f(\xi)$ is periodic). Before we discuss the solution we note that Eq. (14) is the Schrodinger equation with zero energy. In general, it does not have a solution corresponding to given boundary conditions. So the solutions should be found separately for $\xi > 0$ and $\xi < 0$, but not in the region of small ξ where this equation is not applicable. The solutions thus found will not, therefore, match smoothly, but will have a discontinuity in the derivative (a break) at $\xi = 0$.

We consider two exactly solvable model cases that correspond to particular profiles $f(\xi)$ of the reconnecting magnetic field. The first case is $f(\xi) = \tanh(\xi)$. It is the so-called Harris profile [24]. It corresponds to a magnetic field that value does not change in space except for a region of width $\xi \sim 1$, where it reverses its direction. The second solvable case is $f(\xi) = \sin(\xi)$ [50]. In this case the magnetic field changes its strength and reverses its direction on the same scales. The latter case is arguably more relevant for the structures encountered in turbulence, and it is especially convenient for numerical studies as it allows for periodic boundary conditions.

In the first case, $f = \tanh(\xi)$, the solution of Eq. (14) for the magnetic field is (e.g., [51]):

$$\psi(\xi) = Ae^{-\epsilon\xi}\left[1 + \frac{1}{\epsilon}\tanh(\xi)\right], \quad \xi \geq 0, \tag{15}$$

$$\psi(-\xi) = \psi(\xi). \tag{16}$$

The solution for the velocity function $\phi(\xi)$ is then easily found from Eq. (13). In order to match with the inner solution, it is important to know the asymptotic forms of the velocity and magnetic fields for $\xi \ll 1$. Taking into account that $\epsilon \ll 1$, one obtains by expanding the $\tanh(\xi)$ in Eq. (15) that $\phi'(\xi) \sim -A\lambda/\xi^2 + (2A\lambda\xi)/(3\epsilon)$. The second term can be neglected when $\xi^3 \ll \epsilon$. The velocity ϕ' can then be formally expressed in this limit as

$$\phi'(\xi) \sim -\frac{\lambda}{\xi^2}\psi(0). \tag{17}$$

As we will see later, in order to match the magnetic field one can define the tearing parameter

$$\Delta' = \frac{\psi'(\xi) - \psi'(-\xi)}{\psi(0)}, \quad \xi > 0. \tag{18}$$

It is easy to see that in the region $\xi \ll 1$, the tearing parameter approaches a constant value

$$\Delta' = \frac{2}{\epsilon}. \tag{19}$$

In the second case, $f(\xi) = \sin(\xi)$, the solution periodic in $[-\pi, \pi]$ is:

$$\psi(\xi) = A\sin\left[\sqrt{1-\epsilon^2}\left(\xi + \frac{\pi}{2\sqrt{1-\epsilon^2}} - \frac{\pi}{2}\right)\right], \quad \xi \geq 0, \tag{20}$$

$$\psi(-\xi) = \psi(\xi). \tag{21}$$

The derivative of the ϕ function is then found as $\phi' \sim -(A\lambda/\xi^2)(\pi\epsilon^2/4) + (A\lambda\xi/3)$. The second term can be neglected when $\xi^3 \ll \epsilon^2$, in which case the ϕ' function has the asymptotic behavior that is formally identical to that obtained for the first case,

$$\phi' \sim -\frac{\lambda}{\xi^2}\psi(0). \tag{22}$$

Indeed, in this case $\psi(0) = A\pi\epsilon^2/4$. In the region $\xi \ll 1$, the tearing parameter approaches a constant value

$$\Delta' = \frac{8}{\pi\epsilon^2}. \tag{23}$$

Note the different scaling of this parameter with ϵ as compared to the previous result (19).

It is important to check how deeply into the asymptotic region $\xi \ll 1$ the outer solutions can extend. In this region we estimate $f'' \sim f \sim \xi$ for both tanh- and sine-shaped magnetic profiles. From Eq. (14) we have $\psi'' \sim \psi$. Then from Eq. (13) we estimate $\phi'' \sim (\lambda/\xi^3)\psi$. The right-hand sides in Eqs. (11, 12) are, therefore, small if

$$\eta \ll \lambda, \tag{24}$$

$$\lambda^2 \ll \xi^4. \tag{25}$$

IOP Conf. Series: Journal of Physics: Conf. Series **1100** (2018) 012003 doi:10.1088/1742-6596/1100/1/012003

4. The inner equation

In the inner region we need to keep the right hand sides of Eqs. (11) and (12). For that the second derivatives of the fields should be large. For instance, we have to assume $\psi'' \gg \psi$, which holds for $\xi \ll 1$. This allows us to simplify Eqs. (11,12) in the following way:

$$\lambda\psi - \xi\phi = \eta\psi'', \tag{26}$$

$$-\xi\psi'' = \lambda\phi''. \tag{27}$$

By differentiating Eq. (26) twice, we get $\lambda\psi'' = \xi\phi'' + 2\phi' + \eta\psi''''$. We now exclude ψ'' and ψ'''' from this equation by using Eq. (27), $\psi'' = -(\lambda/\xi)\phi''$. A few lines of algebra allow one to cast the resulting equation in the form

$$\lambda^2\phi'' + (\xi^2\phi')' - \lambda\eta\left[\phi'''' - 2\left\{\frac{\phi''}{\xi}\right\}'\right] = 0. \tag{28}$$

This equation can trivially be integrated once. Also, noting that it contains only derivatives of ϕ, we may reduce the order by denoting $Y \equiv \phi'$. We then get:

$$Y'' - \frac{2}{\xi}Y' - \frac{1}{\eta\lambda}\left(\lambda^2 + \xi^2\right)Y = C. \tag{29}$$

Before we analyze this equation further, we note that the first two terms in the left-hand side come from the resistive term in the induction equation. Also, one can directly verify that the term λ^2 in the parentheses would be absent if we used a common approximation treating $\psi(\xi)$ as a constant in the inner region, the so-called "constant-ψ" approximation.

In order to find the constant of integration C we need to match asymptotically the solution of the inner Eq. (29) with the solution of the outer equation. We know that the outer solution exists in $\xi^4 \gg \lambda^2$, see Eq. (25), which, for $\xi < 1$ also implies that $\xi^2 \gg \lambda^2$. What asymptotic does the solution of Eq. (29) have in region (25)? There are two possibilities: $Y \sim \exp\left\{\xi^2/\sqrt{4\lambda\eta}\right\}$ and $Y \sim -C\eta\lambda/\xi^2$. By evaluating different terms in Eq. (29) for these asymptotic solutions, one can check that they hold for $\xi^4 \gg \eta\lambda$, which is less restrictive than Eq. (25).

Obviously, the first asymptotic is not the solution we need, since the outer solution does not have an exponential growth at these scales. We, therefore, are interested in the inner solution with the asymptotic

$$Y \sim -C\frac{\eta\lambda}{\xi^2}. \tag{30}$$

In order to match this asymptotic expression with the outer solution $Y \equiv \phi' = -(\lambda/\xi^2)\psi(0)$, see expressions (17) or (22), we need to choose $C = \psi(0)/\eta$.

We note that the region where we asymptotically matched the two solutions is $\lambda^2 \ll \xi^4 \ll \epsilon^{4/3}$ in the case of the tanh-profile, and $\lambda^2 \ll \xi^4 \ll \epsilon^{8/3}$ in the case of the sine-profile. Our solution, therefore, makes sense only when

$$\lambda^2/\epsilon^{4/3} \ll 1, \quad \text{for tanh-shaped profile,} \tag{31}$$

$$\lambda^2/\epsilon^{8/3} \ll 1, \quad \text{for sine-shaped profile,} \tag{32}$$

which, as can be checked when the solution is obtained, are not restrictive conditions.

So far, we have matched the solutions for the velocity field, the ϕ' functions. To complete the asymptotic matching of the inner and outer solutions, we now need to match the magnetic fields, that is, the ψ' function. This can be done in the following way. From Eq. (27) we get for the inner solution $\psi'' = -(\lambda/\xi)\phi''$. Integrating this equation from $\xi \ll -\sqrt{\lambda}$ to $\xi \gg \sqrt{\lambda}$, which for the inner solution is equivalent to integrating from $-\infty$ to ∞, we obtain

$$-\lambda \int_{-\infty}^{+\infty} \frac{\phi''}{\xi} d\xi = \int_{-\infty}^{+\infty} \psi'' d\xi, \tag{33}$$

which, recalling that $\phi' \equiv Y$, can be rewritten as

$$\int_{-\infty}^{+\infty} \frac{Y'}{\xi} d\xi = -\frac{\psi(0)}{\lambda}\Delta'. \tag{34}$$

This asymptotic matching condition will define the growth rate λ.

It is convenient to change the variables in the following way. Let us introduce a function G such that $Y = (\psi(0)/\lambda)G$, and the independent variable $\zeta = \xi^2/\lambda^2$. Then, the velocity-function equation Eq. (29) takes the form

$$4\zeta G'' - 2G' - \beta^2 (1+\zeta) G = \beta^2, \tag{35}$$

where primes denote the derivatives with respect to ζ, and we have denoted $\beta^2 \equiv \lambda^3/\eta$. The matching condition (34) then takes the form

$$-2\int_0^{\infty} \frac{1}{\sqrt{\zeta}} \frac{\partial G}{\partial \zeta} d\zeta = \lambda\Delta'. \tag{36}$$

We thus need to solve Eq. (35), and then find the tearing growth rate λ from the matching condition (36).

The analytic theory of tearing mode is essentially based on Eqs. (35) and (36). These equations can be solved exactly. Historically, however, a better known case is a simpler case of $\beta \ll 1$, the so-called FKR case [23]. The simplifying assumptions going into the FKR derivation are, however, easier to understand if one knows the exact solution of the problem. Here we, therefore, first concentrate on the exact solution.

5. Solution of the inner equation

Here we present the exact solution of the tearing equation (35). This is an inhomogeneous equation, so its solution is a linear combination of a particular solution of the original inhomogeneous equation (35) and a general solution of the homogeneous equation

$$4\zeta G_0{}'' - 2G_0{}' - \beta^2 (1+\zeta) G_0 = 0. \tag{37}$$

The solution of the homogeneous equation (37) has two possible asymptotics, $G_0 \propto \exp(\pm\beta\zeta/2)$, at $\zeta \to \infty$. We obviously need to consider only the solutions behaving as $G_0 \propto \exp(-\beta\zeta/2)$.

In order to figure out what this boundary condition means in terms of the original function $Y(\xi)$, we need to study the solutions of Eq. (37) in more detail. As can be checked directly, at small ζ the solution has the following asymptotic behavior

$$G_0(\zeta) \sim a_0 \left(1 - \frac{\beta^2}{2}\zeta + \frac{\beta^2}{4}\left\{1 - \frac{\beta^2}{2}\right\}\zeta^2 + \dots\right) + b_0\,\zeta^{3/2}\left(1 + \frac{3}{2}\zeta + \dots\right), \tag{38}$$

where a_0 and b_0 are two arbitrary parameters. The "a_0-part" of the solution, which is a regular function at $\zeta = 0$, corresponds to a solution of the homogeneous version of the original velocity

IOP Conf. Series: Journal of Physics: Conf. Series **1100** (2018) 012003 doi:10.1088/1742-6596/1100/1/012003

equation (29) that is even in ξ, while the singular "b_0-part" corresponds to a solution odd in ξ.[3] Equation (29) is symmetric with respect to $\xi \to -\xi$, therefore, each solution of (29) is a linear combinations of even and odd solutions.

One can see from the asymptotic behavior (38) and from Eq. (37) itself that in the odd solutions, the signs of the first and second derivatives are the same and they do not change on positive or negative axes. This means that every odd solution of the homogeneous version of equation (29) diverges at both $\xi \to \infty$ and $\xi \to -\infty$. In the case of even solutions, the same analysis can be applied to the function $G_0(\zeta)\exp\left(\beta^2\zeta/2\right)$ when $\beta < 1$, from which it follows that every even solution of homogeneous Eq. (29) diverges at both infinity limits as well.[4] Only by choosing a particular relation between a_0 and b_0 can one cancel these divergences either at positive or negative infinity (but not at both).

The method that we will use reproduces all the solutions that decline at $\zeta \to \infty$. Therefore, as we will see, our derived expression for G_0 will contain both even and odd parts, but with a rigid relation between a_0 and b_0, to cancel this divergence. We will need to remove such solutions since, as has been explained, they diverge either at $\xi \to -\infty$ or $\xi \to \infty$. Later, we will use this condition to uniquely define the solution.

In order to find the general solution of Eq. (35) we use the following method. The tearing equation (35) is defined only for positive ζ. We can, however, consider this equation on the whole $\zeta-$axis by formally extending the function G to the negative values of the argument. For that we define

$$G(\zeta) = G_1(\zeta)\theta(\zeta) + G_2(\zeta)\left(1 - \theta(\zeta)\right), \tag{39}$$

Where G_1 and G_2 are solutions of Eq. (35) such that $G_1 \to 0$ at $\zeta \to +\infty$, and $G_2 \to 0$ at $\zeta \to -\infty$, and $\theta(\zeta)$ is the Heaviside step function. These solutions at positive and negative arguments are defined up to arbitrary, declining at infinity solutions of the homogeneous equation, therefore, they can always be chosen so that their amplitudes match at the origin, $G_1(0) = G_2(0)$. This provides a formal extension of the G function to the negative arguments. Note that we match only the values of functions G_1 and G_2 at $\zeta = 0$, but not their derivatives.

If we consider the operator

$$\hat{L} = 4\zeta\frac{\partial^2}{\partial\zeta^2} - 2\frac{\partial}{\partial\zeta} - \beta^2(1+\zeta), \tag{40}$$

we can directly verify that

$$\hat{L}G = \theta\hat{L}G_1 + (1-\theta)\,\hat{L}G_2 = \beta^2. \tag{41}$$

Therefore, the extended function (39) satisfies the same equation (35) on the entire real axis. This function declines at $\pm\infty$, therefore, we can Fourier transform equation (35) using the standard definition

$$G(\zeta) = \frac{1}{2\pi}\int\limits_{-\infty}^{\infty} G(k)e^{ik\zeta}dk. \tag{42}$$

In the Fourier space the equation takes the form

$$\left(4k^2 + \beta^2\right)G' + \left(10k - i\beta^2\right)G = 2\pi i\beta^2\delta(k). \tag{43}$$

[3] This follows from the fact that the velocity function $Y(\xi)$ is analytic at $\xi = 0$.

[4] Indeed, the function $P_0 = G_0\exp\left(\beta^2\zeta/2\right)$ satisfies the equation $4\zeta P_0'' = \left[2 + 4\zeta\beta^2\right]P_0' + \left[\beta^2 - \beta^4\right]\zeta P_0$. Its even solution has the asymptotic behavior $P_0 \sim a_0 + a_0\left(\beta^2/4\right)\left(1-\beta^2\right)\zeta^2$ at small ζ. So, when $\beta < 1$, the function itself and, due to the equation it satisfies, its first and second derivatives have the same sign at $\zeta > 0$. Therefore, the function P_0 diverges at infinity, which means that G_0 must diverge as well.

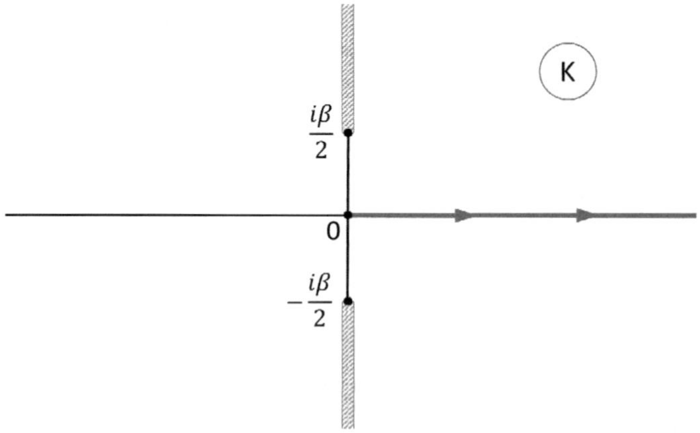

Figure 2. Contour of integration in the k plane in formula (45). Possible branch cuts necessary to define analytic continuations of the integrands in (45) are also shown.

The first-order ordinary differential equation (43) can easily be solved for $k < 0$ and $k > 0$. The solutions are

$$G_{\pm}(k) = 2\pi i A_{\pm}\left[1 + \frac{4k^2}{\beta^2}\right]^{-\frac{5}{4}}\left[\frac{1+\frac{2ik}{\beta}}{1-\frac{2ik}{\beta}}\right]^{\frac{\beta}{4}}, \tag{44}$$

where $A_{\pm}$ are two complex constants, and $\pm$ signs stand for the solutions defined on the positive and negative real k axes, respectively. The function $G(\zeta)$ is real, therefore $A_- = -A_+^*$.

The delta-function in the right-hand side of Eq. (43) implies that the function $G(k)$ is discontinuous on the real k axis at $k = 0$, with the discontinuity condition $A_+ - A_- = 1$. In what follows we will simply denote $A_+ = A$ and $A_- = -A^*$, so that the discontinuity condition reads $A + A^* = 1$. As a result, the function G can be represented as:

$$G(\zeta) = iA\int_0^{+\infty}\left[1 + \frac{4k^2}{\beta^2}\right]^{-\frac{5}{4}}\left[\frac{1+\frac{2ik}{\beta}}{1-\frac{2ik}{\beta}}\right]^{\frac{\beta}{4}} e^{ik\zeta}\,dk - iA^*\int_0^{+\infty}\left[1 + \frac{4k^2}{\beta^2}\right]^{-\frac{5}{4}}\left[\frac{1-\frac{2ik}{\beta}}{1+\frac{2ik}{\beta}}\right]^{\frac{\beta}{4}} e^{-ik\zeta}\,dk. \tag{45}$$

In these integrals the branches of the integrands must be chosen so that they coincide at $k = 0$, and the integration is performed along the positive real line in a complex plane, see Fig.(2). We note that the discontinuity condition defines only the real part of the complex coefficient A, but leaves its imaginary part arbitrary. This reflects the fact the solution is not defined uniquely, but only up to an arbitrary solution of the homogeneous equation (37).[5]

[5] It is easy to see that the solution of the homogeneous equation is given by the integral

$$G_0(\zeta) = A_0\int_{-\infty}^{+\infty}\left[1 + \frac{4k^2}{\beta^2}\right]^{-\frac{5}{4}}\left[\frac{1+\frac{2ik}{\beta}}{1-\frac{2ik}{\beta}}\right]^{\frac{\beta}{4}} \exp(ik\zeta)\,dk, \tag{46}$$

where A_0 is an arbitrary real constant.

For practical calculations, it is convenient to modify Eq. (45) further. In the first integral of Eq. (45) we change the variable of integration

$$k = \left(\frac{\beta}{2i}\right)\frac{q-1}{q+1}, \tag{47}$$

while in the second integral we choose

$$k = -\left(\frac{\beta}{2i}\right)\frac{p-1}{p+1}. \tag{48}$$

Expression (45) now takes the form

$$G(\zeta) = -A\frac{\beta}{4\sqrt{2}}\int_{-1}^{1}\frac{(q+1)^{\frac{1}{2}}}{q^{\frac{5}{4}}}q^{\frac{\beta}{4}}e^{\frac{\beta}{2}\frac{q-1}{q+1}\zeta}\,dq - A^{*}\frac{\beta}{4\sqrt{2}}\int_{-1}^{1}\frac{(p+1)^{\frac{1}{2}}}{p^{\frac{5}{4}}}p^{\frac{\beta}{4}}e^{\frac{\beta}{2}\frac{p-1}{p+1}\zeta}\,dp. \tag{49}$$

The integrals in Eq. (49) look identical. However, they differ by the contours of integration that follow from the changes of variables (47) and (48). If in each integral of Eq. (45), the contours of integration lie along the real axis in the complex k plane, see Fig. (2), then the corresponding contours in the q and p planes are defined as shown in Fig (3).

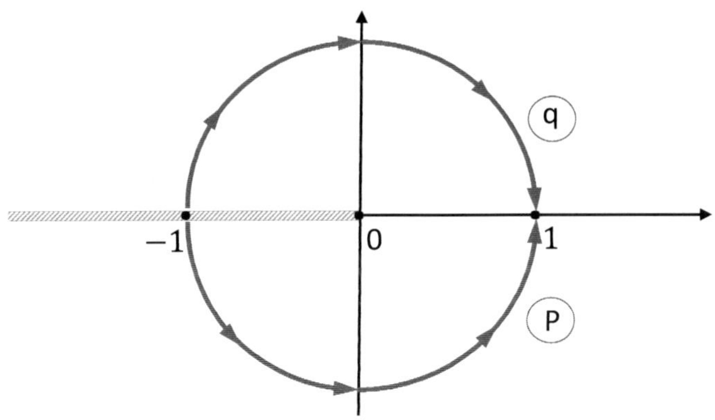

Figure 3. Contours of integration in the q and p complex planes in formula (49).

It is easy to see that for $\zeta > 0$, these integrals will not change if we deform the contours to coincide with the real axis, as shown in Fig (4). This way, one of the contours of integration has to go above the branch cut, and the other one below. It is also useful to integrate by parts once, in order to avoid dealing with too strong a singularity at the origin,

$$G(\zeta) = \frac{A\beta}{1-\beta}\left[1 - \frac{1}{\sqrt{2}}\int_{-1}^{1} q^{\frac{\beta-1}{4}}\frac{d}{dq}\left\{(q+1)^{\frac{1}{2}}\,e^{\frac{\beta}{2}\zeta\frac{q-1}{q+1}}\right\}dq\right]$$
$$+\frac{A^{*}\beta}{1-\beta}\left[1 - \frac{1}{\sqrt{2}}\int_{-1}^{1} q^{\frac{\beta-1}{4}}\frac{d}{dq}\left\{(q+1)^{\frac{1}{2}}\,e^{\frac{\beta}{2}\zeta\frac{q-1}{q+1}}\right\}dq\right]. \tag{50}$$

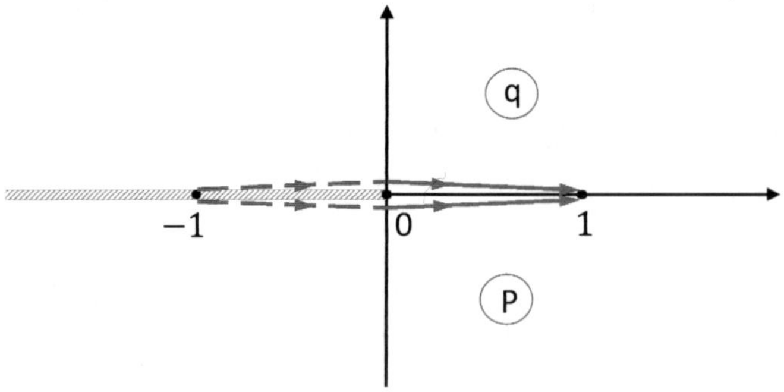

Figure 4. Equivalent contours of integration in the q and p complex planes in formula (50).

In the interval $(0, 1]$ both integrals are the same, and their sum can be simplified since $A + A^* = 1$. In the interval $[-1, 0)$, however, the integrals have different phases. The integrals over the dashed lines add up to $A \exp[i\pi(\beta-1)/4] + A^* \exp[-i\pi(\beta-1)/4]$. Since the imaginary part of A is arbitrary, this sum is arbitrary as well (we assume $\beta < 1$). As a result the solution for $\zeta > 0$ can be represented as

$$G(\zeta) = \frac{\beta}{1-\beta}\left[1 - \frac{1}{\sqrt{2}}\int_0^1 q^{\frac{\beta-1}{4}} \frac{d}{dq}\left\{(q+1)^{\frac{1}{2}} e^{\frac{\beta}{2}\zeta\frac{q-1}{q+1}}\right\} dq\right] + C_0 \int_{-1}^{0} |q|^{\frac{\beta-1}{4}} \frac{d}{dq}\left\{(q+1)^{\frac{1}{2}} e^{\frac{\beta}{2}\zeta\frac{q-1}{q+1}}\right\} dq, \qquad (51)$$

where C_0 is an arbitrary parameter. The first term in this expression is a particular solution of the inner equation (35). It has been originally derived in [28–30], where a different approach involving expansion in the Laguerre polynomials, has been used. The second term, including a free parameter C_0, represents the solution of the homogeneous equation (37), the zero mode.

It is easy to see that the particular solution is an analytic function at $\zeta = 0$, so it describes an even solution of Eq. (29). Also, this solution converges at $\zeta \to \infty$. Therefore, it represents a solution of Eq. (29) that converges at $\xi \to \pm\infty$. The zero mode, on the contrary, is non-analytic at $\zeta = 0$, since its second derivative diverges there. It is therefore a combination of odd and even solutions of homogeneous equation (29). According to what was said in the beginning of this section, a solution of the homogeneous equation diverges at either $\xi \to -\infty$ or $\xi \to +\infty$. We thus have to require $C_0 = 0$, which removes these divergences. The zero mode is therefore absent and the solution is given by the first term in expression (51).

6. Tearing rate in the limit $\beta \ll 1$ (the FKR case)

Now that we have obtained the general solution for the inner region, we can find the tearing mode growth rate by substituting this solution into the matching condition (36). Before considering the general case, however, we discuss the important limit of $\beta \ll 1$, the so-called FKR case [23] (we recall that $\beta^2 = \lambda^3/\eta$). In this limit, we can approximate $q^{(\beta-1)/4} \approx q^{-1/4}$ in Eq. (51). It is possible to show that this approximation is equivalent to the "constant-ψ" approximation discussed after Eq. (29), which demonstrates the equivalence of the constant-ψ approximation to the FKR case. We see that in this case the integral depends on ζ only through the combination $\beta\zeta$. The matching condition (36) now reads

$$C\beta^{3/2} = \lambda\Delta', \tag{52}$$

where the constant C is given by the integral

$$C = \int_0^\infty \sqrt{\frac{2}{x}\frac{d}{dx}} \left[\int_0^1 q^{-\frac{1}{4}} \frac{d}{dq} \left\{ (q+1)^{\frac{1}{2}} e^{\frac{x}{2}\frac{q-1}{q+1}} \right\} dq \right] dx. \tag{53}$$

One can easily do this integral by changing the order of integrations. The answer is expressed through the gamma functions,

$$C = \left(\frac{\pi}{2}\right) \frac{\Gamma\left(\frac{3}{4}\right)}{\Gamma\left(\frac{5}{4}\right)} \approx 1.45. \tag{54}$$

From Eq. (52) we find the growth rate of the tearing mode in the FKR regime of $\beta \ll 1$,

$$\lambda = C^{-4/5}\eta^{3/5}{\Delta'}^{4/5}. \tag{55}$$

In the *dimensional* form, this expression can be rewritten as:

$$\gamma = 2^{4/5}C^{-4/5}\eta^{3/5}k_0^{-2/5}V_A^{2/5}a^{-2} \tag{56}$$

for the tanh-shaped magnetic profile, and

$$\gamma = 8^{4/5}\pi^{-4/5}C^{-4/5}\eta^{3/5}k_0^{-6/5}V_A^{2/5}a^{-14/5} \tag{57}$$

for the sine-shaped magnetic profile. In order to obtain these results we have substituted the expressions (19) and (23) for the corresponding parameters Δ'.

Two important points should be made about the FKR solution. First, since in this limit the inner function $G(\zeta)$ depends on scale only through the combination $\beta\zeta$, this function has a characteristic length scale, $\zeta = 1/\beta$, which is the so-called *inner scale.* In terms of the ξ variable, this scale is $\xi = (\lambda\eta)^{1/4} \gg \lambda$. In dimensional form the inner scale $x = \delta$ is given by

$$\delta = \left(\frac{\gamma\eta a^2}{k_0^2 V_A^2}\right)^{1/4}, \tag{58}$$

where the growth rate γ is given by either (56) or (57) depending on the chosen magnetic profile.

Second, we see that the growth rate and the inner scale formally diverge for $k_0 \to 0$. This is obviously an unphysical behavior. It reflects the fact that the approximation $\beta \ll 1$ is not valid in this case. We will now present the solution for the growth rate in the general case, where the inner function $G(\zeta)$ is given by the exact expression (51). We will see that the growth rate does not diverge, but reaches a maximal value at a certain small value of $k_0 a$.

7. Tearing rate in the general case

Here we analyze the general case, the so-called Coppi case [28–30]. We need to evaluate the integral in the left-hand-side of (36) using the exact expression for the G function given by (51). This can be easily done in the same way as we obtained (54),

$$-2\int_0^\infty \frac{1}{\sqrt{\zeta}}\frac{\partial G}{\partial \zeta}d\zeta = -\frac{\pi}{8}\beta^{3/2}\frac{\Gamma\left(\frac{\beta-1}{4}\right)}{\Gamma\left(\frac{\beta+5}{4}\right)}. \tag{59}$$

The growth rate is found from the transcendental equation [29]:

$$-\frac{\pi}{8}\beta^{3/2}\frac{\Gamma\left(\frac{\beta-1}{4}\right)}{\Gamma\left(\frac{\beta+5}{4}\right)} = \lambda\Delta'. \tag{60}$$

We see that the left-hand-side of this expression is positive when $\beta < 1$, and, therefore, the instability is possible only in this case. In the limit of $\beta \ll 1$, we recover the results discussed in the previous section. The left-hand-side of Eq. (60) is small in this limit. The low-β approximation, however, breaks down at $k_0 \to 0$. As we will see momentarily, in this limit the solution corresponds to $\beta \to 1$. Indeed, equation (60) can be approximated in this case as

$$\sqrt{\pi}\frac{\beta^{3/2}}{1-\beta} = \lambda\Delta'. \tag{61}$$

Recalling now that $\beta^2 \equiv \lambda^3/\eta$, we arrive at the equation for the growth rate

$$\sqrt{\pi}\frac{\beta^{5/6}}{1-\beta} = \eta^{1/3}\Delta'. \tag{62}$$

From the definitions of the dimensionless parameters η and Δ', we see that as $k_0 a \to 0$, the right-hand-side of this equation diverges. This is possible only if $\beta \to 1$ for such solutions. We, therefore, see that $\lambda^3 = \eta$ is the equation that defines the tearing growth rate in this case. In the *dimensional* units, this equation gives

$$\gamma = \eta^{1/3}V_A^{2/3}k_0^{2/3}a^{-2/3}, \tag{63}$$

which is termed the Coppi solution in [32]. The remarkable fact is that as k_0 decreases, the Coppi growth rate decreases as well. This is opposite to the behavior of the FKR growth rate that increases at decreasing k_0. This means that there must exist a maximal growth rate of the tearing instability, γ_*, attainable at a certain wave number k_{0*}. One can *define* this critical wave number as the wave number at which the Coppi growth rate (63) formally matches the FKR growth rate (56) for the tanh-shaped magnetic profile or (57) for the sine-shaped profile.

For the tanh-shaped magnetic profile, a simple algebra gives for the critical wavenumber and the corresponding maximal tearing growth rate:

$$k_{0*} = (2/C)^{3/4}\eta^{1/4}V_A^{-1/4}a^{-5/4}, \tag{64}$$

$$\gamma_* = (2/C)^{1/2}\eta^{1/2}V_A^{1/2}a^{-3/2}. \tag{65}$$

For the sine-shaped profile, the answer is:

$$k_{0*} = (8/\pi C)^{3/7}\eta^{1/7}V_A^{-1/7}a^{-8/7}, \tag{66}$$

$$\gamma_* = (8/\pi C)^{2/7}\eta^{3/7}V_A^{4/7}a^{-10/7}. \tag{67}$$

In the general case, the inner solution $G(\zeta)$ is not a universal function depending only on $\beta\zeta$. However, for $\beta \approx 1$, this function approaches its asymptotic behavior $G(\zeta) \sim 1/\zeta$ at $\zeta \gg 1$. The typical scale (the inner scale) of this solution is therefore $\zeta = 1$, which in terms of the ξ variable reads $\xi = \lambda$. In dimensional variables, the inner scale in this case is

$$\delta = \frac{\gamma a}{k_0 v_A}, \tag{68}$$

where γ is given by Eq. (63), or by expressions (65) or (67) for the fastest growing modes.

8. Tearing rates in the presence of a shear flow

In turbulent systems, magnetic field fluctuations are accompanied by velocity fluctuations, so that both magnetic and velocity shears are present in a turbulent eddy, e.g., [12]. It is therefore useful to comment on the influence of a velocity shear on the tearing instability. A velocity shear across the current layer is expected to be less intense than the magnetic shear, otherwise, such a layer would be destroyed by the Kelvin-Helmholtz instability, e.g., [2, 25]. In MHD turbulence, fluctuations indeed have excess of the magnetic energy over the kinetic energy; the difference between the kinetic and magnetic energies, the so-called residual energy, is negative in the inertial interval, e.g., [52–56].

We assume that, similarly to the magnetic field, the background velocity has the structure $\mathbf{v}_0(x, y) = v_0(x)\hat{\mathbf{y}}$. In the presence of this velocity field, the dimensionless system of equations (9), (10) becomes

$$\lambda\psi = f\phi + \eta\left[\psi'' - \epsilon^2\psi\right] - i\tilde{v}_0\psi, \tag{69}$$

$$\lambda\left[\phi'' - \epsilon^2\phi\right] = -f\left(\psi'' - \epsilon^2\psi\right) + f''\psi - i\tilde{v}_0\left(\phi'' - \epsilon^2\phi\right) + i\tilde{v}_0''\phi, \tag{70}$$

where $\tilde{v}_0 = v_0(x)/V_A$ is the background velocity profile normalized by the Alfvén speed associated with the magnetic profile. Similarly to equations (9), (10), the modified equations can be simplified in the case of small η and λ as:

$$\lambda\psi = f\phi + \eta\psi'' - i\tilde{v}_0\psi, \tag{71}$$

$$\lambda\phi'' = -f\left(\psi'' - \epsilon^2\psi\right) + f''\psi - i\tilde{v}_0\left(\phi'' - \epsilon^2\phi\right) + i\tilde{v}_0''\phi. \tag{72}$$

In the *outer region*, we have from Eqs. (71), (72):

$$\lambda\psi = f\phi - i\tilde{v}_0\psi, \tag{73}$$

$$-f\left(\psi'' - \epsilon^2\psi\right) + f''\psi - i\tilde{v}_0\left(\phi'' - \epsilon^2\phi\right) + i\tilde{v}_0''\phi = 0. \tag{74}$$

A general analysis of the problem is, unfortunately, not very transparent, e.g., [47]. A simplified but quite informative treatment is, however, possible when the velocity profile is similar to that of the magnetic field [44]. We therefore assume that $\tilde{v}_0(x) = \alpha f(x)$, where $-1 < \alpha < 1$.

We see from Eq. (73) that the shear velocity introduces a Doppler shift that competes with the growth rate ($\tilde{v}(\xi)/\lambda = k_0 v(\xi)/\gamma$ in dimensional units). Since in the outer region $v \lesssim V_A$, the Doppler shift dominates, and we can neglect the term containing λ in Eq. (73). Expressing ϕ from Eq. (73), and substituting it into Eq. (74) one obtains after simple algebra [44]:

$$\left(1 - \alpha^2\right)\left\{\psi'' - \epsilon^2\psi - \frac{f''}{f}\psi\right\} = 0. \tag{75}$$

Since $\alpha^2 \neq 1$, the magnetic-field outer equation in this case is identical to the outer equation without a velocity shear (13). The asymptotic behavior of the outer solution for the velocity field is, however, different from Eqs. (17, 22), and it can be found from Eq. (73) to the zeroth order in $\lambda/\tilde{v}_0$:

$$\phi(\xi) \sim i\alpha\psi(\xi), \quad \xi \ll 1. \tag{76}$$

This expression holds in the asymptotic matching region $\delta \ll \xi \ll 1$, where δ is the inner scale.
In the *inner region* $\delta \sim \xi \ll 1$ we have

$$\lambda\psi + i\alpha\xi\psi = \xi\phi + \eta\psi'', \tag{77}$$

$$\lambda\phi'' + i\alpha\xi\phi'' = -\xi\psi''. \tag{78}$$

In order to match the inner solution for the velocity field with the outer solution, we derive from Eq. (78):

$$\lambda \int_{-\infty}^{+\infty} \frac{\phi''}{\xi} d\xi = -i\alpha\phi'|_{-}^{+} - \psi'|_{-}^{+}, \tag{79}$$

where the integral goes from $\xi \ll -\delta$ to $\xi \gg \delta$, and we denote by $|_{-}^{+}$ the jumps of the quantities across the inner layer between the indicated limits. We can use the asymptotic form (76) to evaluate the jump of ϕ'. From Eq. (76) we get

$$\lambda \int_{-\infty}^{+\infty} \frac{\phi''}{\xi} d\xi \sim -\left(1-\alpha^2\right)\psi(0)\Delta'. \tag{80}$$

In order to do the integral in Eq. (80) we need to know the velocity function $\phi(\xi)$ in the inner region. To the best of our knowledge, the exact solution of the inner equations (77, 78) is not available. We may, however, estimate the integral in the left-hand-side of Eq. (80) in the following way. We note that if $\lambda \gg \alpha\delta$, then the shear flow does not affect the inner region of the tearing mode. The inner solution (and the resulting scaling of the growth rate with the Lundquist number $S = aV_A/\eta$ and the anisotropy parameter $k_0 a$) can, therefore, be qualitatively affected by the shear flow only if $\lambda \ll \alpha\delta$, which is essentially the FKR limit. We thus assume this limit in what follows.

Similarly to the case without a flow, the shear-modified FKR limit implies the "constant-ψ" approximation, which reads to the zeroth order in the small parameter $\lambda/(\alpha\delta)$: $\phi(\xi) \sim i\alpha\psi(\xi) \sim$ const at $\xi \ll 1$. This solution trivially satisfies Eqs. (77, 78) to the zeroth order. Obviously, the zeroth order solution does not contribute to the integral in (80), so in order to evaluate this integral we need to go to the first order in $\lambda/(\alpha\delta)$. From Eq. (78) we estimate $\psi'' \sim -i\alpha\phi''$ and substituting this into Eq. (77), we get

$$\lambda\psi(0) + \alpha^2\xi\phi = \xi\phi - i\eta\alpha\phi'', \tag{81}$$

where in the first term in the left-hand-side we have substituted the zeroth order solution. In the region $\xi \gg \delta$, the resistive term is not important, and we obtain the asymptotic form for the first-order velocity field:

$$\phi \sim \frac{\lambda\psi(0)}{(1-\alpha^2)\xi}. \tag{82}$$

This expression does contribute to the integral in Eq. (80). It diverges as ξ decreases, until ξ becomes as small as δ, at which scale the resistive term becomes important and ϕ does not grow anymore. We may then estimate the integral in Eq. (80) as

$$\lambda \int_{-\infty}^{+\infty} \frac{\phi''}{\xi} d\xi \sim 2\lambda \int_{\delta}^{+\infty} \frac{\phi''}{\xi} d\xi \sim -\frac{\lambda^2 \psi(0)}{(1-\alpha^2)\delta^3}, \tag{83}$$

where the inner scale δ is, in turn, estimated from balancing the resistive term in Eq. (81) with the other terms: $\delta^3 \sim \alpha\eta/(1-\alpha^2)$. Substituting this into Eq. (83) and then into Eq. (80) we finally obtain

$$\lambda \sim \left[\eta\, |\alpha| \left(1-\alpha^2\right) \Delta'\right]^{1/2}. \tag{84}$$

This result coincides with the more detailed derivations performed in [44, 47] up to a numerical coefficient of order unity.

In *dimensional* variables, this growth rate takes the following form for the tanh-like magnetic profile:

$$\gamma \sim \kappa^{1/2} \eta^{1/2} V_A^{1/2} a^{-3/2}, \tag{85}$$

while for the sine-like profile we obtain

$$\gamma \sim \kappa^{1/2} \eta^{1/2} V_A^{1/2} a^{-2} k_0^{-1/2}, \tag{86}$$

where we denote $\kappa \equiv |\alpha| \left(1-\alpha^2\right) < 1$. These solutions are shown in Fig. (5), together with the expressions (56), (57), and (63), corresponding to the cases without a flow. We see that a shear flow changes the scaling of the growth rate in the FKR regime, but not in the Coppi regime. In particular, it does not affect the scaling of the fastest growing mode.

Note that the growth rate (85) corresponding to the tanh-like profile is degenerate in that it is independent of the anisotropy parameter $k_0 a$. This explains why the early analysis of [41] based on the Iroshnikov-Kraichnan theory of MHD turbulence, which assumes isotropy of the turbulent fluctuations (i.e, it implies $k_0 a \sim$ const), formally led to the same scaling of the fastest growing mode, $\gamma_* \sim S^{-1/2}$, as the analysis of [16, 17] for the tanh-like magnetic profile, cf. Eq. (56). In the non-degenerate sine-like case, however, the fastest growing rate scales as $\gamma_* \sim S^{-3/7}$ cf. Eq. (57), which is different from $\gamma_* \sim S^{-1/2}$ assumed in [41].

9. Conclusions

We have reviewed the derivation of the anisotropic tearing mode by considering in detail two solvable cases corresponding to the tanh-shaped and sine-shaped magnetic shear profiles. Given large enough anisotropy, the dominating tearing mode has the dimension $\sim 1/k_{0*}$ and grows with the rate γ_*. We see that these parameters depend on the assumed magnetic shear profile, and they are not universal. Their derivation requires one to go beyond the simplified FKR model generally covered in textbooks. We have presented an effective method for solving the inner equation for the current layer in the general case. We have also discussed the influence on the tearing instability of shearing flows that typically accompany magnetic profiles generated by turbulence. Our work provides a detailed and self-contained discussion of the methods required for the study of tearing effects in turbulent systems.

Acknowledgments

SB was partly supported by the NSF grant no. PHY-1707272, NASA Grant No. 80NSSC18K0646, and by the Vilas Associates Award from the University of Wisconsin - Madison. NFL was supported by the NSF-DOE Partnership in Basic Plasma Science and Engineering, award no. DE-SC0016215 and by the NSF CAREER award no. 1654168.

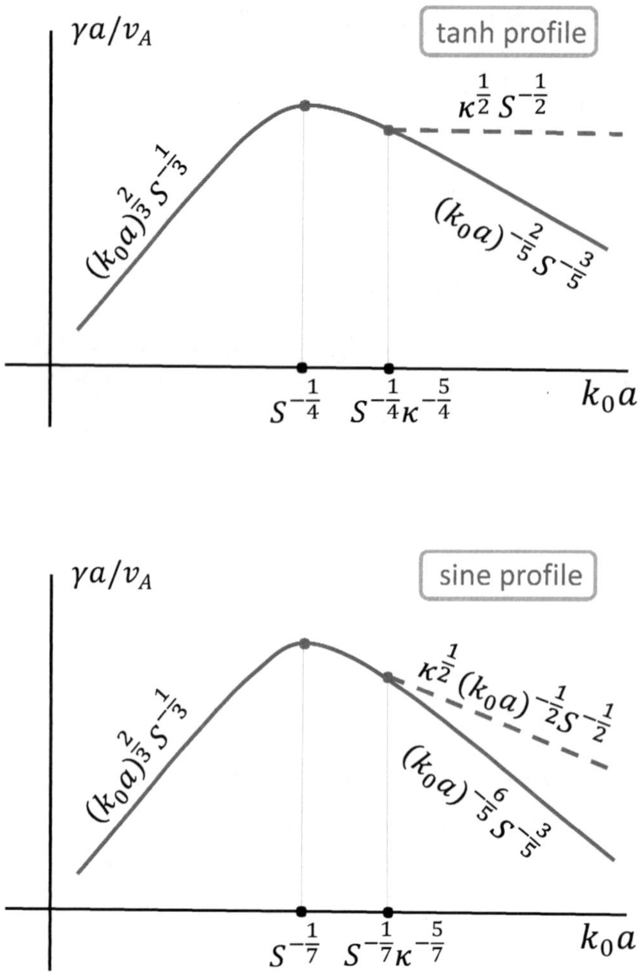

Figure 5. Sketch of the tearing-mode growth rate as a function of (k_0a) for the tanh-like magnetic profile (upper panel) and the sine-like profile (lower panel). The solid lines correspond to the growth rates without a shear flow. The dashed lines show the growth rates in the shear-modified FKR regimes. Here $S = aV_A/\eta$ and $\kappa = |\alpha|\left(1-\alpha^2\right) < 1$.

References

[1] Matthaeus W H and Lamkin S L 1986 *Physics of Fluids* **29** 2513–2534
[2] Biskamp D 2003 *Magnetohydrodynamic Turbulence, Cambridge, UK: Cambridge University Press.* (Cambridge University Press)
[3] Servidio S, Matthaeus W H, Shay M A, Cassak P A and Dmitruk P 2009 *Physical Review Letters* **102** 115003
[4] Servidio S, Dmitruk P, Greco A, Wan M, Donato S, Cassak P A, Shay M A, Carbone V and Matthaeus W H 2011 *Nonlinear Processes in Geophysics* **18** 675–695
[5] Wan M, Matthaeus W H, Servidio S and Oughton S 2013 *Physics of Plasmas* **20** 042307

[6] Zhdankin V, Uzdensky D A, Perez J C and Boldyrev S 2013 *The Astrophysical Journal* **771** 124 (*Preprint* 1302.1460)
[7] Tobias S M, Cattaneo F and Boldyrev S 2013 *MHD Dynamos and Turbulence,* in *Ten Chapters in Turbulence, ed. P.A. Davidson, Y. Kaneda, and K.R. Sreenivasan : Cambridge University Press, p. 351-404.* (*Preprint* 1103.3138)
[8] Zhdankin V, Boldyrev S, Perez J C and Tobias S M 2014 *The Astrophysical Journal* **795** 127 (*Preprint* 1409.3285)
[9] Davidson P A 2017 *An introduction to magnetohydrodynamics* 2nd ed vol 25 (Cambridge university press)
[10] Chen C H K and Boldyrev S 2017 *The Astrophysical Journal* **842** 122 (*Preprint* 1705.08558)
[11] Boldyrev S 2005 *The Astrophysical Journal* **626** L37–L40 (*Preprint* astro-ph/0503053)
[12] Boldyrev S 2006 *Physical Review Letters* **96** 115002 (*Preprint* astro-ph/0511290)
[13] Chen C H K, Mallet A, Schekochihin A A, Horbury T S, Wicks R T and Bale S D 2012 *Astrophysical Journal* **758** 120 (*Preprint* 1109.2558)
[14] Chandran B D G, Schekochihin A A and Mallet A 2015 *The Astrophysical Journal* **807** 39 (*Preprint* 1403.6354)
[15] Mallet A, Schekochihin A A, Chandran B D G, Chen C H K, Horbury T S, Wicks R T and Greenan C C 2016 *Monthly Notices of the Royal Astronomical Society* **459** 2130–2139 (*Preprint* 1512.01461)
[16] Loureiro N F and Boldyrev S 2017 *Physical Review Letters* **118**(24) 245101
[17] Mallet A, Schekochihin A A and Chandran B D G 2017 *Monthly Notices of the Royal Astronomical Society* **468** 4862–4871 (*Preprint* 1612.07604)
[18] Boldyrev S and Loureiro N F 2017 *The Astrophysical Journal* **844** 125 (*Preprint* 1706.07139)
[19] Loureiro N F and Boldyrev S 2017 *The Astrophysical Journal* **850** 182 (*Preprint* 1707.05899)
[20] Mallet A, Schekochihin A A and Chandran B D G 2017 *Journal of Plasma Physics* **83** 905830609 (*Preprint* 1707.05907)
[21] Comisso L, Huang Y M, Lingam M, Hirvijoki E and Bhattacharjee A 2018 *The Astrophysical Journal* **854** 103 (*Preprint* 1802.02256)
[22] Vech D, Mallet A, Klein K G and Kasper J C 2018 *The Astrophysical Journal* **855** L27 (*Preprint* 1803.00065)
[23] Furth H P, Killeen J and Rosenbluth M N 1963 *Physics of Fluids* **6** 459–484 ISSN 1070-6631
[24] Harris E G 1962 *Il Nuovo Cimento* **23** 115–121
[25] Walker J, Boldyrev S and Loureiro N 2018 *Physical Review E* (*Preprint* 1804.02754)
[26] Loureiro N F and Boldyrev S 2018 *ArXiv e-prints* (*Preprint* 1805.09224)
[27] Pucci F, Velli M, Tenerani A and Del Sarto D 2018 *Physics of Plasmas* **25** 032113 (*Preprint* 1801.08412)
[28] Coppi B, Greene J M and Johnson J L 1966 *Nuclear Fusion* **6** 101
[29] Coppi B, Galvão R, Pellat R, Rosenbluth M and Rutherford P 1976 *Soviet Journal of Plasma Physics* **2** 961–966
[30] Ara G, Basu B, Coppi B, Laval G, Rosenbluth M N and Waddell B V 1978 *Annals of Physics* **112** 443–476
[31] Porcelli F 1987 *Physics of Fluids* **30** 1734–1742
[32] Loureiro N F, Schekochihin A A and Cowley S C 2007 *Physics of Plasmas* **14** 100703–100703–4 ISSN 1070664X
[33] Mason J, Cattaneo F and Boldyrev S 2006 *Physical Review Letters* **97** 255002 (*Preprint* astro-ph/0602382)
[34] Mason J, Perez J C, Boldyrev S and Cattaneo F 2012 *Physics of Plasmas* **19** 055902–055902 (*Preprint* 1202.3474)
[35] Perez J C, Mason J, Boldyrev S and Cattaneo F 2012 *Physical Review X* **2** 041005 (*Preprint* 1209.2011)
[36] Chandran B D G, Schekochihin A A and Mallet A 2015 *The Astrophysical Journal* **807** 39 (*Preprint* 1403.6354)
[37] Pucci F and Velli M 2014 *The Astrophysical Journal Letters* **780** L19 ISSN 2041-8205 URL http://iopscience.iop.org/2041-8205/780/2/L19
[38] Uzdensky D A and Loureiro N F 2016 *Physical Review Letters* **116** 105003 (*Preprint* 1411.4295)
[39] Tenerani A, Velli M, Pucci F, Landi S and Rappazzo A F 2016 *Journal of Plasma Physics* **82** 535820501 (*Preprint* 1608.05066)
[40] Dong C, Wang L, Huang Y M, Comisso L and Bhattacharjee A 2018 *ArXiv e-prints* (*Preprint* 1804.07361)
[41] Carbone V, Veltri P and Mangeney A 1990 *Physics of Fluids A* **2** 1487–1496
[42] Iroshnikov P S 1963 *Astron. Zh.* **40** 742
[43] Kraichnan R H 1965 *Physics of Fluids* **8** 1385–1387
[44] Hofmann I 1975 *Plasma Physics* **17** 143–157
[45] Dobrowolny M, Veltri P and Mangeney A 1983 *Journal of Plasma Physics* **29** 393–407
[46] Einaudi G and Rubini F 1986 *Physics of Fluids* **29** 2563–2568
[47] Chen X L and Morrison P J 1990 *Physics of Fluids B* **2** 495–507
[48] Tolman E A, Loureiro N F and Uzdensky D A 2018 *Journal of Plasma Physics* **84** 905840115 (*Preprint*

IOP Conf. Series: Journal of Physics: Conf. Series **1100** (2018) 012003 doi:10.1088/1742-6596/1100/1/012003

1711.09042)
[49] Loureiro N F, Schekochihin A A and Uzdensky D A 2013 *Physical Review E* **87** 013102 (*Preprint* 1208.0966)
[50] Ottaviani M and Porcelli F 1993 *Physical Review Letters* **71** 3802–3805
[51] White R B 1983 *Resistive instabilities and field line reconnection.* (Handbook of plasma physics. Vol. 1: Basic plasma physics I.. A. A. Galeev, R. N. Sudan (Editors).North-Holland Publishing Company, Amsterdam - New York - Oxford. 19+751 pp. (1983).)
[52] Boldyrev S, Perez J C, Borovsky J E and Podesta J J 2011 *The Astrophysical Journal* **741** L19 (*Preprint* 1106.0700)
[53] Wang Y, Boldyrev S and Perez J C 2011 *The Astrophysical Journal* **740** L36 (*Preprint* 1106.2238)
[54] Boldyrev S, Perez J C and Zhdankin V 2012 *American Institute of Physics Conference Series* (*American Institute of Physics Conference Series* vol 1436) ed Heerikhuisen J, Li G, Pogorelov N and Zank G pp 18–23 (*Preprint* 1108.6072)
[55] Boldyrev S, Perez J C and Wang Y 2012 *Numerical Modeling of Space Plasma Slows (ASTRONUM 2011)* (*Astronomical Society of the Pacific Conference Series* vol 459) ed Pogorelov N V, Font J A, Audit E and Zank G P p 3 (*Preprint* 1202.3453)
[56] Chen C H K, Bale S D, Salem C S and Maruca B A 2013 *The Astrophysical Journal* **770** 125 (*Preprint* 1304.7818)

Electron and ion heating due to magnetic reconnection at the heliopause

Iver H Cairns[1] **and S A Fuselier**[2,3]

[1] School of Physics, University of Sydney, Sydney, NSW 2006, Australia
[2] Space Science and Engineering Division, Southwest Research Institute, San Antonio, TX 78228, USA
[3] Southwest Research Institute, San Antonio, TX, Australia

E-mail: iver.cairns@sydney.edu.au,sfuselier@swri.edu

Abstract. Magnetic reconnection is a well-known source of electron and ion bulk heating, as well as energetic particles, in the solar system. Several authors have suggested that reconnection occurs at the heliopause. The paper's primary focus is to predict the amount of electron and ion bulk heating for heliopause reconnection, using the empirical relations of T. Phan and colleagues between the changes in electron and ion temperature and an Alfven speed $V_{A,asym,r}$. This Alfven speed depends on the strengths of the reconnecting magnetic fields and includes asymmetries in the magnetic fields and densities on the two inflowing sides of the reconnection region. For the undisturbed interstellar flow the predicted $V_{A,asym,r} \approx 25$ km/s and the predicted changes in electron and ion temperature are $\Delta T_e 1000$ K and $\Delta T_i 6000$ K. These changes are relatively small and likely not important for dynamics at the heliopause. However, a plasma depletion layer (PDL) is predicted beyond the heliopause, analogous to the PDLs observed sunwards of the magnetopauses of Earth, Mercury, Jupiter, and Saturn. In the PDL, the interstellar (ISM) magnetic field lines drape over the heliopause. Plasma ions and electrons with relatively large parallel temperatures escape along the field, increasing the field strength, decreasing the plasma density, and increasing the Alfven speed. In the region of the PDL where these effects are strong, the expected field and density changes are a factor of 4 and 1/4, respectively, increasing $V_{A,asym,r}$ by a factor close to 3 and the temperature changes by almost a factor of 10. Thus, heliopause reconnection in a strong PDL is predicted to increase the electron and ion temperatures by up to 10^4 K and 8×10^4 K, respectively, corresponding to changes by factors of order 1.5 and 11 compared to the predicted ISM temperature of $\approx$ 7500 K. Thus, the effects of bulk heating in heliopause reconnection regions will be most important for plasma inside or magnetically connected to the strong region of the heliopauses PDL. As an aside, Coulomb collisions appear too slow to relax the ion temperature anisotropies in the PDL beyond the heliopause, different than for the electrons.

1. Introduction

Driven by data from the Voyager, IBEX, and Cassini spacecraft and by theoretical work and simulations, our understanding of the solar wind's interactions with the local interstellar medium continues to evolve (Figure 1. For instance, Voyager-1 and -2 observations demonstrate the existence of the solar wind's termination shock [1, 2], while Voyager-1 data show the existence of a heliopause that separates the shocked solar wind plasma and magnetic fields (in the inner heliosheath) from interstellar plasma and fields [3, 4]. However, IBEX observations provide very strong arguments against an outer bow shock for the solar system, since the inferred fast mode Mach number is very close to 1.0 [5]. The plasma and fields just beyond the heliopause, in the outer heliosheath [6], are not those of the pristine interstellar medium. Instead these are modified by charge-exchange between interstellar plasma and so-called solar wind neutrals (themselves formed by charge-exchange between interstellar neutrals and protons in the solar wind and inner heliosheath) [7], MHD effects associated with the interstellar flow being slowed and diverted around the heliopause, and the heliopause's "plasma depletion layer" or PDL [8–10].

PDLs are observed upstream of Earth's magnetopause [11–13] and at all other planetary magnetopauses where they have been searched for [14, 15]. By analogy with the models for these PDLs, the heliopause's PDL is where interstellar plasma and field are slowed and amplified, respectively, until they can be deflected away from the PDL's center plane and slip around the heliopause obstacle. Maximum slowing and field draping should occur in the plane defined by the pristine interstellar flow velocity $\mathbf{V}_{ISM}$, the pristine interstellar magnetic field $\mathbf{B}_\infty$, and the Sun.

The modern picture of the PDL involves of draping of magnetic fields across the obstacle (e.g., the heliopause), escape of particles with relatively large $v_\parallel$ along $\pm\mathbf{B}$ away from the draping region (thus decreasing the plasma density), increase of $v_\perp$ due to conservation of magnetic moment and B increasing in the draping region, increase of a temperature anisotropy $T_\perp/T_\parallel$ via these effects, wave growth driven by the temperature anisotropy, limitation of the wave growth and temperature anisotropy by moving perpendicular particle energy into the waves and parallel particle energy, and evolution of the system to a state of marginal stability in which particle loss along $\pm\mathbf{B}$, the temperature anisotropy, and wave growth are balanced [8–10, 12, 13, 16]. Thus the PDL in the outer heliosheath should have large-scale, anti-correlated gradients in n and B,

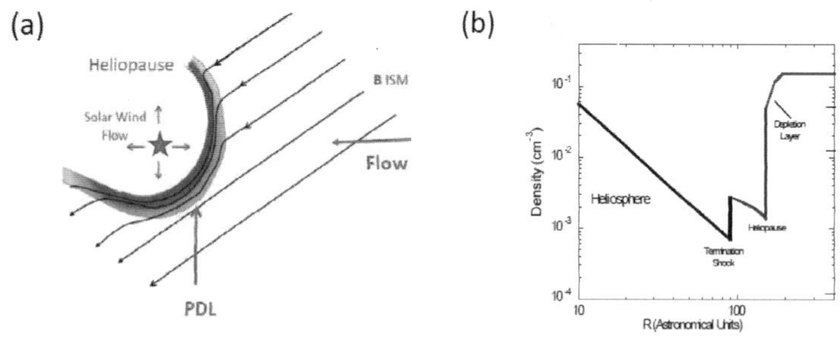

Figure 1. (a) Schematic of the solar system's termination shock, heliopause, and PDL, together with interstellar magnetic field lines that are draped over the heliopause. (b) Schematic of the density as a function of radial distance r from the Sun in a direction close to anti-parallel to the undisturbed interstellar flow direction. This shows the r^{-2} fall-off in the solar wind, the increase at the termination shock, and the increase in density at the heliopause and then in the PDL.

with small n and large B close to the heliopause. These effects are observed [10], providing strong evidence for a PDL beyond the heliopause. The characteristic spatial scale of the PDL is observed to be of order ≈ 2.6 AU [10], somewhat smaller than the estimated $6-12$ AU estimated from a scaling relation based on PDLs observed elsewhere in the solar system [14]; this difference may be due to the heliopause moving outwards due to the effects of solar transients and solar cycle variations [10].

Recently J.R. Jokipii and J. Giacolone [17, 18] and Mostafavi and Zank [19] have argued that the plasma particles should have small mean free paths for Coulomb collisions that are less than 1 AU. This suggests that Coulomb collisions might isotropize the plasma particles and remove the temperature anisotropy required for the foregoing model of the PDL at the heliopause. These aspects are addressed in the Discussion Section below.

When two plasmas with anti-parallel components of $\mathbf{B}$ impinge on one another magnetic reconnection sometimes occurs. This magnetic involves the magnetic field lines reconnecting from one side to the other to change the magnetic topology, the bulk plasma escaping from the reconnection site at speeds close to a suitably defined Alfven speed (see below) perpendicular to the relative velocity vector of the impinging plasmas, and conversion of some magnetic energy (from the anti-parallel field components) into heating of the bulk plasma and the acceleration of energetic particles.

Magnetic reconnection at the heliopause is expected, by analogy for example with Earth's magnetopause, with reconnection between draped interstellar (solar wind) magnetic fields and the inner heliosheath (magnetospheric) fields. This paper does not address reconnection in the distant solar wind or in the inner heliosheath well away from the heliopause or changes in the large-scale topology and the shape of the helioapuse [20, 21]. Similarly, it does not address the conditions under which reconnection should occur at the heliopause [9, 22] or whether reconnection was occurring when Voyager 1 crossed the heliopause [9, 22–24]. It also does not address whether energetic particles are accelerated in reconnection regions at the heliopause.

Here we predict the amount of electron and ion bulk heating expected at the heliopause, both inside and outside the strong region of the PDL. The approach taken is strongly empirical: we apply the empirical relations of Phan et al [25, 26] for bulk heating of electrons and protons, respectively, for appropriate plasma conditions.

The paper proceeds by summarising the empirical relations of Phan and colleagues (Section 2) and applying them to magnetic reconnection at the heliopause, both within and outside the strong region of the PDL (Section 3). The basic result is that reconnection will not produce major heating at the heliopause, with at most factors of 2 and 10 increase for the electrons and ions, respectively, although heating in the strong field region of the PDL will be approximately a factor of 10 larger than in weak regions of the PDL. Given the low plasma temperatures ≈ 7000 K expected in the local ISM. These results are discussed in Section 4, including comparisons with the heating expected from pickup of charge-exchange protons and lower hybrid drive (LHD) beyond the heliopause [6],which are at least 2 orders of magnitude larger, and a quick sketch of other solar system and astrophysical applications for this empirical prediction technique. The conclusions are in Section 5.

2. Summary of Phan et al.'s empirical model

Phan et al. [25] have analysed over 79 crossings of Earth's magnetopause and correlated the variations of the bulk electron heating observed as functions of the plasma and field parameters. As illustrated in Figure 2(a), where the quadratic dependence is clear, their result is

$$k_B \Delta T_e = 0.017\ m_i V_{A,asym,r}^2 \ . \tag{1}$$

Here k_B is Boltzmann's constant, m_i is the ion (proton) mass, and $V_{A,asym,r}$ is the so-called "asymmetric reconnecting Alfven speed" that depends on the possibly different magnetic field

IOP Conf. Series: Journal of Physics: Conf. Series **1100** (2018) 012004 doi:10.1088/1742-6596/1100/1/012004

strengths and plasma densities on either side of the reconnection region and only involves the reconnecting portion of the fields (e.g., not including the guide field) [27–29]. In detail,

$$V_{A,asym,r}^2 = \frac{B_{ih}B_{oh}(B_{ih}+B_{oh})}{\mu_0 m_i(n_{ih}B_{oh}+n_{oh}B_{ih})} \; . \tag{2}$$

Here subscripts ih and oh refer to the inner heliosheath and outer heliosheath, respectively. In the limit that the two sides are symmetric and there is no guide field then (2) reduces to the standard Alfven speed. Including the guide field in B_{ih} and B_{oh} will overestimate $V_{A,asym,r}$ and so the bulk heating.

Equation (1) explains the very small bulk electron heating observed in reconnection regions in the solar wind at 1 AU [25, 30]. In brief, Eq (1) predicts that 1.7% of the of the ion kinetic energy in the reconnection exhaust regions is found in bulk electron heating in the outflow. This result is qualitatively consistent with multiple kinetic simulations[31, 32].

Analyses by Drake et al. and Phan et al. [26, 33] show a very similar empirical relation to Eq. 1 for the bulk ion heating (Figure 2):

$$k_B \Delta T_i = 0.13 m_i V_{A,asym,r}^2 \; . \tag{3}$$

Thus the bulk ion heating is predicted to be approximately 7 times the electron heating, naturally producing strongly non-equilibrium plasma with greater ion temperatures than electron temperatures. At this time we assume the heating is isotropic, rather than anisotropic relative to **B**. Again,kinetic simulations provide some qualitative support for this nodel [31–33].

3. Heating due to reconnection at the heliopause

An initial, order of magnitude, estimate for the bulk hearing expected from magnetic reconnection at the heliopause follows by assuming that $V_{A,asym,r} \approx 20$ km $^{-1}$, a typical value expected in the heliosheath without significant draping effects or a bow shock. Figure 2 immediately predicts that $k_B \Delta T_e < k_B \Delta T_i <$ 10ěV. More precisely Eqs (1) and (3) predict $k_B \Delta T_e \approx 800$ K and $k_B \Delta T_i \approx 7000$ K. In comparison the temperature in the local ISM is believed to be 7500 ± 500 K [5].

Figure 2. (a) After Phan et al.'s [25] Figure N: observations of ΔT_e versus $V_{A,asym,r}$ in terrestrial magnetopause reconnection regions, an associated quadratic fit, and red vertical bars which show the values of 25 and 100 kms^{-1} relevant to the heliopause for a nominal (Voyager 1) and strong PDL, respectively. (b) Similar to (a) but for the ion heating ΔT_i, after Phan et al. [26].

The nominal effect of the PDL is to increase the predicted heating by up to a factor ≈ 4. The reason that is in the strong region of the PDL B is expected to increase by up to a factor of 4 while n decreases by a factor of 4; these changes cause the squared Alfven speed factor in Eqs (1) and (3) to increase by a factor of 4.

Consider now the density and field parameters observed at Voyager 1's crossing of the heliopause [3, 4, 9, 10]: $n_{ih} = 2 \times 10^{-3}$ cm^{-3}, $B_{ih} = 0.25$ nT, $n_{oh} = 0.1$ cm^{-3}, and$B_{oh} = 0.30$ nT, Then $V_{A,asym,r} = 24$ km s^{-1} and so $\Delta T_e \approx 800$ K and $\Delta T_i \approx 6000$ K $\approx T_i$, assuming that the fields are anti-parallel (otherwise these are upper limits. These estimates are slightly smaller than for the nominal value of V due to the asymmetries across the reconnection region between the inner and outer heliosheaths.

Taking these asymmetries across the PDL for a maximally strong region of the PDL (as expected for Voyager 2 [10]) leads to $V_{A,asym,r} = 72$ km s^{-1}. This is a factor of 3 increase, 50% larger than for a symmetric reconnection site, and pointing towards a factor of 9 increase in heating compared with the PDL for Voyager 1. Then Eqs (1) and (3) predict $\Delta T_e \approx 10^4$ K $\approx 1.3 T_e$ and $\Delta T_i \approx 8 \times 10^4$ K $\approx 11 T_i$, respectively.

4. Discussion

The foregoing Section shows that bulk plasma heating at heliopause magnetic reconnection sites should be relatively small (of order a factor of 2 for ions, but $< 20\%$ for electrons) where the PDL is relatively weak, and less than a factor of 10 for ions and a factor of 2 for electrons where the PDL is strong. These estimates are likely overestimates, since they assume that the reconnecting fields are anti-parallel and have no guide field, contrary to realistic expectations [9]. Thus significant bremmstrahlung EUV or X-ray radiation from these regions are not expected.

Future work is necessary to compare these temperature changes in detail with those expected from adiabatic compression in the PDL (less than a factor of 4 [9]) and with the heating expected from pick-up ions and the associated "lower-hybrid drive" (LHD) process [6, 7, 9, 34]. Concerning the latter point, whereas the effective perpendicular temperature of the pick-up ion ring is $\approx m_i v_{sw}^2 / k_B \approx 10^6$ K and the effective parallel electron heating from LHD is $\approx 10^6$ K, the very low pick-up ion fraction $\approx 0.03\%$ suggests that the temperature changes will be less than 10^4 K and so small compared with the heating in heliopause reconnection regions. Future work should consider the importance of these heating mechanisms as the source of the mildly superthermal electrons, acclerated by shocks, that drive Langmuir waves beyond the heliopause [4].

It is clear that this paper's approach can be used to estimate the bulk electron and ion heating expected due to magnetic reconnection in multiple other solar system and astrophysical situations, as well as laboratory experiments. For instance, the characteristic electron and ion temperatures in Earth's plasma sheet, and especially the factor of $\approx 5 - 10$ higher ion temperatures, follow naturally from Eqs (1) and (3) for characteristic Alfven speeds $\approx 500 - 2000$ kms^{-1}. Similar comments are appropriate for solar flares and it would be natural to test Eqs (1) and (3) using solar flare data. In addition, note that the upper limit for $V_{A,asym,r}$ is the speed of light, whence the maximum ensuing bulk electron and ion temperatures are $\approx 0.017 m_i c^2 / k_B = 2 \times 10^{11}$ K and $\approx 0.13 m_i c^2 / k_B = 1.4 \times 10^{12}$ K, respectively. The relevance of these estimates to incoherent UV, X-ray, and radio emission from pulsar winds should be determined.

Another possible line of enquiry is whether analogues to Eqs (1) and (3) exist for the fraction of energy released in energetic particles accelerated in reconnection regions. Data from kinetic simulations and either laboratory could be mined for such relations. These relations may depend on the particular acceleration processes active in the simulations, but would be more powerful if not.

We finish this discussion with some remarks on Coulomb collisions and temperature anisotropies in the PDL at the heliopause. Unexpectedly, recent estimates for the Coulomb collision mean free path beyond the heliopause are mostly less than 1 AU [17–19], suggesting that the temperature anisotropies required for the usual model for the PDL will not develop [17]. In detail the mean free paths for electron-electron, electron-proton, and proton-proton Coulomb collisions in an unmagnetized plasma with temperatures characteristic of the local ISM are $\approx 0.2 - 0.5$ AU, while the proton-electron mean free path is ≈ 20 AU [19]. However, standard theory for temperature isotropisation by Coulomb collisions [35] predicts that

$$\frac{dT_\perp}{dt} = -\frac{dT_\parallel}{dt} = -\nu_T^\alpha (T_\perp - T_\parallel) , \qquad (4)$$

where the rate coefficients for each species are

$$\nu_T^e \approx 8.2 \times 10^{-7} n \lambda T_e^{-3/2} \qquad (5)$$

$$\nu_T^i \approx 1.9 \times 10^{-8} n \lambda T_{e-i}^{-3/2} \qquad (6)$$

in the limit that $T_\perp \approx T_\parallel \approx T_\alpha$, and n and T are measured in cm^{-3} and eV, respectively. Here $\lambda \approx 24$ is the Coulomb logarithim for $n = 0.05$ cm^{-3} and $T = 0.6$ eV. Then $\nu_T^e \approx 2 \times 10^{-6}$ s^{-1} and $\nu_T^i \approx 5 \times 10^{-8}$ s^{-1}, with the corresponding characteristic times $\tau_\alpha = 1/\nu_T^\alpha$ equal to 5×10^5 s and 2×10^7 s for electrons and protons, respectively. The corresponding convection distances at 25 km s^{-1}, the nominal convection speed in the outer heliosheath, are 0.08 AU and 3.3 AU, respectively.

Thus, within the estimated PDL thickness ≈ 2.6 AU observed by Voyager 1, Coulomb collisions are not predicted to be fast enough to isotropise the perpendicular and parallel temperatures of the protons and to remove the temperature anisotropies built by the standard PDL physics. Put another way, the standard ion PDL predicted for the outer heliosheath beyond the heliopause should be robust against the effects of Coulomb collisions. In contrast, Coulomb collisions should be fast enough to remove the corresponding electron temperature anisotropy.

5. Conclusions

Observations near 1 AU provide strong evidence for an empirical model in which the amount of bulk electron and ion heating in reconnection regions is 1.7% and 13% of the ion kinetic energy $1/2 m_i V_{A,asym,r}^2$ in the reconnection outflows [25, 26, 33]. This asymmetric, reconnecting, Alfven speed $V_{A,asym,r}$ is dependent on the strengths of the reconnecting magnetic fields and includes asymmetries in the magnetic fields and densities on the two inflowing sides of the reconnection region [27–29]. Kinetic simulations provide some support for these models [31–33]. Applying this model to reconnection at the nominal heliopause, for instance for the Voyager 1 crossing, yields changes in electron and ion temperatures of only ≈ 800 K and 7000 K, respectively. The nominal ISM temperature is ≈ 7500 K [5]. However, in strong regions of the PDL (e.g., plausibly for the future Voyager 2 crossing), $V_{A,asym,r}^2$ is expected to increase by a factor ≈ 3 and the temperature changes by factors ≈ 11: $\Delta T_e \approx 10^4$ K and $\Delta T_i \approx 8 \times 10^4$ K. Thus reconnection is predicted to produce strong heating at the heliopause in strong regions of the PDL, and so hot electrons and ions in regions magnetically connected to strong regions of the PDL. Multiple future applications of the foregoing approach appear viable for solar system and astrophysical sources, with the maximum electron and ion temperatures estimated to be $\approx 2 \times 10^{11}$ K and 1.4×10^{12} K, respectively, when $V_{A,asym,r} = c$. Finally, Coulomb collisions appear too slow to relax the ion temperature anisotropies in the PDL beyond the heliopause, different than for the electrons.

Acknowledgments

We acknowledge useful discussions at the 2018 AIAA meeting with P. Mostafavi, J. Giacolone, R. Jokipii, M. Shay, T. Terasawa, and G. Zank. Support for this study comes from the IBEX project, part of NASA's Explorer program.

References

[1] Stone E C, Cummings A C, McDonald F B, Heikkila B C, Lal N and Webber W R 2005 *Science* **309** 2017–2020

[2] Stone E C, Cummings A C, McDonald F B, Heikkila B C, Lal N and Webber W R 2008 *Nature* **454** 71–74

[3] Burlaga L F, Ness N F and Stone E C 2013 *Science* **341** 147–150

[4] Gurnett D A, Kurth W S, Burlaga L F and Ness N F 2013 *Science* **341** 1489–1492

[5] McComas D J, Alexashov D, Bzowski M, Fahr H, Heerikhuisen J, Izmodenov V, Lee M A, Möbius E, Pogorelov N, Schwadron N A and Zank G P 2012 *Science* **336** 1291

[6] Cairns I H and Zank G P 2002 *Geophysical Research Letters* **29** 1143

[7] Zank G P 1999 *Space Science Reviews* **89** 413–688

[8] Fuselier S A and Cairns I H 2013 *The Astrophysical Journal* **771** 83

[9] Fuselier S A and Cairns I H 2015 *Journal of Physics Conference Series* (*Journal of Physics Conference Series* vol 642) p 012010

[10] Cairns I H and Fuselier S A 2017 *The Astrophysical Journal* **834** 197

[11] Zwan B J and Wolf R A 1976 *Journal of Geophysical Research* **81** 1636–1648

[12] Anderson B J and Fuselier S A 1993 *Journal of Geophysical Research* **98** 1461–1479

[13] Anderson B J, Fuselier S A, Gary S P and Denton R E 1994 *Journal of Geophysical Research* **99** 5877–5891

[14] Gershman D J, Slavin J A, Raines J M, Zurbuchen T H, Anderson B J, Korth H, Baker D N and Solomon S C 2013 *Journal of Geophysical Research (Space Physics)* **118** 7181–7199

[15] Masters A, Phan T D, Badman S V, Hasegawa H, Fujimoto M, Russell C T, Coates A J and Dougherty M K 2014 *Journal of Geophysical Research (Space Physics)* **119** 121–130

[16] Gary S P, Anderson B J, Denton R E, Fuselier S A and McKean M E 1994 *Physics of Plasmas* **1** 1676–1683

[17] Jokipii J and Giacolone J 2017 private communication

[18] Jokipii J and Giacolone J 2018 private communication

[19] Mostafavi P and Zank G P 2018 *The Astrophysical Journal Letters* **854** L15

[20] Opher M, Drake J F, Swisdak M, Zieger B and Toth G 2017 *The Astrophysical Journal Letters* **839** L12 (*Preprint* `1702.06178`)

[21] Pogorelov N V, Heerikhuisen J, Roytershteyn V, Burlaga L F, Gurnett D A and Kurth W S 2017 *The Astrophysical Journal* **845** 9 (*Preprint* `1706.09637`)

[22] Fuselier S A and Cairns I H 2017 *Journal of Physics Conference Series* (*Journal of Physics Conference Series* vol 900) p 012007

[23] Swisdak M, Opher M, Drake J F and Alouani Bibi F 2010 *Astrophyscal Journal* **710** 1769–1775 (*Preprint* `1001.0589`)

[24] Swisdak M, Drake J F and Opher M 2013 *Astrophysical Journal Letters* **774** L8 (*Preprint* `1307.0850`)

[25] Phan T D, Shay M A, Gosling J T, Fujimoto M, Drake J F, Paschmann G, Oieroset M, Eastwood J P and Angelopoulos V 2013 *Geophysical Research Letters* **40** 4475–4480

[26] Phan T D, Drake J F, Shay M A, Gosling J T, Paschmann G, Eastwood J P, Oieroset M, Fujimoto M and Angelopoulos V 2014 *Geophysical Research Letters* **41** 7002–7010

[27] Cassak P A and Shay M A 2007 *Physics of Plasmas* **14** 102114

[28] Swisdak M and Drake J F 2007 *Geophysical Research Letters* **34** L11106 (*Preprint* `0704.2758`)

[29] Birn J, Borovsky J E, Hesse M and Schindler K 2010 *Physics of Plasmas* **17** 052108

[30] Gosling J T 2007 *The Astrophysical Journal Letters* **671** L73–L76

[31] Shay M A, Haggerty C C, Phan T D, Drake J F, Cassak P A, Wu P, Oieroset M, Swisdak M and Malakit K 2014 *Physics of Plasmas* **21** 122902 (*Preprint* 1410.1206)

[32] Haggerty C C, Shay M A, Drake J F, Phan T D and McHugh C T 2015 *Geophysical Research Letters* **42** 9657–9665 (*Preprint* 1504.02110)

[33] Drake J F, Cassak P A, Shay M A, Swisdak M and Quataert E 2009 *The Astrophysical Journal Letters* **700** L16–L20

[34] McBride J B, Ott E, Boris J P and Orens J H 1972 *Physics of Fluids* **15** 2367–2383

[35] Huba J D 2016 *NRL Plasma Formulary* (Washington: Naval Research Laboratory)

How Nanoflares Produce Kinetic Waves, Nano-Type III Radio Bursts, and Non-Thermal Electrons in the Solar Wind

H. Che

University of Maryland, College Park, MD, 20742, USA
NASA Goddard Space Flight Center, Greenbelt, MD, 20771, USA

Abstract. Observations of the solar corona and the solar wind discover that the solar wind is unsteady and originates from the impulsive events near the surface of the Sun's atmosphere. How solar coronal activities affect the properties of the solar wind is a fundamental issue in heliophysics. We report a simulation and theoretical investigation of how nanoflare accelerated electron beams affect the kinetic-scale properties of the solar wind and generate coherent radio emission. We show that nanoflare-accelerated electron beams can trigger a nonlinear electron two stream instability, which generates kinetic Alfvén and whistler waves, as well as a non-Maxwellian electron velocity distribution function, consistent with observations of the solar wind. The plasma coherent emission produced in our model agrees well with the observations of Type III, J and V solar radio bursts. Open questions in the kinetic solar wind model are also discussed.

1. Introduction

The origin of the solar wind is one of the most important unsolved problems in heliophysics. The concept of solar wind and the first steady hydrodynamic solar wind model were proposed by Parker in 1958 [1]. The Parker model describes the solar wind as a continuous plasma outflow from the solar corona, maintained by the stationary expansion of the Sun's atmosphere. As pointed out by Parker in 1965 [2], the steady solar wind model only concerns the general dynamical principles but the solar corona is actively heated to maintain the outflow. It is expected that direct observations of the corona and the solar wind will allow more details to be incorporated into the solar wind model. Over the past decades, significant improvements have been made in solar and solar wind probes, and thanks to these improvements, it has become increasingly clear that the solar wind is indeed not steady and is associated with small-scale impulsive events ubiquitously occurring near the surface of the photosphere. On the theoretical front, with the help of more and more powerful computer simulations, it has become possible to reach a better understanding of the detailed physical processes in the solar corona and how these processes affect the properties of the solar wind [3, 4, 5, 6, 7, 8, 9, 10, 11].

In this paper, we report some recent progress in the understanding of how the electron beams accelerated by nanoflares shape the solar wind non-thermal electron velocity distribution function (VDF), generate kinetic waves, and produce nano-Type III radio bursts.

IOP Conf. Series: Journal of Physics: Conf. Series **1100** (2018) 012005 doi:10.1088/1742-6596/1100/1/012005

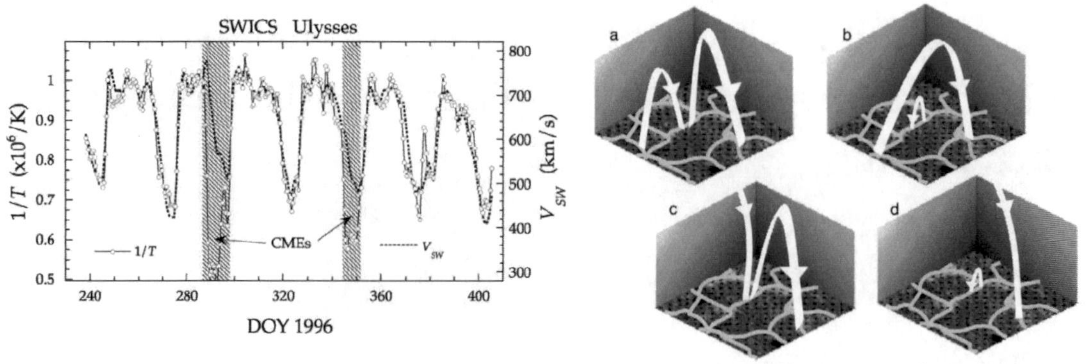

Figure 1: **Left:** Time variations of the inverse of the electron temperature $1/T$ (open circles) and of the solar wind proton bulk speed V_{sw} (dotted curves) during a 166-day time period (27 August 1996 to 9 February 1997), observed with the SWICS on Ulysses [22]. **Right:** An illustration of the reconnection of loops and open field lines from Fisk (2003) [23]. (a) The foot points of two loops move with convective velocities along the lanes separating the granular and supergranular cells on the solar surface. (b) Two of the foot points of the loops have reconnected to form a new larger loop and a small secondary loop will subduct back into the photosphere. (c) The foot points of a loop and an open field line move along the lanes. (d) A foot point of the loop and the open field line have reconnected, the open field is displaced to lie over the location of another foot point of the loop, and a small secondary loop is again formed that should subduct back into the photosphere.

1.1. Evidence for the Connection Between Solar Wind and Nanoflares

Recent observations of the corona and the solar wind elemental composition and the frozen-in electron temperature strongly suggest that the solar wind is associated with impulsive events close to the surface of the Sun [12, 13, 14, 15, 16, 17, 18, 19, 20, 9]. It is discovered that 1) the frozen-in electron temperature of the fast wind is $\sim 8 \times 10^5$ K, and its nearly photospheric composition suggests that the fast wind may originate from coronal holes; 2) the frozen-in electron temperature of the slow wind is $\sim 1.5 \times 10^6$ K, and its lower coronal composition suggests that slow wind may originate from the quiet Sun. Using data from the Solar Wind Ion Composition Spectrometer (SWICS) onboard *Ulysses* [21], obtained over an entire solar cycle and full latitude range, Gloeckler, Zurbuchen & Geiss discovered that the solar wind speed is anti-correlated with the electron temperature derived from the density ratio O^{7+}/O^{6+} [22]. This discovery implies that the plasma heating process in the lower coronal may affect the acceleration of the solar wind. Fisk [23, 24] proposed that the solar wind was produced by photosphere-rooted small-scale loop-loop magnetic reconnections (MR), and loop-open field-line MR, also known as interchange MR (Fig. 1). The MR process releases Poynting flux and mass flow that has been heated from the loops into the corona. Assuming a constant mass ejection from the source region, the mass flow escapes to interplanetary along open field lines powered by Poynting flux produces the anti-correlation between the solar wind and the local corona temperature.

The observations [19, 18] and Fisk's theoretical picture imply that the small-scale impulsive events that occur everywhere in the quiet Sun, including corona holes have similar properties as the nanoflare proposed by Parker [25]. Recent high-resolution observations of the Sun from sounding rockets, spacecrafts such as SDO, IRIS, and NuSTAR, are providing a increasingly detailed picture of nanoflares [26, 27, 28, 29, 30]. The estimated occurrence rate of nanoflares with energy release $\sim 10^{24}$ erg is $\sim 10^{6}$ s^{-1} for the whole Sun.

1.2. Coronal Weak Type III Bursts and Nanoflare-Accelerated Electron Beams

Similar to flares, nanoflares can accelerate particles and the characteristic energy of nanoflare-accelerated electrons is in keV range [31]. The accelerated electron beams can trigger electron two-stream instability (ETSI), generate Langmuir waves, and produce type III radio bursts. Indeed, Observations [32, 33] have found in the solar corona a new kind of type III radio bursts whose brightness temperature $T_b \sim 10^{6}$ K is about 9 orders of magnitude lower than the flare-associated Type III bursts $T_b \sim 10^{15}$ K and are far more abundant, implying these bursts very possibly originate from nanoflares (see Fig. 2). The high occurrence rate of these "nano type III bursts" indicates that electron beams and ETSI are common in the solar corona.

A group of solar radio bursts including Type III, J, and V radio bursts shows the same characteristics: their frequencies are close to the local electron plasma frequency $\omega_{pe} \sim (4\pi n_e e^2/m_e)^{1/2}$, which changes as the electron beams travel along magnetic field lines, as first discovered by wild in 1940s [35, 36]. Type III radio bursts propagate along open magnetic field lines and can escape from the corona and enter the interplanetary space if the bursts are strong enough, these bursts are called interplanetary bursts; otherwise, bursts are called coronal radio bursts. Type J and V bursts propagate along closed field lines and thus belong to the coronal bursts class (Fig. 3). Nano-Type III radio bursts are weak coronal radio bursts.

If the origin and acceleration of solar wind are associated with the physical processes of nanoflares, the plasma heating produced by electron beams should affect the solar wind properties. Our recent work has addressed this question and is presented in § 2.

1.3. Observational Problems of Non-thermal Electron Velocity Distribution Function and Kinetic Turbulence in the Solar Wind

Observations of electron VDFs at heliocentric distances from 0.3 to 1 AU show a prominent "break" or a sudden change of slope at a kinetic energy of a few tens of electron volts as shown in Fig. 4. The electron VDF below the break is dominated by a Maxwellian known as the *core* while the flatter wing above the break is called the *halo* [38, 39]. How to naturally produce the nearly isotropic halo population that can be described by an approximate Maxwellian function has been a long-standing puzzle in heliophysics [40]. The isotropic nature of the halo suggests that halo formation needs strong turbulence scattering and is likely related to the kinetic turbulence in the solar wind [40]. In addition, the *strahl* – an anisotropic tail-like feature skewed with respect to the magnetic field direction is found in the electron VDF of fast solar wind with speed ~ 400 km s^{-1}. In the slow solar wind coming from the sector boundary with speed < 400 km s^{-1}, the strahl is nearly invisible and the isotropic core-halo feature dominates.

Kinetic models show that magnetic focusing effect can produce a strahl-like tail at a minimum heliocentric distance of 10 $R_\odot$ [41, 42], but the different modes of solar wind turbulence are unable to produce the isotropic halo, suggesting that stronger scattering – probably caused by kinetic instabilities – is required [43, 40, 44, 45]. Observations of ion charge states of the solar wind imply a nonthermal tail in the electron VDF of the lower corona, suggesting the coronal origin of the electron halo [46, 47, 48, 49]. Whistler waves [50, 51] and other possible kinetic waves [52, 53] in the solar corona are found to be able to effectively scatter the electrons to form the electron halos, but the processes that generate these waves are unknown.

IOP Conf. Series: Journal of Physics: Conf. Series **1100** (2018) 012005 doi:10.1088/1742-6596/1100/1/012005

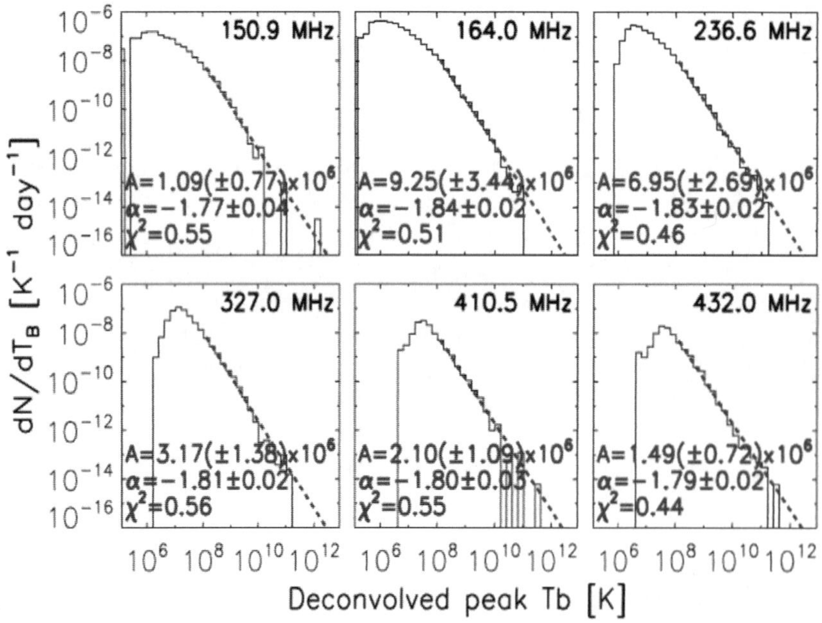

Figure 2: Histogram of deconvolved (i.e., assuming both observed and true sources are Gaussian-shaped) peak brightness temperature ($T_b \sim 10^6$ K) of all type III bursts (The typical T_b of solar flare Type III bursts is about 10^{15} K). The dashed blue line is a power-law $dN/dT_b = AT_\alpha$ B fit to the data, using the C-statistic (Cash 1979[34]), technically better suited than Poisson statistics for data sets with small number of counts per bin (as is our case for the high-value bins, but in this case leading to negligible differences). The associated best-fit parameters are in the lower left corner. Statistical survey of 10,000 type III radio bursts observed by the Nancy Radioheliograph from 1998 to 2008 found associated with nanoflares (Saint-Hilaire et al, ApJ, 2013 [33]).

To understand what wave scattering processes produce the observed isotropic electron halo, we need to understand the cause of the kinetic-scale turbulence in the corona and the solar wind. Observations of the solar wind turbulence (Fig. 5) have shown that as scales approaching the ion inertial length where wave-particle interactions become important, the power-spectrum of magnetic fluctuations, which in the inertial range follows the Kolmogorov scaling law $B_k^2 \propto k^{-5/3}$, is replaced by a steeper anisotropic scaling law $B_{k_\perp}^2 \propto k_\perp^{-\alpha}$, where $\alpha > 5/3$. Spectral index $\alpha \sim 2.7$ is found in observations but can vary between 2 and 4. Magnetic fluctuations with frequencies much smaller than ion gyro-frequency propagating nearly perpendicularly to the solar wind magnetic field are identified as kinetic Alfvén waves (KAWs) [54, 55, 56, 57, 58, 59] and the break frequencies of the magnetic power-spectra from 0.3 to 1 AU suggest the break likely corresponds to the ion inertial length [60, 61]. In the past decades, extensive studies of solar wind kinetic-scale turbulence have focused on the idea that the kinetic-scale spectrum is due to the cascade of large-scale turbulence and dissipation on kinetic-scales. However, there are concerns that the energy in the solar wind large-scale turbulence may not be enough to cascade and support the observed kinetic-scale turbulence and heating [62].

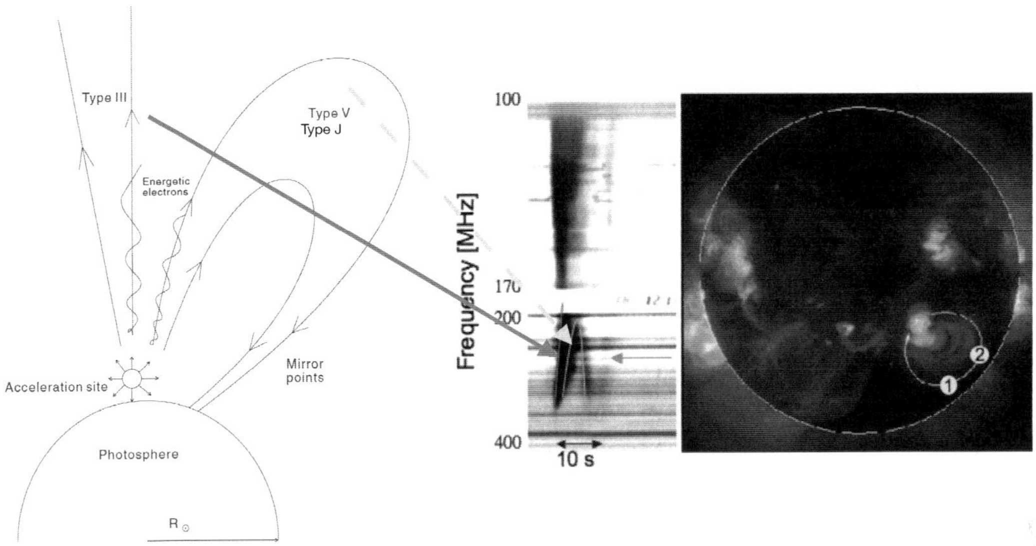

Figure 3: **Left**: an illustration of propagation paths for Type III, J and V radio bursts (from Tang et. al. 2013 [37]). **Right**: The spectrum of a meter wave type III burst followed by a Type V burst. The bursts are observed with the *Nanay Radioheliograph.* (Credit: Potsdam Astrophysical Institute.)

In the following section, we show that the electron beams produced by nanoflares can produce the observed core-halo structure and KAWs and whistler waves.

2. Electron Two-stream Instability and the Common Origin of Kinetic Turbulence, Nonthermal Electron VDFs and Nano-Type III Radio Bursts

In the classical Kolmogorov turbulence scenario, the energy is injected from the large scale. The balance between energy input and its final absorption is controlled by a nonlinear forward cascade from long wavelengths to dissipation-dominated short wavelengths, resulting in a universal energy cascade power-law with index $-5/3$. In plasmas, the source of instability is often beams of charged particles that inject energy on kinetic scales. For example, electron beams inject energy on electron inertial length c/ω_{pe} through electron two steam instability (ETSI). Different from Kolmogorov turbulence, at shorter wavelengths the natural candidate to provide the sink of wave energy is Landau damping. However, nonlinear disparate-scale wave interactions which follow from the direct calculation of basic three-wave coupling can only lead to inverse cascades (to longer wavelengths) through modulational instability [64], and away from the Landau damping region of the spectrum. The eventual nonlinear process capable of overriding this inverse cascade was suggested by Zakharov, namely Langmuir collapse (LC), which is analogous to a self-focusing of the Langmuir wave packets, or cavitons [65].

How does the ETSI driven by nanoflares shape the kinetic properties of the solar wind? Our recent particle-in-cell (PIC) simulations clearly demonstrate that the nonlinear effects of

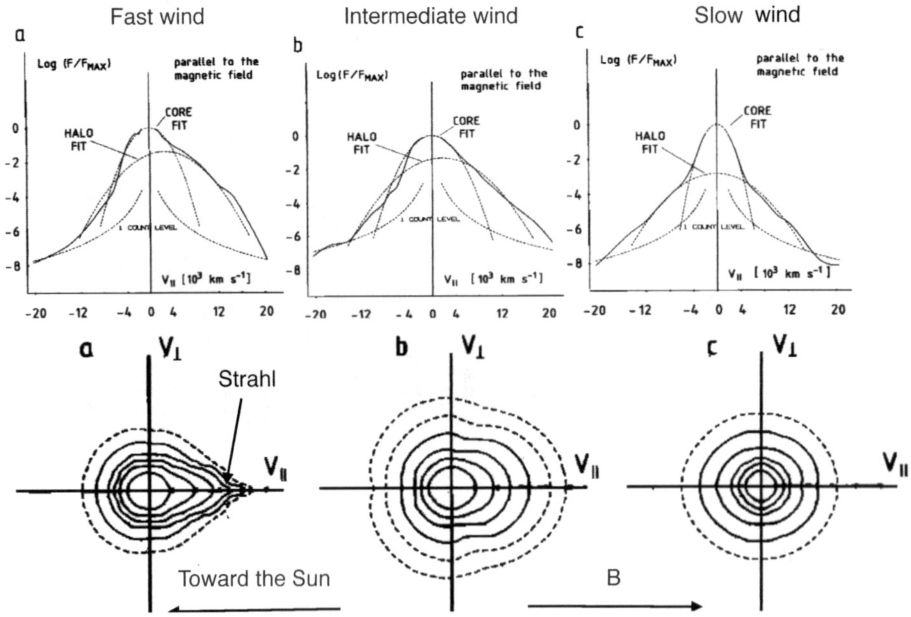

Figure 4: The electron VDFs of solar wind at 1AU from Pilipp et. al. (1987) [38]. The top panels show the 1D cuts parallel to the magnetic field crossing the centers of 2D contours displayed below. Slow wind is define with velocity $v_{wind} < 400$ km s^{-1} and fast wind with $v_{wind} > 700$ km s^{-1}. The speed of the intermediate wind is between these two.

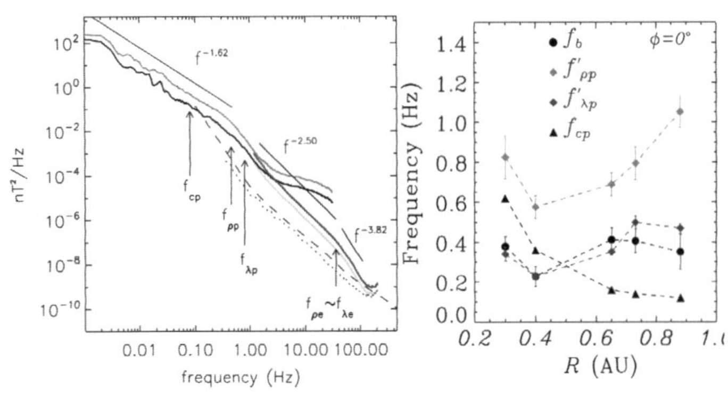

Figure 5: **Left:** The observed solar wind magnetic turbulence power spectrum at 1AU [56]. **Right:** The evolution of magnetic power spectra break from 0.3-1AU [63], where f_b is the observed magnetic power spectral break frequency, $f_{\rho p,e}$ corresponds to the proton (electron) gyro-radius, and $f_{\lambda p,e}$ corresponds to the proton (electron) inertial length. The spectrum is observed in the solar wind co-moving frame.

ETSI can naturally explain the origin of the observed non-thermal electron VDFs in the solar wind, and the electron beam's contribution to the kinetic scale turbulence is non-negligible [3, 4]. Coherent plasma emission produced by ETSI is found to be able to last for more than five orders of magnitude longer than its linear saturation time, and this long duration possibly resolves the so-called "Sturrock dilemma" [7]. We briefly summarize the major results below.

2.1. Electron Two-stream Instability

ETSI is an electrostatic instability and shows different physical evolutions in cold plasma and warm plasma. The details can be found in a recent review [66] and references therein.

The plasma is cold if the wave phase speed v_p is much larger than the electron thermal speed v_{te}, i.e. $v_p \gg v_{te}$. In the cold plasma limit the phase speed of the fastest growing mode of ETSI is $v_p = (n_b/2n_0)^{1/3} v_d$, where v_d is the drift of electron beams, n_0 is the background electron density and n_b is the beam density. In cold plasma ETSI grows with a rate $\gamma \sim \sqrt{3}/2(n_b/2n_0)^{1/3}\omega_{pe}$, and the fastest growing mode is $k_f = \omega_{pe}/v_d$. During the linear growth, most of the kinetic energy of the beams is converted into the growth of electric field $\delta E \propto e^{\gamma t} \sim ek_f m_e(v_d - v_{te})^2/2$. The linear growth time-scale $1/\gamma$ is short and comparable to $1/\omega_{pe}$. If $v_d > 2v_{te}$ and $\delta E > ek_f m_e v_{te}^2/2$ then the electric field can trap more electrons with velocity $< v_{te}$ and develop electron holes. The trapping and de-trapping of electrons by electron holes can efficiently heat the plasma [67].

In warm plasma $v_p \sim v_{te}$, the thermal effects must be considered and the kinetic theory is required to describe the ETSI. Different from cold plasma, Langmuir waves are produced and Landau damping becomes the dominant process. Let $\omega = \omega_r + i\gamma$ where $\gamma \ll \omega_r \sim \omega_{pe,0}$, the dispersion relation can be found as for two beams 1 & 2[68, 66]:

$$\frac{\omega_r^2}{\omega_{pe,0}^2} = 1 + 3k^2\lambda_{D0}^2, \tag{1}$$

$$\frac{\gamma}{\omega_{pe,0}} = \sqrt{2\pi}\left[-\frac{\omega_{pe,1}^2\omega_r}{k^3 v_{t1}^3} e^{-\omega_r^2/2k^2v_{t1}^2} + \frac{\omega_{pe,2}^2(v_d - \omega_r/k)}{k^2 v_{t2}^3} e^{-(\omega_r/k - v_d)^2/2v_{t2}^2}\right]. \tag{2}$$

Eq. (1) is the classical dispersion relation of Langmuir wave in a warm plasma and Eq. (2) is the growth rate of Landau damping.

When the electron beam is strong, ETSI can heat the plasma and cause the plasma to become warm before the kinetic energy in the beam gets exhausted. It is difficult to obtain an analytic solution for what occurs during the transition from cold to warm plasma due to the strong nonlinear effects, particularly, the effects caused by wave coupling and wave-particle interactions. Instead, the complete nonlinear evolution of ETSI is demonstrated using PIC simulations as we describe below.

2.2. PIC Simulations of Electron Two-stream Instability

We have carried out 2.5D massive parallel PIC simulations to study the nonlinear evolution of ETSI in a uniformly magnetized plasma with equal ion and electron temperature [3, 4, 69]. The initial physical parameters resemble the typical physical condition in the solar corona. The initial density ratio of beam and core is 10% and the drift of the electron beams is about 10 times larger than the thermal velocity, thus the ETSI starts from a cold plasma and ends as warm plasma.

The ETSI experience four phases: linear growth, nonlinear growth, saturation and nonlinear decay till turbulent equilibrium. The time-scale of the linear growth phase is about tens of ω_{pe}^{-1} while the time-scale of the nonlinear evolution of ETSI is $\sim 10^4\omega_{pe}^{-1}$. The complete evolution of ETSI is shown in Fig. 6. The linear growth stage of ETSI can be well described by the cold plasma

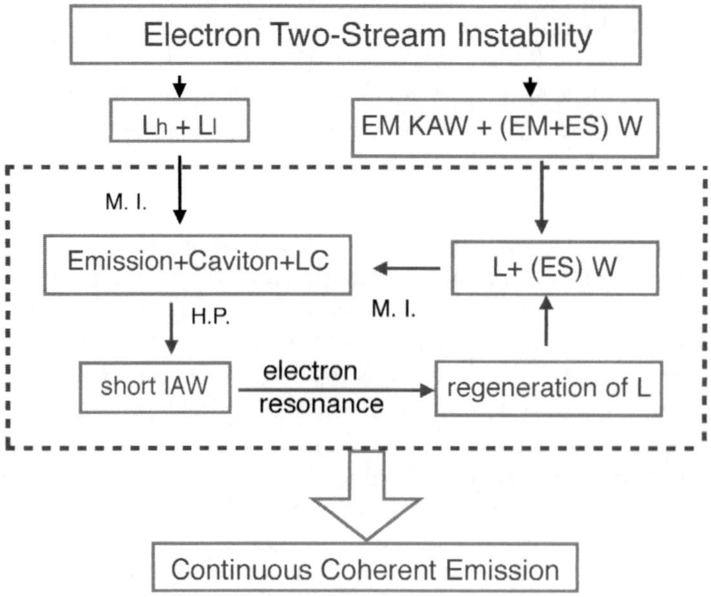

Figure 6: The main processes during the nonlinear evolution of ETSI which includes linear growth, nonlinear growth, saturation and nonlinear decay stages [7]. The following abbreviations are used: L_h: High frequency langmuir wave produced in the background; L_l: Low frequency langmuir wave produced inside the electron holes; EM: Electromagnetic; ES: Electrostatic; L: Langmuir wave; W: Whistler waves; EVDF: electron VDF, M. I.: modulational instability; LC: Langmuir Collapse; IAW: ion acoustic wave. Electron holes form in the nonlinear growth stage. Modulational instability, which occurs at the saturation stage through the nonlinear decay stage, leads to Langmuir collapse and electron heating that fills in cavitons. The high pressure is released via the excitation of a short wavelength IAW that is damped by electrons and re-excites small-scale Langmuir waves—this process closes a feedback loop (outlined by blue dashed lines) that maintains the continuous coherent emission.

limit. Then electron holes form in the nonlinear growth stage and produce fast electron heating, the plasma becomes warm and enters the nonlinear evolution stage. Two parallel processes are found: the generation of low frequency KAWs and whistler waves through bi-directional energy cascades, and the generation of high-frequency Langmuir waves. The plasma emission is continuously maintained through the repeating modulational instability driven by the disparate-scale wave coupling between Langmuir waves and low-frequency waves (shown in the blue box in Fig. 6). The wave-wave and wave-particle interactions eventually lead to the balance of energy exchange, and the turbulence stays at a nonthermal equilibrium in which the non-Maxwellian electron VDF and kinetic waves co-exist, a self-consistent solution of Vlasov equation [5]. These relics of the violent dissipation through ETSI can be carried out into interplanetary with the solar wind along open field lines [4].

2.3. ETSI and the Common Origin of Kinetic Turbulence and Non-thermal Electron VDFs

ETSI injects beam kinetic energy on scales close to the Debye length, and the energy cascade is quickly stopped by strong electron heating. On the other hand, we found that the coupling

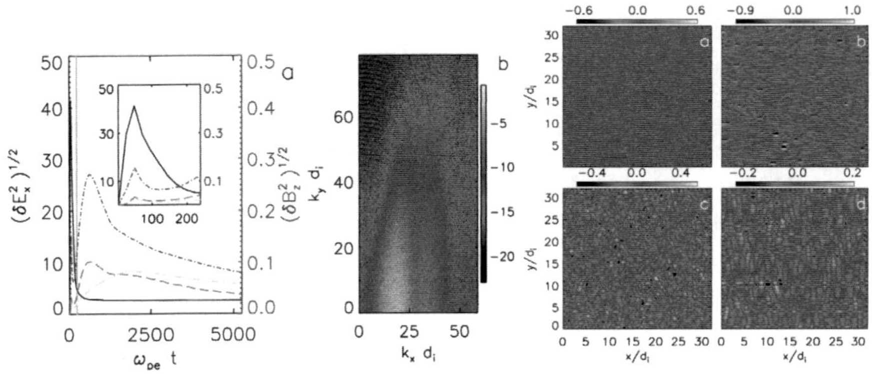

Figure 7: **Left:** Time evolution of turbulent energies in the PIC simulation: $\langle \delta E_x^2 \rangle$ (black line), $\langle \delta B_x^2 \rangle$ (red line), $\langle \delta B_y^2 \rangle$ (green line), $\langle \delta B_z^2 \rangle$ (blue line). The embedded is an expanded view of the time evolution from $\omega_{pe}t = 0 - 230$. The orange line indicates $\omega_{pe}t = 230$. **Middle:** Power spectrum of $|\delta E_x(k_x, k_y)|^2$ at the linear phase of ETSI, shown in logarithmic scale. **Right:** B_z/B_0 at (**a**) the linear phase, (**b**) the nonlinear growth phase, (**c**) the nonlinear decay phase, and (**d**) the turbulence equilibrium. (reference by Che et al [?])

between KAW and whistler waves can inversely transfer the energy to large scales and develop kinetic-scale turbulence. Simultaneously, the waves scatter the hot electron tail that lies along the magnetic field into an isotropic population superposed over the Maxwellian core electron VDF, forming the electron halo in the solar wind [3, 4].

The linear growth of ETSI lasts about $\omega_{pe}t = 20$ (Fig. 7), and $\sim 10\%$ of the kinetic energy in the beam is converted into magnetic energy at the nonlinear growth phase, while nearly 90% is converted into the thermal motion of trapped electrons. The fast growth of the electric field induces a magnetic field, and the electric current density j_{ex} produced by the inductive magnetic field becomes as important as the displacement current when the ETSI starts to decay. The current j_{ex} then drives a Weibel-like instability that generates nearly non-propagating transverse electromagnetic waves. The fast decay of the localized j_{ex} breaks up the transverse waves and produces randomly propagating KAWs and whistler waves. The wave-wave interactions drive a bi-directional energy cascade. The perpendicular KAW energy is transferred from the electron inertial scale up to the ion inertial scale. The parallel whistler wave energy is transferred from the ion inertial scale down to the electron inertial scale. Eventually, magnetic power is concentrated in two branches in the energy spectrum: the nearly perpendicular branch with $k_x d_i < 1$, and the parallel branch with $k_y d_i < 2$ (Fig. 7). Around $\omega_{pe}t = 10000$, the energy exchange between particles and waves reaches a balance. The turbulence reaches its new steady state with $P^2 + B^2/8\pi = constant$, where P is the total pressure of ions and electrons.

The amplitude ratio of the magnetic fluctuations to the background magnetic field is ~ 0.2, which agrees with observations of solar wind kinetic turbulence. The 1D magnetic fluctuation power spectrum is shown in Fig. 8. With $1 < k_y d_i < 2$, which corresponds to the range of wavelengths current instruments can probe, both KAWs and whistler waves are important. The perpendicular power spectrum is fitted with a power-law with an index of -2.2 which falls within the observed range. The perpendicular power spectrum terminates at the ion inertial length, is also consistent with observations [62, 60, 61]. A ubiquitous observable feature is a spectral break at the electron scale caused by energy injection. This model also predicts the existence of

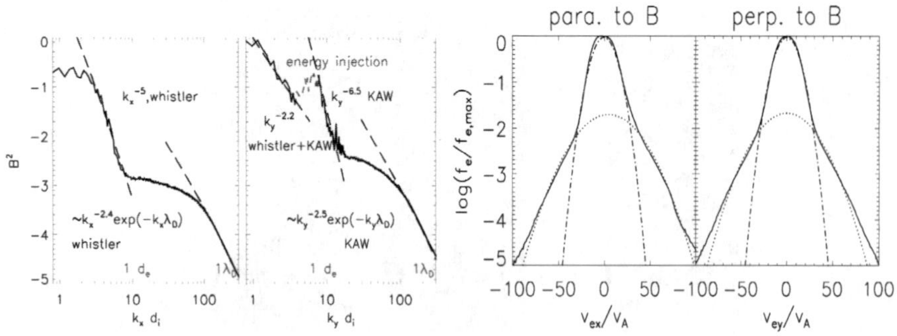

Figure 8: **Left:** The 1D spectra of $\delta B^2(k)$ in x and y directions [3]. The blue short-dashed line shows the wavenumber where magnetic energy injection occurs. **Right:** The 1D electron VDF cuts parallel and perpendicular to magnetic field when ETSI simulation reaches turbulence equilibrium [4]. The dot-dashed lines delineate the core Maxwellian VDF and the dashed lines represent the halo VDF. They are plotted in the same manner as in Pilipp et. al. (1987) [38].

whistler waves, and the cutoff of the parallel power spectrum at the ion gyro-radius [3].

The steady-state electron VDF in our simulation agrees with the observed core-halo structure in the solar wind (Fig. 8). This is expected if the beam heated plasma escapes from the inner corona and advects into interplanetary medium along open field lines, forming the solar wind and preserving its kinetic properties. This nonlinear heating process predicts that the core-halo temperature ratio T_h/T_c of the solar wind is insensitive to the initial conditions in the corona but is linearly correlated to the core-halo density ratio of the solar wind n_c/n_h:

$$\frac{T_h}{T_c} \approx \frac{n_c}{n_h}\frac{1-C_T}{C_T} + 4, \tag{3}$$

where C_T is the rate at which kinetic energy of electron beams converts to heat, and $C_T \sim 0.9$ is found in our simulations. If the core and halo experience similar temperature evolutions when traveling from the Sun to 1AU, the temperature ratio can be approximately preserved. In fast wind where the strahl is strong, the halo temperature can be replaced by the mean temperature of halo and strahl $T_{hot} = n_{strahl}T_{strahl}/n_c + n_h T_h/n_c$ and halo density be replaced by the total density of both strahl and halo $n_{hot} = n_{strahl} + n_h$ because the energy and density are approximately conserved during scattering [70]. Thus we have:

$$\frac{T_{hot}}{T_c} \approx \frac{n_c}{n_{hot}}\frac{1-C_T}{C_T} + 4, \tag{4}$$

The break point dividing the core and halo in electron VDF, which is a useful quantity in observations, satisfies:

$$v_{brk} \approx [\ln(T_h/T_c) - \ln(n_h/n_c)^2]^{1/2} v_{te,c}. \tag{5}$$

In addition, the relative drift between the core and halo is close to the core thermal velocity – a relic of the ETSI saturation.

Our simulations [3] show that when the kinetic turbulence fully saturates, the ratio of parallel to perpendicular electric field fluctuations $\langle|\delta E_{\|}|/|\delta E_{\perp}|\rangle$ is enhanced by the relic parallel electric field by a factor of $\sim 2-3$, consistent with observations that the parallel turbulent electric field is larger than the perpendicular turbulent electric field but contrary to what is expected if the turbulent fluctuations are dominated by KAWs [71]. The enhanced electric field might be caused by electrostatic whistler waves and Langmuir waves.

2.4. ETSI and Continuous Coherent Plasma Emission

The continuous plasma coherent emission is produced while the electron halo and the KAWs and whistler waves develop as have been shown in detail in Che et. al. (2017) [7]. Here we present a basic picture.

Ginzburg and Zhelezniakov [72] in 1958 first proposed a basic framework for Type III radio bursts. The essence of the scenario is that the ETSI, driven by electron beams, generates Langmuir waves that are converted into plasma coherent emission via nonlinear three-wave coupling (e.g., One Langmuir wave, one ion acoustic wave (IAW) produce one photon). However, the deacceleration time of ETSI in solar corona is $\sim 10^{-7}$s estimated by the growth rate in warm plasma shown in Eq. 2[73], which is more than five orders of magnitude shorter than the duration of radio bursts. This long-standing problem was first pointed out by Sturrock and is known as the "Sturrock's dilemma" [73].

Several theoretical models have been proposed to refine the Ginzburg & Zhelezniako model to address this problem [74, 75, 76, 77, 78]. However, there are two major problems in these models. 1) All the models focus only on the regeneration of Langmuir waves and ignored the generation of the low-frequency waves. These models assume that IAWs are present in the background which is not always true in the realistic environment. On the other hand, the coupling between the two waves requires the waves to be in phase, a condition that cannot be guaranteed if the IAWs are not self-consistently produced in the process. 2) All the models are based on quasi-linear theory. But it is found that Langmuir collapse is often associated with Type III radio bursts [78, 79]. Langmuir collapse is a strong turbulence process, which contradicts to the conditions required for the quasi-linear theory, implying the regenerated Langmuir waves by these model must be much weaker than what occurs in the observations. With PIC simulations we show our strong ETSI model can overcome these two problems and provide a self-consistent mechanism to continuously generate coherent radio emission.

In the aforementioned PIC simulation of the ETSI, the coherent plasma emission is not present until the saturation stage when the plasma becomes warm due to the heating by electron holes. At this stage, $\omega_{pe,0}t = 320 - 420$, the high-frequency Langmuir wave L_h is generated in the background plasma and the low-frequency Langmuir wave L_l is generated in the trapped electrons inside the electron holes. These two Langmuir waves satisfy the following dispersion relation (normalized by the initial $\omega_{pe,0}$ and $\lambda_{De,0}$):

$$\frac{\omega}{\omega_{pe,0}} = \left(\frac{n_e^2}{n_0^2} + \frac{T_{ce}n_e}{T_{ce,0}n_0}\gamma k_x^2\lambda_{De,0}^2\right)^{1/2}, \tag{6}$$

where $\gamma = 3$ as the electron heating caused by the solitary wave is nearly adiabatic [67].

The coupling between the two Langmuir waves $L_h + L_l \to T$ (where T is transverse emission) drives modulational instability and produces the first coherent emission with frequency about $1.6\omega_{pe,0}$ (Fig. 9). The emission propagates forward much stronger than backward and satisfies the dispersion relation:

$$\frac{\omega}{\omega_{pe,0}} = \left(\frac{n_e^2}{n_0^2} + \frac{c^2}{v_{te}^2}k_x^2\lambda_{De,0}^2\right)^{1/2}. \tag{7}$$

The maximum growth rate for modulational instability is $\gamma_m = \omega_{pi}(\langle E_L^2\rangle/8\pi n_0 T_e)^{1/2}$ and the critical condition for Langmuir collapse (LC) is [65]:

$$\frac{E_L^2}{8\pi n_e T_e} > \frac{1}{4}k_x^2\lambda_{De}^2. \tag{8}$$

As the modulation instability grows and the critical condition in Eq. (8) is satisfied, Langmuir collapse (LC) occurs [65, 64]. LC leads to the contraction of the modulated Langmuir envelope

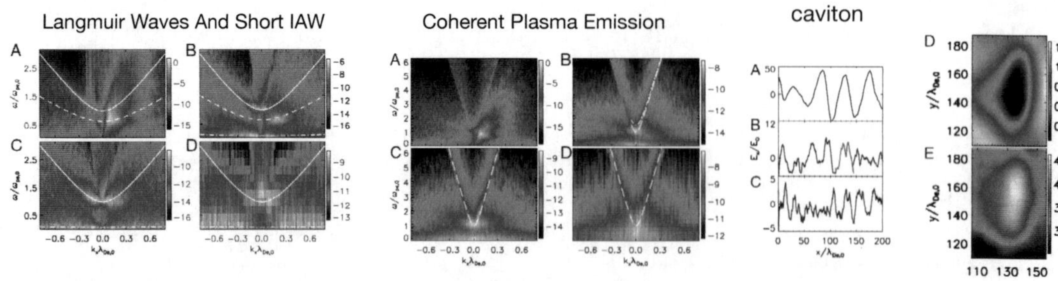

Figure 9: **Left**: The $\omega/\omega_{pe,0} - k_x\lambda_{De,0}$ diagrams of the parallel propagating high frequency electric field component E_x at four time intervals: (A): $\omega_{pe,0}t = 0 - 100$; (B): $\omega_{pe,0}t = 320 - 420$; (C): $\omega_{pe,0}t = 2880 - 2980$; and (D): $\omega_{pe,0}t = 10560 - 10580$. Also shown are the Langmuir wave dispersion relations of the background electrons (solid lines), trapped electrons (dashed lines), and the short wavelength IAW (dash-dotted lines). **Middle**: The $\omega/\omega_{pe,0} - k_x\lambda_{De,0}$ diagrams of parallel propagating high frequency E_y are shown for the same four time intervals for Langmuir waves. Dashed lines: dispersion relation of plasma emission with base frequency $\sim \omega_{pe,0}$. **Right**: Parallel electric field E_x for $x \in [0, 200\lambda_{De,0}]$ and $y = 100\lambda_{De,0}$ at (A) $\omega_{pe,0}t = 72$ when the ETSI nearly saturates and hot electrons excite Langmuir waves, (B) $\omega_{pe,0}t = 320$ when the modulational instability grows, LCs start and cavitons form, and (C) $\omega_{pe,0}t = 648$ LCs continue. The ion density map (D) and the electron temperature map (E) of the caviton that corresponds to the Langmuir envelope (in red) in panel (B) are also shown. Figures from the paper by Che et al.[7].

and the formation of ion density cavitons. We plot an example of the parallel electric field E_x in Fig. 9 at three moments: $\omega_{pe,0}t = 72,\ 320,\ 680$. At $\omega_{pe,0}t = 72$, the solitary waves with wavelengths near the fastest growing mode reach the peak. The critical condition for LC $E^2/8\pi n_0 T_e > \frac{1}{4}k_x^2\lambda_{De}^2$ is satisfied since $(E^2/8\pi n_0 T_e)^{1/2} \sim 0.4$ is larger than the fastest growing mode of the ETSI $k\lambda_{De}/2 \sim v_{te,0}/2v_{db,0} \sim 0.05$. At $\omega_{pe,0}t = 320$, the modulated wave envelopes decrease from 50 to 30 $\lambda_{De,0}$ and ion density cavitons form. Contraction of the Langmuir wave envelopes efficiently dissipates the Langmuir wave energy into electron thermal energy.

Consequently, LC results in the destruction of the cavitons and the release of hot plasma that was inside the caviton. The process generates intermediate short ion acoustic waves (IAWs). Short IAWs resonate with the electrons and regenerate Langmuir waves. The emission is maintained by the cyclic couplings between the regenerated Langmuir waves and the electrostatic component of the whistler waves (Fig. 6). The frequency of the intermediate short IAW is of the same order as the ion plasma frequency ω_{pi}, and satisfies a dispersion relation similar to that of Langmuir waves:

$$\omega^2 \approx \omega_{pi}^2 \frac{1+\sqrt{3T_i/T_e}}{4} + \sqrt{\frac{3T_e}{T_i}} k^2 v_{ti}^2. \qquad (9)$$

The Langmuir waves, short IAWs, and emissions at four different evolution stages are shown in Fig. 9. For each cycle, the regeneration of the Langmuir wave leads to a small frequency shift and a reduced amplitude of the Langmuir waves. The emission continues beyond the saturation of turbulence.

In our simulations, the ETSI nonlinear saturation time is $\sim 1.5 \times 10^4 \omega_{pe}^{-1}$. Since the

Table 1: Model Predictions and Observational Evidence

Model Predictions	Observations	References
In the solar corona emission duration $\sim 10^5\omega_{pe}^{-1} \sim 1-10$ ms.	Coronal Type J & U radio bursts Weak Coronal Type III radio bursts	[82, 81, 33]
Langmuir waves & whistler waves	Interplanetary Type III radio bursts	[83, 84, 85, 86]
Langmuire collapse & short wavelength IAW	Interplanetary Type III radio bursts	[87, 83, 88]

modulational instability nearly dominates the entire process, the nonlinear saturation time is approximately proportional to $(m_i/m_e)^{1/2}$, and for real mass ratio, the ETSI nonlinear saturation time is translated to $\sim 10^5\omega_{pe}^{-1}$, which is significantly longer than the ETSI linear saturation time $(n_0/n_b)^{1/3}\omega_{pe}^{-1} \sim 2\omega_{pe}^{-1}$. Note that our simulation assumes instantaneous injection of the electron beam, while in the corona the electron-acceleration time is finite and the beam will propagate out of the region of initial generation. The acceleration time also affects the actual duration of the bursts [75, 80]. The overall scenario is that coronal bursts produce nonthermal electrons that escape into space and produce interplanetary bursts [81] with accompanying waves. Our simulation assumes the beam energy is about 100 times the coronal thermal energy. For nanoflares, the beam energy is about 1 keV and the corona temperature is ~ 10 eV. We estimate the emission power in our simulation is $\sim 10^{-4} - 10^{-6}$ of the Langmuir wave power. Such small energy loss is negligible dynamically. The emission mechanism we discussed provides a self-consistent solution to the long-standing "Sturrock's dilemma" [73]. In Table 1 we provide an incomplete sample of literature in which the observations of various types of radio bursts are consistent with our model predictions.

3. Summary and Open Questions

To summarize, we have shown that nanoflare-accelerated electron beams can trigger ETSI, which generates kinetic turbulence as well as a non-Maxwellian electron VDF, consistent with observations of the solar wind [4]. The major attraction of this finding is that it can account for the origin of both the electron VDF and kinetic turbulence in a unified picture, while past studies treat these two phenomena as unrelated. The link between the solar wind and nanoflares directly relates solar wind properties to photospheric dynamics and puts useful constraints on kinetic processes in both the solar corona and the solar wind. The plasma coherent emission produced in our model agrees well with the radio observations of nano-Type III, J & V solar radio bursts. The model also predicts features that can be tested with current and future solar and solar wind probes. One of the most important predictions of our model is the correlation between the temperature of core and halo of solar wind electron VDF, and this correlation is confirmed in a recent analysis of 12 years of *WIND* data [89]. Recent SDO observations of the corona also suggest that the plasma heating is associated with open and close field [90]. Using CHIANTI data base, it is found electron VDF is non-Maxwellian [49] and electron beams form in the lower corona [91].

The upcoming Parker Solar Probe (PSP) and Solar Orbiter (SO) spacecrafts will provide unprecedented *in situ* observations of the solar wind and multi-bands remote observations of corona activities. The *in situ* observations of the solar wind around 10 solar radii can test whether the electron halo and strahl in the electron VDF are of coronal origin [9, 10, 92]. The observations from 10 solar radii to 1AU will enable us to investigate the evolution of electron and ion VDFs in the inner heliosphere. The simultaneous X-ray and radio observations will provide us more information on the particle acceleration in solar flares.

How the particle heating and acceleration in the corona affect the properties of the solar wind is the core science for both PSP and SO. Nanoflares [93, 28, 30] and plasma waves are two dominant sources of coronal heating [94, 95, 8, 11]. Therefore, *in situ* observation of the solar wind opens a window to nanoflare heating.

There are several unsolved problem related to nanoflares induced particle energization: 1) MR is believed to be the engine of particle energization in nanoflares. The interchange MR is essential for slow solar wind to escape from the lower corona to interplanetary space. How MR accelerates and transports particles in the solar corona is still not very well understood and is a subject of active study [96, 97, 98, 99, 100, 101]. Current observations [102, 103] and theoretical/numerical models [104, 105, 106] of interchange MR still cannot provide sufficient details on how beams and energetic particles are produced; 2) The ion VDF of the solar wind is also non-Maxwellian [107]. How the ion beams in the corona impact the solar wind ion VDF has not been closely investigated; 3) Energetic particle propagation is very important in space weather applications. In the past, the study of energetic particle transport in the heliosphere focuses on particle scattering and acceleration by Alfvénic wave turbulence, and the role of kinetic scale turbulence on solar wind has not drawn sufficient attention, particularly the impact on solar wind electrons such as the evolution of the anisotropic strahl [92].

How to incorporate kinetic processes into the global solar wind model is a profound theoretical and observational challenge. The study of the impact of nanoflares on the solar wind's properties on both large and kinetic scales will certainly be enlightening to the pursuit of a complete understanding of the origin and interactions of the solar wind. Such understanding will have broader implications for astrophysical winds and outflows beyond the solar system.

4. Acknowledgement

HC would like to thank all collaborators on this project: Drs. M. Goldstein, P. Diamond, R. Sagdeev and A. Vinãs. HC would like to thank Dr. G. Zank for the kind invitation to the *17th Annual International Astrophysics Conference* and inspiring discussions during the meeting. HC also like to thank the participants at this meeting for the interesting discussions. HC is partly supported by NASA Grant No.NNX17AI19G. The simulations and analysis were supported by the NASA High-End Computing (HEC) Program through the NASA Advanced Supercomputing (NAS) Division at Ames Research Center.

References

[1] Parker E N 1958 The Astrophysical Journal **128** 664
[2] Parker E N 1965 Space Sci. Rev. **4** 666–708
[3] Che H, Goldstein M L and Viñas A F 2014 Phys. Rev. Lett. **112** 061101
[4] Che H and Goldstein M L 2014 The Astrophysical Journal Letters **795** L38
[5] Kim S, Yoon P H, Choe G S and Wang L 2015 The Astrophysical Journal **806** 32
[6] Vocks C, Dzifčáková E and Mann G 2016 Astronomy and Astrophysics **596** A41
[7] Che H, Goldstein M L, Diamond P H and Sagdeev R Z 2017 *Proceedings of the National Academy of Science* **114** 1502–1507
[8] Zank G P, Adhikari L, Hunana P, Shiota D, Bruno R and Telloni D 2017 The Astrophysical Journal **835** 147
[9] Cranmer S R, Gibson S E and Riley P 2017 Space Sci. Rev. **212** 1345–1384
[10] Dudík J, Dzifčáková E, Meyer-Vernet N, Del Zanna G, Young P R, Giunta A, Sylwester B, Sylwester J, Oka M, Mason H E, Vocks C, Matteini L, Krucker S, Williams D R and Mackovjak Š 2017 Solar Physics **292** 100
[11] Zank G P, Adhikari L, Hunana P, Tiwari S K, Moore R, Shiota D, Bruno R and Telloni D 2018 The Astrophysical Journal **854** 32
[12] Axford W I 1977 The three-dimensional structure of the interplanetary medium *Study of Travelling Interplanetary Phenomena* (*Astrophysics and Space Science Library* vol 71) ed Shea M A, Smart D F and Wu S T pp 145–163
[13] Wang Y M and Sheeley Jr N R 1990 The Astrophysical Journal **365** 372–386
[14] Deforest C E, Hoeksema J T, Gurman J B, Thompson B J, Plunkett S P, Howard R, Harrison R C and Hassler D M 1997 Solar Physics **175** 393–410

[15] Zurbuchen T H, Hefti S, Fisk L A, Gloeckler G and von Steiger R 1999 Space Sci. Rev. **87** 353–356
[16] Wang Y M and Sheeley Jr N R 2003 The Astrophysical Journal **587** 818–822
[17] Woo R, Habbal S R and Feldman U 2004 The Astrophysical Journal **612** 1171–1174
[18] Tu C Y, Zhou C, Marsch E, Xia L D, Zhao L, Wang J X and Wilhelm K 2005 Science **308** 519–523
[19] Feldman U, Landi E and Schwadron N A 2005 J. Geophys. Res. **110** A07109
[20] Brosius J W, Daw A N and Rabin D M 2014 The Astrophysical Journal **790** 112
[21] Geiss J, Gloeckler G, von Steiger R, Balsiger H, Fisk L A, Galvin A B, Ipavich F M, Livi S, McKenzie J F, Ogilvie K W and Wilken B 1995 *Science* **268** 1033–1036
[22] Gloeckler G, Zurbuchen T H and Geiss J 2003 J. Geophys. Res. **108** 1158
[23] Fisk L A 2003 J. Geophys. Res. **108** 1157
[24] Fisk L A and Zurbuchen T H 2006 *Journal of Geophysical Research (Space Physics)* **111** A09115
[25] Parker E N 1988 The Astrophysical Journal **330** 474–479
[26] Winebarger A R, Walsh R W, Moore R, De Pontieu B, Hansteen V, Cirtain J, Golub L, Kobayashi K, Korreck K, DeForest C, Weber M, Title A and Kuzin S 2013 The Astrophysical Journal **771** 21
[27] Viall N M and Klimchuk J A 2013 The Astrophysical Journal **771** 115 (*Preprint* 1304.5439)
[28] Testa P, De Pontieu B, Allred J, Carlsson M, Reale F, Daw A, Hansteen V, Martinez-Sykora J, Liu W, DeLuca E E, Golub L, McKillop S, Reeves K, Saar S, Tian H, Lemen J, Title A, Boerner P, Hurlburt N, Tarbell T D, Wuelser J P, Kleint L, Kankelborg C and Jaeggli S 2014 Science **346** 1255724 (*Preprint* 1410.6130)
[29] Hannah I G, Grefenstette B W, Smith D M, Glesener L, Krucker S, Hudson H S, Madsen K K, Marsh A, White S M, Caspi A, Shih A Y, Harrison F A, Stern D, Boggs S E, Christensen F E, Craig W W, Hailey C J and Zhang W W 2016 The Astrophysical Journal Letters **820** L14
[30] Klimchuk J A 2015 *Philosophical Transactions of the Royal Society of London Series A* **373** 20140256–20140256
[31] Gontikakis C, Patsourakos S, Efthymiopoulos C, Anastasiadis A and Georgoulis M K 2013 The Astrophysical Journal **771** 126 (*Preprint* 1305.5195)
[32] Thejappa G, Gopalswamy N and Kundu M R 1990 Solar Physics **127** 165–183
[33] Saint-Hilaire P, Vilmer N and Kerdraon A 2013 The Astrophysical Journal **762** 60 (*Preprint* 1211.3474)
[34] Cash W 1979 The Astrophysical Journal **228** 939–947
[35] Wild J P, Roberts J A and Murray J D 1954 Nature **173** 532–534
[36] Wild J P, Smerd S F and Weiss A A 1963 Annual Review of Astronomy and Astrophysics **1** 291
[37] Tang J F, Wu D J and Tan C M 2013 The Astrophysical Journal **779** 83
[38] Pilipp W G, Muehlhaeuser K H, Miggenrieder H, Montgomery M D and Rosenbauer H 1987 J. Geophys. Res. **92** 1075–1092
[39] Pilipp W G, Muehlhaeuser K H, Miggenrieder H, Rosenbauer H and Schwenn R 1987 J. Geophys. Res. **92** 1103–1118
[40] Marsch E 2006 Living Reviews in Solar Physics **3** 1
[41] Smith H M, Marsch E and Helander P 2012 The Astrophysical Journal **753** 31
[42] Landi S, Matteini L and Pantellini F 2012 The Astrophysical Journal **760** 143
[43] Pierrard V, Maksimovic M and Lemaire J 1999 J. Geophys. Res. **104** 17021–17032
[44] Marsch E 2012 Space Sci. Rev. **172** 23–39
[45] Vocks C 2012 Space Sci. Rev. **172** 303–314
[46] Ko Y K, Fisk L A, Gloeckler G and Geiss J 1996 Geophys. Res. Lett. **23** 2785–2788
[47] Esser R and Edgar R J 2000 The Astrophysical Journal Letters **532** L71–L74
[48] Feldman U, Ralchenko Y and Landi E 2008 The Astrophysical Journal **684** 707–714
[49] Dzifčáková E, Dudík J, Kotrč P, Fárník F and Zemanová A 2015 The Astrophysical Journal supplement **217** 14
[50] Vocks C and Mann G 2003 The Astrophysical Journal **593** 1134–1145
[51] Vocks C, Salem C, Lin R P and Mann G 2005 The Astrophysical Journal **627** 540–549
[52] Cranmer S R, Field G B and Kohl J L 1999 The Astrophysical Journal **518** 937–947
[53] Laming J M 2004 The Astrophysical Journal **604** 874–883
[54] Leamon R J, Matthaeus W H, Smith C W, Zank G P, Mullan D J and Oughton S 2000 The Astrophysical Journal **537** 1054–1062
[55] Bale S D, Kellogg P J, Mozer F S, Horbury T S and Reme H 2005 Phys. Rev. Lett. **94** 215002 (*Preprint* arXiv:physics/0503103)
[56] Sahraoui F, Goldstein M L, Robert P and Khotyaintsev Y V 2009 Phys. Rev. Lett. **102** 231102
[57] Kiyani K H, Chapman S C, Khotyaintsev Y V, Dunlop M W and Sahraoui F 2009 Phys. Rev. Lett. **103** 075006 (*Preprint* 0906.2830)
[58] Salem C S, Howes G G, Sundkvist D, Bale S D, Chaston C C, Chen C H K and Mozer F S 2012 The

Astrophysical Journal Letters **745** L9
[59] Podesta J J 2013 Solar Physics
[60] Perri S, Carbone V and Veltri P 2010 The Astrophysical Journal Letters **725** L52–L55
[61] Bourouaine S, Alexandrova O, Marsch E and Maksimovic M 2012 The Astrophysical Journal **749** 102
[62] Leamon R J, Smith C W, Ness N F and Wong H K 1999 J. Geophys. Res. **104** 22,331–22,344
[63] Boldyrev S and Perez J C 2012 The Astrophysical Journal Letters **758** L44 (*Preprint* 1204.5809)
[64] Rudakov L I and Tsytovich V N 1978 Physics Reports **40** 1–73
[65] Zakharov V E 1972 *Soviet Journal of Experimental and Theoretical Physics* **35** 908
[66] Che H 2016 *Modern Physics Letters A* **31** 1630018-163
[67] Che H, Drake J F, Swisdak M and Goldstein M L 2013 Phys. Plasma **20** 061205 URL http://link.aip.org/link/?PHP/20/061205/1
[68] Bohm D and Gross E P 1949 *Physical Review* **75** 1864–1876
[69] Che H 2016 Common origin of kinetic scale turbulence and the electron halo in the solar wind - Connection to nanoflares *American Institute of Physics Conference Series* (*American Institute of Physics Conference Series* vol 1720) p 030001 (*Preprint* 1603.00549)
[70] Maksimovic M, Zouganelis I, Chaufray J Y, Issautier K, Scime E E, Littleton J E, Marsch E, McComas D J, Salem C, Lin R P and Elliott H 2005 J. Geophys. Res. **110** A09104
[71] Mozer F S and Chen C H K 2013 The Astrophysical Journal Letters **768** L10 (*Preprint* 1304.1189)
[72] Ginzburg V L and Zhelezniakov V V 1959 On the mechanisms of sporadic solar radio emission *URSI Symp. 1: Paris Symposium on Radio Astronomy* (*IAU Symposium* vol 9) ed Bracewell R N p 574
[73] Sturrock P A 1964 *NASA Special Publication* **50** 357
[74] Papadopoulos K, Goldstein M L and Smith R A 1974 The Astrophysical Journal **190** 175–186
[75] Goldstein M L, Smith R A and Papadopoulos K 1979 The Astrophysical Journal **234** 683–695
[76] Freund H P and Papadopoulos K 1980 Phys. Fluid **23** 1546
[77] Goldman M V 1983 Solar Physics **89** 403–442
[78] Robinson P A 1997 *Reviews of Modern Physics* **69** 507–573
[79] Sinclair Reid H A and Ratcliffe H 2014 *Research in Astronomy and Astrophysics* **14** 773–804 (*Preprint* 1404.6117)
[80] Ratcliffe H, Kontar E P and Reid H A S 2014 Astronomy and Astrophysics **572** A111
[81] Aschwanden M J 2002 Space Sci. Rev. **101** 1–227
[82] Aschwanden M J, Benz A O, Dennis B R and Schwartz R A 1995 The Astrophysical Journal **455** 347
[83] Lin R P, Levedahl W K, Lotko W, Gurnett D A and Scarf F L 1986 The Astrophysical Journal **308** 954–965
[84] Kellogg P J, Goetz K, Lin N, Monson S J, Balogh A, Forsyth R J and Stone R G 1992 Geophys. Res. Lett. **19** 1299–1302
[85] MacDowall R J, Hess R A, Lin N, Thejappa G, Balogh A and Phillips J L 1996 Astronomy and Astrophysics **316** 396–405
[86] Ergun R E, Malaspina D M, Cairns I H, Goldman M V, Newman D L, Robinson P A, Eriksson S, Bougeret J L, Briand C, Bale S D, Cattell C A, Kellogg P J and Kaiser M L 2008 *Physical Review Letters* **101** 051101
[87] Lin R P, Potter D W, Gurnett D A and Scarf F L 1981 The Astrophysical Journal **251** 364–373
[88] Kellogg P J, Goetz K, Howard R L and Monson S J 1992 Geophys. Res. Lett. **19** 1303–1306
[89] Macneil A R, Owen C J and Wicks R T 2017 *Annales Geophysicae* **35** 1275–1291 URL https://www.ann-geophys.net/35/1275/2017/
[90] Orange N B, Chesny D L, Gendre B, Morris D C and Oluseyi H M 2016 The Astrophysical Journal **833** 257
[91] Dzifčáková E, Dudík J and Mackovjak Š 2016 Astronomy and Astrophysics **589** A68
[92] Graham G A, Rae I J, Owen C J and Walsh A P 2018 The Astrophysical Journal **855** 40
[93] Benz A O 2004 Nanoflares and the Heating of the Solar Corona *Stars as Suns : Activity, Evolution and Planets* (*IAU Symposium* vol 219) ed Dupree A K and Benz A O p 461
[94] De Pontieu B, Rouppe van der Voort L, McIntosh S W, Pereira T M D, Carlsson M, Hansteen V, Skogsrud H, Lemen J, Title A, Boerner P, Hurlburt N, Tarbell T D, Wuelser J P, De Luca E E, Golub L, McKillop S, Reeves K, Saar S, Testa P, Tian H, Kankelborg C, Jaeggli S, Kleint L and Martinez-Sykora J 2014 Science **346** 1255732
[95] van Ballegooijen A A, Asgari-Targhi M and Voss A 2017 The Astrophysical Journal **849** 46 (*Preprint* 1710.05074)
[96] Drake J F, Swisdak M, Che H and Shay M A 2006 Nature **443** 553–556
[97] Zank G P, le Roux J A, Webb G M, Dosch A and Khabarova O 2014 The Astrophysical Journal **797** 28
[98] Zank G P, Hunana P, Mostafavi P, Le Roux J A, Li G, Webb G M, Khabarova O, Cummings A, Stone E and Decker R 2015 The Astrophysical Journal **814** 137

[99] Guo F, Li X, Li H, Daughton W, Zhang B, Lloyd-Ronning N, Liu Y H, Zhang H and Deng W 2016 The Astrophysical Journal Letters **818** L9
[100] Borovikov D, Tenishev V, Gombosi T I, Guidoni S E, DeVore C R, Karpen J T and Antiochos S K 2017 The Astrophysical Journal **835** 48
[101] Dahlin J T, Drake J F and Swisdak M 2017 *Physics of Plasmas* **24** 092110
[102] Aschwanden M J 2005 *Physics of the Solar Corona. An Introduction with Problems and Solutions (2nd edition)*
[103] Benz A O 2017 *Living Reviews in Solar Physics* **14** 2
[104] Antiochos S K, DeVore C R, Karpen J T and Mikić Z 2007 The Astrophysical Journal **671** 936–946 (*Preprint* 0705.4430)
[105] Antiochos S K, Mikić Z, Titov V S, Lionello R and Linker J A 2011 The Astrophysical Journal **731** 112
[106] Higginson A K, Antiochos S K, DeVore C R, Wyper P F and Zurbuchen T H 2017 The Astrophysical Journal Letters **840** L10
[107] Marsch E, Schwenn R, Rosenbauer H, Muehlhaeuser K H, Pilipp W and Neubauer F M 1982 J. Geophys. Res. **87** 52–72

Observational Analysis of Small-scale Magnetic Flux Ropes from Ulysses In-situ Measurements

Yu Chen[1], Qiang Hu[1,2], Jakobus le Roux[1,2], Jinlei Zheng[1]

[1] Department of Space Science, The University of Alabama in Huntsville, USA
[2] Center for Space Plasma and Aeronomic Research, The University of Alabama in Huntsville, USA

E-mail: [1]yc0020@uah.edu

E-mail: [2]qh0001@uah.edu

Abstract. Small-scale magnetic flux ropes, which have similar magnetic field configuration as their large-scale counterparts (i.e., magnetic clouds), but with different sizes and origin, constitute an important element of solar wind structures. They are also considered to be associated with local particle energization and other related processes. In this report, we apply the Grad-Shafranov (GS) reconstruction method to detect these small-scale flux ropes with a set of quantitative criteria by utilizing data from the Ulysses spacecraft measurements for the first time. We conduct full range automatic detection for years 1994, 1996, 2004 and 2005 during the solar minimum periods. Based on solar wind speed/helio-latitude ranges, these periods are categorized into two groups: one with high solar wind speeds at high latitudes (1994 and 1996) and the other with low solar wind speeds at low latitudes (2004 and 2005). Through mainly statistical analysis of the results from these four years worth of Ulysses data, we have obtained the following findings: (1) Alfvénic structures occur more frequently at higher latitudes or in high speed solar wind (1994 and 1996). (2) Small-scale flux ropes at lower latitudes tend to align with the nominal Parker spiral direction. (3) The scale sizes of small-scale flux ropes are in the same range for different heliocentric distances. Both scale size and duration distributions seem to obey power laws, similar to the analysis results at 1 astronomical unit (AU). (4) The waiting time distribution (WTD) is fitted well by an exponential function rather than a power law. (5) The power law fitting is applied to the wall-to-wall time distribution with the break point at $\sim$ 200 min which is 3$\sim$4 times the result at 1 AU.

1. Introduction

Magnetic flux ropes, with helical structures and winding magnetic field lines, can be categorized into two types: the large-scale magnetic flux ropes or magnetic clouds which have been extensively studied, and the small-scale ones or magnetic islands which are still under investigation. Compared to the former, the small-scale flux ropes have similar magnetic field configuration but different sizes (0.001 $\sim$ 0.01 AU at 1 AU while large-scale ones have $\sim$ 0.1 AU to 0.4 AU) and perhaps different origination and evolution processes. The existence of small-scale flux ropes in the solar wind was first introduced by Moldwin et al. [1]. They reported a small-scale magnetic flux rope with scale size $\sim$ 0.05 AU observed by the Ulysses spacecraft at around 5 AU at low latitudes. With characteristics of high proton temperature, density, and plasma beta, etc., they interpreted this type of events as being generated from multiple magnetic reconnection events at the heliospheric current sheet (HCS). Later, Moldwin et al. [2] discovered more small-scale flux ropes at 1 AU from the IMP 8 and Wind spacecraft measurements. Features of these small helical field structures including bimodal size distribution, different behavior of plasma parameters such as the proton temperature, and scale size similar to the HCS thickness were presented. Therefore, they concluded that these small helical field structures are signatures of magnetic reconnection across the HCS.

However, it is still debatable what factors triggered or formed these structures. Feng et al. [3] identified magnetic flux ropes including small- and intermediate-sized structures from 1995 to 2001 from the Wind spacecraft measurements. Based on their estimate of the energy contents of flux ropes, the authors suggested that the small- and intermediate-sized magnetic flux ropes are the interplanetary manifestations of small coronal mass ejections (CMEs), like magnetic clouds which are related to the large interplanetary coronal mass ejections (ICMEs). Moreover, Feng et al. [4] presented the annual number and distribution of the axial orientations of small- and intermediate-sized magnetic flux ropes and showed that these characteristics are similar to those of magnetic clouds. Based on these and other related features, they suggested that small- and intermediate-sized magnetic flux ropes have the same origin as magnetic clouds, i.e., solar eruptions. More supporting evidence that some small-scale flux ropes originate in the solar corona was also presented by investigating the associated counterstreaming suprathermal electron (CSE) signatures [5].

On the other hand, Cartwright and Moldwin [6] compared small-scale flux ropes with magnetic clouds by applying a semi-automated routine to 11 years of Wind data. The bimodal size distribution was confirmed again, and they concluded that small and large-scale flux ropes have different source mechanisms. Furthermore, they presented a comprehensive survey of small-scale flux ropes using multiple spacecraft measurements between 0.3 and 5.5 AU, including the Helios 1, Helios 2, IMP8 spacecraft missions, etc. By examining the statistical results, they found that the occurrence rate of small-scale flux ropes has a weak dependence on the solar cycle, and the generation of these structures from magnetic reconnection across the HCS is consistent with their results.

Small-scale flux ropes are thought to be connected to energization of particles in both theory, simulations and observations. Theory and simulations predict that merging and contraction of magnetic islands generate electric fields which accelerate quasi-trapped particles in a quasi-2D non-resonant interacting process [7; 8]. When suprathermal ions traverse a region of numerous contracting and merging magnetic islands, they can be accelerated to hard power-law spectra [8; 9]. When diffusive shock acceleration (DSA) of energetic particles is combined with acceleration by downstream contracting and merging magnetic islands, theoretical predictions suggest that energetic particle fluxes peak behind the shock instead of at the shock as predicted by DSA only [10; 11]. Thus, this provides an alternative, and novel explanation for downstream peaks in energetic particle fluxes, not solely relying on time-dependent DSA effects. Observational support for the latter theoretical prediction can be found in [10], [12], [13], and [14]. Figure 1 illustrates the magnetic reconnection associated with merging magnetic islands downstream of a shock.

Until recently, small-scale flux rope identification in spacecraft data was hampered by a limited number of events and was narrow in scope by only focusing on temporal profiles. Accurate and automatic detection of small-scale flux ropes from *in-situ* spacecraft measurements is essential for better categorizing and understanding how they originate and interact with other processes. The Grad-Shafranov (GS) reconstruction technique, which can recover two-dimensional (2-D) magnetic field structures without assuming the force-free condition and axisymmetric cross sections from one-dimensional (1-D) *in-situ* spacecraft data, was developed and applied to study magnetic flux ropes [16–18]. It is based on the assumption of 2-D magnetohydrostatic equilibrium, and the solution of magnetic flux function over the cross section of a small-scale flux rope can be obtained by examining the calculated transverse pressure as

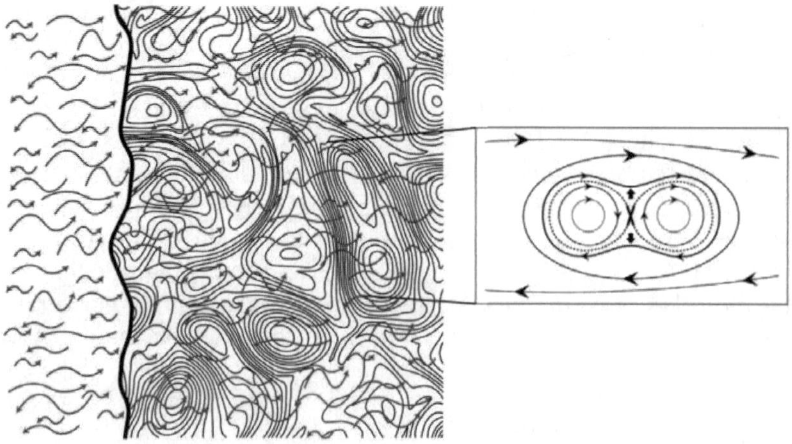

Figure 1. Schematic of merging magnetic islands downstream of a shock. The heavy black wavy line is a shock. Alfvénic-like waves are shown by red wavy lines. Magnetic reconnection involving a pair of merging magnetic islands is illustrated in the right panel. Image credit: Zank et al. [10].

a function of the magnetic flux function using initial data from spacecraft measurements. This technique, therefore, enables us to develop the automated detection of small-scale flux ropes. A database containing over 70,000 small-scale flux ropes identified via the GS method from the Wind spacecraft data has been constructed [19]. By analyzing this database, Zheng and Hu [19] showed that the wall-to-wall time distribution and the non-Gaussian probability density function of the axial current density distribution correspond well to observations and numerical simulation results of magnetohydrodynamic (MHD) turbulence in the solar wind.

Although a number of small-scale flux ropes at 5 AU has been studied and hence resulted in suggestions of possible origination [1], quantitative modeling and analysis of small-scale flux ropes by applying the GS reconstruction to both the inner and outer heliosphere, and especially higher latitude regions, have not been attempted. In this study, we apply the Grad-Shafranov (GS) reconstruction method to the Ulysses *in-situ* spacecraft measurements and present statistical analysis results for four years during the solar minimum periods over a range of heliographic latitudes.

The *in-situ* data from the Ulysses spacecraft and the Grad-Shafranov (GS) equation will be introduced in section 2. The steps of the Grad-Shafranov (GS) reconstruction and a new set of criteria for the automated detection will be introduced as well. In section 3, we present the statistical results of small-scale flux ropes based on the automated detection from the Ulysses data. They include various properties of small-scale flux ropes, the waiting time distribution (WTD) analysis, wall-to-wall time distribution, partial variance of increments (PVI) statistic, the angular change ($\Delta\theta$) of magnetic field across flux rope boundaries, and the probability density function (PDF) of the axial current density. We also present two preliminary case studies of the particle energization signatures associated with the small-scale flux ropes. In the last section, we summarize the key findings of this study and discuss future work involving the GS reconstruction method.

IOP Conf. Series: Journal of Physics: Conf. Series **1100** (2018) 012006 doi:10.1088/1742-6596/1100/1/012006

2. Data and Method

In this section, the *in-situ* data from the Ulysses spacecraft are introduced first. Then, the standard Grad-Shafranov equation is presented. With this equation, the Grad-Shafranov reconstruction method can be implemented, and the reconstruction steps are explained. The automated detection procedure of small-scale flux ropes, based on this technique, is described with a new set of criteria needed for analysis of Ulysses data.

2.1. Data from Ulysses In-situ Measurements

The analysis covers four years during solar minima. The solar wind flow had both high and low speed intervals during these times which were well separated by heilo-latitude [20]. Besides, by combining flux-rope data at $\sim$ 1 AU, mostly obtained in low latitudes near the ecliptic plane, with flux-rope events near the ecliptic at larger heliocentric distances, we are able to analyze the dependence of flux-rope characteristics on heliocentric distance. For this purpose, the years 1994, 1996, 2004 and 2005 were selected in the Ulysses database.

Ulysses has various instruments on board. In this study we use data from Vector Helium Magnetometer (VHM), Solar Wind Observations Over the Poles of the Sun (SWOOPS), Heliosphere Instrument for Spectra, Composition and Anisotropy at Low Energies (HI-SCALE), and Cosmic Ray and Solar Particle Investigation (COSPIN). The data files from the first two instruments contain date, time, magnetic field data with 1 min. resolution in the RTN (radial, tangential and normal) coordinates. They also provide the field magnitude, proton density, three components of solar wind velocity in the RTN coordinates, and the proton temperature in terms of T_{large} and T_{small}. These two kinds of proton temperature are measured in two ways: determination of T_{large} involves the integral of the proton distribution over all energy channels, which may lead to overestimation of temperature, while T_{small} is calculated at a fixed energy resulting in underestimation for very cold plasma. Following most studies, T_{small} is regarded as proton temperature in this study. Since the electron temperature is unavailable from the Ulysses spacecraft, the thermal pressure p calculated in the first step of the GS reconstruction only contains proton temperature. The proton β, which partly defines a magnetic flux rope, is considered to be equivalent to the plasma β. Data files of the last two instruments which contain energetic proton flux measured from the low energy telescope (LET), and ion data from the low energy magnetic spectrometer at 30° relative to the spacecraft spin axis (LEMS30), are applied to case studies.

2.2. The Grad-Shafranov Equation

Using the assumption of two-dimensional (2D) magnetohydrostatic equilibrium and four field line invariants, i.e., the thermal pressure p, the transverse current density j_t, the axial component of the magnetic field B_z, and the magnetic flux function A, the Grad-Shafranov (GS) equation, describing non-force-free magnetohydrostatic structures in space plasmas, is given by:

$$\frac{\partial^2 A}{\partial x^2} + \frac{\partial^2 A}{\partial y^2} = -\mu_0 \frac{dP_t(A)}{dA} = -\mu_0 j_z(A). \tag{1}$$

Here j_z is the axial component of the current density, and P_t is defined as the total transverse pressure $P_t = p + (B_z{}^2/2\mu_0)$ which is a single-variable function of A only.

From the solution to the standard GS equation (1), we obtain the magnetic flux function $A(x, y)$, i.e., the x-y plane, of a cross section and the axial current density j_z over the cross section of a small-scale flux rope. Each flux surface or field line is of a distinct A value.

2.3. The Grad-Shafranov (GS) Reconstruction

As aforementioned, the GS reconstruction is a tool that employs 1-D data from *in-situ* spacecraft measurements to recover 2-D magnetohydrostatic or magnetic field structures, i.e., the cross

Table 1. The criteria of small-scale magnetic flux rope detection for Ulysses.

Duration (minutes)	R_{dif}	R_{fit}	*Walén test slope*
$45 \sim 2165$	≤ 0.12	≤ 0.14	≤ 0.5

section of a small-scale flux rope.

Following Hu [21], for example, the basic steps of the GS reconstruction are:

1. Select a data interval with magnetic field $\mathbf{B}$ and plasma parameters including the proton temperature T_p, proton number density N, the electron temperature T_e if available, and the solar wind velocity $\mathbf{V}_{SW}$. The thermal pressure p, therefore, can be computed by

$$p = Nk(T_e + T_p). \tag{2}$$

2. In order to satisfy the requirement of quasi-static equilibrium from the GS equation, we have to find the co-moving frame of reference. The deHoffmann-Teller (HT) frame is usually adopted [18]. Next, the frame velocity $\mathbf{V}_{HT}$ is determined by minimizing the residual electric field in a frame moving with unknown velocity $\mathbf{V}_F$. The residual electric field will reach its minimum when $\mathbf{V}_F = \mathbf{V}_{HT}$.

3. The minimum-variance analysis of magnetic field vector data (MVAB) provides three eigenvalues: the minimum, the intermediate and the maximum variance and their corresponding eigenvectors. We take the intermediate eigenvector as the initial trial z-axis and perform the trial-and-error process which is based on the requirement that $P_t(x,0)$ versus $A(x,0)$ be single-valued to find the optimal z-axis.

4. Once the optimal z-axis is determined, the reconstruction frame can be established. The x-axis is projected along the spacecraft path, and following the right-handed orthogonal coordinate system, the direction of the y-axis is determined.

5. With the spacecraft data of magnetic field along the x-axis as initial value, the magnetic flux function $A(x,0)$ along the x-axis can be calculated by integrating $-B_y$:

$$A(x,0) = \int_0^x \frac{\partial A}{\partial \xi} d\xi = \int_0^x -B_y(\xi,0) d\xi, \tag{3}$$

where $d\xi = -\mathbf{V}_{HT} \cdot \hat{\mathbf{x}} dt$, obtained from the previous steps, and dt is the sampling time increment. By far, $P_t(x,0) = p(x,0) + B_z^2(x,0)/2\mu_0$ versus $A(x,0)$ is ready.

6. Having determined $P_t(x,0)$ versus $A(x,0)$ for data points along the x-axis, the transverse pressure as a function of the magnetic flux, $P_t(A)$, can be obtained by minimizing $\sum_i [P_t(x_i,0) - P_t(A(x_i,0))]^2$, where $P_t(x,0)$ is calculated from spacecraft measurements while $P_t(A(x_i,0))$ is found from a functional fitting.

7. With the determined $P_t(A)$ function and the initial data provided by the spacecraft, we can solve the GS equation and finally obtain the solution of $A(x,y)$ which describes the magnetic field configuration of a flux rope, together with the axial field $B_z(A)$.

2.4. Automated Detection

Using the Grad-Shafranov (GS) reconstruction, Zheng and Hu [19] completed the automated detection of small-scale flux ropes utilizing data from the Wind spacecraft in the vicinity of 1 AU. We applied their method to Ulysses data using a new set of appropriate detection criteria which is presented in Table 2.1.

The basic search criterion is based on the double-folded behavior of $A(x,0)$ along the spacecraft path. Since the spacecraft passes through the same set of magnetic field lines from one edge to the center as it passes from the center to the other edge but in reversed order, we can split this path into two branches: the first and the second halves which let $P_t(A)$ curve be double-folded with an inflection or turning point. Then we apply the set of metrics given in Table 2.1 to filter event candidates:

1. *Duration:* In order to find candidate events fulfilling the criteria quantitatively and qualitatively, multiple iterations are adopted for detecting flux ropes: the searching window time interval is varied between (45, 80), (70, 105), to an upper bound of 2165 minutes. Taking (45, 80) min as an example, event candidates which fall within the limits of the width of sliding window set at 80 minutes maximum and the lower limit 45 minutes while exhibiting a double-folded $P_t(A)$ are recorded by the detection algorithm. Moreover, we allow slight overlaps, 10 min, to exist between adjacent search windows in order to identify the maximum number of qualified candidate events.

2. *Residues:* In order to check the double-folding quality, the difference residue and the fitting residue are both used [18; 22]. Separated by the turning or inflection point, $P_t(A)$ values of the two branches are calculated to obtain the difference residue R_{dif} while the fitting residue R_{fit} is determined through the deviation of P_t values between the measurements and the fitting function. Because P_t may vary from one individual flux rope to another, normalization by the difference between the maximum and minimum is necessary. Therefore, the smaller these residues are, the more accurately the z-axis is determined:

$$R_{dif} = [\frac{1}{2N}\sum_{i=1}^{N}((P_t)_i^{1st} - (P_t)_i^{2nd})^2]^{\frac{1}{2}}/|max(P_t) - min(P_t)|, \quad (4)$$

and

$$R_{fit} = [\frac{1}{L}\sum_{i=1}^{L}(P_t(x_i,0) - P_t(A(x_i,0)))^2]^{\frac{1}{2}}/|max(P_t) - min(P_t)|, \quad (5)$$

where the maximum P_t value is usually reached in the center of flux ropes.

3. *Walén slope threshold:* Walén slope is the slope of linear regression between the remaining flow $\mathbf{V} - \mathbf{V}_{HT}$ and the local Alfvén velocity $\mathbf{V_A}$. A small slope threshold excludes Alfvénic structures or waves and therefore ensures the satisfaction of the requirement of quasi-static equilibrium. In this study, a value 0.5 is set as the threshold in order to achieve sufficient event statistics while fulfilling the quasi-static assumption.

3. Results

3.1. Statistical Results of Occurrence Rate

Figure 2 presents the polar plot of solar wind speed as a function of helio-latitude for the four years: 1994, 1996, 2004 and 2005. Both 1994 and 1996 data were in high speed solar wind conditions, and for most of the time during these two years, the Ulysses was at higher latitudes. The Ulysses was at lower latitudes or near the ecliptic plane for most of 2004 and 2005. Also, the average solar wind speed is below 500 km/s for these two years. Therefore, we can categorize these four years of data into two groups based on the aforementioned information: Group 1 (1994 and 1996) corresponds to higher solar wind speed/latitudes, and Group 2 (2004 and 2005) refers to lower solar wind speed/latitudes, respectively.

Table 3.1 illustrates the parameters for these four years along with the flux-rope occurrence rates for different Walén slope thresholds. The data files of average solar wind speed, heliocentric distance and HelioGraphic Inertial (HGI) latitude are provided by Ulysses from COHOweb 1 hour merged data products. Occurrence rates or counts of small-scale flux ropes with different Walén slope thresholds are obtained from automated detection as described in the last section. Because this slope represents the ratio between the remaining flow speed and the local Alfv́en speed, the smaller Walén slope threshold we set, the more Alfvénic structures and waves will

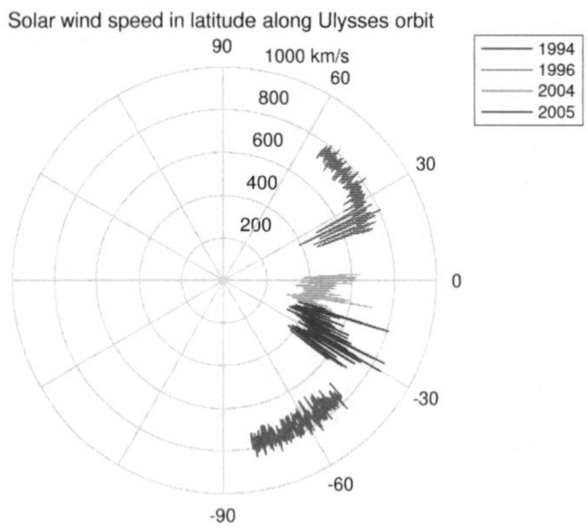

Figure 2. Polar plot of solar wind speed as a function of latitude during 1994 (blue line), 1996 (orange line), 2004 (gold line) and 2005 (purple line). HelioGraphic Inertial (HGI) latitude and plasma flow speed in RTN coordinates are provided by Ulysses from COHOweb 1 hour merged data files.

Table 2. The statistical results of small-scale flux ropes from Ulysses measurements. From left to right the columns are: Year, average solar wind speed in units of km/s, heliocentric distance (AU), latitude, counts with Walén slope (WS) ≤ 0.5, counts with Walén slope ≤ 0.3, and counts with Walén slope ≤ 0.1.

Year	$<V_{SW}>$	Distance (AU)	Latitude (°)	WS $\leq$ 0.5	WS $\leq$ 0.3	WS $\leq$ 0.1
1994	762	1.56 $\sim$ 3.83	-80.2 $\sim$ -44.2	1585	1418	609
1996	712	3.05 $\sim$ 4.71	19.1 $\sim$ 53.1	1729	1706	1171
2004	445	5.31 $\sim$ 5.41	-15.7 $\sim$ 2.6	1166	1164	1108
2005	494	4.48 $\sim$ 5.31	-37.1 $\sim$ -15.7	1204	1199	1161

be excluded. According to the results in Table 3.1, it is clear that with decreasing threshold values, the counts of small-scale flux ropes significantly decrease in Group 1, but not in Group 2. Therefore, we can conclude that Alfvénic structures (or waves) occur more frequently at higher latitudes under higher speed solar wind speed conditions, while the quasi-static small-scale flux ropes appear more dominantly at lower latitudes where lower solar wind speeds prevail.

3.2. Statistical Properties of Small-scale Flux Ropes

The histogram of average solar wind speed V_{sw} is shown in Figure 3. Group 1 data (years 1994 and 1996) is represented by the blue curve. A peak occurs near V_{sw} = 760 km/s close to the median V_{sw} = 753 km/s. Group 2 data (years 2004 and 2005) is denoted by the red curve. It also has high solar wind speed, such as the maximum V_{sw} = 798 km/s. But Group 2 can still be regarded as low solar wind speed since the mode is 429 km/s and the median is 462 km/s, which are all less than 500 km/s. Note that Ulysses went to lower latitudes (down to 19.1° north) during year 1996 and higher latitudes (up to -37.1° south) during year 2005. Thus, there is an intersection of two groups and the minimum solar wind speed of Group 1 is 482 km/s which is really close to the mean value of Group 2, i.e., 480 km/s, and the difference between two maximum values is just 36.51 km/s. Fortunately, because low speed solar wind only occurred

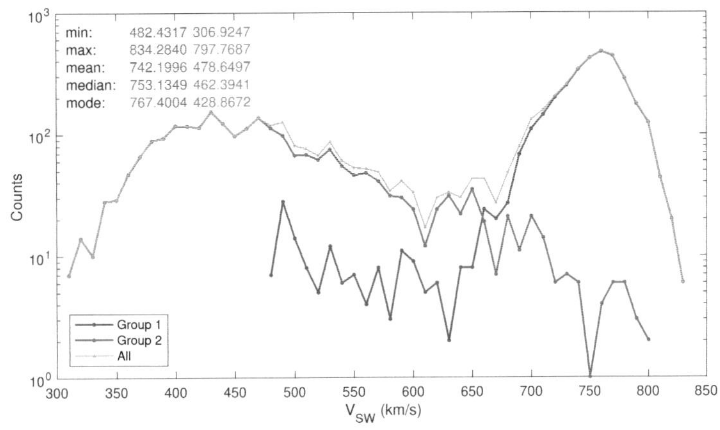

Figure 3. Histogram of average solar wind speed with 10 km/s as bin size.

for part of year 1996 (Figure 2), our categorization of two groups based on solar wind speed is still valid.

Figure 4 presents the angular axial orientations of small-scale magnetic flux ropes in the RTN coordinate system. The blue and yellow bars represent two different groups: Group 1 data is associated mainly with high solar wind speed while Group 2 data refers mainly to relatively low solar wind speed as mentioned in Table 3.1. In the histogram of polar angle θ, the angle between the z axis (given in the RTN coordinates) and N direction, the largest count of small-scale flux rope occurrence is at $80 \sim 90°$ bin which indicates that most cases for both groups tend to lie parallel to the local RT plane. The histogram of the azimuthal angle ϕ, the angle between the projected flux-rope z-axis and the T direction of Group 2 shows that two peaks are at $100 \sim 120°$ and $280 \sim 300°$ which are separated by 180° from each other. It indicates that the projection of the z axis onto the RT plane tends to align with the nominal Parker spiral direction on the ecliptic plane. However, Group 1 does not have clear tendency, although there might be two peaks located at $160 \sim 170°$ and $300 \sim 310°$.

Figure 5 presents the histogram of duration of small-scale flux ropes for these four years. The gold line represents the entire event set. The counts for Group 1 and 2 are shown by blue and red lines, respectively. Based on the criteria of automated detection, the minimum duration investigated is 45 minutes for both data groups, which guarantees the minimum duration of double-folding $P_t(A)$. The largest search window limits are between (1450, 2165) minutes. The distributions of both groups obey a power law, similar to the results for flux ropes identified at 1 AU, although the slopes may be different. Plus, as shown by both lines, the same tendencies indicate that there is no significant difference in duration between the high and low latitudes among the identified flux rope events except for the slopes in the power-law distributions. Both the Group 1 and Group 2 data exhibit the same trend of plateaus for flux-rope durations less than ~ 80 min, and a negative power-slope for durations more than ~ 100 minutes. However, Group 1 data forms a steeper power-law with a lower mean duration, which is partly due to Group 1 data being associated with higher solar wind speeds.

Figure 6 shows the distribution of scale sizes which are the spatial extent of the cross section of small-scale flux ropes along the spacecraft path. The maximum flux rope scale size is ~ 0.43 AU at least while the minimum is within the range of scale sizes observed at 1 AU. Similarly, the distributions of scale sizes of two groups seem to be close to each other which also indicates that small-scale flux ropes occurring at high or low latitudes or with different solar wind speed

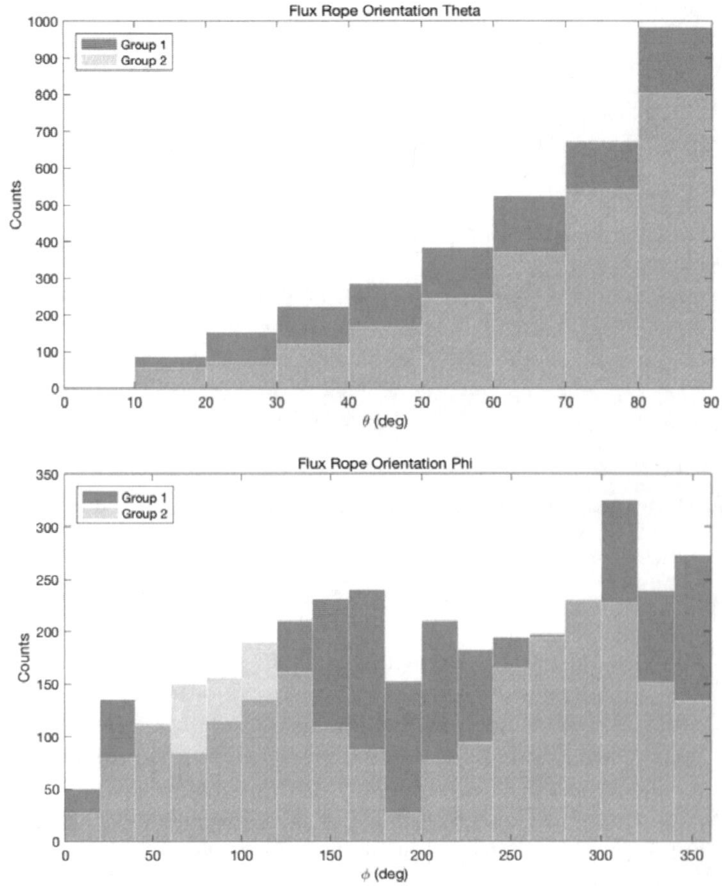

Figure 4. The top panel is histogram of the polar angle θ of the flux rope axial orientation with bin size 10°. In our searching process, the theta angle has been limited to be within 0 $\sim$ 90°. That is if the theta angle exceeds 90°, we simply flip the z axis. The bottom panel is histogram of azimuthal angle ϕ with 20° bin size. Blue and yellow bars represent Group 1 and 2, respectively.

maintain similar characteristics. Again, the distributions appear to follow power laws but with different power-law indices.

In addition, we compare the distributions of scale size at different heliocentric distances in the ecliptic plane using Group 2 (2004 and 2005) Ulysses data at $\sim$ 5 AU and Wind data at $\sim$ 1 AU for the same time periods. The occurrence rate from Wind observations was obtained for a duration range $\sim$ 9 - 361 minutes [19]. In order to make this comparison more consistent, we set the duration range from 5 to 355 minutes for the automated detection for Ulysses. The results are presented in Figure 7. One can see that two lines representing results from different spacecraft have a similar trend in that more event counts occur at smaller scale size and there are larger scale sizes continuously distributed through a range that may be defined as intermediate-scale. Although the scale size of the cross section of small-scale flux ropes is not dramatically different at different latitudes (Figure 6), the differences are somewhat more pronounced at different heliocentric distances. The largest count of scale size in the Wind database occurs at $\sim$ 0.002 AU which is twice that from the Ulysses data. Moreover, the difference between counts

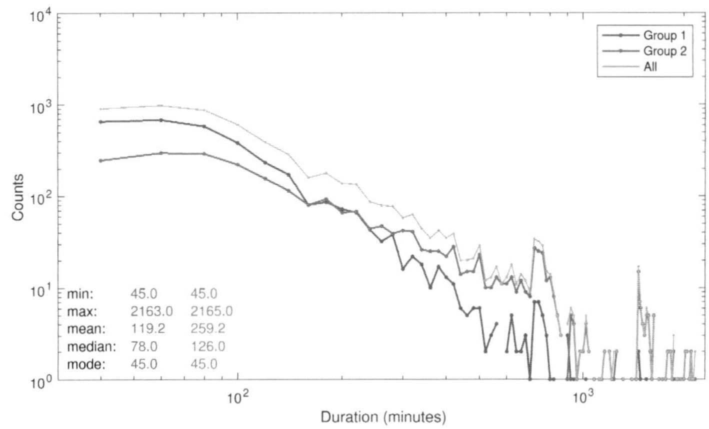

Figure 5. The histogram of duration of small-scale flux ropes in minutes which corresponds to the length of each flux rope interval. The range is from 45 minutes to 2165 minutes with 20 minutes bin size.

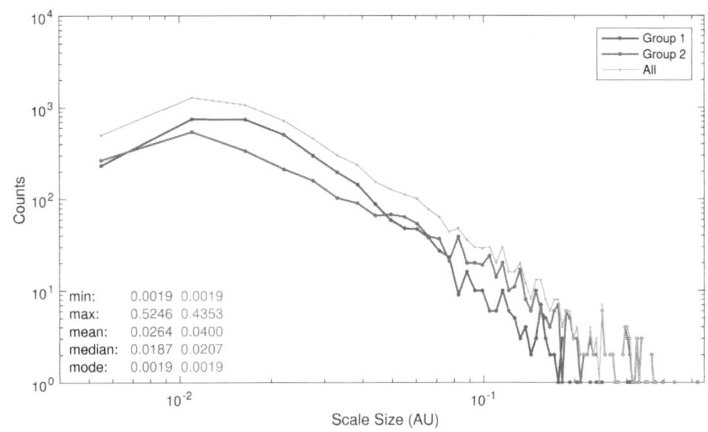

Figure 6. The histogram of scale size of small-scale flux ropes cross section with bin size 0.0055 AU.

of small-scale flux ropes between these two spacecraft measurements becomes larger at scale sizes greater than the line intersection value of $\sim$ 0.01 AU based on the power law fitting curves. Beyond the intersection point, Ulysses data tend to yield more events with relatively larger scale sizes.

In Figure 8 the histograms of average magnetic field magnitude, maximum magnetic field magnitude, proton temperature and plasma β are displayed. T_{small} of proton temperature data is used as in previous studies. The distribution of proton temperature peaks at higher temperature for Group 1 compared to Group 2. This indicates that small-scale flux ropes with lower solar wind speed are inclined to have low proton temperature and high proton temperature persists in high speed solar wind flux ropes, consistent with results at 1 AU. The histogram of average magnetic field shows that Group 1 has relatively higher peak values than Group 2.

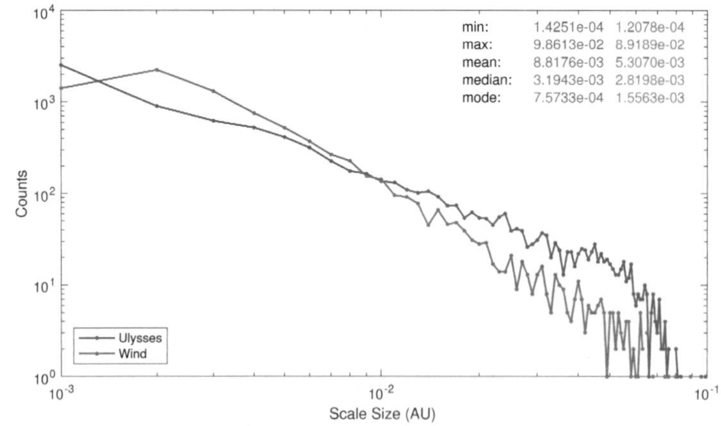

Figure 7. The histogram of scale size of small-scale flux ropes with data from Ulysses (blue line) and Wind (red line) in years 2004 and 2005.

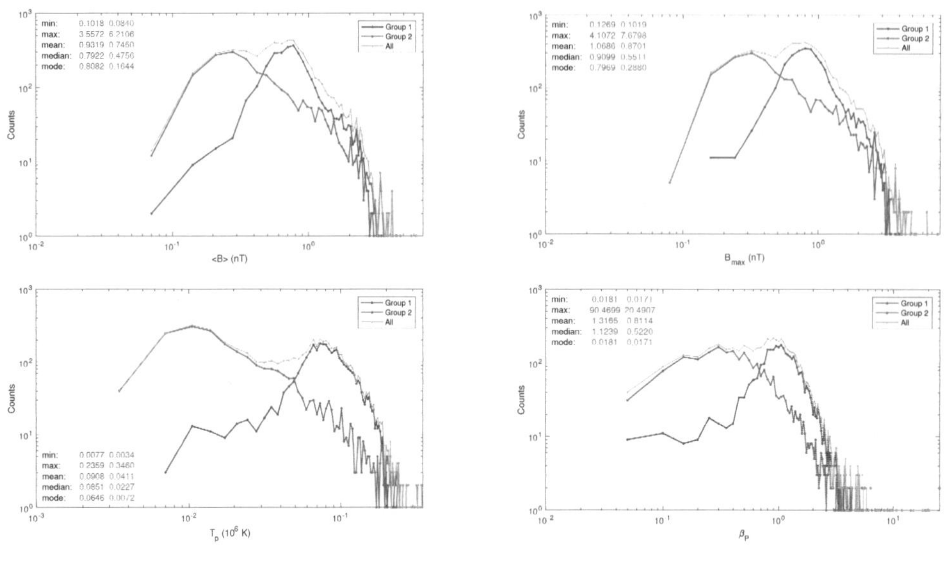

Figure 8. From the top to the bottom panels: the histograms of average magnetic field magnitude with bin size 0.07 nT, the maximum magnetic field with 0.08 nT bin size, the proton temperature T_{small} with 0.0035×10^6 K bin size, and the plasma β with 0.05 as bin size.

Except for the minimum value, the maximum, mean, median and mode of average magnetic field are all greater under high speed solar wind conditions.

Since the plasma β is the ratio between the plasma pressure NkT_p and magnetic pressure $B^2/2\mu_0$, its distribution must contain features of aforementioned parameters. Beyond the value 0.07×10^6 K on the distribution of proton temperature plot, small-scale flux ropes under high speed solar wind conditions dominate since the gold line which represents the entire data set, then overlaps the blue line of Group 2 data. This tendency is also clear in distribution of plasma β. Most β-values are less than 1 and the occurrence of flux ropes drops strongly with increasing

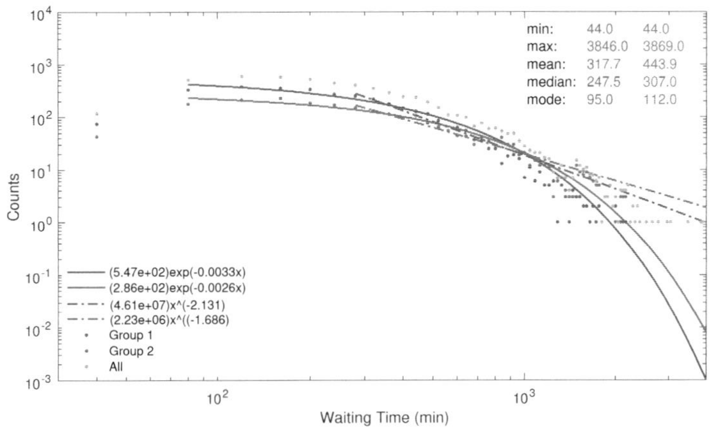

Figure 9. The waiting time distribution (WTD), which is defined as the separation time between the starting times between adjacent flux-rope structures (dotted curves). The WTDs are fitted both by an exponential (solid curves) and a power law (dash-dotted curves).

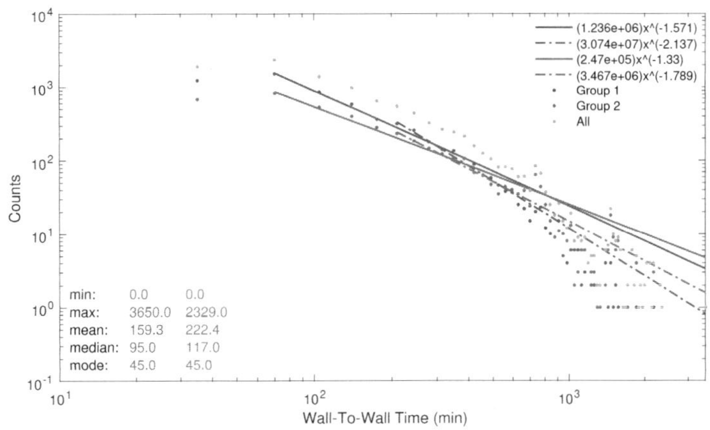

Figure 10. The distribution of wall-to-wall time which is defined as the separation between boundaries of small-scale flux ropes.

β-values beyond $\beta \simeq 1$, thus confirming that small-scale flux ropes are non-linear structures often associated with strong magnetic fields in force balance with significant plasma pressure gradient. An abnormally large β-values of 90.46 was detected in the Group 1 data set, which may be due to strong current sheet structure.

The waiting time distribution (WTD), which indicates if discrete events occur independently, is presented by Figure 9. The WTD in this study is defined by the separation time between the starting times of adjacent flux ropes. The minimum therefore has to be $\sim$ 45 min. The WTD of each group of data is fitted both by an exponential function and a power law function, respectively. It is clear that the former yields a better fit to both WTDs.

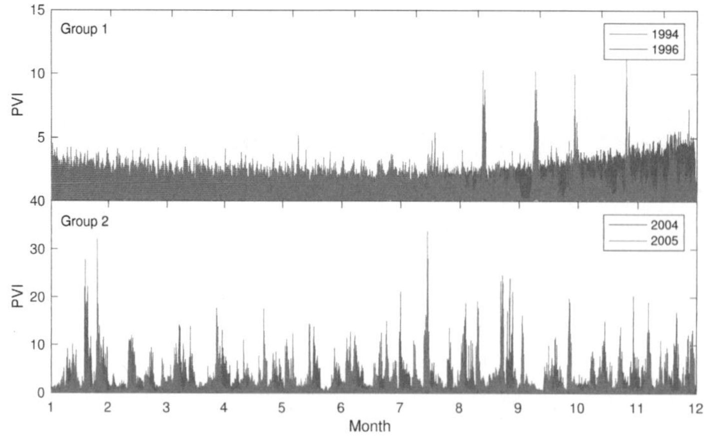

Figure 11. Values of partial variance of increments (PVI) for these four years. The increment is 40 min while the average is taken over the whole year. The first panel is Group 1 data including 1994 (blue line) and 1996 (red line). The second panel is Group 2 data including 2004 (blue line) and 2005 (red line).

3.3. Wall-to-wall Time Distribution Analysis

Figure 10 illustrates the distribution of wall-to-wall time. We assume that small-scale flux ropes are bounded by current sheets or magnetic field jumps with negligible thickness. Thus, current sheets, i.e., walls, exist at the start and end points of each flux rope interval. A power law fit is applied to each wall-to-wall time distribution. Each distribution can be thought of as double power law with a breaking point separating the two slopes, indicating separation between the inertial range (harder slope) and the energy containing range (softer range) of flux rope cross sections. The breaking point is located at $\sim$ 200 min which is 3 - 4 times the observational result at 1 AU (not shown). Compared to Greco et al. [23] who observed a power law index -1.23 $\pm$ 0.03 for 1 AU *in-situ* observations in the inertial range, we find an index of -1.57 below the breaking point and -2.1 beyond the breaking point for Group 1 data, and for Group 2 data, these indices are -1.33 and -1.78. So the results from Group 2 are closer to those at 1 AU of Greco et al. [23], within the inertial range. Since Group 1 may contain a lot of Alfvénic structures or waves (Table 3.1), the result of wall-to-wall time analysis at higher latitudes or in high speed solar wind might be affected, and is different from the result in low latitudes.

3.4. Partial Variance of Increments (PVI) Statistics

The partial variance of increments (PVI), according to Greco et al. [24; 25], is defined as

$$\Im = \frac{|\Delta \mathbf{B}|}{\sqrt{< |\Delta \mathbf{B}|^2 >}} = \frac{|\mathbf{B}(t+\tau) - \mathbf{B}(t)|}{\sqrt{< |\mathbf{B}(t+\tau) - \mathbf{B}(t)|^2 >}}. \quad (6)$$

Any sharp changes in the magnetic field will be detected via large PVI values. This is a quick method to locate the boundaries of structures and also can be used for comparing signatures of observations with results of simulations.

Values of calculated partial variance of increment (PVI) for the selected time periods are illustrated in Figure 11. The first panel is Group 1 data with two separate lines for 1994 and 1996, respectively. There are no apparent peaks for most high latitudes during mostly high speed solar wind conditions. But peaks occur for the last 4 months in 1996 when Ulysses was

IOP Conf. Series: Journal of Physics: Conf. Series **1100** (2018) 012006 doi:10.1088/1742-6596/1100/1/012006

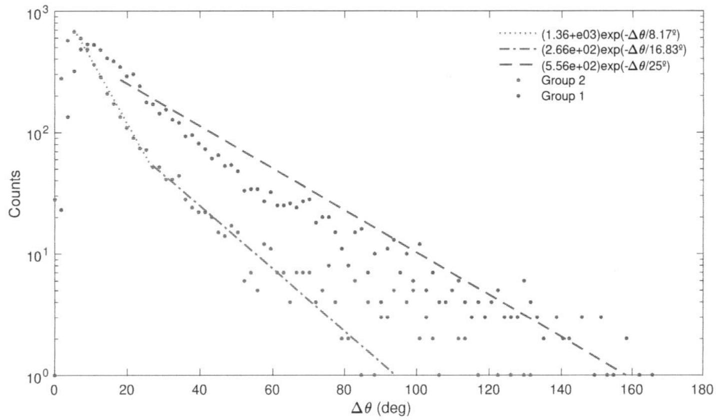

Figure 12. The histogram of the angular change in **B** across the small-scale flux rope boundaries, which is calculated with increment of 2 minutes before and after each boundary point. Data and fitting curve by an exponential function for Group 1 are marked in blue, while the red dots, red dotted line and the dash-dotted line are for Group 2.

transitioning into low latitudes and low speed solar wind streams. Such peaks are also visible in the predominantly low latitude, slow solar wind Group 2 data in the second panel. Group 2 data has about at least two strong peaks per month, in addition to relatively weak peaks that occur more often throughout the year. During solar minimum, the heliospheric current sheet (HCS) occurs more frequently in the equatorial plane while tending to occur more frequently at high latitudes at solar maximum [26]. All four years of data analyzed in this study were observed during solar minima. Therefore, it is reasonable that Group 2 data near the ecliptic plane contains more recurring peaks probably corresponding to HCS crossings.

3.5. $\Delta\theta$: Angle Across Small-scale Flux Rope Boundaries

The distribution of deflection angle across small-scale flux rope boundaries is presented with exponential fits in Figure 12. Miao et al. [28] identified about 28,000 current sheets in Ulysses 1-2 second resolution magnetic field measurements, and showed a breaking point around $\Delta\theta = 72°$. The largest count for the deflection angle occurred at around 25°. In this study, Group 2 data can be compared directly with the results of Miao et al. [28] since in both cases the observations were made near ecliptic plane or at low latitudes. The distributions were fitted with double exponential functions with characteristic decay values of $1/8.17°$ and $1/16.83°$, respectively, separated by a break point around $\Delta\theta = 30°$. Although the main features of our distributions seem agree with those of Miao et al. [28], both two characteristic exponential decay values and location of the breaking point differ substantially. Note that Miao et al. [28] detected 28214 events and their analysis used much higher resolution data while in this study we consider only about 4 years of data with much less events. Therefore, our statistical results may be affected by the relative lack of sample size and data resolution. Moreover, we did not impose a limit on the magnetic field magnitude as one of the criteria for the automated detection, which means our statistical result might contain small fluctuations and other structures, which needs to be investigated further.

The axial electric current density can be determined by the derivative of the transverse pressure with respect to the magnetic flux function A: $j_z = dP_t/dA$. The distribution of j_z is shown in Figure 13 normalized to the standard deviation of j_z, $\sigma = 3.2 \times 10^{-12}\ A/m^2$.

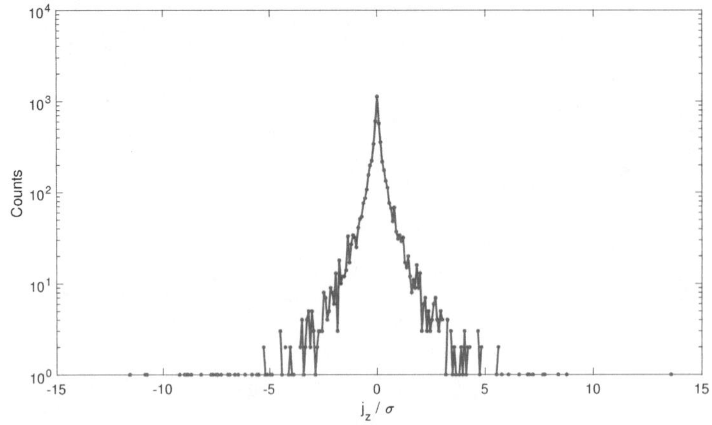

Figure 13. The distribution of the axial electric current density j_z. The x-axis is normalized by the standard deviation $\sigma = 3.2 \times 10^{-12}\ A/m^2$.

The distribution is non-Gaussian distribution with pronounced tails in agreement with the 1 AU observational result (not shown), thus confirming the accuracy and unique outcome of combining the Grad-Shafranov reconstruction method with automated detection in small-scale flux rope identification and characterization.

3.6. Applications: Small-scale Flux Ropes and Particle Energization
Observation evidence of particle acceleration near the HCS indicated local acceleration related to merging and contraction of small-scale magnetic islands [12]. Particles, magnetically confined inside cavities, may experience additional acceleration up to 1 $\sim$ 1.5 MeV if they are pre-accelerated [15]. They can also be trapped after acceleration via magnetic reconnection during the initial process and re-accelerated up to $\sim$ 5 MeV via small-scale magnetic flux rope dynamics [13].

The Grad-Shafranov reconstruction method has been applied to various data of spacecraft missions, such as the Wind, Advanced Composition Explorer (ACE) and Ulysses, to detect small-scale magnetic flux ropes automatically. Analysis of results and further applications are presented in the following case studies. Case 1 is an analysis combining flux-rope and energetic ion data from the ACE and the Solar and Heliospheric Observatory (SOHO) spacecraft. Case 2 is an application based on the detection result of flux rope and energetic ions in Ulysses data. The idea is to look for evidence that small-scale flux-ropes coincide with enhancements in energetic ion fluxes, thus proving potential evidence that these structures are responsible for ion accerleration in the solar wind.

3.6.1. Case 1: Observations During January 10 - 11, 2002. Case 1 consists of data from both SOHO and ACE since SOHO provides more channels of particles energies up to 22 MeV while the highest energy of ions is 4.8 MeV from ACE/EPAM. Because of their proximity in location, it is considered feasible for double-checking the same structures in both ACE and SOHO spacecraft data. We applied our automated detection to both data sets to analyze structures in the vicinity of interplanetary shocks with a special emphasis on the occurrence of small-scale flux ropes downstream of these shocks. Figure 14 shows the automatic identification of small-scale flux ropes downstream (up to 15 hours after the shock transition) of the shock at January 10, 2002 that occurred just before 16:00 UT. The presence of flux ropes coincides with particle flux

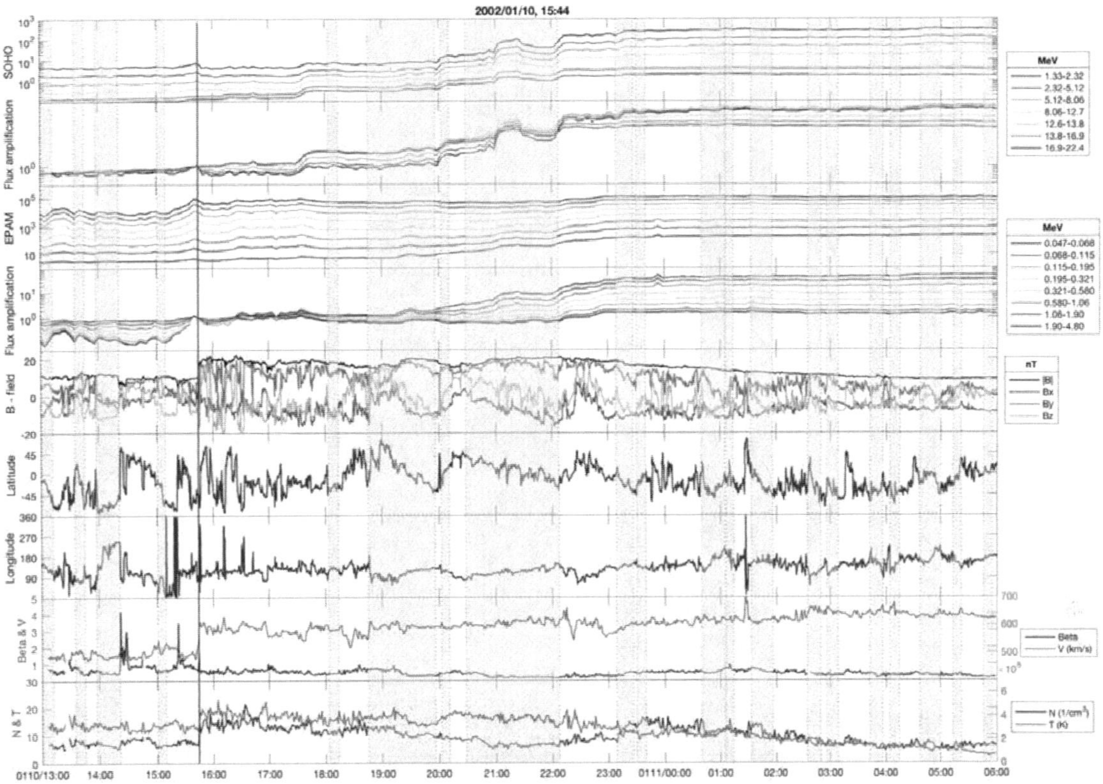

Figure 14. Example of small-scale flux-rope detection during the period from January 10, 2002, 13:00 UT to January 11, 2002, 06:00 UT. An interplanetary shock is shown by the solid vertical line and 20 identified flux rope intervals are shaded in grey. The top two panels are proton intensity data for the energy range, 1.33 MeV to 22.4 MeV, and its flux amplification normalized at the shock from SOHO. The third and fourth panels are the corresponding ion data from ACE EPAM (LEMS 120; 5 minute averages). The magnetic field data in the GSE coordinates are shown in the next three panels. The plasma parameters including the proton β, density, temperature, and the solar wind bulk speed are shown in the last two panels.

enhancements for flux rope intervals suggesting energetic ion energization by flux ropes during these intervals. The particle flux enhancements were observed at both spacecraft, but what flux-rope accerleration mechanism is responsible for the continuous acceleration associated with the flux enhancement starting from around January 10, 2002, 22:00 UT and persisting for quite a long period of at least 8 hours far downstream, still needs to be investigated.

3.6.2. Case 2: Observations During February 12 - 25, 2004. Case 2 represents our detection results based on Ulysses data. PVI analysis is applied, and different energy levels of proton flux from the COSPIN and LEMS30 telescopes are presented. Ulysses provides three RTN coordinate components of solar wind velocity. We present the radial component V_r only since the other two components are relatively small compared to V_r. We figure that the second continuous particle flux enhancement which occurs at around late February 16th to early 17th (day 47-48), may be associated with small-scale flux ropes, but this preliminary conclusion needs further

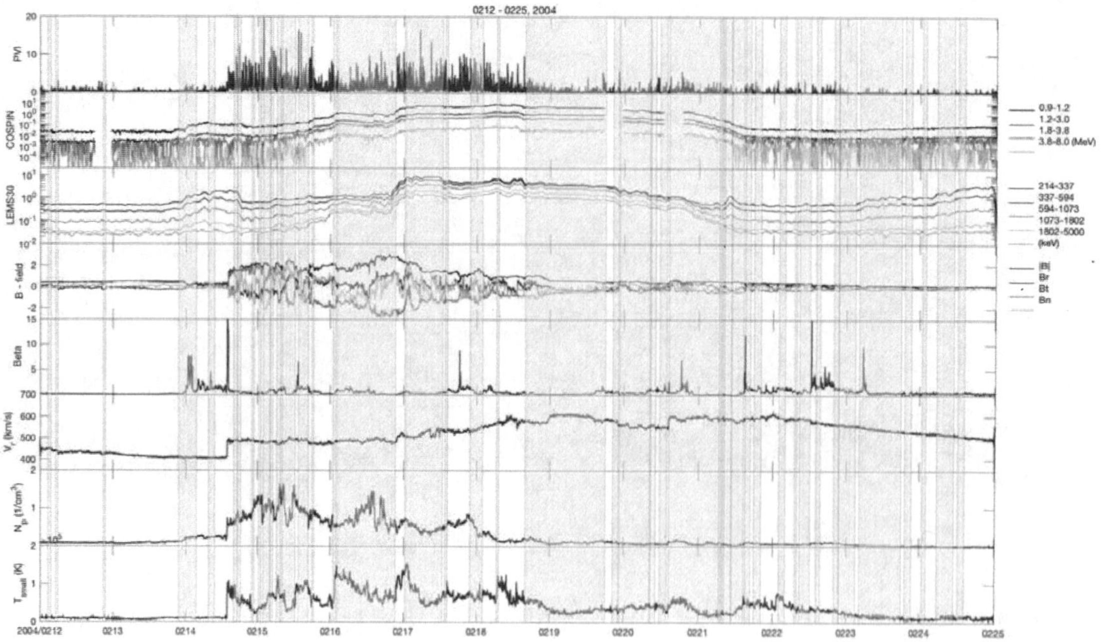

Figure 15. Example of small-scale flux ropes application from February 12, 2004, 00:00 UT to February 25, 2004, 060:00 UT at Ulysses (courtesy of Lingling Zhao). The total number of identified flux rope (intervals shaded in grey) is 43. All data are provided by the Ulysses *in-situ* measurements. The panels are, from the top to the bottom: calculated PVI values (same method as Figure 11) with time, proton flux from 0.9 to 8 MeV from COSPIN, proton flux from 214 to 5000 keV from LEMS30, magnetic field in RTN coordinate system and field magnitude (black line), plasma beta, the radial component of solar wind velocity, the proton number density and the proton temperature T_{small}.

investigation (Figure 15).

4. Summary and Discussion

We have applied a combination of Grad-Shafranov reconstruction with automated detection to Ulysses *in-situ* measurements at high helio latitudes and low latitudes at $\sim$ 5 AU to perform a statistical analysis of small-scale flux ropes for years 1994, 1996, 2004 and 2005. The results are compared for some statistical properties with the corresponding 1 AU observational results. The main results of this study are summarized as follows:

1. Significant drop of occurrence rate of small-scale flux ropes with decreasing Walén slope threshold indicates that at higher latitudes or under high speed solar wind conditions during 1994 and 1996, there are considerably more Alfvénic structures (or waves) than at low latitudes. This tendency may affect the statistical results of small-scale flux ropes during 1994 and 1996.

2. Most small-scale flux ropes identified in this study tend to lie parallel to the local RT plane, and in low speed solar wind near the ecliptic plane, they are likely aligned with the nominal Parker spiral magnetic field direction.

3. Statistical properties of small-scale flux ropes, such as the distributions of duration, scale size, proton temperature, etc., remain similar for both high and low latitudes. The scale sizes for different heliocentric distances in the ecliptic plane are also in the same range. The distributions are mainly power laws with different slopes at high latitudes as compared to low latitudes.

4. The waiting time distribution (WTD) of flux ropes are best fitted by an exponential function for both high and low speed winds in Ulysses data while 1 AU statistical results have good fits both by a power law and an exponential function.

5. Fitting with a double power law worked best for the observed wall-to-wall time distribution of flux ropes, yielding a breaking point at $\sim$ 200 min. The power law indices for Group 2 data (low latitude Ulysses data at $\sim$ 5 AU) correspond approximately to the low latitude indices at 1 AU while for Group 1 high latitude data, containing more Alfvénic structures or waves, the power-law indices are distinctly different.

6. Preliminary results combining Grad-Shafranov reconstruction method with energetic ion fluxes data from ACE, SOHO and Ulysses provide the potential evidence that enhancements in energetic ion fluxes can be attributed partly to small-scale flux ropes.

In retrospect, the limit of magnitude of magnetic field applied in the previous study was not applied in this paper. Therefore, our statistical results might contain small magnetic field turbulences. As shown by Table 3.1, more Alfvénic structures could be included by using a larger Walén slope threshold, e.g., 0.5, especially for high speed solar wind in high latitudes. Thus, Grad-Shafranov reconstruction can be extended in a future study to take into account significant field-aligned flow [29], corresponding to cases with significant Walén slopes.

5. Acknowledgments

All of Ulysses data in this paper are from Coordinate Data Analysis Web (CDAWeb) of NASA GSFC. We would like to thank people behind this website and scientists working for Ulysses spacecraft. In addition, we acknowledge NASA grants NNX15AI65G and NNX17AB85G, subawards NRL N00173-14-1-G006 and SAO SV4-84017, and NSF grant AGS-1650854 for support. We thank our colleagues at CSPAR/UAH, and Dr. O. Khabarova for stimulating discussions.

References

[1] Moldwin M B, Phillips J L, Gosling J T, Scime E E, McComas D J, Bame S J, Balogh A and Forsyth R J 1995 *J. Geophys. Res.* **100** 19903–19910

[2] Moldwin M B, Ford S, Lepping R, Slavin J and Szabo A 2000 *Geophys. Res. Lett.* **27** 57–60

[3] Feng H Q, Wu D J and Chao J K 2007 *Journal of Geophysical Research (Space Physics)* **112** A02102

[4] Feng H Q, Wu D J, Lin C C, Chao J K, Lee L C and Lyu L H 2008 *Journal of Geophysical Research (Space Physics)* **113** A12105

[5] Feng H Q, Zhao G Q and Wang J M 2015 *Journal of Geophysical Research (Space Physics)* **120** 10

[6] Cartwright M L and Moldwin M B 2008 *Journal of Geophysical Research (Space Physics)* **113** A09105

[7] Drake J F, Swisdak M, Che H and Shay M A 2006 **443** 553–556

[8] Zank G P, le Roux J A, Webb G M, Dosch A and Khabarova O 2014 *Astrophys. J.* **797** 28

[9] le Roux J A, Webb G M, Zank G P and Khabarova O 2015 *Journal of Physics: Conference Series* **642** 012015 URL `http://stacks.iop.org/1742-6596/642/i=1/a=012015`

[10] Zank G P, Hunana P, Mostafavi P, Le Roux J A, Li G, Webb G M, Khabarova O, Cummings A, Stone E and Decker R 2015 *Astrophys. J.* **814** 137

[11] le Roux J A, Zank G P, Webb G M and Khabarova O V 2016 **827** 47

[12] Khabarova O, Zank G P, Li G, le Roux J A, Webb G M, Dosch A and Malandraki O E 2015 *Astrophys. J.* **808** 181 (*Preprint* 1504.06616)

[13] Khabarova O V and Zank G P 2017 *Astrophys. J.* **843** 4

[14] Zheng J, Hu Q, Chen Y and le Roux J 2017 **900** 012024

[15] Khabarova O V, Zank G P, Li G, Malandraki O E, le Roux J A and Webb G M 2016 *Astrophys. J.* **827** 122

[16] Sonnerup B U Ö and Guo M 1996 *Geophys. Res. Lett.* **23** 3679–3682

[17] Hu Q and Sonnerup B U Ö 2001 *Geophys. Res. Lett.* **28** 467–470

[18] Hu Q and Sonnerup B U Ö 2002 *Journal of Geophysical Research (Space Physics)* **107** 1142

[19] Zheng J and Hu Q 2018 *Astrophys. J. Lett.* **852** L23 (*Preprint* 1801.01771)

[20] McComas D J, Gosling J T and Skoug R M 2000 *Geophys. Res. Lett.* **27** 2437–2440

[21] Hu Q 2017 *Science China Earth Sciences* **60** 1466–1494 ISSN 1869-1897 URL https://doi.org/10.1007/s11430-017-9067-2

[22] Hu Q, Smith C W, Ness N F and Skoug R M 2004 *Journal of Geophysical Research (Space Physics)* **109** A03102

[23] Greco A, Matthaeus W H, Servidio S and Dmitruk P 2009 *Physical Review* **80** 046401

[24] Greco A, Chuychai P, Matthaeus W H, Servidio S and Dmitruk P 2008 *Geophys. Res. Lett.* **35** L19111

[25] Greco A, Matthaeus W H, Servidio S, Chuychai P and Dmitruk P 2009 *Astrophys. J. Lett.* **691** L111–L114

[26] Richardson J D, Wang C and Burlaga L F 2004 *Adv. Space Res.* **34** 150–156

[27] Greco A, Matthaeus W H, Perri S, Osman K T, Servidio S, Wan M and Dmitruk P 2018 *Space Science Reviews* **214** 1

[28] Miao B, Peng B and Li G 2011 *Annales Geophysicae* **29** 237–249

[29] Teh W L 2018 *Earth, Planets and Space* **70** 34

IOP Conf. Series: Journal of Physics: Conf. Series **1100** (2018) 012007 doi:10.1088/1742-6596/1100/1/012007

On the magnetic field topology in reconnection region

G Consolini, V Quattrociocchi and M F Marcucci

INAF-Istituto di Astrofisica e Planetologia Spaziali, Roma, Italy

E-mail: giuseppe.consolini@iaps.inaf.it

Abstract. Magnetic reconnection is generally accompanied by large and quasi-turbulent spatio-temporal fluctuations from the MHD scales down to the kinetic ones. The study of these turbulent fluctuations is generally based on the investigation of spectral and scaling features. In a recent paper by Consolini et al. [1] it has been shown how the statistics of geometrical invariants of coarse-grained gradient tensor of plasma velocity is able to provide valuable information on the topology of the turbulent structures. Here, we present a preliminary study of the topology of magnetic field structures at kinetic scales in the X-line/dissipation region of a reconnection event observed by MMS constellation [2]. The analysis evidenced vortex sheet Ohmic-dissipation.

1. Introduction

Several space environments display complex dynamics characterised by multi-scale processes, such as fluid and magnetoydrodynamic (MHD) turbulence. In the framework of heliospheric plasmas, turbulence is a phenomenon widely documented in several different regions: interplanetary medium, Earth's magnetosheath and tail central plasma sheet, etc. Furthermore, the occurrence of turbulent spatio-temporal fluctuations has been observed also in the course of fast dynamical processes such as the magnetic reconnection and/or current disruption events.

The understanding and characterisation of spatio-temporal turbulent fluctuations are very important for modelling plasma heating, acceleration and transport across boundary regions. In this framework of a great importance is the study of the scaling features and intermittent nature of such fluctuations down to the kinetic domain, where plasma acceleration and heating is mainly taking place in the course of fast relaxation processes.

The traditional approach to space plasma turbulence is based on Fourier-spectral analysis and statistical methods assuming an Eulerian point-of-view. Although using such an approach much has been gathered on the general spectral and statistical features of these fluctuations, less is known about the topology of the relevant magnetic field and plasma structures responsible for them.

In this framework, information on the relevant topologies involved in the spatio-temporal turbulence observed in several space plasma regions can be obtained by investigating the statistical features of the geometrical invariants of the coarse-grained gradient tensor of velocity and magnetic field. Recent multi-satellite missions such as the ESA-Cluster and the NASA-MMS, due to their quasi-tetrahedral configuration, allow the computation of the magnetic and velocity field gradient tensor, $A_{ij} = \partial_i u_j$ and $X_{ij} = \partial_i b_j$, at the tetrahedron scale (the so called *coarse-grained gradient tensor*), providing more detailed information on the spatial features of the observed fluctuations and their evolution with the scale.

Recently, Consolini et al. [1] have shown the potentiality of the investigation of coarse-grained velocity gradient tensor and its geometrical invariants at small-scales for a case study of space plasma turbulence using the ESA-CLUSTER mission, following the same methodology used in the case of fluid turbulence [3, 4, 5]. In this work they have analysed the joint statistics $P(R, Q)$ of the second and third geometric invariants, Q and R, of the coarse-grained velocity gradient tensor, showing the presence of a good similarity to what is found for the low end of the inertial range in fluid turbulence. Clearly, a full comprehension of the role that structure topology plays in magnetised turbulent plasmas would also require the investigation of magnetic field gradient tensor features [6].

This work presents a preliminary analysis of magnetic field topologies in the kinetic domain (i.e., below the ion-cyclotron frequency) in the X-line/dissipation region of a reconnection event observed by MMS constellation [2] by evaluating the statistics $P(R_X, Q_X)$ and $P(Q_j, Q_K)$ of the geometric invariants of the coarse-grained magnetic field gradient tensor.

2. Magnetic field gradient tensor and geometrical invariants: a brief introduction

In turbulent media the analysis of the structures involved in the dissipation and in the generation of intermittency is central, and can be achieved by studying the statistics of the geometrical invariants of velocity field gradient tensor, $A_{ij} = \partial_i u_j$, which provides an information of the role that nonlinear self-stretching plays in generating small-scale large fluctuations [7, 8, 9].

The same can be done for the magnetic field gradient tensor, $X_{ij} = \partial_i b_j$ ($\mathbf{X} = \nabla \mathbf{b}$), whose geometrical invariant quantities can be useful to characterize the local geometrical features of the magnetic field lines [6].

On the basis of the Cayley-Hamilton Theorem the characteristic polynomial of the magnetic field gradient tensor,

$$|| \mathbf{X} - \lambda_i \mathbf{I} ||= \lambda_i^3 + Q_X \lambda_i + R_X = 0 \tag{1}$$

is invariant under SO3 group, so that the quantities,

$$Q_X = -\frac{1}{2} X_{im} X_{mi} \tag{2}$$

and

$$R_X = -\frac{1}{3} X_{im} X_{mk} X_{ki}, \tag{3}$$

are geometrical invariants under rotations and reflections. In particular, depending of the values of Q_X and R_X we get different topologies of the magnetic field lines. Thus, as in the fluid case, we can use two invariants, Q_X and R_X to classify the most relevant topological structures of magnetic field lines using the (R_X, Q_X) plane and investigating the joint statistics $P(R_X, Q_X)$ of the invariant quantities. In detail, the (R_X, Q_X) plane can be divided into four sectors using the $R_X = 0$ axis and the discriminant line $D = 0$, i.e.,

$$D = \frac{27}{4} R_X^2 + Q_X^3 = 0, \tag{4}$$

so that different topologies of the field lines can be associated to the different sectors: *i)* unstable vortex compressing $[D, R_X > 0]$; *ii)* stable vortex stretching $[D > 0$ and $R_X < 0]$; *iii)* stable tube-like structures $[D, R_X < 0]$ *iv)* unstable sheet-like structures $[D < 0$ and $R_X > 0]$.

Furthermore, by decomposing the magnetic field gradient tensor into its symmetric and anti-symmetric parts other invariants can be introduced:

$$Q_K = -\frac{1}{2} Tr(\mathbf{K}^2), \tag{5}$$

$$R_K = -\frac{1}{3} Tr(\mathbf{K}^3) \tag{6}$$

and

$$Q_j = -\frac{1}{2} Tr(\mathbf{J}^2) = \frac{1}{4} j^2. \tag{7}$$

Here, the first two are related to the symmetric strain rate tensor **K**, while the last one is associated to the current density j.

Among these last invariant quantities the two scalars Q_j and Q_K can be used to evaluate the relative importance of rotational part of the energy and the energy associated to the strain. In this framework, the joint probability distribution $P(Q_j, -Q_K)$, defined on the plane $(Q_j, -Q_K)$, allows to identify the topologies that are mainly involved in the Ohmic dissipation [6].

Here, we investigate the two joint probabilities $P(R_X, Q_X)$ and $P(Q_j, -Q_K)$ to get an information of the most relevant topologies involved in the spatio-temporal turbulent fluctuations observed in the magnetic reconnection dissipation region.

3. Data and Methods

To study the topological features of the magnetic field spatio-temporal fluctuations at the non-MHD scales in the dissipation region of a magnetic reconnection event we consider the NASA-MMS encounter of the dissipation region already studied by Burch et al. [2] on October, 16, 2015 from 13:06:45 UT to 13:07:15 UT. During the selected time interval the MMS constellation crosses the reconnection X line/dissipation region, where jet reversals, intense currents and strong plasma heating is observed [see Burch et al. [2] for a more detailed description]. Furthermore, the MMS satellite configuration resembles a quasi-perfect tetrahedral shape (pseudo-sphere), being the tetrahedral shape characterised by a *planarity coefficient* $P = 0.19$, an *elongation factor* $E < 0.01$ and a *characteristic scale* (radius of the circumscribed sphere) $L \sim 10$ km [10]. This characteristic scale L is intermediate between the electron inertial length $\eta_e \sim 2$ km and the proton inertial length $\eta_p \sim 70$ km, so that MMS configuration allows to investigate the magnetic field topologies at non-MHD (kinetic) scales.

To evaluate the coarse-grained gradient tensor the dataset used in our analysis consists in the magnetic field measurements from FGMs and SCM experiments and satellite attitude data. Data come from NASA-CDAWeb (https://cdaweb.sci.gsfc.nasa.gov/index.html). FGMs and SCM data have been combined so to get a resolution of 1024 samples per second following a general scheme reported in Ref. [11].

Figure 1 shows a sample of the obtained power spectral densities (PSD) of the three magnetic field FGMs-SSC combined components relative to satellite MMS1. No anomalous effects of the data merging are visible at the Nyquist frequency ($f_N = 64$ Hz) corresponding to FGM sampling rate. Furthermore, the observed PSDs show a common power-law behaviour, $S(f) \sim f^{-\beta}$, with $\beta \sim -8/3$ over nearly two orders of magnitude above the proton cyclotron frequency ($f_\Omega \sim 0.4$ Hz) computed using the magnetosheath local mean magnetic field.

To compute the coarse-grained gradient tensor of the magnetic field at the tetrahedron scale we make use of the TETRAD method introduced by Chertkov et al. [3]. Indicating with the $\tilde{X}_{ij}$ the coarse-grained gradient tensor $X_{ij} = \partial_i b_j$ evaluated on the volume Γ, i.e.

$$\tilde{X}_{ij} = \frac{1}{\Gamma} \int_\Gamma \partial_i b_j d^3x, \tag{8}$$

it results,

$$\tilde{X}_{ij} = \left(\hat{\rho}^{-1}\right)_{ik} \hat{b}_{kj} - \frac{\delta_{ij}}{3} \mathrm{Tr}\left(\hat{\rho}^{-1}\hat{b}\right), \tag{9}$$

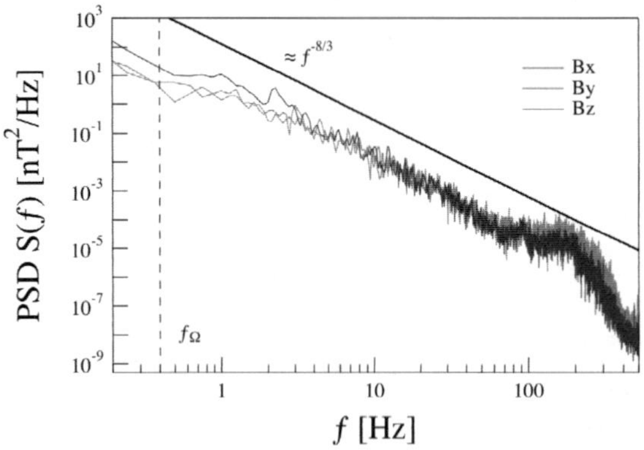

Figure 1. The PSDs $S(f)$ of the three magnetic field FGMs-SSC combined components relative to satellite MMS1. Solid line refers to power-law behaviour, $S(f) \sim f^{-\beta}$, with $\beta \sim -8/3$, while the vertical dashed line indicates the average proton cyclotron frequency, $f_\Omega \sim 0.4$ Hz.

where $\hat{\rho}$ and $\hat{b}$ are the distance matrix tensor and the magnetic field one, computed as described in Refs. [3, 1] using the four measurements at the vertexes of the tetrahedron. In Eq. (9) the second term on the right hand side is the constraint of a divergence-free field, i.e., $\nabla \cdot b = 0$. The quantity, $\tilde{X}_{ij}$, represents the average gradient tensor of the magnetic field at the scale of the characteristic size L of the tetrahedron, and is a coarse-grained approximation of the local magnetic field gradient tensor.

Once the magnetic field coarse-grained gradient tensor, $\tilde{X}_{ij}$, is computed, all the main four geometrical invariants, $[Q_X,\ R_X,\ Q_K,\ Q_j]$, can be evaluated using the expressions of Eqs. (2, 3, 5) and (7). The estimation of the associated joint probabilities $P(R_X, Q_X)$ and $P(Q_j, -Q_K)$ allows the investigation of the most relevant magnetic field line topologies at the observational scale L.

The computation of the joint probabilities $P(R_X, Q_X)$ and $P(Q_j, -Q_K)$ is done by applying a kernel-based method as described in Kaiser & Schreiber [12]. The used kernel is a Gaussian one, whose width is twice the resolution scale adopted to compute the joint PDFs.

4. Results

Following the procedure described in the previous Section 3 we have computed point-by-point the four geometrical invariants, $[Q_X,\ R_X,\ Q_K,\ Q_j]$ and successively estimated the associated joint probabilities $P(R_X, Q_X)$ and $P(Q_j, -Q_K)$ for the X-line/dissipation region crossing.

Figure 2 shows the computed joint probability $P(R_X, Q_X)$ of the first two geometrical invariants, Q_X and R_X. The joint PDF $P(R_X, Q_X)$ shows a cigar-like shape and is mainly concentrated in upper part of the (R_X, Q_X) plane, i.e., above the discriminant zero-line $D = 0$. This suggests that the most relevant topologies of the magnetic field line structures are typically vortical structures. In particular we observe both stable vortex stretching and unstable vortex compressing, while there is a less evidence for the formation of sheet-like and tube-like structures, which are associated to the lower part of the $[R_X, Q_X]$ plane below the discriminant zero-line. Furthermore, the distribution core is quite well in agreement with the one observed in MHD decaying turbulence simulations by Dallas & Alexakis [6].

Apart from the distribution core we can observe some spotty increments of the probability in the upper-left $[R_X, Q_X]$ plane. These probability spots are associated with the occurrence of extreme events that corresponds with those points where strong electron heating is observed (data not-shown: please refer to Burch et al. [2]).

Figure 3 reports the joint probability $P(Q_j, -Q_K)$ of the other two invariants. As already said in Section 2 the distribution $P(Q_j, -Q_K)$ on the $(Q_J, -Q_K)$ plane characterises the dissipative structures, i.e., the structures where the Ohmic dissipation is mainly concentrated [6]. In our case we can find that the PDF is mainly aligned along the bisector of the $(Q_J, -Q_K)$ plane.

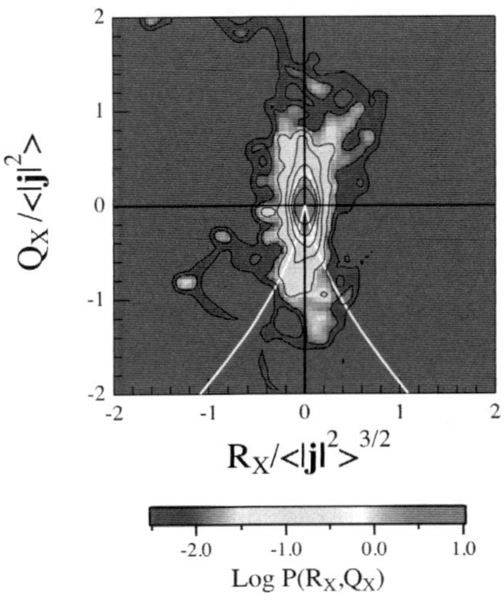

Figure 2. The joint probability $P(R_X, Q_X)$ of the first two geometrical invariants. The dashed red line is the *discriminant zero-line*. The two invariants are normalized using the mean-squares value of the current density, **j**, i.e., $j^2_{RMS} = \langle | \mathbf{j} |^2 \rangle$.

This result, which is in agreement with previous numerical study by Dallas & Alexakis [6] in a MHD decaying turbulence regime, suggests that the Ohmic dissipation mainly occurs in current layers located in vortex sheets. The same it occurs also for those extreme events identified in the (R_X, Q_X) plane.

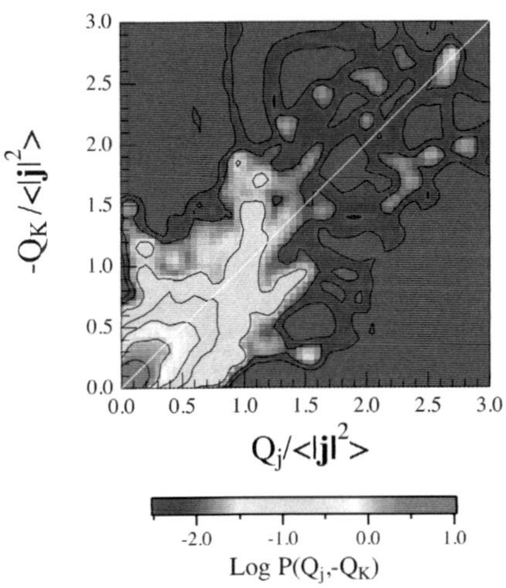

Figure 3. The joint probability $P(Q_j, -Q_K)$ of the other two geometrical invariants. The white line is the *bisector line*. The two invariants are normalised using the mean-squares value of the current density, **j**.

5. Summary and Conclusions

In this work we have presented a preliminary study of the geometrical/topological features of the magnetic field turbulent fluctuations at non-MHD/kinetic scales in a reconnection dissipation region performed by the analysis of the coarse-grained gradient tensor topological scalars. The principal aim of this work is to stress the potentiality of this multi-point method in unveiling the most relevant topologies that play a fundamental role in the observed turbulent spatio-temporal fluctuations.

Our analysis of the most relevant topologies has essentially shown that:

- the magnetic field streamlines are mainly elliptic, i.e. characterised by swirling streamlines associated with stable vortex stretching and unstable vortex compressing;
- the Ohmic dissipation concentrates in current layers in vortex sheets;
- there is a good correlation between electron heating and Ohmic dissipation in vortex sheets.

The emerging scenario from our analysis on the topologies in the dissipation region of a reconnection site is that of a complex domain, characterised by turbulent multi-scale structures. We believe that this complex and turbulent topology can strongly affect the reconnection phenomenon and plasma heating as already discussed in some previous works [see e.g. Refs. [13, 14, 15, 16] and references therein].

Clearly, more observational and theoretical studies, capable of describing the evolution of coarse-grained gradient tensor of the magnetic field from MHD scales down to kinetic ones, are necessary to explain the interrelationship between the observed turbulent structures and the occurrence of strong plasma acceleration events at kinetic scales.

Acknowledgments
We thank the NASA-CDAWeb (https://cdaweb.sci.gsfc.nasa.gov/index.html) and the FGM and SCM instruments PIs (Burch, C.T. Russell, W. Magnes, O. Le Contel) and teams for providing data used in this work. This research has received funding from the Italian Space Agency under Grant agreement No. ASI-INAF 2015-039-R.O. "Missione M4 di ESA: Partecipazione Italiana alla fase di assessment della missione THOR".

References

[1] Consolini G, Materassi M, Marcucci MF and Pallocchia G 2015 *Astrophys. J.* **812** 84
[2] Burch JL, Torbert RB, Phan TD, et al 2016 *Science* **352** 1189, id.aaf2939
[3] Chertkov M, Pumir A and Shraiman BI 1999 *Phys. Fluids* **11**, 2394.
[4] Celani A, Mazzino A and Pumir A 2003, in *The Kolmogorov Legacy in Physics*, Livi R and Vulpiani A eds., *Lect. Notes Phys.*, **636**, 173.
[5] Chevillard L and Meneveau C 2006 *Phys. Rev. Lett.* **97**, 174501.
[6] Dallas V and Alexakis A 2013 *Phy. Fluid* **25**, 105106
[7] Vieillefosse P 1984 *Physica A* **125**, 150
[8] Cantwell BJ 1992 *Phys. Fluids A* **4**, 782.
[9] Cantwell BJ 1993 *Phys. Fluids A* **5**, 2008.
[10] Robert P, Roux A, Harvey CC et al 1998 in *Analysis Methods for Multi-Spacecraft Data*, Paschmann G and Daly PW eds, *ISSI Sci Rep*, **SR-001**, 323.
[11] Fisher D, Magnes W, Hagen C, et al 2016 *Geosci. Instrum. Method. Data Syst.* **5**, 521
[12] Kaiser A and Schreiber T 2002 *Physica D* **166**, 43.
[13] Yankov VV 1997 *Phys. Plasmas 11*, 1287.
[14] Lazarian A and Vishniac ET 1999 *Astrophys. J. 217*, 700.
[15] Materassi M and Consolini G 2007 *Phys. Rev. Lett. 99*, 175002.
[16] Valentini F, Perrone D, Stabile S, et al 2016 *New J. Phys. 18*, 125001.

What causes the variability in the properties of energetic storm particle (ESP) events?

M A Dayeh[1], M I Desai[1], R W Ebert[1], H Elliott[1], A Farahat[2], K Kozarev[3], G Li[4]

1: Southwest Research Institute, United States of America;
2: King Fahd University of Petroleum and Minerals, Saudi Arabia;
3: Institute of Astronomy, Bulgarian Academy of Sciences, Bulgaria;
4: University of Alabama in Huntsville, United States of America

Correspondance: maldayeh@swri.edu

Abstract. Energetic storm particle (ESP) events are enhancements above ~0.05 MeV/nucleon ions near 1 AU in association with the passage of an interplanetary (IP) coronal mass ejection (ICME). The primary candidate of producing these enhancements is diffusive shock acceleration (DSA). ESPs can produce significant increases in the near-Earth particulate radiation and pose severe hazards to astronauts and hardware in space. Physical parameters thought to affect ESP production include IP shock properties (e.g., speed, strength, obliquity) and conditions upstream of the propagating shock (e.g., seed population, ambient plasma conditions). Several theoretical and observational studies tried to relate ESP production to these drivers; however, reliable prediction of ESP properties (e.g. intensities, spectra, abundances), including their event-to-event variability, has so far proven elusive. This indicates an incomplete understanding of how ICME-driven IP shocks accelerate ESPs.

Using instruments onboard ACE, we investigate the relations between a large set of parameters (28) that characterize (i) ESP properties, (ii) IP shock and ICME properties, and (iii) the upstream and downstream conditions measured across the IP shock. Ee present a comprehensive correlation matrix between all parameters and those of the ESP properties, in an attempt to identify the dominant parameters that influence ESP production and variability. We find that within the selected dataset, (1) spectral and compositional relations strongly indicate a rigidity-dependent acceleration mechanism, (2) correlations between observations and DSA predictions are poor, (iii) ICME sheath temperature appears to play a role in determining the ESP peak intensities.

1. Introduction

Fast interplanetary (IP) coronal mass ejections (CMEs) often drive shock waves through the corona and the IP medium. These shocks are sometimes accompanied by enhancements in the intensities of energetic ions above ~0.05 MeV/nucleon (e.g., Armstrong et al. 1985; Richter et al. 1985; Scholer 1985; Kennel et al. 1986; Reames 1999) known as energetic storm particle (ESP) events. ESPs can last for several hours and their associated particle enhancements might arrive ahead or behind the shock (e.g., Cohen 2006, Giacalone 2012). A classic ESP event (see Lario et al. 2005 for ESP categorization) is illustrated in Figure 1.

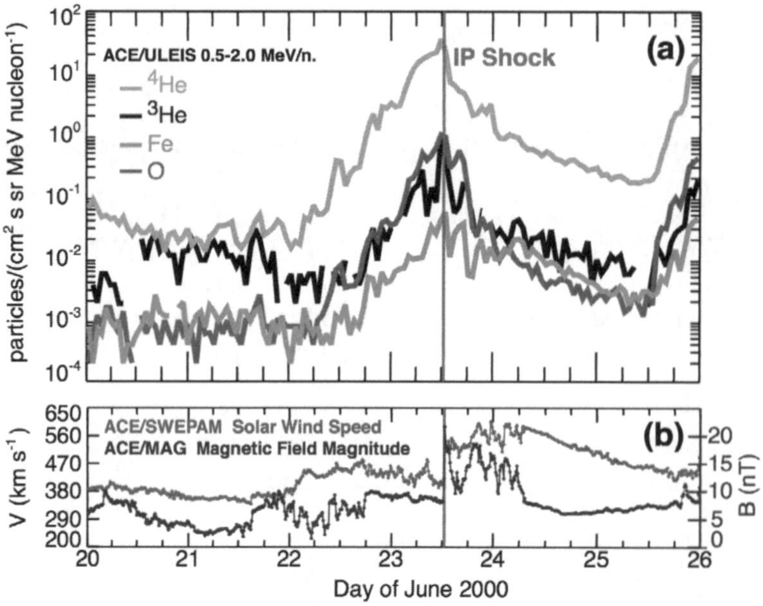

Figure 1: A classic ESP event observed at ACE. (a) Intensity-time profiles of 0.5–2.0 MeV/nucleon ^{3}He, ^{4}He, O, and Fe nuclei. (b) 5 minute average of the magnetic field magnitude B (in blue) and the 5 minute average of the solar wind speed from 2000 June 20 through 26. Purple vertical line marks the arrival of the interplanetary shock at ACE at 1227 UT on 2000 June 23.

The primary candidate for accelerating energetic particles at CME-driven shocks is thought to be the diffusive shock acceleration (DSA), comprising the shock-drift mechanism at quasi-perpendicular shocks (Decker, 1981), and the first-order Fermi mechanism at quasi-parallel shocks (Lee, 1983). DSA theory lays some expectations for ESP properties and how they relate to their IP shock properties and ambient conditions. For instance, (i) the acceleration of a mono energetic seed population results in a power-law with a spectral index, γ, that is independent of ion species and determined solely by the IP shock density compression ratio, (e.g. Van Nes et al. 1984). The stronger the compression, the flatter (i.e., lower spectral index) the spectrum. Tylka et al. (2005) suggested that particle abundances at the shock have a dependence on shock obliquity, with quasi-perpendicular shocks being more likely to produce heavy ion enhancements compared to quasi-parallel shocks. This ties into the effects that limit the acceleration processes, such as shock geometry (width, curvature), escape of ions from shock, and/or finite acceleration time, to produce a characteristic exponential rollover with e-folding energy E_0 that is species-dependent (e.g., Jones & Ellison 1991; Li et al. 2003). E_0 in this case depends on the diffusion coefficient, which increases with ion rigidity, such that higher rigidity ions (e.g., Fe) are accelerated less efficiently than lower rigidity ions (e.g., O). Furthermore, Zank et al. (2006) suggested that quasi-parallel shocks accelerate protons to higher energies more efficiently than quasi-perpendicular shocks, and that the latter are more likely to accelerate particles out of a pool of pre-energized source material than from to the solar wind, thereby producing ESP events with different composition than those generated by parallel shocks. These authors suggested that the maximum energy that particles can attain is a function of the upstream turbulence for perpendicular shocks and the generation of self-excited waves by the accelerated particles for parallel shocks. For the latter case, Giacalone et al. (2003, 2005) argued that sometimes the self-excited waves are not necessary for the particle acceleration. Other approaches have included detailed treatment of field geometry and shock evolution (e.g., Lee 2005), coupling particle acceleration, transport, and wave generation (Vainio & Laitinen 2007, 2008; Ng & Reames 2008), parameterizing the particle source at the shock and comparing with observations (e.g., Lario et al. 1998; Aran et al. 2005), and finally, a hybrid approach that combines analytical, MHD, and numerical simulations (Zank 2000; Li et al. 2003; Rice et al. 2003).

Observational evidence is undoubtedly a decisive way to test the ESP theoretical predictions. Measurements at 1 AU have shown that ESP properties exhibit significant event-to-event variations. Reames et al. (2012; see also Mäkelä et al. 2011) found that ESP proton intensities strongly correlate with the IP shock speeds, but these intensities varied broadly for similar shock speeds, as shown in figure 2a. This indicates that the physical processes responsible for producing this variability are likely dependent on the interplay of several parameters associated with ESP events.

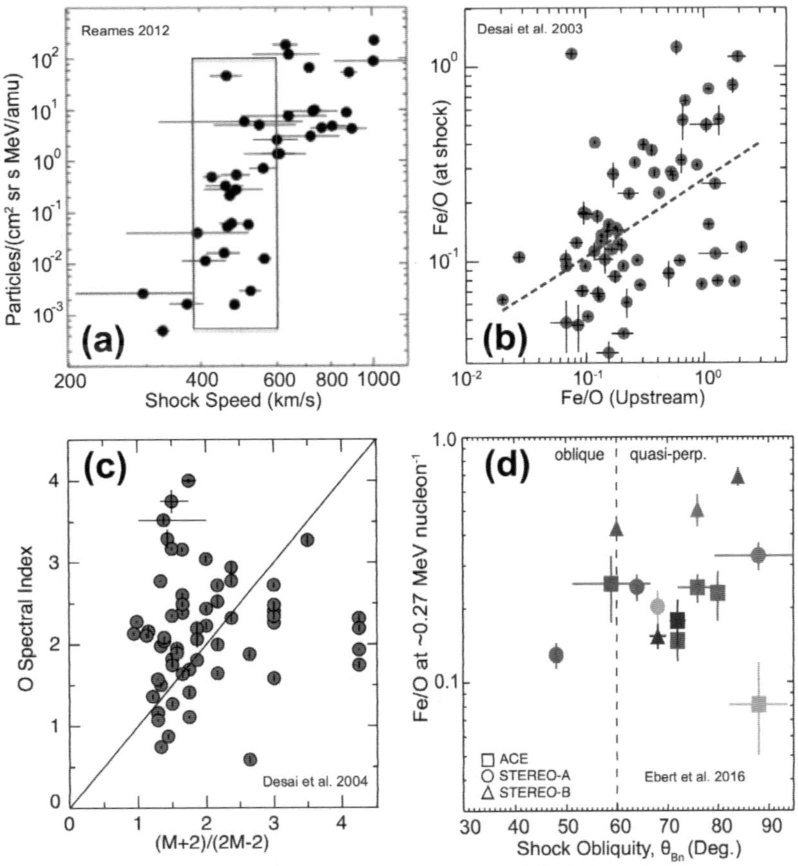

Figure 2: **(a)** Large intensity variations for similar shock speeds, suggesting complex inter-connected physical processes at play. **(b)** Upstream Fe/O is related to the ratio present at the shock. **(c)**, **(d)** Predictions of the DSA theory do not often agree with observations when numerous events are involved. Predicting ESP properties from DSA theory has proven to be difficult.

ESP abundance ratios and spectral properties have also proven difficult to predict due to their high variability when correlated with various parameters. Desai et al. (2003; 2004) surveyed ~72 IP shocks and associated ESP events that occurred during the ascending phase of solar cycle 23 and found that the Fe/O ratio exhibited strong energy-dependent behavior only in ~25% of these events. The authors also found that the 0.5-2 MeV/n. Fe/O ratio during the ESP event is well correlated with that measured upstream prior to the event (see figure 2b), the 0.1-0.5 MeV/n. oxygen spectral indices in ESP events at 1 AU are poorly correlated with the steady state prediction of the indices using the magnetic compression ratio at the shock (figure 2c), and that there is poor agreement between the spectral break energy and the local shock speed and obliquity. The latter results were confirmed by a multi-spacecraft study performed by Ebert et al. (2016) using ESP events measured at longitudinally-separated spacecraft, as illustrated in figure 2d. Finally, Lario et al. 2005 performed a detailed ESP case-study in context of the DSA theory predictions and found that although many of the event signatures were consistent with the predictions of DSA, heavy-ion elemental abundances were indeed poorly correlated with the solar wind composition.

To summarize, numerous theoretical and observational studies showed that IP shock and ESP properties exhibit large event-to-event variability and non-conclusive linkage, but the associated parameters provide strong insights into the physical mechanisms that produce ESPs. A detailed systematic examination of these parameters, along with modeling, is thus critical to characterize the ESP event-to-event variability, sources, acceleration, and transport processes and to understand and ultimately be able to predict ESP properties at 1 AU.

In this article, we perform a comprehensive analysis using several instruments onboard the Advanced Composition Explorer (ACE; Stone et al. 1998a) spacecraft, and examine the linear correlations between a large set of parameters that characterize (i) ESP properties, (ii) IP shock and ICME properties, and (iii) the upstream and downstream IP conditions measured near the IP shock.

2. Data Selection and Observations

We use energetic ion, solar wind plasma, and magnetic field data from the Ultra–Low-Energy Isotope Spectrometer (ULEIS; Mason et al. 1998), Solar Isotope Spectrometer (SIS; Stone et al. 1998b), the Solar Wind Electron Proton Alpha Monitor (SWEPAM; McComas et al. 1998), and the magnetometer (Smith et al. 1998) on board ACE. To select the events in this study, we started with a list of 72 CME-driven IP shocks that were associated with ESP enhancements identified in Desai et al. 2003 between 1997 October 1 and 2002 September 30. The selection criteria for the events were constrained by the following conditions, (1) The 0.5–2.0 MeV/n. ^{4}He, O, and Fe intensities should increase by at least a factor of 5 within a 24 hour period centered on the arrival of the shock. (2) The intensity-time profiles of different species should track each other, indicating a common acceleration and transport history for all species. (3) The associated Fe-group ions in the 0.3–3.0 MeV nucleon energy range should not exhibit velocity dispersion during onsets, indicating that the ions are accelerated in Interplanetary space near 1 AU.

We then visually examined all events and eliminated those where the IP shock appeared to be a driveless event, meaning that the event was not driven by a ICME or a corotating interaction region (CIR) [e.g., Gopalswamy et al., 2009]. The presence of a magnetic ejecta and a sheath region were used as signatures to determine the ICME boundaries (see Richardson and Cane 2010). These criteria left us with 62 ICME-associated ESP events. For each identified event, we then derived a comprehensive list of parameters that are descriptive of the properties and conditions of the IP shock, the ICME, the ESP event, and the ambient environment near the shock. Because the energetic particle event (i.e., the ESP component) evolves continuously with the traveling shock, whereas the shock itself is a discontinuity and many of its properties are very localized, the inferred parameters were calculated using different sampling time periods. Parameters of interest and a brief description of how they were determined are detailed in the following subsections.

2.1 ESP properties

ESP properties include the spectral and composition properties of heavy ions (^{3}He, ^{4}He, C, O, Fe), following the procedure illustrated in Desai et al. 2003, 2004. Spectral indices are determined by fitting a power-law of the form $j(E) = j_o E^{-\gamma}$ (0.1-0.5 MeV/n.) and the e-folding energies (E_0) are determined from fitting the spectral function $j(E) = j_o E^{\gamma} \exp(-E/E_o)$, at ~0.1-5 MeV/n. Peak O intensity (J^{O}_{max}) is the maximum hourly averaged 0.5–2.0 MeV/n. O intensity measured during the shock-associated time intervals. The ^{3}He/^{4}He and Fe/O ratios are respectively calculated from the 0.5–2.0 MeV/n. helium mass histograms (for examples, see Desai et al. 2001, Dayeh et al. 2017) and the average 0.5–2.0 MeV/n. Fe and O intensities measured during the ESP event time intervals (differ for each event).

2.2 IP shock and interplanetary environment properties

Measurable quantities upstream and downstream of the shock are determined over a fixed analysis interval of eight minutes, following the procedure described in Kilpua et al. 2015. The number of data points within the analysis intervals varied with the data resolution and availability. A minimum of three points was required to determine an average. If a particular data product is not available for the shock time period (e.g., data gap, saturation), the value of the respective parameter gets flagged and excluded from the analysis via pairwise deletion. Values were cross-checked with the ip shocks database hosted at (http://ipshocks.fi; see also Kilpua et al. 2015). Inferred parameters include, upstream and downstream magnetic field magnitude, proton number density, speed, and temperature (B|U, $B|_U$, $n|_U$, $n|_U$, $V|_U$, $V|_U$, $T|_U$, $T|_U$ respectively), in addition to the shock speed and obliquity (V_{shock}, θ_{B_n}). Finally, Fe/O upstream and downstream were also calculated from the average 0.5–2.0 MeV nucleon^{-1} Fe and O intensities 3 hours before ($Fe/O|_U$) and after ($Fe/O|_D$) the shock arrival time.

2.3 ICME shock properties

Focus here is on the plasma parameters in the sheath region, namely, the sheath pressure (P_{sheath}), temperature (T_{sheath}), and number density (n_{sheath}). Values are calculated using the corresponding data within three hours after shock arrival (i.e., downstream).

A total of 28 parameters were inferred for each of the selected 62 events. We then performed a Pearson correlation analysis between all parameters and those of the ESP properties, as indicated in table 1. This table shows a heatmap that is color-coded between -1 and +1, and provides the correlation coefficient (r) between ESP properties and each of the parameters characterizing (i) ESP properties, (ii) IP shock properties, and (iii) ICME properties. Values of r between +0.7 to +1 indicate a strong positive correlation, between +0.4 to +0.7 indicate a moderate positive correlation, and below +0.4 poor or no correlation. Similar classification is adopted if r is negative. Overlaid numbers are those corresponding to the values of r in each cell. Bold numbers indicate that the correlation is statistically significant within 0.05 significance level. We emphasize here that these correlation coefficients are descriptive of the linearity correlation only, which is not necessarily the only relation, since different quantities could be related by non-linear relationships and still show some linearity. Inspection of the scatter plot is thus necessary in all cases. However, the work reported here is meant to only examine the linearity between different parameters and full analysis is beyond the scope of this article.

IOP Conf. Series: Journal of Physics: Conf. Series **1100** (2018) 012008 doi:10.1088/1742-6596/1100/1/012008

		ESP								
		γ_C	E_o^C	γ_O	E_o^O	γ_{Fe}	E_o^{Fe}	J^O_{max}	$^3He/^4He$	Fe/O
ESP	γ_C	1.00	-0.29	0.98	-0.34	0.69	-0.06	-0.48	0.39	0.29
	E_o^C	-0.29	1.00	-0.31	0.86	-0.44	0.55	0.05	-0.04	0.58
	γ_O	0.98	-0.31	1.00	-0.34	0.76	-0.07	-0.54	0.37	0.23
	E_o^O	-0.34	0.86	-0.34	1.00	-0.43	0.47	-0.01	-0.11	0.41
	γ_{Fe}	0.69	-0.44	0.76	-0.43	1.00	-0.10	-0.56	0.07	-0.15
	E_o^{Fe}	-0.06	0.55	-0.07	0.47	-0.10	1.00	-0.12	-0.03	0.26
	J^O_{max}	-0.48	0.05	-0.54	-0.01	-0.56	-0.12	1.00	-0.17	-0.15
	$^3He/^4He$	0.39	-0.04	0.37	-0.11	0.07	-0.03	-0.17	1.00	0.55
	Fe/O	0.29	0.58	0.23	0.41	-0.15	0.26	-0.15	0.55	1.00
Upstream and downstream of the IP shock	$Fe/O\|_U$	0.03	0.39	0.01	0.23	-0.12	0.15	0.00	0.42	0.53
	$Fe/O\|_D$	-0.10	0.76	-0.11	0.61	-0.21	0.21	-0.09	0.30	0.64
	$B\|_U$	0.08	0.08	0.04	-0.08	-0.10	-0.10	0.20	0.03	0.21
	$B\|_D$	-0.12	0.47	-0.18	0.31	-0.32	0.01	0.31	-0.08	0.22
	$B\|_D / B\|_U$	-0.26	0.60	-0.29	0.55	-0.29	0.12	0.13	-0.13	0.08
	$V\|_U$	-0.18	0.27	-0.16	0.27	-0.22	0.39	0.10	0.16	0.15
	$V\|_D$	-0.33	0.32	-0.35	0.32	-0.32	0.36	0.29	-0.02	0.07
	Δ_V	-0.33	0.19	-0.38	0.20	-0.26	0.11	0.37	-0.24	-0.08
	$n\|_U$	-0.03	0.41	-0.02	0.31	-0.15	-0.12	0.06	-0.03	0.23
	$n\|_D$	0.00	0.31	0.00	0.18	-0.08	-0.16	-0.01	-0.09	0.07
	$n\|_D / n\|_U$	0.03	-0.12	0.02	-0.10	0.10	-0.15	-0.09	-0.17	-0.22
	$T\|_U$	-0.13	-0.01	-0.14	0.01	-0.17	-0.05	0.17	-0.06	-0.01
	$T\|_D$	-0.31	0.21	-0.31	0.29	-0.23	0.29	0.23	-0.16	-0.07
	$T\|_D / T\|_U$	-0.24	0.20	-0.26	0.27	-0.14	0.27	0.02	-0.13	-0.03
	θ_{B_n}	0.02	0.01	0.02	-0.11	-0.04	-0.26	-0.07	0.01	0.02
	V_{shock}	-0.34	0.34	-0.40	0.28	-0.38	0.39	0.52	-0.16	0.07
ICME	P_{sheath}	-0.10	0.13	-0.11	0.20	-0.12	-0.13	0.02	-0.16	0.05
	T_{sheath}	-0.39	0.40	-0.41	0.40	-0.41	0.28	0.44	-0.17	-0.02
	n_{sheath}	0.10	-0.09	0.11	-0.12	0.09	-0.29	-0.08	-0.14	-0.07

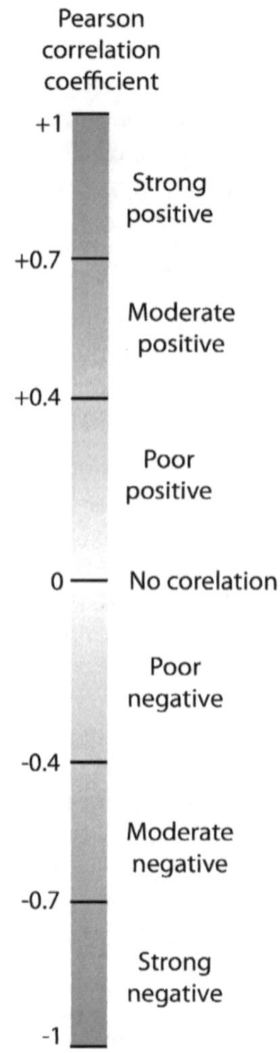

Table 1: Correlation heatmap of all inferred parameters with those of ESP properties.

3. Discussion

Linear correlations between parameters associated with ESP events are presented. In the following subsections, we comment on some of these correlations in context of their driving physical mechanisms.

3.1 ESP vs. ESP correlations

Top 9 x 9 cells in table 1 show the correlation results for the spectral and compositional properties of the 62 ESP events. The correlation between parameters varies from poor to high. Figure 3a plots the Oxygen peak intensity at ~1 MeV/n. versus the Oxygen spectral index. A moderate negative correlation (r = -54) is shown. This trend indicates that the most intense ESP events tend to have harder (i.e., flatter) spectra. Similarly, smaller ESP events would be characterized with softer (steeper) spectra.

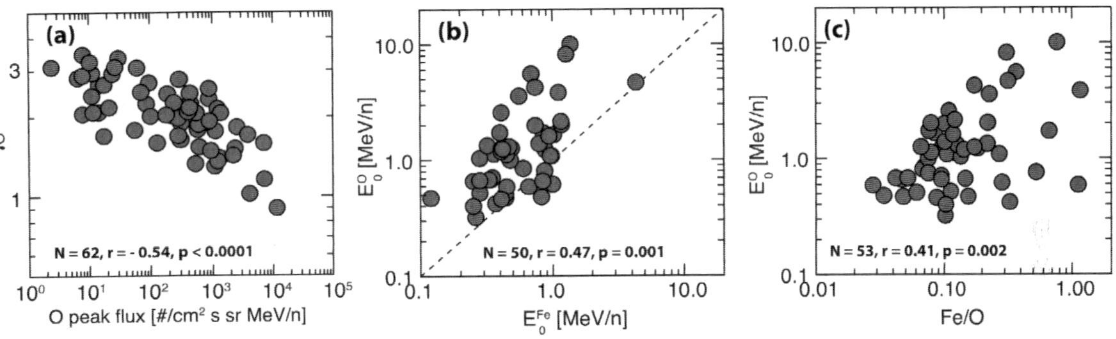

Figure 3: (a) power-law spectral indices of O vs. the maximum O peak flux at ~1 MeV/nucleon. (b) e-folding energies of O vs. that of Fe. The solid line has slope = 1 and represents a 1:1 correlation. (c) Scatter plot of Fe/O ratio and the e-folding energy of O. In all panels, N, r, and p denote the number of points, the linear correlation coefficient, and its statistical significance.

Figure 3b shows the relation between the e-folding energies of Fe and O. A moderate positive correlation (r=0.47) exits between the two quantities. Except for the two events at both extremes of E_0^{Fe} (at ~0.12 and ~4.5 MeV/n.), the clustering of points shows that the relation between the two parameters is monotonic but not a 1:1 relation. Instead, it is obvious that E_0^{O} extends to higher energy values (between ~0.3 and ~11 MeV/n.) compare to Fe (between ~0.2 to 2 MeV/n.). This could be explained by the rigidity-dependent acceleration mechanisms, where ions with higher charge-to-mass ratio (M/Q), such as Fe, are accelerated less efficiently than those with lower M/Q, such as O, leading to a faster roll over in the Fe spectrum before that of Oxygen, and producing the observed behavior. Figure 3c shows the relation between Fe/O ratio and E_0^{O}. A moderate correlation exists (r=0.41) and is also expected from a rigidity-dependent acceleration mechanism.

This part of table 1 also shows other moderate or strong correlations including,

- C and O spectral indices are strongly correlated, with the Fe correlation slightly weaker. This is probably not at odds assuming that C and O have very similar Q/M values, compared to Fe.
- $^3He/^4He$ ratio is moderately correlated with Fe/O ratio (r=0.55). One mechanism that could produce behavior is the reacceleration of impulsive flare material by the IP shock. However, a closer examination of this correlation shows that that it is driven by 4 events, which have consistently high Fe/O and $^3He/^4He$ ratios. The remaining events show no correlation.

3.2 ESP vs ICME properties

Figure 4 shows the relation between the downstream sheath temperature and the O peak flux at ~1 MeV/n. Sheath temperature is calculated as the 3-hour average downstream of the shock. While the total linear correlation coefficient is moderate (r=0.44), the trend behavior is quite interesting. For low sheath temperatures (below ~2 . 10^5 K), O peak fluxes do not exceed ~ 100 particle/(cm^2 sr s MeV/n.); for higher temperatures (above ~2 . 10^5 K), there seem to be a consistent trend of increasing peak fluxes with increasing temperature. This behavior suggests that there might be a threshold above which the ESP peak fluxes increase monotonically with the ICME sheath temperature. Such a trend could be a manifestation of many effects. For instance, it may be related to the shock geometry: temperature jumps at a quasi-parallel ($\theta_{B_n} \leq 45$) can differ significantly from a quasi-perpendicular shocks ($\theta_{B_n} > 45$) depending on plasma beta (e.g., Leroy et al., 1982; Burgess et al., 2005). Across a quasi-parallel shock, the jump of magnetic field is small; therefore the kinetic energy of the flow converts mostly to downstream thermal energy. At a quasi-perpendicular shock, however, the magnetic field jump is significant, so that part of the flow kinetic energy is converted to downstream magnetic field energy. This suggests that for shocks of similar compression ratio with similar upstream medium, the downstream temperature is higher for a quasi-parallel shock. Preliminary analysis of examining the effect of geometry on the observed trend of figure 4 does not confirm this (not shown). This, however, is mainly because in our sample, the number of quasi-parallel events is too little (~10) compared to the quasi-perpendicular events (~50), a larger sample is thus required to test the possible geometry dependence hypothesis. Furthermore, one should also compare the upstream plasma properties between quasi-perpendicular and quasi-parallel shocks.

Indeed, the observed trend could also be a result of an elevated upstream temperature for those large ESP events, a factor that could well be related to the accelerated particle intensities at IP shocks (e.g., Lario et al. 2017). Finally, the turbulence levels associated with all shocks involved might also contribute the observed trend, since turbulence is more intense behind quasi-parallel shocks than behind the quasi-perpendicular shock [Lucek et al., 2005]. Streaming protons can excite Alfven waves in the upstream region, which are then transmitted into the downstream region. This process is much more efficient at a quasi-parallel shock than at a quasi-perpendicular shock. In the latter case, the inability of particles to stream away from the shock results in the very inefficient excitation of the upstream waves (Lee 1980, 2005; Zank et al 2000, 2006; Li et al. 2003, 2005).

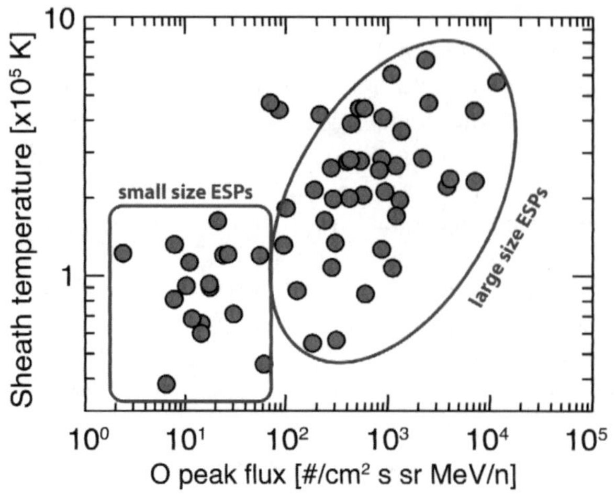

Figure 4: Sheath temperature (T_{sheath}) versus O peak flux at ~1 MeV/nucleon. Small and large ESP evens appear to behave differently as a function of T_{sheath}.

3.3 ESP vs IP shock properties

Figure 5a plots the shock obliquity and the O spectral indices averaged over two-hour intervals centered on the IP shock. Figure 5b shows the relation between the magnetic compression ratio and the O spectral indices. In both plots, the correlation is poor and quantities appear to be uncorrelated. The one-dimensional steady state theory predicts that density compression ratio and the spectral index are correlated (e.g., Van Nes et al. 1984, Parker and Zank 2012, Zank et al. 2006, Giacalone 2012). Results shown indicate that the acceleration is likely less efficient than the theoretical predictions, or the simplified form of the DSA is just not applicable for comparison will all types of ESP events. Several studies (e.g., Lario et al. 2004; Ho et al. 2005) attempted to relate observed shocks of varying obliquity to the particle spectra expected from DSA theory. While DSA appears to be the most feasible explanation for particle acceleration at shocks, unfortunately, there are very few examples of one-to-one correspondence between DSA predictions and observations (e.g., Giacalone 2012 and references therein).

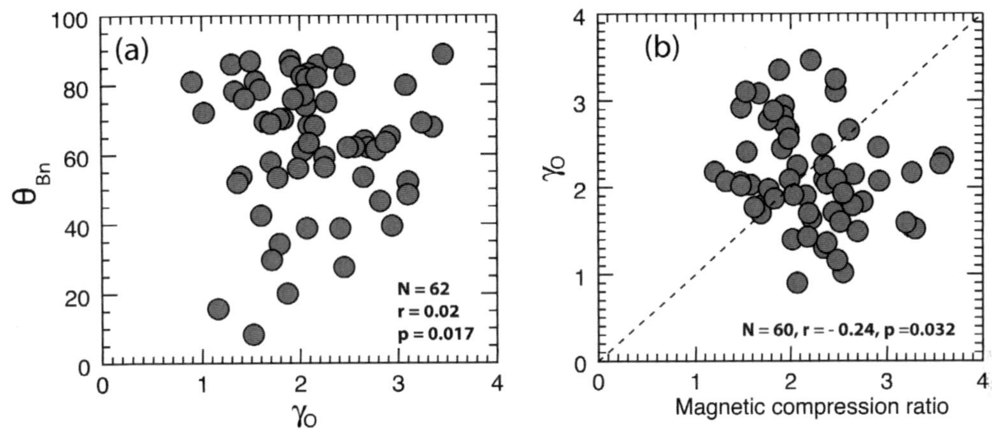

Figure 5: (a) shock obliquity versus the O spectral indices. (b) O spectral indices versus the magnetic compression ratio.

Furthermore, we note that table 1 indicates a high correlation between the magnetic compression ratio (B_d/B_u) and the roll-over energies of O and C in table 1. Upon closer investigation, the correlation appears to be driven by few events only. Finally, Shock speed (Vs_{hock}) shows moderate/poor correlation with all ESP properties.

In summary, using instruments onboard ACE, we investigate the relations between a large set of parameters (28) parameters that characterize (i) ESP properties, (ii) IP shock and ICME properties, and (iii) the upstream and downstream IP conditions measured near the IP shock. We present a comprehensive correlation matrix between all parameters and those of the ESP properties. We find that within the selected dataset, the (1) spectral and compositional relations strongly indicate a rigidity-dependent acceleration mechanism, (2) correlations between observations and the DSA predictions are poor, (iii) ICME Sheath temperature appears to play a role in determining the ESP peak intensities.

4. Acknowledgements.

We acknowledge using data from the Heliospheric Shock Database, generated and maintained at the University of Helsinki. Events are listed in Desai et al. 2003, 2004 and all associated data are available for download from the ACE Science Center at (www.srl.caltech.edu/ACE/ASC/. This work is partially supported by NSF grant AGS-1460118 and NASA grant NNX17AI17G. Author AF acknowledges the support provided by the Deanship of Scientific Research (DSR) at the King Fahd University of Petroleum and Minerals (KFUPM) through project IN151005. Solar wind analysis was partially supported by ACE-SWEPAM MO&DA funding NNX13AQ01G.

5. References

[1] Aran A, Sanahuja B, Lario D 2005 Ann. Geophys. 23, 3047
[2] Armstrong T P, Pesses M E, Decker R B 1985 geo. mono. series, pp 271–285. doi:10.1029/GM035p0271
[3] Burgess D, Lucek E A, Scholer M, Bale S D, Balikhin M A, Balogh A, et al. 2005 Space Sci. Rev., 118, 205–222
[4] Cohen C M S 2006 Geophys. Monogr. Ser., 165, 275–282
[5] Dayeh M A et al. 2017 Astrophys J., 835, 155
[6] Decker R B 1981 J. Geophys. Res., 86, 4537
[7] Desai M I et al. 2003, Astrophys J., 588, 2 1149, 2003
[8] Desai M I et al. 2004, Astrophys J., 611, 2, 1156, 2004
[9] Desai M I, et al. 2001, Astrophys J. Lett, 553, L89, 2001
[10] Ebert R W et al. 2016, Astrophys J., 831 (2), 153
[11] Giacalone J 2012 Astrophys. J. 761, 28
[12] Giacalone J 2005 Astrophys J., 628, L37
[13] Giacalone J 2003 Plan. Sp. Sci., 51, 659
[14] Giacalone J and J R Jokipii 2012 Astrophys. J., 751, L33
[15] Gopalswamy N, Makela P, Xie N H, Akiyama S and Yashiro S 2009 J. Geophys. Res., 114, A00A22, doi:10.1029/2008JA013686.
[16] Ho G C, Roelof E C and Mason G M 2005, Astrophys. J., 621, L141
[17] Jones F C and Ellison D C 1991 Space Sci. Rev., 58, 259
[18] Kennel C F, Coroniti F V, Scarf F L, Livesey W A, Russell C T and Smith E J 1986 J. Geophys. Res. 91:11917–11928. doi:10.1029/JA091iA11p11917
[19] Kilpua E K J, Lumme E, Andreeova K, Isavnin A and Koskinen H E J 2015 J. Geophys. Res., 120, 4112–4125, doi:10.1002/2015JA021138
[20] Leroy M M, Winske D, Goodrich C C, Wu C S and Papadopoulos K 1982 J. Geophys. Res., 87, 5081
[21] Lario D, Decker R B, Roelof E C, Viñas A F, Wimmer-Schweingruber R F and Berger L 2017 J. Phys.: Conf. Ser. 900 012012
[22] Lucek E A, Constantinescu D and Goldstein M L, et al. 2005, SSRv, 118, 95
[23] Lario D, Livi S, Roelof E C, et al. 2004, J. Geophys. Res., 109, A09S02
[24] Lario D, Sanahuja B and Heras A M 1998, Astrophys. J., 509, 415
[25] Lario D, et al. 2005 J. Geophys. Res., 10, A09S11, doi:10.1029/2004JA010940
[26] Lee M A 2005, Astrophys. J. Supp., 158, 38
[27] Lee M A 1983 J. Geophys. Res., 88, 6109
[28] Li G, Zank G P and Rice, W K M 2003, J. Geophys. Res., 108, 1082
[29] Li G, Zank G P and Rice, W K M 2005, J. Geophys. Res. A, 110, 06104
[30] Mäkelä P et al. 2011 J. Geophys. Res., 116, A8, A08101
[31] Mason G M, Gold R E, Krimigis S M, Mazur J E, Andrews G B, Daley K A, Dwyer J R, Heuerman K F, James T L, Kennedy M J,LeFevere T,Malcolm H,Tossman B, Walpole P H 1998 Rev. Space Phys., 86, 1–4, pp 409–448
[32] McComas D J, Bame S J, Barker P, Feldman W C, Phillips J L, Riley P and Griffee J W 1998 Rev. Space Phys. 86, 1–4, pp 563–612
[33] Ng C K, Reames D V 2008 Astrophys. J. Lett. 686, L123
[34] Parker L N and Zank G P 2012 Astrophys. J., 757:97, 11pp
[35] Reames D V 1999 Space Sci Rev 90:413–491 doi:10.1023/A:1005105831781
[36] Reames D V 2012, Astrophys. J., 757, 93
[37] Rice W K M, Zank G P and Li G 2003, J. Geophys. Res., 108, A10
[38] Richardson I G and Cane H V 2010, Sol. Phys., 264
[39] Stone E C, Frandsen A M, Mewaldt R A, Christian E R, Margolies D, Ormes J F and Snow F 1998 Rev. Space Phys. 86, 1–4

[40] Stone E C, Cohen C M S, Cook W R, Cummings A C, Gauld B, Kecman B, Leske R A, Mewaldt R A and Thayer M R 1998 Rev. Space Phys. 86, 1–4, pp 357–408
[41] Smith C W, L'Heureux J, Ness N F, Acuña M H, Burlaga L F, Scheifele J 1998 Rev. Space Phys., 86, 1–4, pp 613–632
[42] Tylka A J et al. 2005 Astrophys. J., 625, 1, 474
[43] van Nes P et al. 1984 J. Geophys. Res., 89, 2122
[44] Vainio R and T Laitinen 2007 Astrophys. J., 658, 622–630
[45] Vainio R, and T Laitinen 2008 J. Atmos. Sol. Terr. Phys., 70,467–474
[46] Zank G P, Rice W K M and Wu C C 2000, J. Geophys. Res., 101, 17093
[47] Zank G P, Li G, Florinski V, et al. 2006 J. Geophys. Res., 111, A06108
[48] Zank G P, Li G, Florinski V, Hu Q, Lario D, Smith C W 2006 J. Geophys. Res., 111, A06108, doi:10.1029/2005JA011524
[49] Zank G P et al. 2000 J. Geophys. Res., 105, A11
[50] Zank G P et al. 2006 J. Geophys. Res., 111, A06108, 25079

Particle Acceleration in Interacting Magnetic Flux Ropes

Senbei Du[1], Gary P. Zank[1,2], Fan Guo[3,4], Xiaocan Li[3], Adam Stanier[3]

[1] Department of Space Science, University of Alabama in Huntsville, Huntsville, AL 35899, USA

[2] Center for Space Plasma and Aeronomic Research (CSPAR), University of Alabama in Huntsville, Huntsville, AL 35805, USA

[3] Los Alamos National Laboratory, Los Alamos, NM 87545, USA

[4] New Mexico Consortium, Los Alamos, NM 87544, USA

E-mail: sd0050@uah.edu, garyp.zank@gmail.com, fanguo.ustc@gmail.com

Abstract. Magnetic flux ropes (or magnetic islands) are ubiquitous space plasma structures. Recent observations suggest that they are often associated with the acceleration of charged particles, but detailed acceleration mechanisms remain unclear. In this study, we present PIC simulations studying particle acceleration due to magnetic flux ropes. We consider a simple 2D configuration of two-magnetic-island coalescence. Some electrons and protons are found to be accelerated to more than 10 times their initial kinetic energies at the end of the simulation. We use a particle tracing technique on the high-energy particles to clarify the associated acceleration mechanisms. We find that reconnection electric field and Fermi-type acceleration due to magnetic island contraction can explain the particle energy gain, which is consistent with previous simulation studies. Our results also suggest that electrons are more responsive to the island contraction mechanism compared to ions. An effective island contraction rate is derived from the simulation data. Finally we briefly discuss a statistical description of particle acceleration associated with interacting magnetic flux ropes, and how it can be connected to simulations.

1. Introduction

Magnetic flux ropes are structures characterized by helical magnetic field lines. They are referred to as magnetic islands when viewed in 2D, where magnetic field lines form closed-loop-like structures. Magnetic flux ropes are frequently observed in space plasmas, such as in the Earth's magnetosphere [1, 2] and in the solar wind [3–5].

Magnetic reconnection, which is a sudden change in magnetic field configuration, plays a crucial role in the formation and evolution of magnetic flux ropes. Numerous simulations suggest that magnetic islands are formed in an elongated reconnecting current sheet. Multiple magnetic islands may be generated in a 2D turbulent reconnecting flow, and they undergo dynamic evolution such as coalescence and merging [e.g., 5–8]. Observational studies appear to support the view that small scale flux ropes in the solar wind are generated by the local magnetic reconnection and turbulent processes [3–5].

An interesting aspect of magnetic flux ropes is their potential to accelerate charged particles. Despite Gosling et al. [9] find that direct magnetic reconnection is not an efficient particle accelerator in the solar wind, recent observations suggest that magnetic flux ropes may be responsible for particle acceleration in the heliosphere. Khabarova et al. [5] discuss the crossing of the heliospheric current sheet (HCS) by interplanetary shocks. They find that atypical energetic particle events downstream of the shocks can be associated with small-scale magnetic flux ropes. Khabarova & Zank [10] reexamined the "Gosling event" [9] and 126 related events over a larger time and more extended spatial range and energies. They found evidence of an energetic particle population, most likely accelerated by turbulence/magnetic island related structures generated by the initial reconnection events. In these regions, magnetic flux ropes undergo dynamic interactions through magnetic reconnection. Particles that are trapped in such regions can be accelerated more efficiently than in an isolated reconnection exhaust.

Numerical simulations have suggested several mechanisms for particle acceleration. The basic processes include first order Fermi acceleration due to magnetic island contraction, and direct acceleration by the reconnection electric field generated during the merging of two adjacent magnetic islands [11–15]. Zank et al. [16] develop a theoretical particle transport equation that describes the particle acceleration in a sea of interacting magnetic islands, incorporating the abovementioned basic acceleration mechanems. le Roux et al. [17] derive a more sophisticated transport equation from a quasi-linear approach, where the energization is associated with guiding center drift motions. The two equations discuss the same energization mechanisms but in terms of field line contraction and guiding center motions. However, both theories need to be tested against more careful simulations and observations. In return, detailed simulation studies will enable us to better understand the acceleration processes, and thus guide the development of more sophisticated transport models.

In this study, we perform fully kinetic particle-in-cell (PIC) simulations of the coalescence of two magnetic flux ropes. We use a particle tracing technique to study the acceleration of both electrons and ions. We report some findings regarding the acceleration mechanisms, recognizing the limitation of a 2D model, and discuss the connection between simulations and a statistical transport theory.

2. Simulation Model

The simulation uses the VPIC code [18], which solves Maxwell's equations for electromagnetic fields and the relativistic equations of motion for particles. The simulation setup is reported in an earlier work (Du et al., in preparation). The initial configuration contains two magnetic islands embedded in a reconnecting current sheet, illustrated in the top left panel of Figure 1. A similar setup has also been used in several previous simulation studies [e.g., 19–21]. The magnetic field is given by

$$B_x = \frac{B_0 \sinh(z/L)}{\cosh(z/L) + \varepsilon \cos(x/L)}; \; B_z = \frac{B_0 \sin(x/L)}{\cosh(z/L) + \varepsilon \cos(x/L)}; \; B_y = \frac{B_0\sqrt{1-\varepsilon^2}}{\cosh(z/L) + \varepsilon \cos(x/L)},$$

where L is the distance between the two islands, which also determines the system size, and ε is a measure of the island size. This setup ensures that the initial condition is force free, i.e., $\mathbf{J} \times \mathbf{B} = (\nabla \times \mathbf{B}) \times \mathbf{B} = 0$. The simulation box is set to $x \in [-2\pi L, 2\pi L]$, $z \in [-\pi L, \pi L]$. We set $L = 2d_i$ and $\varepsilon = 0.4$ in the simulation, where $d_i = c/\omega_{pi}$ is the ion inertial length. A periodic boundary condition is applied to the x direction, whereas conducting field/reflective particle boundaries are used in the z direction. We consider only an electron–proton plasma, and the mass ratio m_i/m_e is set to 25. The initial electron and proton temperatures are uniform throughout the simulation domain with $kT_e = kT_i = 3.75 \times 10^{-3} m_e c^2$. The magnetic field strength B_0 is determined by the ω_{pe}/Ω_{ce} value, where $\omega_{pe} = \sqrt{4\pi n_0 e^2/m_e}$ and $\Omega_{ce} = m_e c/eB_0$ are the electron plasma frequency and electron gyrofrequency, respectively. We set $\omega_{pe}/\Omega_{ce} = 2$ in our simulations. The particle number density profile is also uniform initially. The initial plasma beta is then given by $\beta_e = \beta_i = n_0 kT_e/(B_0^2/2) = 0.03$. About 400 macro-particles per cell are used for each species.

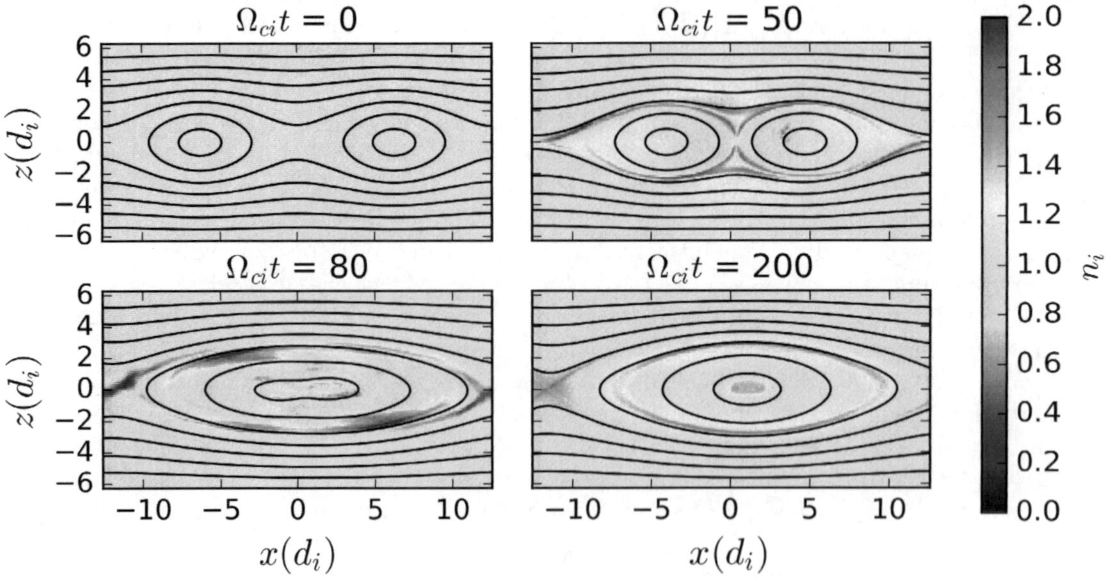

Figure 1. Snapshots at four normalized time, showing the ion number density, on which in-plane magnetic field lines are superimposed.

3. Results

3.1. Methodology

Figure 1 shows several snapshots of the ion number density on which in-plane magnetic field lines are superimposed. The general evolution and the energy conversion of the system is addressed in a previous study (Du et al., 2018, in preparation), so we do not discuss them here. In this paper, we focus on particle energization using a particle tracing technique. We present a detailed analysis of tracer particle trajectories and discuss the acceleration mechanisms for both electrons and ions.

A total of 4×10^8 macro-particles (including both electrons and ions) are used in the simulation, and we select 2×10^4 particles as tracers during the first run. To save storage space, the tracer trajectories are not well resolved in time. We then select the highest-energy tracers and run the simulation a second time to obtain better time-resolved particle trajectories. The energy selection criteria is $E_e \gtrsim 7.8 \times 10^{-2} m_e c^2$ for electrons, and $E_i \gtrsim 5.3 \times 10^{-2} m_e c^2$ for ions, where E_e and E_i are the kinetic energy of the tracer particles at the end of the simulation. We find a total of 131 electrons and 86 ions that satisfy the criteria, and obtain their trajectories.

As discussed in Zank et al. [16], the basic particle acceleration mechanisms associated with flux ropes include island contraction and reconnection electric field. A simple way to model the island contraction mechanism is by using adiabatic invariants. For a contracting magnetic island, assuming a compressible process, the conservation of first and second adiabatic invariants, as well as the magnetic flux yield the rate of change in particle momentum

$$\frac{\mathrm{d}p_\parallel}{\mathrm{d}t} = \eta_c p_\parallel; \quad \frac{\mathrm{d}p_\perp}{\mathrm{d}t} = \frac{1}{2}\eta_c p_\perp. \tag{1}$$

On the other hand, if we assume the coalescence of magnetic islands is an incompressible process, i.e., the total area of the islands remains the same after coalescence, then the rate of change of particle momentum becomes

$$\frac{\mathrm{d}p_{\parallel}}{\mathrm{d}t} = \eta_m p_{\parallel}; \quad \frac{\mathrm{d}p_{\perp}}{\mathrm{d}t} = -\frac{1}{2}\eta_m p_{\perp}. \tag{2}$$

Here, η_c and η_m are the associated island contraction rates for the compressible and incompressible magnetic island contraction cases, respectively. We adopt this simple model to analyze the simulation data.

3.2. Electron Acceleration

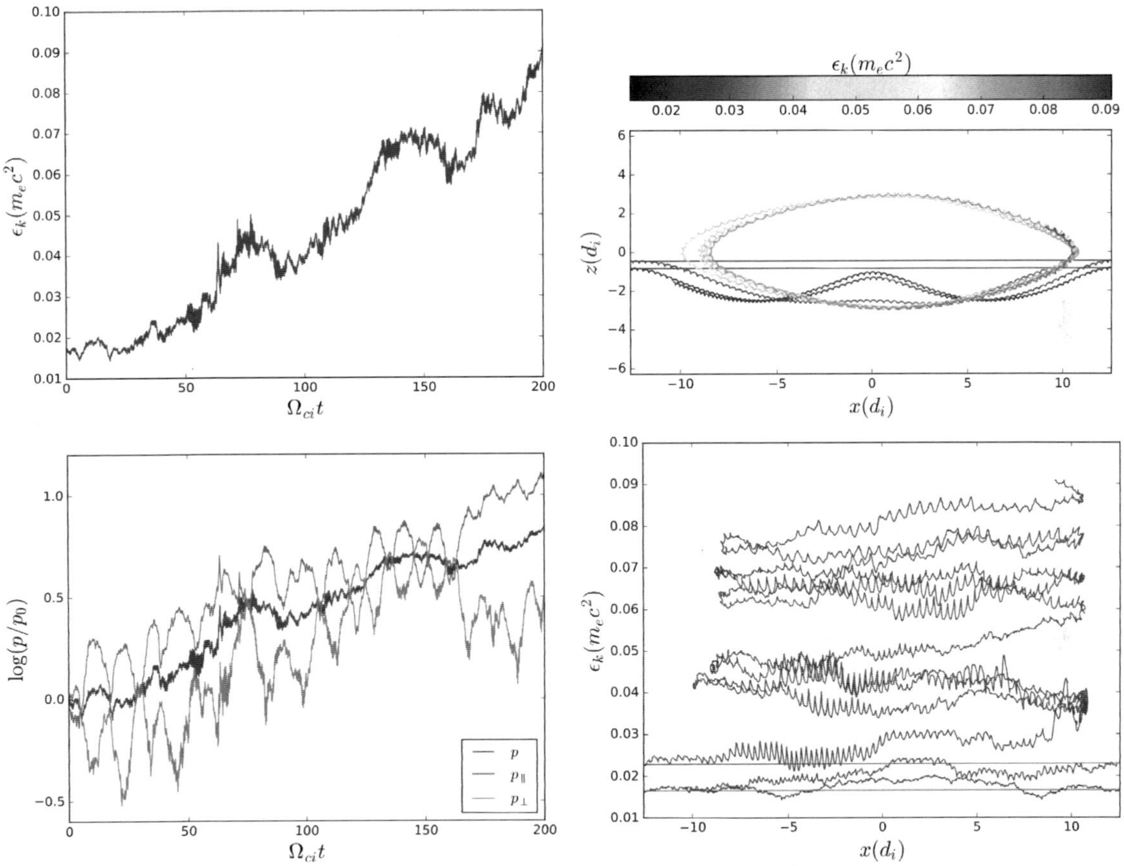

Figure 2. An example of an electron tracer. Top left: particle kinetic energy vs. time; bottom left: logarithm of normalized momentum (parallel, perpendicular and total) vs. time; top right: particle trajectory in the x-z plane; bottom right: particle kinetic energy vs. x-coordinate.

Some examples of electron trajectories are shown in Figures 2–5. To illustrate the acceleration mechanisms, we plot the kinetic energy vs. time (top left), momentum vs. time (bottom left), trajectory in the x-z plane (top right), and kinetic energy vs. x-coordinate (bottom right). Particle trajectories are color-coded by the corresponding kinetic energy. Since we apply a periodic boundary condition in x direction, a horizontal line appears in the trajectories and energy vs. time plots (right panels) when a particle crosses the x-boundaries. The momentum plots include both the parallel and perpendicular momenta, and also the total momentum. We plot them logarithmically as $\log(p/p_0)$, where p_0 is the momentum (parallel, perpendicular,

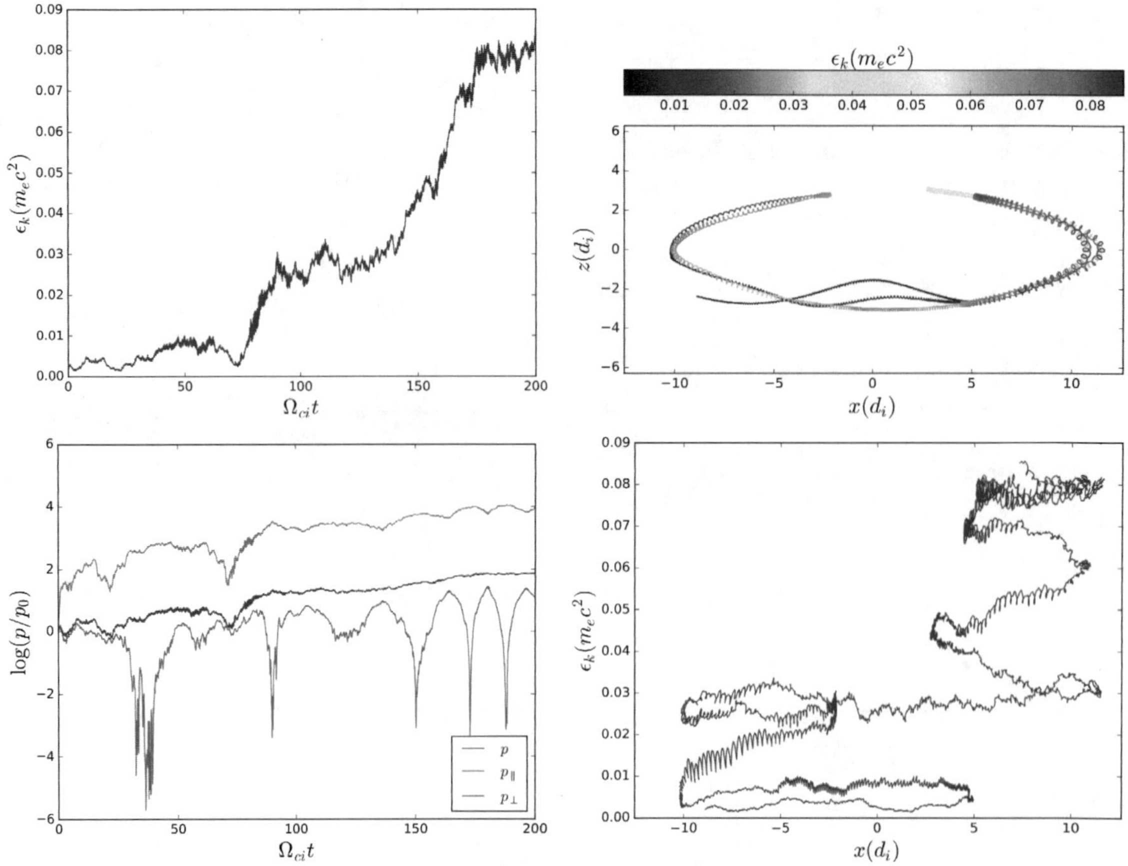

Figure 3. An example of an electron tracer. The format is the same as Figure 2

or total) at the initial time. As discussed below, this helps us understand the acceleration mechanisms. The momentum plots have been smoothed using a running average.

Figure 2 shows a tracer electron that undergoes acceleration due to magnetic island contraction. The particle energy increases as it orbits the flux ropes. As seen in the bottom left panel, both parallel and perpendicular momenta increase until nearly the end of the simulation when the perpendicular momentum drops. The linear increase in the particle momentum log(p) indicates that this is a first-order Fermi mechanism (as discussed later).

Figure 3 is another example of first-order Fermi acceleration, although the particle behaves differently than in the previous case. Here the particle is reflected several times inside the flux ropes, due to its small parallel momentum. The particle gains energy during the process because the length of the field lines is shortening.

Besides the island contraction (or Fermi acceleration) mechanism, we also see direct acceleration by the reconnection electric field. In Figure 4, the particle energy and momentum show a sudden increase at $\Omega_{ci}t \sim 75$. The bottom right panel shows that the sudden increase happens when the particle is near the boundary in the x direction. The trajectory also suggests that the particle is not trapped by the flux ropes before merging, but it is then trapped inside the single merged island at later time. Since the acceleration happens at the primary reconnection location, we attribute the main acceleration mechanism to the electric field generated by the primary reconnection event. Figure 5 is another example of a sudden energy increase at

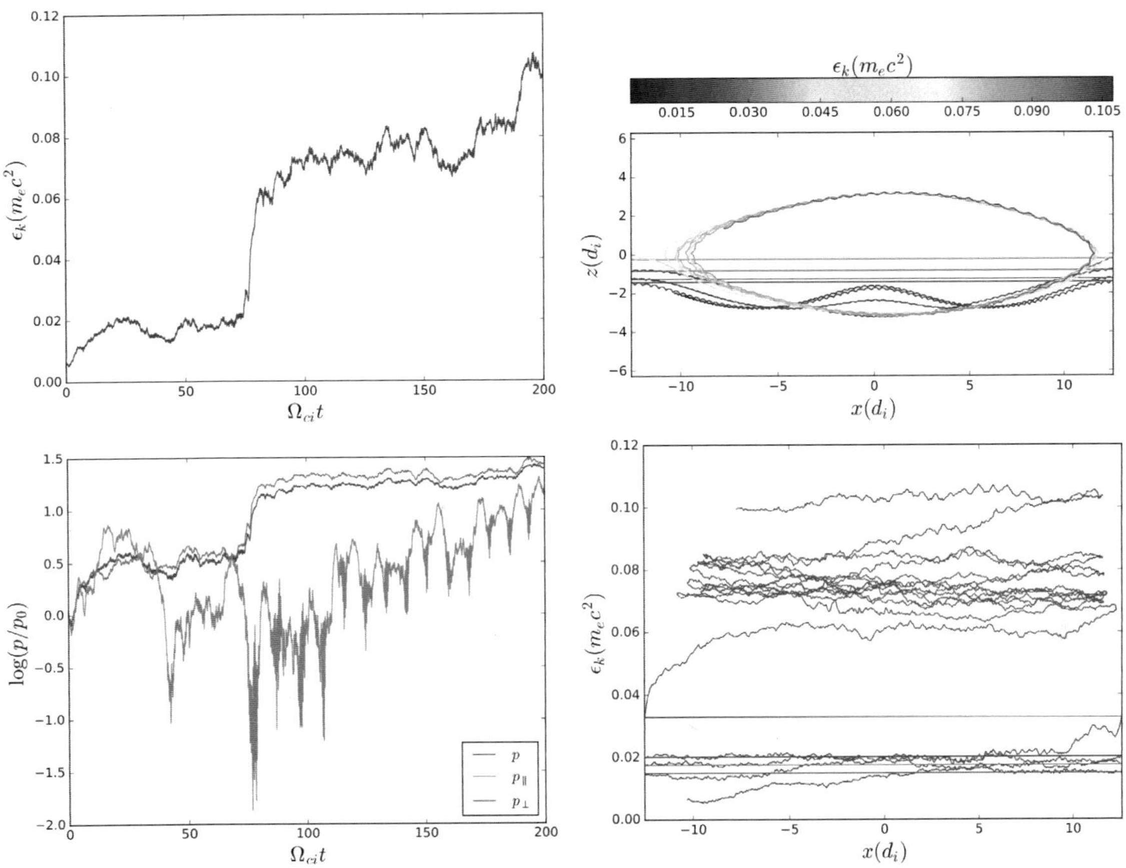

Figure 4. An example of an electron tracer. The format is the same as Figure 2

$\Omega_{ci}t \sim 75$. However, the acceleration site is near $x = 0$, where the island coalescence happens. Therefore, the reconnection electric field generated during coalescence is likely to be responsible for the particle energy gain. Since this electric field is in the opposite direction to the electric field generated by the primary reconnection event (near the x boundary of the simulation domain), it is called an anti-reconnection electric field in some literature [14, 16]. We also notice that in both examples that illustrate direct acceleration by reconnection electric fields, the particle energy continue to increase after the main merging phase. This suggests that the single merged magnetic island undergoes contraction, which leads to addition Fermi acceleration of particles, although the effect is not as large as the electric field energization for these two particular particles.

3.3. Ion Acceleration

Now we consider ion tracers. We find that the basic acceleration mechanisms are still the same—magnetic island contraction and reconnection electric field. Here we show examples of ion acceleration in Figures 6–8 in the same format as the previous plots.

In Figure 6, the ion energy increases suddenly at about $\Omega_{ci}t \sim 55$. From the trajectory, the main acceleration site is near the center of the simulation domain, where island coalescence occurs. This is consistent with direct acceleration by the anti-reconnetion electric field. The particle energy continues increasing after merging, due to island contraction.

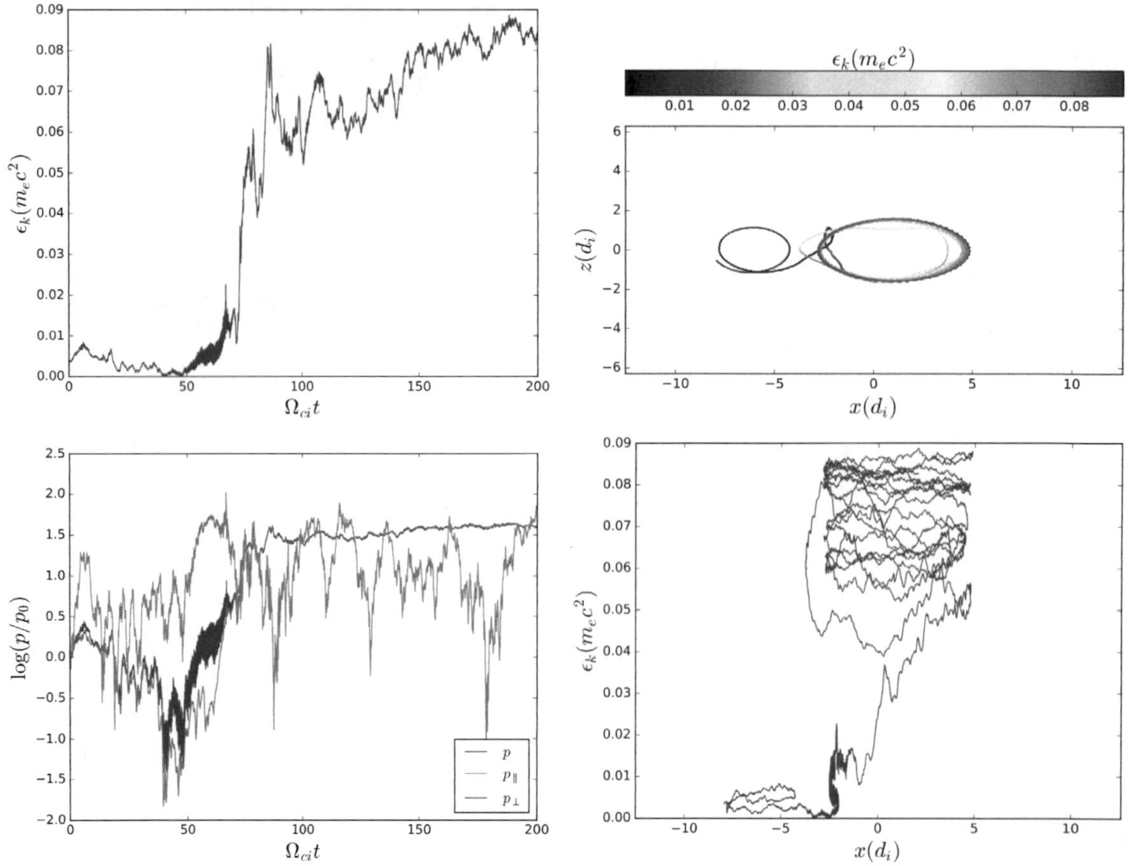

Figure 5. An example of an electron tracer. The format is the same as Figure 2

As a second example, in Figure 7, the ion gains energy at about $\Omega_{ci}t \sim 75$, near the location of primary reconnection. The particle is then trapped in the merged island, and undergo a reflection while gaining energy. The acceleration after island merging can still be interpreted as a first-order Fermi process.

Another interesting finding is that the kinetic energy does not necessarily increase after the main coalescence phase ends. Some of the ion tracers appear to undergo significant deceleration during later times of the simulation. An example is shown in Figure 8. In this case, the particle is accelerated to a high energy before $\Omega_{ci}t \sim 75$, which is likely due to the anti-reconnection electric field. It then experiences a gradual deceleration, and loses about half of its maximum kinetic energy by the end of the simulation. The deceleration seems to suggest a expanding magnetic island, which is inconsistent with the examples discussed before, so the mechanism is still not clear.

4. Discussion and Conclusions

4.1. Differences between Electron and Ion acceleration

We now discuss differences between the electron and ion acceleration. Since ions are much heavier than electrons, their gyroradii are much larger than those of electrons, especially after they are energized. As indicated by the tracer trajectories, the ion gyroradii approach scales comparable to the magnetic island size or the length scale on which the field varies. The large

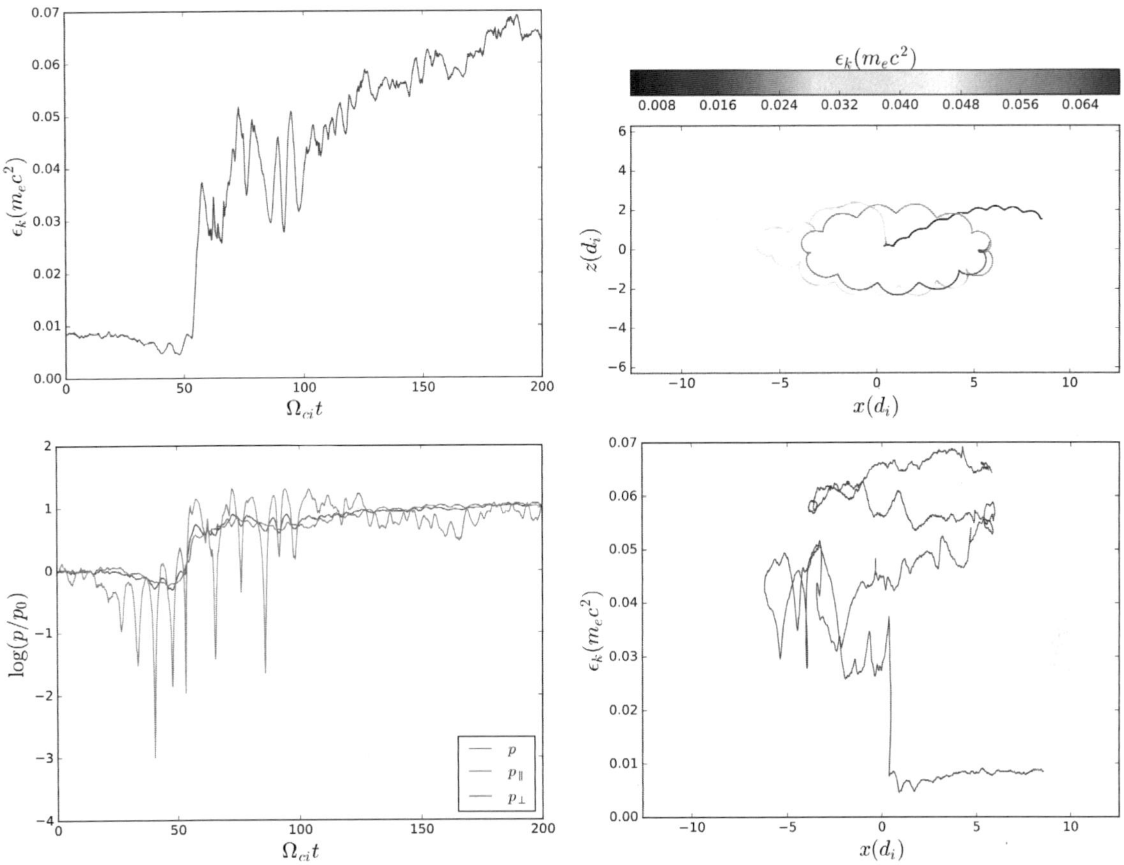

Figure 6. An example of an ion tracer. The format is the same as figure 2

gyroradii and gyroperiod may violate the conservation of adiabatic invariants, which may pose a problem on estimating the acceleration rate from the simple Zank et al. model [16]. We caution that the current simulation size is relatively small compared to the scale size of magnetic flux ropes observed in the solar wind. For example, recent observations [22] find that the scale size of observed small-scale magnetic flux ropes is $\sim 0.001 - 0.01$AU at a heliocentric distance of 1AU, which corresponds to $\sim 2\times10^3 - 2\times10^4 d_i$ on assuming a proton number density of $n_i = 10\text{cm}^{-3}$. The proton gyroradii could be much smaller compared to the flux rope sizes, and may behave like electrons. However, the plasma beta in the solar wind is usually larger than the value used in the simulation, which will result in a larger gyroradius.

A noticeable difference between ion and electron behavior is that ions are mostly accelerated during the main merging phase. This can be seen in the evolution of the total particle energy of the system (Figure 9). Electrons, while being accelerated at the merging time, continue to gain energy after the two islands merge. Since island contraction is the main acceleration mechanism after merging, we may infer that this mechanism seems to have a bigger impact on electrons than ions. Also we note that the simulation produces more high-energy electrons than ions, and the highest energy achieved by an electron tracer ($0.153 m_ec^2$) is slightly larger than that achieved by an ion tracer ($0.140 m_ec^2$). This difference may also be a consequence of the large ion gyroradii or the small simulation size. Interestingly, our previous study (Du et al., 2018, in preparation) shows that the macroscopic behavior is not much different in larger-size simulations

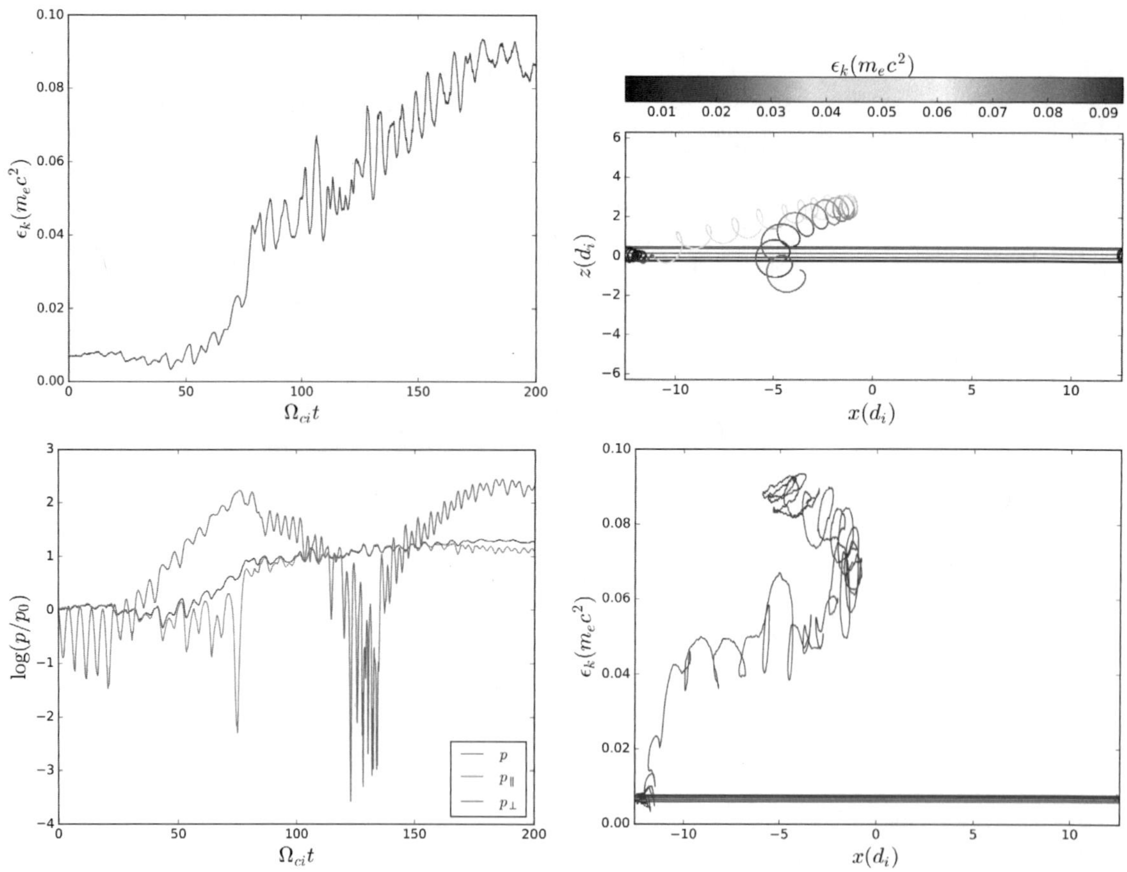

Figure 7. An example of an ion tracer. The format is the same as figure 2

(up to $\sim 100d_i$). A particle tracing study with larger simulations may be helpful.

Since magnetic island contraction is responsible for the particle acceleration during the later half of the simulation, we attempt to obtain an acceleration rate from the simulation data. We adopt the simple Zank et al. model [16], i.e., equations 1 and 2. Upon defining momentum logarithms $\xi_\parallel = \log(p_\parallel/p_{\parallel,0})$ and $\xi_\perp = \log(p_\perp/p_{\perp,0})$ where $p_{*,0}$ $(* = \parallel, \perp)$ are suitable constants, the equations can be rewritten in terms of new variables as

$$\frac{\mathrm{d}\xi_\parallel}{\mathrm{d}t} = \eta_c; \quad \frac{\mathrm{d}\xi_\perp}{\mathrm{d}t} = \frac{1}{2}\eta_c \tag{3}$$

for a compressible process, and

$$\frac{\mathrm{d}\xi_\parallel}{\mathrm{d}t} = \eta_m, \quad \frac{\mathrm{d}\xi_\perp}{\mathrm{d}t} = -\frac{1}{2}\eta_m \tag{4}$$

for an incompressible process. These expressions show that the slope of a $\log(p)$ vs. t curve gives an acceleration rate (or island contraction/merging rate). We observe from the $\log(p)$ vs. t plots that either the parallel or perpendicular momentum often includes large fluctuations. Therefore, we consider only the total momentum evolution. By transforming the variables $\xi_\parallel$ and $\xi_\perp$ to a total momentum logarithm $\xi = p/p_0$ and a pitch angle cosine $\mu = p_\parallel/p$, it is straightforward to

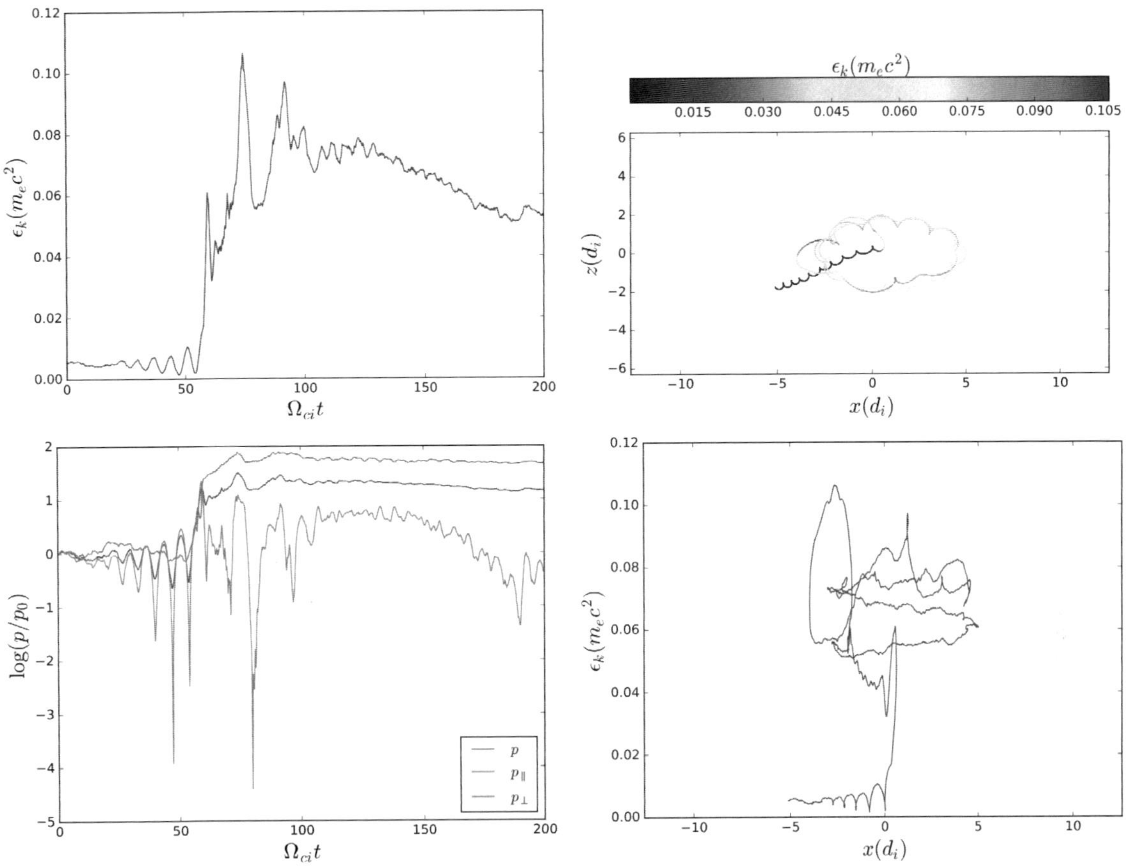

Figure 8. An example of an ion tracer. The format is the same as figure 2

convert equations 3 and 4 to

$$\frac{\mathrm{d}\xi}{\mathrm{d}t} = \frac{1}{2}\eta_c(1+\mu^2); \quad \frac{\mathrm{d}\mu}{\mathrm{d}t} = \frac{1}{2}\eta_c\mu(1-\mu^2); \tag{5}$$

$$\frac{\mathrm{d}\xi}{\mathrm{d}t} = \frac{1}{2}\eta_m(3\mu^2-1); \quad \frac{\mathrm{d}\mu}{\mathrm{d}t} = \frac{3}{2}\eta_m\mu(1-\mu^2), \tag{6}$$

for compressible and incompressible processes, respectively. Assuming the contraction/merging rates $\eta_{c/m}$ are independent of ξ and μ, we can obtain the rate of total momentum change after averaging the above equations over particle pitch angles. The resulting equations are formally written as

$$\frac{\mathrm{d}\xi}{\mathrm{d}t} = \tilde{\eta}_{c/m} \tag{7}$$

where $\tilde{\eta}_{c/m}$ is an effective contraction/merging rate that has absorbed the average of the pitch angle part of Equations (5) and (6). Note that in both compressible and incompressible cases, the changing rate of pitch angle cosine $d\mu/dt$ has the same sign as μ itself. This means that the pitch angle will increase in time for positive μ, and decrease for negative μ, aligning particle motions with the magnetic field, which may lead to an anisotropic particle distribution. However, suppose the particle distribution remains isotropic for some reasons (for example, due to particle scattering by background turbulence or self-generated waves), one concludes that $\tilde{\eta}_m = 0$ because

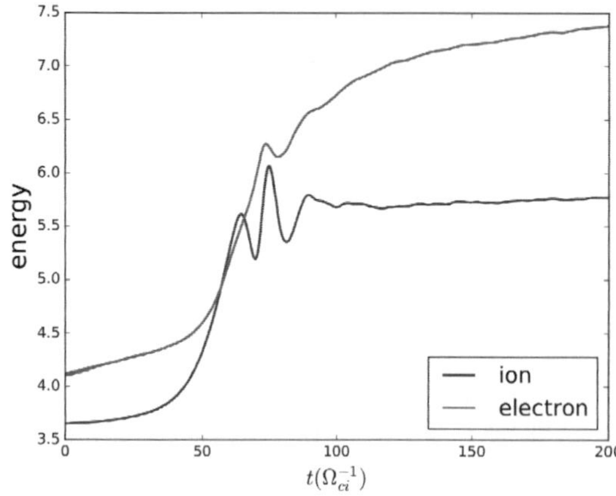

Figure 9. Total particle energy evolution of the system for ions and electrons. This plot includes all particles and not just the tracer particles.

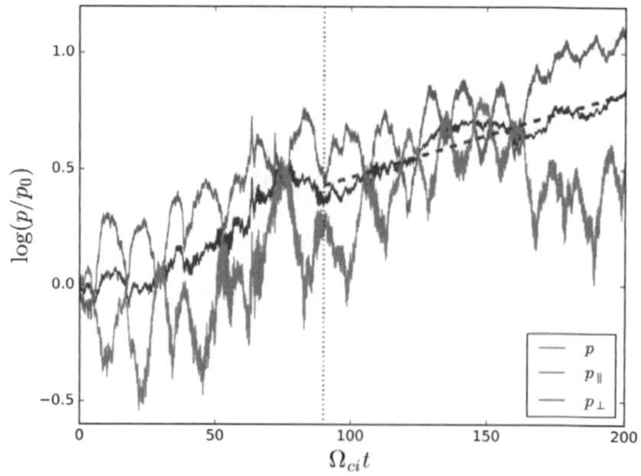

Figure 10. A linear fit to the log(p)-t curve, corresponding to the same tracer electron as Figure 2. The dotted blue line indicates the starting time of the fit $\Omega_{ci}t = 90$, and the dashed blue line is the best fit to the solid blue curve (total momentum).

the integration of $3\mu^2 - 1$ is zero in Equation 6. More generally, we can expand the particle distribution in terms of Legendre polynomials [16]. Since the second-order Legendre polynomial is $P_2(x) = (3x^2 - 1)/2$, the only contributing term to the zeroth-order distribution function will be the second-order term. On the other hand, $\tilde{\eta}_c$ always has the same sign with η_c regardless of the particle distribution, as suggested by equation 5. Whether it is positive or negative depends on the dynamics of the flux rope: it may either be contracting or expanding. In this sense, one may call the compressible energization (η_c effect) first-order Fermi acceleration, and the incompressible energization (η_m effect) second-order Fermi acceleration or stochastic acceleration. In reality, one may expect both compressible and incompressible effects contribute to the acceleration of particles. We simply use Equation 7 in our analysis, and denote the acceleration rate as $\tilde{\eta}_c$, though it may be a combination of the η_c and η_m effects.

As discussed in the previous section, at early simulation times, particle acceleration is strongly affected by the reconnection electric field, so we apply a linear fit to log(p)-t plots from the later part of the simulation, $\Omega_{ci}t = 90$ to $\Omega_{ci}t = 200$. Figure 10 illustrates our fitting procedure, where the same tracer electron as in Figure 2 is plotted. The fitting procedure is applied to all tracer electrons and ions, and a slope is obtained for each tracer. In Figure 11, we plot the pitch angle cosine at $\Omega_{ci}t = 90$, denoted as μ_1, against the acceleration rate $\tilde{\eta}_c$. Electron and ion tracers are plotted separately in the left and right panel. A mean value and a standard deviation are calculated and included in the figure. The fitting results show that electrons experience clear acceleration during the considered time interval, while ions are less strongly accelerated. This is consistent with Figure 9 and our previous discussion. Note that the fitting is only applied to the highest energy tracer particles, and the selection criteria is higher for electrons, so we caution about the potential bias in the analysis. One may also notice that electron pitch angles are strongly clustered near $\mu = -1$. This is mostly due to the initial setup of the simulation, because the initial force-free condition is achieved by introducing a particle drift motion along the magnetic field.

Tracer trajectories show that all the high energy electrons are trapped within the flux rope, although some of them are not trapped initially (such as the example shown in Figure 2). The trapping and escape of particles is very important for the acceleration process. For a particle to

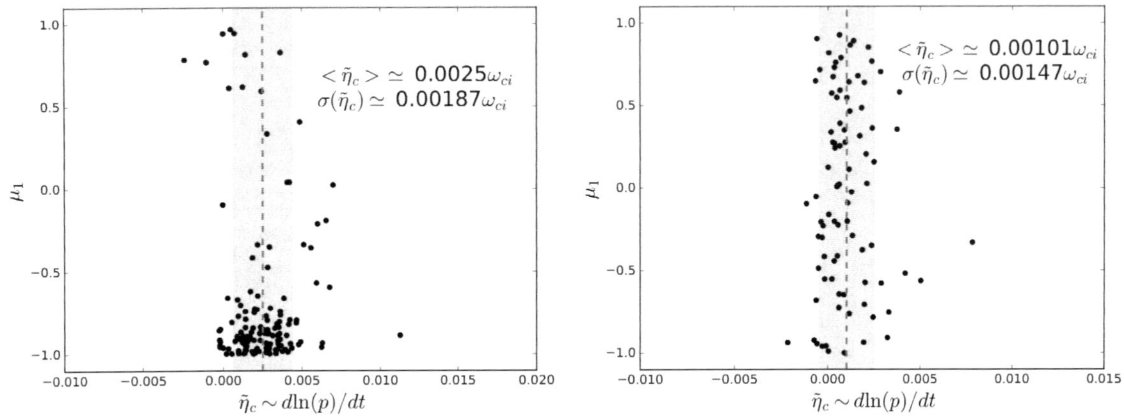

Figure 11. Scatter plots of acceleration rates for electrons (left) and ions (right). μ_1 is the pitch angle cosine at the start time of the fitting ($\Omega_{ci}t = 90$), and $\tilde{\eta}_c$ is the slope of $\log(p/p_0)$-t curves. The red dashed vertical line shows the mean value of $\tilde{\eta}_c$, and the shaded area shows the range $< \tilde{\eta}_c > \pm\sigma(\tilde{\eta}_c)$.

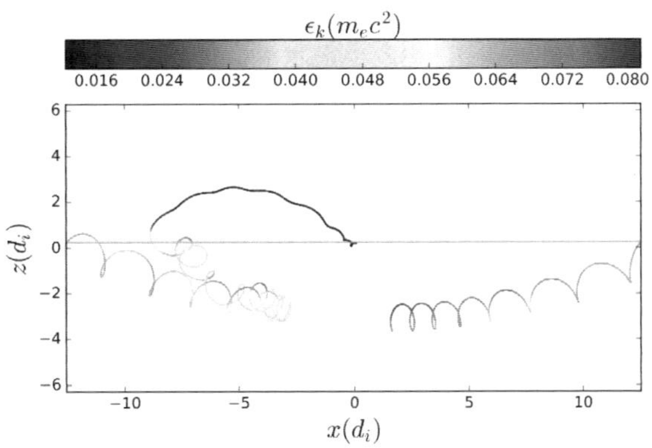

Figure 12. The trajectory of an ion tracer that escaped a flux rope.

gain energy in a contracting magnetic island via field line shortening, it needs to be trapped in the island. Escape is important for the formation of a power law energy spectrum by stochastic acceleration. In our simulation, high-energy electrons do not escape once they are trapped. This is likely due to the 2D nature of the simulation, as indicated by previous simulations [23]. Despite the large ion gyroradii, most of the ion tracers are also trapped within the flux rope. We find one case where an initially trapped ion seems to be scattered out of the flux rope after merging (shown in Figure 12). However, the escape mechanism is still not clear. We will return to this point later when we discuss the associated transport formalisms.

4.2. A Statistical Description of Particle Acceleration

In this section, we discuss briefly a statistical description of particle acceleration associated with multiple magnetic flux ropes. Similar to our previous analysis, we consider the island contraction mechanism only, and use the simple relation (Equation (7)) to estimate the particle momentum rate of change. The problem of stochastic acceleration can be treated as a random walk in momentum space. The classical approach to study the problem is to use the Fokker–Planck

(FP) equation

$$\frac{\partial f(\xi,t)}{\partial t} = -a\frac{\partial f(\xi,t)}{\partial \xi} + D\frac{\partial^2 f(\xi,t)}{\partial \xi^2}. \tag{8}$$

By using the FP equation, we essentially assume that the random walk is a Markov process, and the steps have a finite mean and variance, so that the coefficients are given by $a =< \Delta\xi > /\Delta t$ and $D =< \Delta\xi^2 > /2\Delta t$, where $< \Delta\xi >$ and $< \Delta\xi^2 >$ are the mean and variance of the momentum increment, and Δt is the time increment. Using equation (7), we can estimate the coefficients as

$$a \simeq < \eta_c >; \quad D \simeq < \eta_c^2 > \tau \tag{9}$$

where $< \eta_c >$ and $< \eta_c^2 >$ is the mean and variance of the acceleration rate, and τ is a characteristic time scale.

A transport equation of the FP type is applied to the flux rope acceleration problem by Bian & Kontar [24]. A more general approach is the so called continuous time random walk (CTRW). Depending on the form of the distribution of the acceleration rate and waiting time, the CTRW approach yields different forms of the transport equation (see, for example, Metzler & Klafter [25] for a comprehensive review). In the simplest case, the classical FP equation can be recovered, while in some other circumstances, a fractional differential equation (FDE) is more suitable. The idea of CTRW is that the step size (or jump length) and the waiting time between two steps (jumps) are drawn from a joint probability density function (pdf). If we further assume that the jump length is independent of the waiting time, each following a pdf $\lambda(\xi)$ and $\psi(t)$, then the governing equation for the particle distribution function can be written as a modified Chapman–Kolmogorov equation,

$$f(\xi,t) = \int_{-\infty}^{\infty} d\xi' \int_0^t dt' f(\xi',t')\lambda(\xi-\xi')\psi(t-t') + f_0(\xi)\int_t^{\infty} dt'\psi(t'). \tag{10}$$

It is often more convenient to work with the Fourier and Laplace transformed equation,

$$\hat{f}(k,u) = \frac{1-\hat{\psi}(u)}{u}\frac{\hat{f}_0(k)}{1-\hat{\lambda}(k)\hat{\psi}(u)}, \tag{11}$$

where we introduce a Fourier transform $\xi \to k$, and a Laplace transform $t \to u$ [25]. Considering only the island contraction mechanism, from Equation (7), the momentum change during a given time interval Δt is approximately

$$\Delta\xi = \eta_c \Delta t, \tag{12}$$

where we have dropped the tilde over η_c for notational simplicity. Suppose the effective contraction rate η_c has a pdf $g(\eta_c)$, then the jump length pdf is connected to η_c according to

$$\lambda(\Delta\xi)d\Delta\xi = g(\eta_c)d\eta_c. \tag{13}$$

On combining the two equations above,

$$\lambda(\xi) = \frac{1}{\Delta t} g\left(\frac{\xi}{\Delta t}\right). \tag{14}$$

Equation (14) provides a direct relation between the jump length pdf required for a CTRW process and the island contraction rate.

From our discussion, one sees that a theoretical description of stochastic acceleration is based on the distribution of acceleration rates $g(\eta_c)$. Our particle tracing technique provides a tool for obtaining such information, as the effective contraction rate $\tilde{\eta}_c$ associated with a flux rope can be obtained from tracer particles. Currently we don't have any statistical information of magnetic islands in our current simulation, because there is only one coalescence event. However, we expect that in a large-scale multiple island simulation, a similar technique may be useful. Besides the acceleration rate, the waiting time pdf $\psi(t)$ is also important for determining the shape of the transport formalism in CTRW. Another important parameter is the acceleration time Δt in Equation (14), which may be interpreted as the characteristic trapping time of a particle by a contracting flux rope. Information about the waiting time and trapping time will be difficult to obtain, since in the current simulation most energetic particles are very efficiently trapped by flux ropes and rarely escape. This is where 3D vs. 2D simulations may be important. A future study of the trapping and escape mechanism will be important.

In summary, we carried out a 2D PIC simulation of the coalescence of two magnetic flux ropes. The analysis of tracer particle trajectories shows that the basic acceleration mechanisms are the reconnection electric field and magnetic island contraction. The primary particle acceleration mechanism is magnetic island contraction at late times. However, the reconnection or anti-reconnection electric field is important at early times when the islands coalescence happens, and it may contribute to most of the ion energy increase and a large portion of the electron energy increase. The island contraction mechanism seems to be more important for electrons than for ions, but this conclusion may depend on parameters such as the island size distribution and in terms of simulations, the box size. Effective island contraction rates are obtained from tracer particle data. Although we are not able to derive a full particle transport equation from the current simulation, we expect that the particle tracing technique will be useful for future theoretical developments of the transport formalism.

Acknowledgments

G.P.Z. and S.D. acknowledge the partial support of an NSF DOE grant PHY - 1707247 and a NASA grant SV4 - 84017. This material is based also in part upon work supported by the NSF EPSCoR RII-Track-1 Cooperative Agreement OIA-1655280. F. G. and X. L. acknowledge the support by NASA under grant NNH16AC60I, DOE OFES, and the support by the DOE through the LDRD program at LANL. F.G.s contributions are also partly supported by the U.S. Department of Energy, Office of Fusion Energy Science, under Award Number DE-SC0018240 and by the National Science Foundation under grant No. 1735414. G.P.Z. acknowledges the generosity of the International Space Science Institute (ISSI) in supporting him through the 2017 Johannes Geiss Fellowship. This work is also partly supported by the International Space Science Institute (ISSI) in the framework of International Team 504 entitled Current Sheets, Turbulence, Structures and Particle Acceleration in the Heliosphere. We thank Jakobus le Roux for the useful discussion. The simulations were performed at Los Alamos National Laboratory, and this research used resources of the National Energy Research Scientific Computing Center, a DOE Office of Science User Facility supported by the Office of Science of the U.S. Department of Energy under Contract No. DE-AC02-05CH11231.

References

[1] Wang R, Lu Q, Nakamura R, Huang C, Du A, Guo F, Teh W, Wu M, Lu S and Wang S 2016 *Nature Physics* **12** 263–267

[2] Huang S Y, Retino A, Phan T D, Daughton W, Vaivads A, Karimabadi H, Zhou M, Sahraoui F, Li G L, Yuan Z G, Deng X H, Fu H S, Fu S, Pang Y and Wang D D 2016 *J. Geophys. Res.* **121** 205–213

[3] Cartwright M L and Moldwin M B 2010 *J. Geophys. Res.* **115** A08102

[4] Zheng J and Hu Q 2018 *Astrophys. J.* **852** L23 (*Preprint* 1801.01771)

[5] Khabarova O, Zank G P, Li G, le Roux J A, Webb G M, Dosch A and Malandraki O E 2015 *Astrophys. J.* **808** 181 (*Preprint* 1504.06616)

[6] Servidio S, Matthaeus W H, Shay M A, Cassak P A and Dmitruk P 2009 *Phys. Rev. Lett.* **102** 115003

[7] Greco A, Matthaeus W H, Servidio S, Chuychai P and Dmitruk P 2009 *Astrophys. J.* **691** L111–L114

[8] Zank G P, Adhikari L, Hunana P, Shiota D, Bruno R and Telloni D 2017 *Astrophys. J.* **835** 147

[9] Gosling J T, Skoug R M, Haggerty D K and McComas D J 2005 *Geophys. Res. Lett.* **32** L14113

[10] Khabarova O V and Zank G P 2017 *Astrophys. J.* **843** 4

[11] Drake J F, Swisdak M, Che H and Shay M A 2006 *Nature* **443** 553–556

[12] Drake J F, Swisdak M, Schoeffler K M, Rogers B N and Kobayashi S 2006 *Geophys. Res. Lett* **33** L13105

[13] Drake J F, Swisdak M and Fermo R 2013 *Astrophys. J.* **763** L5 (*Preprint* 1210.4830)

[14] Oka M, Phan T D, Krucker S, Fujimoto M and Shinohara I 2010 *Astrophys. J.* **714** 915–926 (*Preprint* 1004.1154)

[15] Le A, Karimabadi H, Egedal J, Roytershteyn V and Daughton W 2012 *Phys. Plasmas* **19** 072120

[16] Zank G P, le Roux J A, Webb G M, Dosch A and Khabarova O 2014 *Astrophys. J.* **797** 28

[17] le Roux J A, Zank G P, Webb G M and Khabarova O 2015 *Astrophys. J.* **801** 112

[18] Bowers K J, Albright B J, Yin L, Bergen B and Kwan T J T 2008 *Phys Plasmas* **15** 055703

[19] Stanier A, Simakov A N, Chacón L and Daughton W 2015 *Phys. Plasmas* **22** 101203

[20] Stanier A, Daughton W, Chacón L, Karimabadi H, Ng J, Huang Y M, Hakim A and Bhattacharjee A 2015 *Phys. Rev. Lett.* **115** 175004

[21] Stanier A, Daughton W, Simakov A N, Chacón L, Le A, Karimabadi H, Ng J and Bhattacharjee A 2017 *Phys. Plasmas* **24** 022124 (*Preprint* 1611.05933)

[22] Zheng J, Hu Q, Chen Y and le Roux J 2017 *Journal of Physics: Conference Series* **900** 012024

[23] Dahlin J T, Drake J F and Swisdak M 2017 *Phys. Plasmas* **24** 092110 (*Preprint* 1706.00481)

[24] Bian N H and Kontar E P 2013 *Phys. Rev. Lett.* **110** 151101 (*Preprint* 1302.6090)

[25] Metzler R and Klafter J 2000 *Phys. Reports* **339** 1 – 77 ISSN 0370-1573

Reflections on a 50-year career as a theoretician in heliospheric and solar physics

L A Fisk
Department of Climate and Space Sciences and Engineering, University of Michigan, 2455 Hayward St, Ann Arbor, MI 48109-2143, USA

E-mail: lafisk@umich.edu

Abstract. This paper is the story of what has motivated my approach to research throughout my career, and a statement of my belief that in order to maintain our scientific discipline as vibrant, dynamic and worth supporting, it is essential that we always seek and welcome new concepts and paradigm shifts.

1. Introduction

This paper is based on the talk I gave at the session of the 17th Annual International Astrophysics Conference that acknowledged my 75 years on this Earth, and my 50-year career as a theoretician in heliospheric and solar physics. The paper is not intended to be a review of the science that has been conducted during the 50 years of my career, by me or by others, but rather is simply the story of what has motivated my approach to research throughout my career, and a statement of my belief that in order to maintain our scientific discipline as vibrant, dynamic and worth supporting, it is essential that we always seek and welcome new concepts and paradigm shifts.

I entered the field of heliospheric research in 1968, when I published my first paper in the Journal of Geophysical Research [1]. My Ph.D. degree is from the University of California, San Diego, in 1969, and was followed by a postdoctoral appointment at the NASA Goddard Space Flight Center, and a civil service position in 1971, working for Frank McDonald, who will appear several times in this story.

The community of space researchers at that time was small, particularly the theoreticians. You could count on two hands your competitors or colleagues. Everything was still new. Much was unknown. We were data starved. New ideas were welcomed and encouraged. There was no conventional wisdom yet.

Theoreticians in some cases had yet to prove their worth. Many of the important scientists were of the opinion that a good scientist can build the instrument, analyze the data, and interpret the results. What value added is there for a theoretician?

Ambitious theoreticians usually took a two-prong approach. You needed some problems to work on and publish that people were interested in. In my case it was cosmic ray modulation. However, you were also always on the lookout for a new observation that demanded a new explanation, for which you could become famous.

We all knew the so-called Parker moment. When Gene Parker developed his theory for the acceleration of the supersonic solar wind in 1958 [2], he was roundly criticized as wrong, only to be vindicated when Mariner 2 in route to Venus in 1962 observed the supersonic wind. This was a goal for us. Have a theory that is controversial and challenged, which you defend against criticism, and then are ultimately proven correct. If you have a few Parker moments in your career, your success is assured.

2. My first Parker moment – Anomalous Cosmic Rays

My first opportunity for a Parker moment came in 1973 with the discovery of what became known as anomalous cosmic rays, or ACRs. As shown in Figure 1, Hovestadt et al. in 1973 [3] and McDonald et al. in 1974 [4] discovered a new component of cosmic rays at energies of a few MeV per nucleon, which had a very unusual composition. In was greatly enhanced in oxygen, but not in carbon, and also had some extra helium [5]. I, Reuven Ramaty and Ben-Zion Kozlovsky looked at this discovery and came up with what ultimately was established as the origin of the ACRs [6]. The composition is the same as interstellar neutral gas, which flows into the heliosphere. It is then ionized by charge exchange and photoionization, picked up by the solar wind, and convected into the outer heliosphere. All you had to do was imagine that there is an acceleration process somewhere in the outer heliosphere that raises the particle energies from 1 keV per nucleon, the energy at which they are first picked up, to 10 MeV per nucleon, the energy at which they are observed. Moreover, the ACR particles would be singly charged, and therefore, particularly for oxygen, of high rigidity, which would allow them to propagate back into the inner heliosphere, maintaining the steep spectrum seen in figure 1.

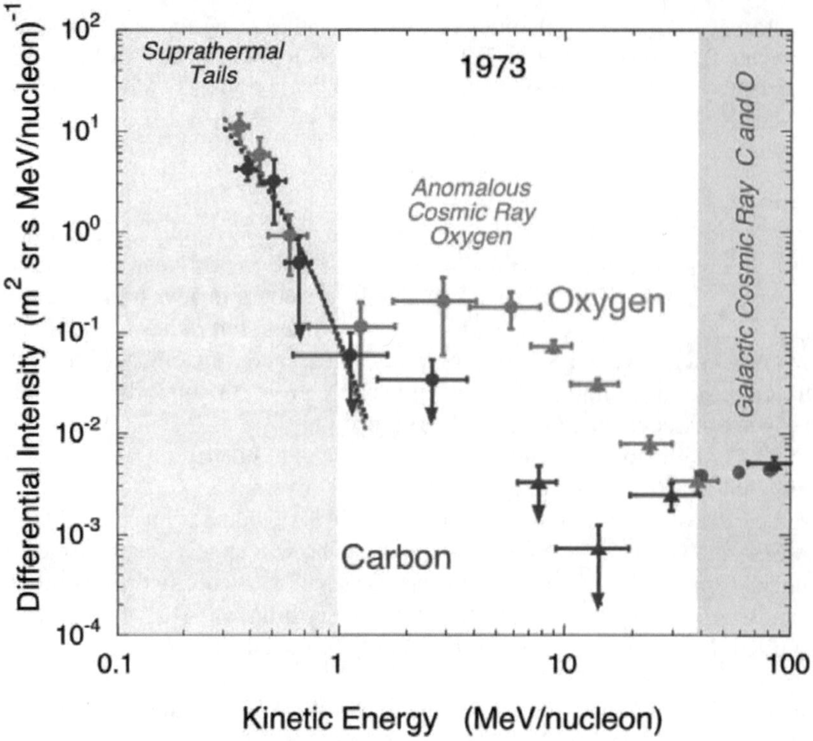

Figure 1. Differential intensity spectra of carbon (blue symbols) and oxygen (red symbols) measured by Hovestadt et al. [3] (filled circles) and McDonald et al. [4] (filled triangles). The oxygen spectrum has an extra component labeled "Anomalous Cosmic Rays" in the energy interval from about 1 to 30 MeV/nuc not seen in the carbon spectrum where only upper limits are shown.

This theory had all the ingredients of being my Parker moment. Frank McDonald, who I worked for, hated this theory. He thought he had discovered a new component of galactic cosmic rays, from some exotic object. In fact, some major astrophysicists, e.g. Hoyle and Clayton [7], published a theory arguing that the ACRs could be from white dwarfs in binary star systems. And here was this young scientist, in Frank McDonald's employ, depriving him of this major discovery.

Frank tried to suppress the publication and redirect my research into other more acceptable areas. However, this was my Parker moment, and I left Goddard in 1977 for the faculty of the University of New Hampshire (UNH).

3. My second Parker moment – He^3-rich solar flares

Almost immediately upon arriving at UNH, I had another opportunity for a Parker moment: ^{3}He-rich solar flares. The story was the same. There was a new unanticipated observation. In the mid 1970s, observations by Caltech, Goddard and the University of Chicago [8,9,10] found that there are small impulsive solar flares where ^{3}He is enhanced by up to a factor of 10,000 over cosmic abundances. To many astrophysicists, when you say ^{3}He they think of nuclear processes, and major astrophysicists like Colgate, Audouze, and Fowler [11] proposed theories that involved extreme conditions of temperatures and densities in solar flares. I went another direction [12] and noted that ^{3}He is a unique isotope, in that it is the only commonly available isotope whose cyclotron frequency lies between $^4He^{+2}$ and H^{+1}. Thus, if you can excite waves with frequencies in between the cyclotron frequencies of $^4He^{+2}$ and H^{+1}, you can preferentially heat the $^3He^{+2}$ and preferentially inject it into the flare acceleration process. I used electrostatic ion cyclotron waves, driven by current instabilities.

This theory was not really a Parker moment, because there was no real resistance to it. Other researchers developed more complete models, used other instabilities, but the general principle of using the unique cyclotron frequency of ^{3}He was common to subsequent models. Thus, I considered this a victory.

4. Other significant events in the 1970s

In the 1970s there were also several events and activities that have been very important for my career. Perhaps of most importance was the beginning of what has been a career-long collaboration with George Gloeckler. George was at the University of Maryland and recognized that there was a region of the energetic particle spectrum, lying between solar wind energies and a few MeV per nucleon, that was basically unexplored. And therefore, it was virgin territory for a new experimental group that did not compete with the powerhouse cosmic ray groups at Goddard and at the University of Chicago. I was at Goddard, anxious to be involved, and largely being ignored in this ambition. So, George and I formed a collaboration that has been essential to my career. Through his projects he has supported my research. We have coauthored more than 100 papers together, and are still to this day enjoying doing research together.

In the 1970s I also decided that I would like to be influential and help initiate and sell a new major space mission: the joint NASA-ESA dual spacecraft International Solar Polar mission. With a gravitational assist from Jupiter, the NASA spacecraft was to orbit in one direction over the poles of the Sun. The ESA spacecraft was to orbit in the opposite direction over the solar poles. Great mission. I put a lot of effort into selling this mission in the U.S., and we received a new start in 1978.

However, in 1980 Ronald Reagan was elected President of the United States, and his Director of the Office of Management and Budget, David Stockman, slashed the NASA budget and cancelled the U.S. spacecraft. We fought hard to keep the U.S. spacecraft, but to no avail. The European spacecraft, of course, flew and became the Ulysses mission, which had a number of U.S. instruments aboard, including George Gloeckler's SWICS instrument. Ulysses was an excellent, revolutionary mission, but not as good as it could have been had we had the two spacecraft.

5. The administration years

The end of the dual spacecraft Solar Polar mission was discouraging. So, I looked around for something else to do, and decided to try my hand at administration, which to my great surprise I turned out to be good at. In 1983 I became the Director of Research at UNH, a dean's level position. By 1985 I was the Vice President for Research and Financial Affairs.

Then in 1986 the Challenger accident happened and NASA cleaned out its entire upper management. So, NASA, in fact Frank McDonald, then NASA's chief scientist, came looking for an Associate

IOP Conf. Series: Journal of Physics: Conf. Series **1100** (2018) 012010 doi:10.1088/1742-6596/1100/1/012010

Administrator for Space Science and Applications who was a scientist recognized by the space science community, with senior administration experience. That combination was hard to find, and I was offered the position and moved my family to Washington, DC, in April of 1987.

The years at NASA in the late 1980s and early 1990s were good years. The NASA budget doubled, as did the budget for space science. We had major new starts, increases in the Explorer program and in Research & Analysis funding, etc. However, around 1992 things went sour. The growth in the NASA budget stopped. Dan Goldin became the NASA administrator and drove out the management talent at NASA Headquarters. Within about three years, only two of the senior leaders that were there when Dan arrived were left, and I was among the casualties.

I thought about what to do next. I decided that all I really wanted to administrate was NASA, and with that option closed, I decided to return to my research roots. The University of Michigan gave me a nice offer. In July of 1993 I became a Professor of Space Science at Michigan and started my brain again for research.

6. The community of researchers I returned to

I noted some changes in our research community that had occurred in the approximately 10 years while I was otherwise engaged. The community of heliospheric and solar researchers was very much larger. They had been diligently working over the ten years, and considered many issues settled, particularly in the heliosphere. There were details to work out but the big picture was known.

This was also the beginning of the development of major numerical modeling groups, well-funded with professional programmers. I had neither the interest nor the capability to start a numerical modeling group, so I chose to pursue my research as I had done before I became an administrator: Look for a new observation, which has the potential for resulting in a paradigm shift in our understanding. Run with it. Look for another Parker moment.

7. My third opportunity for a Parker moment

The opportunity for another Parker moment came in 1995. Ulysses, over the south pole of the Sun, observed 26-day increases in the intensity of a few hundred keV electrons and 0.5 MeV protons, which gave every evidence of having originated from corotating interaction regions within 30 degrees of the solar equatorial plane [13, 14]. How can this be? Low-energy electrons in particular follow field lines, and the Parker spiral heliospheric magnetic field, lying on cones of constant latitude, provides no means for electrons to be transported from low to high heliographic latitudes. Time for a new thought.

That new thought is illustrated in figure 2 [15]. The year 1995 was solar minimum, and the heliospheric magnetic field throughout much of the heliosphere originates from the polar coronal hole, which is offset from the solar rotation axis. Coronal holes are observed to rigidly rotate at the equatorial rotation rate of the Sun. The Sun differentially rotates, slower at the poles than at the equator. There is a nonradial expansion of the magnetic field from the polar coronal hole. When you put all this together [15], you find that the location from where the heliospheric magnetic field is convected radially outward with the solar wind moves dramatically in latitude and the field in the heliosphere that results provides a direct magnetic connection from low to high heliographic latitudes. This introduction of a polar component to the heliospheric magnetic field is hard to observe at any one location, since it is a small deviation. However, it is a systematic deviation, and the connection for low-energy particles from low to high latitudes establishes its existence.

This concept of the heliospheric magnetic field was reasonably well accepted in the heliospheric community. However, the real consequences of this theory were for solar physics. Ulysses also established that there is a single current sheet separating two hemispheres of opposite magnetic polarity in the heliosphere, and at solar minimum it lies near the equatorial plane [16]. The open magnetic flux of the Sun, the component of the solar magnetic field that opens into the heliosphere, is difficult to eliminate. You have to reconnect two oppositely directed open magnetic field lines at one location, at the current sheet, inside the Alfven point. Since there is very little evidence that such reconnection occurs, as you transport all this open flux down to the current sheet (see figure 2), you need to transport the open magnetic field around the Sun. As is discussed in Fisk and Schwadron [17] and Fisk [18], this transport can be accomplished by diffusion due to reconnection with coronal loops. Then, when you consider that there is global transport of open magnetic flux on the Sun, you can develop a comprehensive theory for how the solar magnetic field behaves and how the solar wind is accelerated.

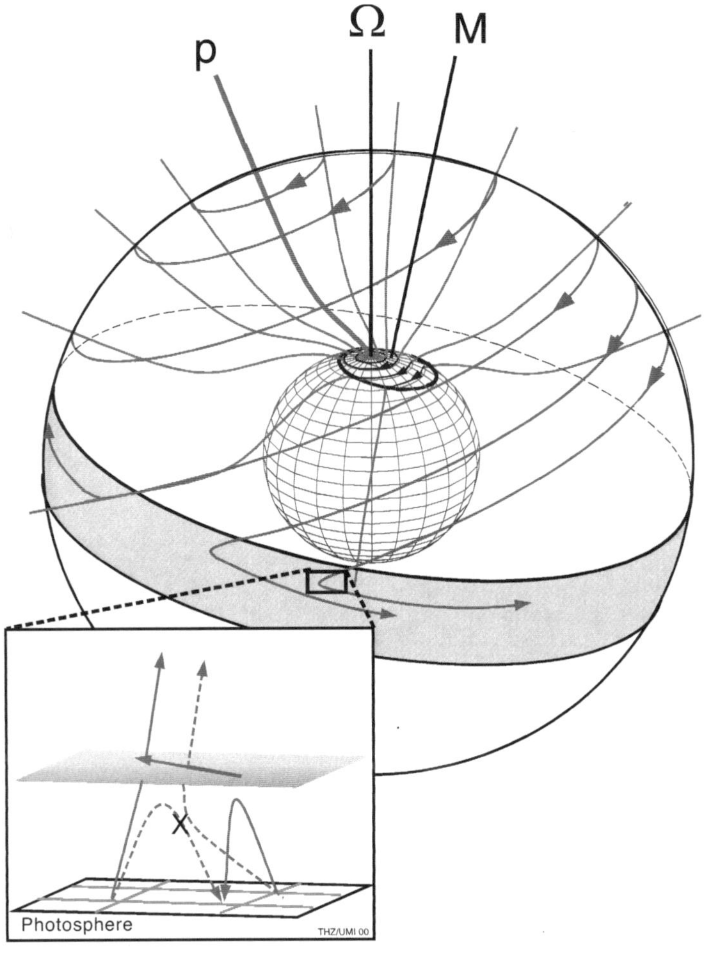

Figure 2. A schematic of the motions (in red) of open magnetic field lines (in green) from the polar coronal hole resulting from a combination of: (1) the offset of the polar coronal hole, which is observed to rigidly rotate; (2) differential rotation of the Sun; and (3) the nonradial expansion of the magnet field from the polar coronal hole [15]. The insert illustrates the diffusive motions of open field lines that can result from reconnection with coronal loops.

The schematic flow chart in figure 3 is meant to illustrate the various interlocking components of a comprehensive theory, which was developed together with Nathan Schwadron and Thomas Zurbuchen, for how the solar magnetic field behaves and the solar wind is acceleration. The processes associated with each component are discussed in one or more of the many papers that were written about this model [e.g., 17-25]. The model is based on three basic concepts. There is emerging magnetic flux on the Sun in the form of loops. The Sun differentially rotates and the open magnetic flux is organized into two regions separated by a single current sheet.

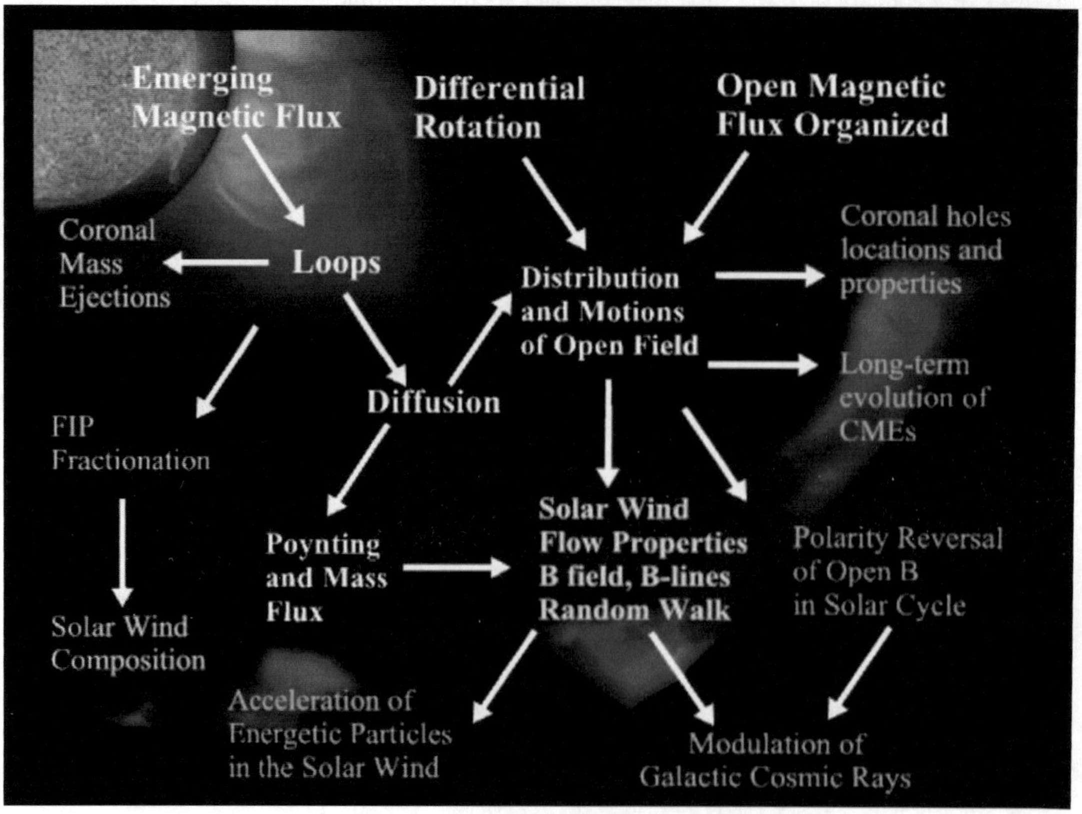

Figure 3. A flow chart of the interlocking processes that control the evolution of the solar magnetic field and the acceleration of the solar wind in the model of Fisk and colleagues [17–25].

The emerging loops are responsible for diffusion of the open flux and, together with differential rotation and the constraint that the open flux is organized, distributes the open flux around the Sun. It accumulates at the solar poles at solar minimum and this yields the polar field that drives the solar dynamo. In fact, you can derive that open flux will accumulate and form coronal holes in regions where the emergence of new magnetic flux is a local minimum, a prediction that was confirmed through solar observations by Abromenko et al. [24]. The diffusion by reconnection of open field lines with loops releases mass and results in a Poynting flux, which can be shown to be consistent with the resulting mass flux and flow energy in the solar wind. And so on.

So, was this a Parker moment? To be a Parker moment, you have to ultimately be proven to be correct. Unfortunately, when this model was first introduced, many solar physicists chose to ignore this work on the grounds that they didn't need new concepts for the solar magnetic field.

Spiro Antiochos's group has been working on aspects of how open flux becomes embedded in closed magnetic fields and the consequences for the solar wind, and their approach has been evolving to be similar to mine [e.g., 26, 27]. A very recent paper by Gary Zank both acknowledges my earlier work and develops a comprehensive model for the interaction of small loops and open field lines, and the resulting heating [28]. Thus, perhaps in time the concepts contained in my work, and possibly even the broader concepts for how the solar magnetic field behaves, will find their way into mainstream solar physics.

8. My fourth opportunity for a Parker moment – the common spectral shape of accelerated particles

As my work on the solar magnetic field was coming to an end, there was a new observation and another opportunity for a Parker moment. Voyager 1 finally crossed the termination shock. It did not observe that the ACRs were accelerated at the termination shock, but as it traveled downstream from the termination shock, it observed something remarkable. The spectrum of low-energy particles that were somehow being accelerated locked onto a power law spectrum with a specific spectral index, a differential intensity spectrum in energy, with spectral index of -1.5 [29, 30]. If we express the spectrum as a distribution function in velocity, the spectral index is -5. Remarkably, when Voyager 2 crossed the termination shock four years later, the exact same spectrum was observed. And much later, when Voyager 1, now deep in the heliosheath, observed the full accelerated ACR spectrum, it also had a -5 spectrum when expressed as a distribution function (also referred to as phase space density), as seen in figure 4 [31].

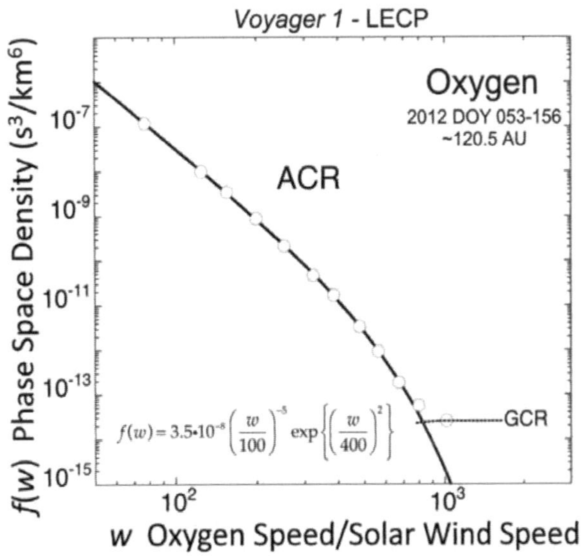

Figure 4. Phase space density of ACR oxygen as a function of oxygen speed/solar wind in the prime acceleration region at ~120 AU. The solar wind speed is taken to be 100 km/s. The data is from the LECP instrument on *Voyager* 1. The equation shown in the figure is the formula for the common spectrum and provides an excellent fit to the observations (from [31]).

The initial Voyager observations caused George Gloeckler to go back and look at the spectrum of much lower energy suprathermal tails from Ulysses and from ACE, and he found with a careful transformation into the frame of the solar wind that there were -5 spectra everywhere: Downstream from shocks, in regions where there were no shocks around. Everywhere there is a region of acceleration, where particles are being accelerated locally, the spectra of the suprathermal tails had a spectral index of -5, as illustrated in figure 5 and with other observations summarized in [32].

IOP Conf. Series: Journal of Physics: Conf. Series **1100** (2018) 012010 doi:10.1088/1742-6596/1100/1/012010

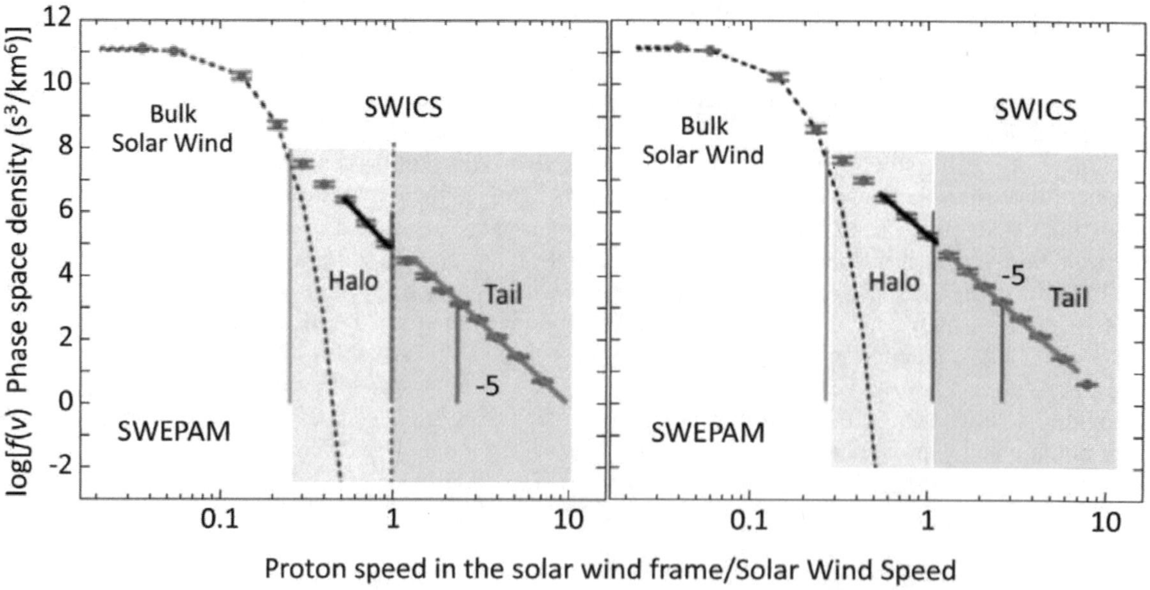

Figure 5. One-hour averaged solar wind frame velocity distribution functions showing the proton bulk solar wind, the halo and the tail segments during hour 11 of 12 August 2001, during which the strong (compression ratio of 3.85 ±0.15) shock passed *ACE* (left panel) and during the hour of peak tail density that was observed one hour downstream of the shock (right panel) (from [32]).

There was no acceleration theory known at that time that automatically yields a single common spectral shape for the tails on the distribution function, a power law spectrum with spectral index of -5. Diffusive shock acceleration yields power law spectra, but with variable spectral index depending on the compression ratio at the shock. Besides, the -5 spectra can be observed with no shock nearby. Stochastic acceleration is more likely to yield exponential spectra. This is exciting. We need a new acceleration mechanism.

So, you think. You note first that the acceleration is most likely to be strong in subsonic regions, in the heliosheath and downstream from shocks. Subsonic regions contain moderate scale compression regions and so the new acceleration mechanisms should involve compressions and expansions, and it should include an irreversible process. You note second that we have an equation that describes particle acceleration by compressions and expansions, with irreversible spatial diffusion, the Parker transport equation, written in terms of the distribution function f as a function of particle speed v [32]:

$$\frac{\partial f}{\partial t} + \delta \mathbf{u} \bullet \nabla f = \frac{\nabla \bullet \delta \mathbf{u}}{3} v \frac{\partial f}{\partial v} + \nabla \bullet \nabla \left(\vec{\kappa} \cdot \nabla f \right), \tag{1}$$

where $\delta \mathbf{u}$ is the random convective velocity of the compressions and expansions, and $\vec{\kappa}$ is the spatial diffusion tensor.

The Parker transport equation describes acceleration and deceleration in individual compression and expansion regions. We need to average this equation to determine the net acceleration due to multiple compressions and expansion regions. However, we need to be careful. Some of usual tricks of solving for the average distribution function of accelerated particles do not work here. We cannot treat this as a Markovian process, in which particles at different energies behave independently of each other. There is a cohesiveness in this problem that we have to capture. All particles in a compression or expansion region are either accelerated or decelerated. All compression regions must be surrounded by expansion regions.

To build in this cohesiveness, as is discussed in detail in [32, 34], we need to solve for the distribution function that defines the mean pressure, f_o, and then show that this distribution function can to a very good approximation be compared to observations. This is the equation that results. It describes the time evolution of the spectrum of the accelerated particles.

$$\frac{\partial f_o}{\partial t} = \frac{1}{v^4}\frac{\partial}{\partial v}\left(\frac{\delta u^2}{9\kappa}v\frac{\partial}{\partial v}\left(v^5 f_o\right)\right), \tag{2}$$

where κ is the cross-field diffusion coefficient.

The new acceleration mechanism is a pump acceleration process. It is a first order acceleration, and thus competitive with diffusive shock acceleration. The solutions to this equation fit the observations. For example, the excellent fit to the full ACR spectrum shown in figure 4 is from a solution to this equation.

So, was this a Parker moment? Well, not yet. There are many in our field who are firmly attached to diffusive shock acceleration as the preferred and dominant acceleration mechanism, and are reluctant to consider this new acceleration process, or, in the case of theorists, to extend and improve it. There are others now who are finding the -5 spectra when they consider acceleration with the same constraints as apply for the pump acceleration. There may be new observations that will be available from the upcoming IMAP mission that can verify the pump acceleration mechanism. And so, we will have to wait to see if this is a Parker moment.

9. My fifth opportunity for a Parker moment – the interaction of the heliosphere with the local interstellar medium

As my work on this new acceleration theory was winding down, there was yet another new observation and another opportunity for a Parker moment. As Voyager 1 penetrated further into the heliosheath, amazing new conditions and regions were discovered, which are summarized with references in [35]. First, the dominant pressure in the heliosheath is in mobile ACRs and pickup ions. Second, the solar wind flow, which on Voyager 1 is measured by observing the convective anisotropies of low-energy energetic particles, slows systematically down in the radial direction, and also in the polar and azimuthal directions, and creates what the Voyager investigators refer to as a stagnation region. Third, at about 122 AU, the particles accelerated in the heliosheath abruptly disappear and the galactic cosmic ray intensity increases. Fourth, Gurnett et al. [36] observe plasma waves, which provide a direct measure of the electron density, in the vicinity of ~122 AU and beyond. The density is ~0.1 per cc, much larger than the density that Voyager 2, with its working plasma detector, was observing closer to the termination shock, and comparable to the density expected in the interstellar medium.

As the Voyager observations became public, George Gloeckler and I undertook to develop a model that could explain the observations. Our first model involved the solar wind flowing in supersonic jets out the sides of the heliosheath [31]. However, as this paper was being completed, the density observations of Gurnett et al. became available and were inconsistent with our first model. Hence, we developed an entirely new model, which is illustrated in figure 6 [35]. The model is intended to be consistent with every Voyager observation known at that time.

The streamlines of the solar wind cross the termination shock, and then turn initially azimuthal, so the solar wind can flow out of the nose region of the heliosheath. The azimuthal flow creates a centerline region, which causes the flow speed to decrease. The azimuthal speed goes to zero first, and then the streamlines flow radially outward at a very low speed, across 122 AU, which we refer to as the heliocliff, out to the actual heliopause, which lies out at 140 AU or more. The solar wind flows across the heliopause, which is a rotational discontinuity.

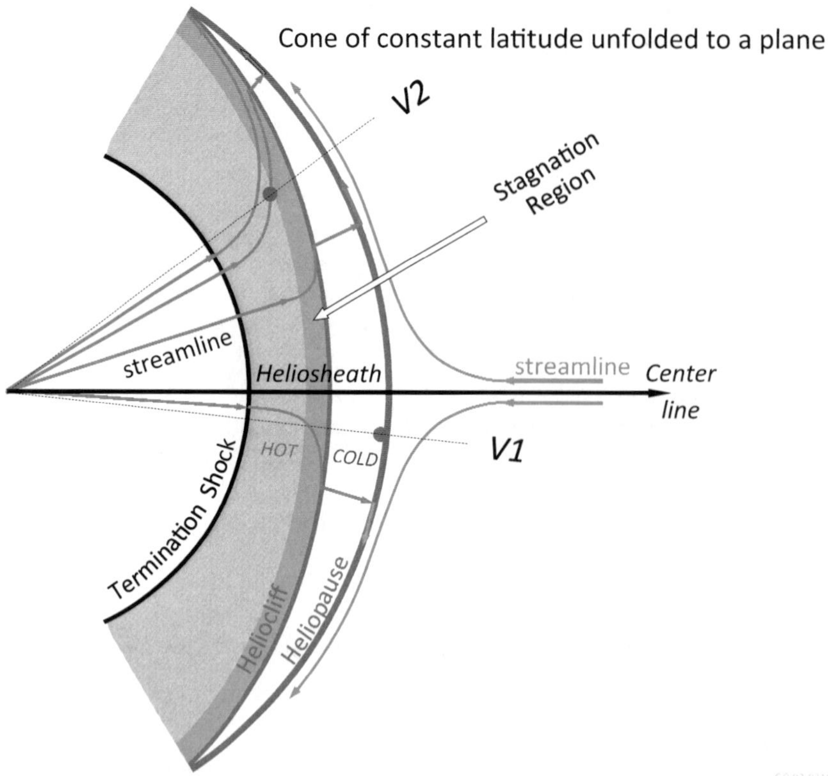

Figure 6. A simplified schematic of the basic features of the Fisk and Gloeckler model for the nose region of the heliosheath [35].

The new model is consistent with all Voyager observations. It naturally yields the stagnation region, where the solar wind flow becomes small in all directions. The solar wind is compressed and heated in this region, consistent with IBEX ENA observations [37]. Beyond 122 AU, where the solar wind and energetic particles can escape along the azimuthal magnetic field across the actual heliopause, the solar wind is cold and dense.

As we were completing this model, the Voyager investigators decided that Gurnett's density measurements [36], densities comparable to what is expected in the interstellar medium, is in fact conclusive evidence that the heliocliff is the heliopause, and Voyager 1 had crossed into the interstellar medium.

But wait. The Voyager investigators are the ones that said there is a stagnation region just inside the heliocliff where the solar wind flow velocity becomes small in all directions. And the mass conservation equation clearly states if a flow is compressed in all directions the density increases. In fact, subsequent IBEX observations of the ENA spectrum along the line of sight of the Voyager 1 trajectory can be understood only if the solar wind inside the heliocliff is compressed and heated [37].

And if you take away the argument that the only explanation for the Gurnett et al. density observations is that they are observing interstellar gas, the argument that Voyager 1 has passed into the interstellar medium dissolves. However, the press conference was held, and the public was thrilled to learn that Voyager 1 is now in the interstellar medium.

So, will this turn out to be a Parker moment, when I am ultimately proven to be correct? There may be some measurements possible from Voyager 2 that will support my position, but I expect that I will have to wait until an interstellar probe flies through the heliosheath and shows that indeed the heliopause is not at 122 AU, and there is a fascinating new region of compressed solar wind lying between 122 AU and the actual heliopause. I hope someone lets my great-grandchildren know.

10. Concluding remarks

Working on these last three theories has brought me enjoyment and satisfaction. It has been stimulating and challenging to seek and to recognize new observations that are significant, in that they cannot be understood with existing theories. And then to develop a new theory that can account for the observations. And then to pursue the broader ramifications for our understanding of the Sun and the heliosphere that result from this new theory. There has been a certain thrill in risking one's reputation by defending the new theory against other theoreticians and experimentalists. There has been satisfaction in recognizing that it is essential for the future of our field that we continuously develop and consider new concepts, and paradigm shifts.

Indeed, we cannot allow our field to evolve to where new concepts, where paradigm shifts, are neither looked for nor welcome. Where we have become complacent that everything we observe now or in the future can be understood with concepts and physical processes that were developed years ago. Should this be the fate of our field, we will cease to be a scientific discipline worth supporting.

As scientists, we should always be in search of the real truth. Not simply the validation of our preconceptions. As theorists we should always recognize that our theories will be judged as to whether they can account for observations. And all of us should embrace change as essential to a dynamic scientific discipline. Cherished theories, models and explanations that are found to be wanting should be quickly improved or replaced.

And we should recognize that change should come from the young. It has been disappointing to me that there were very few if any young people who have worked on explaining the observations on which these last three theories were based. I am in my mid 70s. Much of this work was done when I was in my 60s or late 50s. Where were the young people who viewed these observations as their opportunity for a Parker moment?

11. References

[1] Fisk L A and Axford W I 1968 Effects on energy changes on solar cosmic rays *J. Geophys. Res.* **73** 4397

[2] Parker E N 1958 Dynamics of the interplanetary gas and magnetic fields *Astrophys. J.* **128** 664

[3] Hovestadt D, Vollmer O, Gloeckler G and Fan C Y 1973 Differential energy spectra of low-energy (<8.5 MeV per nucleon) heavy cosmic rays during solar quiet times *Phys. Rev. Letts.* **31** 650

[4] McDonald F B, Teegarden B J, Trainor J H, von Rosenvinge T T and Webber W R 1974 The anomalous abundance of cosmic-ray nitrogen and oxygen at low energies *Astrophys. J.* **187** L105

[5] Garcia-Munoz M, Mason G M and Simpson J A 1973 The anomalous 1972 low-energy galactic cosmic ray proton and helium spectrum *Astrophys. J.* **18** L81

[6] Fisk L A, Kozlovsky B and Ramaty R 1974 An interpretation of the observed oxygen and nitrogen enhancements in low-energy cosmic rays *Astrophys. J.* **190** L35

[7] Hoyle F and Clayton D D 1974 Nucleosynthesis in white-dwarf atmospheres *Astrophys. J.* **191** 705

[8] Garrard T L, Stone E C and Vogt R E 1973 The isotopes of H and He in solar cosmic rays *High Energy Phenomena on the Sun Symp. Proc.*, ed R Ramaty and R G Stone (NASA SP-342) p 341

[9] Serlemitsos A T and Balasubrahmanyan V K 1975 Solar particle events with anomalously large relative abundance of ^{3}He *Astrophys. J.* **198** 195
[10] Anglin J D 1975 The relative abundance and energy spectra of solar-flare-accelerated deuterium, tritium, and Helium-3 *Astrophys. J.* **198** 733
[11] Colgate S A, Audouze J and Fowler W A 1977 Possible interpretations of the isotopic composition of hydrogen and helium in solar cosmic rays *Astrophys. J.* **213** 849
[12] Fisk L A 1978 ^{3}He-rich flares: A possible explanation *Astrophys. J.* **224** 1048
[13] Simnett G M, Sayle K A, Tappin S J and Roelof E C1995 Corotating particle enhancements out of the ecliptic plane *Space Sci. Rev.* **72** 327
[14] McKibben R B, Connell J J, Lopate C, Simpson J A and Balogh A 1995 Ulysses out-of-ecliptic observations of 27-day variations in high energy cosmic ray intensity *Space Sci. Rev.* **72** 403
[15] Fisk L A 1996 Motion of the footpoints of heliospheric magnetic field lines at the Sun: Implications for recurrent energetic particle events at high heliographic latitudes *J. Geophys. Res.* **101** 15,547-15,554
[16] Balogh A, Smith E J, Tsurutani B T, Southwood D J, Forsyth R J and Horbury T S 1995 The heliospheric magnetic field over the south polar region of the Sun *Science* **269** 1007
[17] Fisk L A and Schwadron N A 2001 The behavior of the open magnetic field of the Sun *Astrophys. J.* **560** 425-438
[18] Fisk L A 2005 The open magnetic flux of the Sun 1: Transport by reconnections with coronal loops *Astrophys. J.* **626** 563-573
[19] Fisk L A, Zurbuchen T H and Schwadron N. A. 1999 On the corona magnetic field: Consequences of large-scale motions *Astrophys. J.* **521** 868-877
[20] Fisk L A, Zurbuchen T H and Schwadron N A 1999 Coronal hole boundaries and their interactions with adjacent regions *Space Sci. Rev.* **87** 43-54
[21] Fisk L A, Schwadron N A and Zurbuchen T H 1999 Acceleration of the fast solar wind by the emergence of new magnetic flux *J. Geophys. Res.* A **104** 19,765-19,772
[22] Fisk L A 2003 Acceleration of the solar wind as a result of the reconnection of open magnetic flux with coronal loops *J. Geophys. Res.* A **108** DOI 10.1029/2002JA009284
[23] Fisk L A and Zurbuchen T H 2006 Distribution and properties of open magnetic flux outside of coronal holes *J. Geophys. Res.* **111** A09115
[24] Abramenko V I, Fisk L A and Yurhyshyn V B 2006 The rate of emergence of magnetic dipoles in coronal holes and adjacent quiet-sun regions *Astrophys. J. Letts.* **642** L65
[25] Fisk L A and Zurbuchen T H 2006 Distribution of open magnetic flux outside of coronal holes *Proc. of Solar Wind 11* vol SP-592 (Noordwijk: ESA) p 227
[26] Higginson A K, Antiochos S K, DeVore C R, Wyper P F and Zurbuchen T H 2017 Formation of heliospheric arcs of slow solar wind *Astrophys. J. Letts.* **840** L10
[27] Higginson A K, Antiochos S K, DeVore C R, Wyper P F and Zurbuchen T H 2017 Dynamics of coronal hole boundaries *Astrophys. J.* **837** 113
[28] Zank G P, Adhikari L, Hunana P, Tiwari S K, Moore R, Shiota D, Bruno R and Telloni D 2018 Theory and transport of nearly incompressible magnetohydrodynamic turbulence. IV. Solar coronal turbulence *Astrophys. J.* **854** 32
[29] Decker R B, Roelof E C, Krimigis S M and Hill M E 2006 Low-energy ions near the termination shock *Physics of the Inner Heliosheath*, *AIP Conf. Proc.* vol 858 (Melville: AIP) pp 73–78
[30] Gloeckler G, Fisk L A, Mason G M and Hill M E 2008 Formation of power law tail with spectral index -5 inside and beyond the heliosphere *Particle Acceleration and Transport in the Heliosheath and Beyond*, *AIP Conf. Proc.* vol 1039 (Melville: AIP) pp 367–374
[31] Fisk L A and Gloeckler G 2013 The global configuration of the heliosheath inferred from recent Voyager 1 observations *Astrophys. J.* **776** 79
[32] Fisk L A and Gloeckler G 2014 The case for a common spectrum of particles accelerated in the heliopshere: Observations and theory *J. Geophys. Res.* **119** 8733

[33] Parker E N 1965 The passage of energetic charged particles through interplanetary space *Planet. Space Sci.* **13** 9
[34] Fisk L A and Gloeckler G 2012 Particle acceleration in the heliosphere: Implications for astrophysics *Space Sci. Rev.* **173** 433-458
[35] Fisk L A and Gloeckler G 2014 On whether or not Voyager 1 has crossed the heliopause *Astrophys. J.* **789**, 41
[36] Gurnett D A, Kurth W S, Burlaga L F and Ness N F 2013 In situ observations of interstellar plasma with Voyager 1 *Science* **341** 1489
[37] Gloeckler G and Fisk L A 2015 IBEX observations provide strong evidence that Voyager 1 is still in the heliosheath *J. Phys.: Conf. Ser.* **642** 012011

Acknowledgements
I would like to express my profound gratitude to Gary Zank for organizing this celebration of my 75 years on this Earth and my 50-year career as a theoretician in heliospheric and solar physics.

IOP Conf. Series: Journal of Physics: Conf. Series **1100** (2018) 012011 doi:10.1088/1742-6596/1100/1/012011

Magnetically-Aligned Interstellar Dust Grains at the Heliosphere

P C Frisch[1]

[1] Department of Astronomy and Astrophysics, University of Chicago, Chicago IL 60637 USA

E-mail: frisch@oddjob.uchicago.edu

Abstract. Light from nearby stars becomes linearly polarized while traversing a dichroic interstellar medium. Polarization data which reveal heliosphere features are summarized. Polarizing dust grains trace a magnetic field direction consistent with the field affecting the warm breeze of secondary interstellar helium found by IBEX. Polarization data also trace the field direction that dominates the configuration of the IBEX ribbon of energetic neutral atoms (ENA). The polarizations that echo heliospheric features appear to arise from magnetically-aligned submicron interstellar dust grains that are excluded from the inner heliosphere by large charge-to-mass ratios, but, due to long collision times, retain their alignment with respect to the interstellar magnetic field draping over the heliosphere.

1. Introduction

The low density interstellar material around the heliosphere sustains a magnetic field that aligns interstellar dust grains and creates a dichroic medium capable of polarizing the light of background stars. These linear polarizations are parallel to the direction of the interstellar magnetic field (ISMF) producing the alignment [1]. The polarizations thus map the direction of the ISMF. The high sensitivities now achievable with modern polarimeters, capable of detecting polarization strengths of 0.001% or less, permit mapping of the ISMF formed in the local interstellar medium (ISM) where column densities are low, $N(H^{o}) < 10^{18.5}$ [2, 3, 4, 5, 6, 7, 8, 9, 10, 11, 12, 13, 14, 15, 16, 17, 18, 19, 20, 21].

A survey of polarized starlight tracing the local interstellar magnetic field (ISMF) is underway [22] in order to connect the ambient interstellar magnetic field with the ISMF that shapes the heliosphere, orders the ribbon of energetic neutral atoms [23, 24, 25], and deflects the ions giving rise to the warm breeze of secondary neutral atoms [26, 27, 28] discovered by IBEX. [1] Results of the study reveal an imprint of heliosphere geometry on the polarization data. One feature is a filament of aligned interstellar dust grains with a magnetic pole corresponding to the upwind direction of the warm breeze of neutral He° discovered by IBEX (Figure 1, [15]). A second feature is a band of statistically probable magnetic field directions that echos the heliosphere geometry [22].

The interstellar cloud surrounding the heliosphere provides the source of the "interstellar wind" feeding interstellar gas and dust into the inner heliosphere, and is termed here as the

[1] These features are ordered by the B-V plane, formed by the directions of the interstellar magnetic field B from the IBEX "ribbon" of interstellar neutral atoms and the velocity of the flow of primary interstellar neutral atoms through the heliosphere [28].

IOP Conf. Series: Journal of Physics: Conf. Series **1100** (2018) 012011 doi:10.1088/1742-6596/1100/1/012011

circumheliospheric interstellar material (CHM). The name "Local Interstellar Cloud" (LIC) is used to describe the cloudlet in the 15-cloud model of Redfield and Linsky [29]. The nose of the heliosphere, located 16° northeast of the galactic center [30], is outside of the LIC as defined in the 15-cloud model (Figure 19 in [29]).

The context for the ISMF surrounding the heliosphere is provided by radio continuum [31] and H° 21-cm data [32] that show giant shells associated with the Loop I superbubble. We assume spherical shells, which places the Sun in the rim of the Loop I superbubble. The bulk motion of the local interstellar clouds through the Local Standard of Rest (LSR) is directed away from the center of the Loop I superbubble [33, 34]). IBEX data on the interstellar He° velocity in the inner heliosphere, and the ISMF direction from the center of the IBEX ENA ribbon, indicate that the velocity of the CHM through the LSR is perpendicular to the orientation of the interstellar magnetic field [35].

2. Interstellar Boundary Conditions of the Heliosphere

Determining the interstellar boundary conditions of the heliosphere requires identifying a star with an absorption component at the CHM velocity projected onto the stellar sightline. Interstellar absorption lines in that component then provide data on the composition and ionization of the pristine CHM. Neutral atoms from the CHM enter the heliosphere, become ionized and reprocessed, to produce the pickup ions and anomalous cosmic ray particle populations. These populations provide boundary conditions for radiative transfer models of the surrounding interstellar gas, after accounting for the filtration of neutral interstellar atoms in heliosheath regions [36]. Slavin and Frisch (2008) found that the predicted ionization levels of H, He, N, O, and Ne depend in detail on the interstellar radiation field in the range 500A–1500A, the total interstellar cloud opacity to penetrating radiation, and the properties of the soft X-ray background component that maintains the He and Ne ionization. The composition of interstellar dust grains in the CHM is inferred from the atoms not seen in the gas phase.

In situ measurements of interstellar He° atoms in the inner heliosphere yield the CHM velocity vector, which relates the CHM to interstellar clouds observed through the absorption lines they create in the spectra of nearby stars. Reconstruction of the trajectories of the atoms through the heliosphere, after taking into account gravitational forces and weak ionization effects, yields the upwind direction for the flow of CHM through the heliosphere. In the inertial frame of the heliosphere the interstellar helium gas velocity arrives from the direction $\ell = 3.66^\circ \pm 0.94^\circ$, $b = 15.05^\circ \pm 1.22^\circ$, with a velocity of 25.4 ± 1.1 km s^{-1} and temperature of $7,260 \pm 270$ K [30, 37, 38]. [2]

The velocity vector of interstellar grains flowing through the heliosphere has been determined from Ulysses dust data to have a similar upwind direction (ℓ=8.0° ± 10°, b=13.5° ± 5° [40]) and velocity ($24.5^{+0.11}_{-0.12}$ km s^{-1} [41]) as the gas.

The LIC component observed toward the "Dog Star", Sirius[3], is the interstellar cloud containing the CHM. Projecting the CHM velocity toward the direction of Sirius, 137° away from the nose of the heliosphere, yields a velocity of -18.6 ± 0.8 km s^{-1}. The projected CHM velocity coincides with the velocity of the LIC component in the spectrum of Sirius (-17.6 ± 1.5 km s^{-1} [43]), which firmly links the CHM to the LIC cloud in front of Sirius.

The boundary conditions of the heliosphere have been modeled by Slavin and Frisch (2008, [36]) using radiative transfer models of the ionization of the LIC component that is measured toward epsilon CMa [44], located 12.6° away from Sirius. Both Sirius and the more distant star epsilon CMa show the same two interstellar neutral clouds. The LIC radiative transfer

[2] It is a curiosity that for the relative Sun-CHM velocity of 25.4 km s^{-1} the Sun will have traveled over 1334 AU (or 0.006 pc) through the CHM since Sir William Herschel first determined the proper motion of the Sun through space [39].

[3] Sirius has been referred to as the Dog Star since the days of Homer [42].

IOP Conf. Series: Journal of Physics: Conf. Series **1100** (2018) 012011 doi:10.1088/1742-6596/1100/1/012011

models utilize a range of assumptions, and include predictions of the neutral helium densities and temperatures at the heliosphere boundaries in agreement with IBEX and Ulysses data [38, 37]. The observed temperature of the LIC component toward Sirius (7000–8500 K) represents interstellar gas over the integrated sightline through the cloud, and is consistent with the interstellar He° temperature K derived from in situ IBEX measurements [30, 37, 38].

Reconstructed cloud abundances from the radiative transfer models [36] give the composition of the interstellar dust grains in the CHM based on the "missing mass" principle, where atoms not found in the gas phase are assumed to be captured by the interstellar dust grains in the same cloud. These models indicate that the dust grains in the LIC are composed primarily of olivines, and that Mg, Si, and Fe are strongly depleted onto the LIC interstellar grains, O is moderately depleted, and C and S are not present in the grains[4] (also see [33]).

The LIC dust grain composition inferred from the radiative transfer models of the LIC cloud toward epsilon CMa [36] are in agreement with the LIC interstellar dust grain composition measured by the CDA instrument on Cassini in orbit around Saturn [46]. Both investigations find the interstellar dust grains in the cloud around the heliosphere lack a carbonaceous component, and are dominated by silicate oxides. Cassini measured 36 interstellar grains and found magnesium-rich grains with a silicate and oxide composition and iron-rich inclusions. Similar grain compositions were found from the radiative transfer models. The models show that the amount of carbon predicted by cosmic abundances is found in the interstellar gas, within nothing left for dust grains. The high abundances of refractory elements in interstellar dust grains is attributed to the processing of interstellar grains through shock fronts that erode the volatile grain components [47, 40] and also decrease the average size distribution of dust grains.

3. Extinction and Small Grains

The spectral dependence of the linear polarization strengths provide a diagnostic of the dimensions of the polarizing grains. Marshall et al. [18] have shown that the maximum polarization towards nearby stars tends to occur at shorter (bluer) wavelengths than found for more obscured distant stars, giving wavelength of maximum polarization $\lambda_{\rm max} \sim 0.315$ μm. The smaller grains implied by this value, radius $a \sim 0.05$ μm, are consistent with shocked grains in the low density local interstellar material. Fortunately polarization position angles are independent of grain size [48].

Czechowski & Mann [49] have shown that the ratio of the Larmor radius of the grain to the thickness of the outer heliosheath governs the trajectories of interstellar dust grains through the heliosheath regions, with only a very small fraction of the smallest grains likely to reach the inner-most heliosphere [50, 51, 52]. Figure 3 in Frisch et al. (1999) [40] indicates that carbonaceous grains with radii ~ 0.05 μm have gyroradii ~ 10 AU in the heliopause region. Increasing grain density by 50% to account for the probable silicate composition still results in the exclusion of the smallest interstellar dust grains.

Silicate grains with radii of ~ 0.05 μm have high surface potentials in the outer heliosheath region ($\sim 5-10$ V) that will lead to tight coupling between the ISDGs and the deflected ISMF in the outer heliosheath region [53]. For comparison, this distance interval is significantly smaller than the thickness of the inner heliosheath in the directions of Voyager 1 (27 AU) and Voyager 2 (larger than 31 AU), and significantly smaller than the 2000 AU scale of the outer heliosheath derived from the properties of the magnetic turbulence detected by Voyager 1 [54].

The interstellar dust population interacting with the heliosphere thus contains ample abundances of micron-sized dust grains that both are excluded from entering the heliosphere

[4] This argument for grain composition breaks down when grain gyroradii are larger than the cloud. Gruen and Landgraf [45] show that large interstellar dust grains, with masses larger than 10^{-15} gr, have large gyroradii and couple to interstellar clouds over length scales of 100-1000 pc.

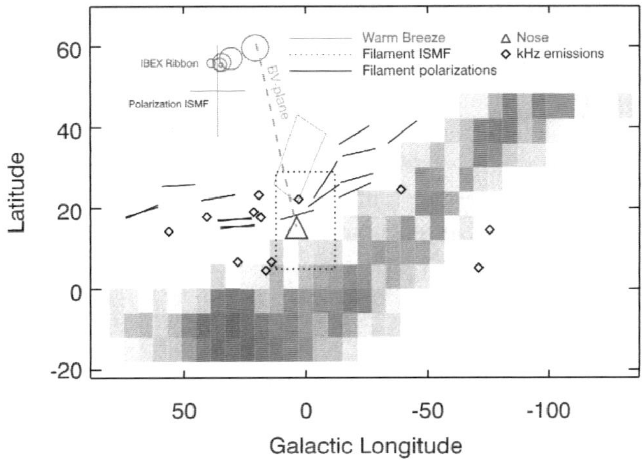

Figure 1. A filament of interstellar dust grains interacting with the heliosphere has been identified in the polarizations of stars within 40 pc [15]. The best-fitting ISMF to the polarizations of the stars forming the polarization filament is directed toward ℓ=357.3°, b=17.0° (±11.2°) (black cross). and magnetic turbulence appears to account for 9.6° of the direction uncertain. The axis of the polarization filament makes an angle of 80°± 14° with respect to the fundamental B-V symmetry plane of the heliosphere , for surrounding magnetic field direction B and inflowing interstellar velocity V. The confluence of the ISMF traced by the polarization filament and the warm breeze of secondary interstellar He° discovered by IBEX [57, 26, 27] strongly suggest that interstellar dust grains interacting with the heliosphere are responsible for the polarization filament.

(Section 6) and also have compositions consistent with those of magnetically aligned grains elsewhere (e.g. [55]).

4. Geometric Imprint of the Heliosphere on Polarized Starlight

Linearly polarized starlight provides the only means of detecting the ISMF direction over short sightlines through very low density interstellar material such as surrounds the heliosphere. We have undertaken to study the local interstellar magnetic field through the collection and analysis of linear polarization data of stars within 40 pc [56, 8, 14, 15, 17, 20, 18, 19].

Linear polarization data for each star trace the projection of the three-dimensional polarization vector onto the plane of the sky so that a statistical combination of the polarization data reconstructs the missing radial component of the polarizations. Combining multiple measurements of polarization position angles thus allows the directions of the magnetic field to be mapped throughout the sky. A surprising outcome of our polarization survey is that the interaction between the heliosphere and interstellar medium leaves an imprint on the linear polarizations of stars within ~ 15 pc, including a polarization filament [15] and polarization band [22].

The low densities of interstellar gas near the Sun [33] indicate an absence of dense clouds that often show strongly polarized starlight. Thus the low local densities also enable the detection of weak polarization features arising from the interaction between the heliosphere and the interstellar medium [22].

ISMF direction of IBEX Ribbon: The direction of the interstellar magnetic field traced by the center of the IBEX Ribbon [23, 25] was recovered from measurements of linearly polarized starlight for stars within 40 pc, based on a merit function that statistically minimized the sine of polarization position angles. The best-fitting ISMF direction was directed toward ℓ=36.2°, b=49.0° (±16°). The IBEX Ribbon ISMF direction was recovered after omitting a set of data that trace an independent filament of polarizations (below). For comparison, the weighted mean center of the IBEX Ribbon is located at $\ell = 36.7° \pm 2.1°$, $b = 56.0° \pm 0.6°$ [58].

Filament of Polarization: A polarization filament has been identified with a best-fitting ISMF direction aligned with the direction of the warm breeze of magnetically deflected helium atoms (Figure 1, [15]). The coincidence between the filament ISMF direction and the warm breeze of secondary interstellar He°, which forms from magnetically deflected helium ions [57, 26, 27], is the basis for concluding that the filament is formed by interstellar dust grains in the upwind outer heliosheath. The filament star polarizations trace a magnetic field oriented toward ecliptic coordinates λ=255.0° ±7.0°, β=7.9° ±8.5°, in agreement with the warm He° breeze direction of λ,β=251°, 12.0° ±7.8°. The warm He° breeze arises from secondary helium atoms created from charge-exchange between the parent interstellar helium ions that are magnetically deflected in the outer heliosheath regions and subsequently reneutralized before propagating to the inner heliosphere. Both the filament magnetic pole and warm breeze upwind direction appear to trace the magnetic deflection of charged particles in the magnetic field in outer heliosheath regions.

Polarization band: A "band" of overlapping polarization position angles has been identified by a great circle on the sky that traverses the heliospere nose and has a best-fitting ISMF direction toward the ecliptic poles. The properties of this band are under study [22].

Low frequency radio emissions: Figure 1 shows the locations of the dozen low frequency ($\sim 2 - 3$ kHz) radio emissions originating in the ionized interstellar gas beyond the heliopause, and located by the triangulation of radio signals received by Voyagers 1 and 2 [59]. Half of the 3 kHz emissions are located in the directions of the polarization filament (Figure 1). The ISMF traced by the filament polarizations may provide for the propagation of Langmuir waves into the upstream interstellar electron plasma where the oscillation frequency of the plasma at density n_e=0.08 cm^{-3} produces the 3 kHz emissions [60, 61].

5. Retention of Magnetic Alignment of ISDGs Interacting with Heliosphere

The polarization data discussed here suggest that the magnetic alignments of the charged, optically asymmetric, interstellar grains that are trapped in the ISMF interacting with the heliosphere must be preserved during the interaction so that grains retain their alignment. The IBEX ribbon of enhanced ENA fluxes [58] provides a remote diagnostic of the orientation of the ISMF interacting with the heliosphere. The ribbon is seen in directions where the ISMF that is draping over the heliosphere is perpendicular to the radial sightlines, as viewed from the 1 AU orbit of IBEX [24, 25]. The existence of the IBEX ribbon requires that the ISMF interacting with the heliosphere smoothly varies in the outer heliosheath in the upwind directions.

6. Filament Polarizations and Exclusion of Submicron Grains from Heliosphere

The dusty component of the LIC contains grains with masses ranging from $10^{-15} - 10^{-9}$ gr [40, 53, 62], or larger if micro-meteorites detected by radar are included [63]. In situ measurements of interstellar dust grains by Ulysses and Galileo (Section 6) show a steep decrease in the numbers of grains as grain mass decreases, including a lack of grains with radii less than 0.18 μm compared to a nominal MRN [64] interstellar distribution [40, 65, 62]. The smallest grains are deflected away from the dust flow at the heliopause region, where the grain potentials are increased by interactions with heliosheath plasma that include the enhanced ejection of secondary electrons from the smallest particles embedded in the million degree plasma (see Figure 2 in Slavin et al. 2012). Grain charging and deflection will increase if the interstellar grains are porous and not compact [66]. Orbits of the smallest dust grains deflected around the heliosphere by Lorentz forces are shown in [67, 53]. These small deflected grains are a plausible origin for the polarizing dust grains traced by the polarization filament.

In situ grain measurements by Ulysses and Galileo give data on the grain trajectories and mass distribution [68, 69, 40, 63, 51, 62]. The dominant features of the grain mass distributions are the deficit of low mass grains and grains with larger massses than predicted by the typical MRN interstellar grain mass distributions. For silicate grains the deficit sets in for grain radii below ~ 0.3 μm and the excess appears for grain radii larger than ~ 1.0 μm [40, 62]. The missing grain mass varies with solar cycle phase due to the filtration of submicron interstellar grains interacting with the solar wind.

The gas-to-dust mass ratios determined from the in situ measurements appears to vary over time, possibly due to variations in the filtration of the smallest grains. The gas-to-dust mass ratio, $R_{g/d}$, found for the grains detected by Ulysses and Galileo during the decade of the 1990's is 94 (+46, -38), based on an interstellar gas density of 0.3 cm^{-3} [40, 51]. The entire sixteen years of the Ulysses data set (1992–2008) yielded $R_{g/d}$=193 (+85 -57), based on a slightly lower interstellar density (0.247 cm^{-3}) and a CHM velocity of 26 km s^{-1}.[5] In order to compare the $R_{g/d}$ value for the 1990's and the $R_{g/d}$ value for the full 16 years of data, the latter value should be adjusted upwards by 45% to account for the differences in the assumed gas densities and for the velocity dependence of the Ulysses impact ionization detector [62]. There remains a significant difference between the values for $R_{g/d}$ found during the measurements in the 1990's and the total sixteen year of measurements.

The difference between the $R_{g/d}$ for the 1990's and full 16 years is likely due to variations in the polarity of the solar wind between the 1990's and the first decade of the 21st century. During the 1990's the solar cycle magnetic polarity acted to "focus" interstellar dust grains toward the plane of the ecliptic [65], which has the effect of decreasing the gas-to-dust mass ratios obtained from data at 1 AU during that decade, compared to later values based on both phases of solar magnetic polarities [62]. The signature of the solar magnetic cycle on the gas-to-dust ratios provides additional support to the view that the small grains are filtered from the interstellar flow in the outermost heliosphere and heliosheath where the deflection of polarizing interstellar dust grains around the heliosheath would occur.

Slavin et al. [51] predict that time-variable submicron-sized grains are deflected in the outer heliosphere and heliosheath regions and form plumes around the heliosphere for the defocusing phase of the solar cycle. Sterken et al. [52] suggested that additional filtration of small grains in the outer heliosheath is needed to explain the Ulysses observations of grains. The deflection of small grains with large charge-to-mass ratios at the heliopause is supported by the in situ measurements of interstellar dust (also see review of Mann [70]).

7. Closing Remarks

The similar interstellar magnetic field directions traced by the polarization filament [15] and warm breeze of neutral helium flowing through the heliosphere [26], and the imprint of the heliospheric geometry on a statistically defined polarization band [22], indicate that magnetically aligned interstellar dust grains interacting with the heliosphere produce visible polarizations. The polarization data indicate that submicron-sized dust grains with an interstellar origin are trapped on the interstellar magnetic field lines draped over the heliosphere (Figure 1).

Magnetometers on board Voyager 1 have measured the ISMF in the outer heliosheath[6] and found a strength of 5 μG and laminar flow [71]. During quiet intervals, outside of disturbances from solar storms, the magnetic field direction was displaced toward the center of the IBEX ribbon [28]. These characteristics in the inner regions of the outer heliosheath suggests an

[5] The grain velocity affects the interpretation of the grain mass distribution) [62].

[6] Voyager 1 is presently located about 30° to the ecliptic north of the heliosphere nose.

orderly magnetic field. Further out, Voyager 1 found an outer scale of turbulence of ~ 0.01 pc, which is thought to represent the distance to the undisturbed interstellar medium [54]. The outer scale of turbulence corresponds to the maximum allowed thickness for the location of the aligned dust grains that create the polarization filament.

The orderly nature of the ISMF allows the polarizing ISDGs to retain their alignment during their approach through the outer heliosheath. The extremely low volume densities of the CHM gas, < 0.3 cm^{-3}, means that grain alignement is disrupted by collisions with gas over timescales of millions of years.

The results showing that aligned interstellar dust grains trace the interaction between the heliosphere and interstellar magnetic field suggests that the long-term monitoring of weakly polarized light of nearby stars can monitor the outer heliosheath regions and possibly provide the first evidence of changes in the galactic environment of the Sun.

Acknowledgements. The author is grateful for support by the Interstellar Boundary Explorer mission as a part of the NASA Explorer Program.

References

[1] Andersson B G, Lazarian A and Vaillancourt J E 2015 *ARA&A* **53** 501–539
[2] Magalhaes A M, Rodrigues C V, Margoniner V E, Pereyra A and Heathcote S 1996 *Polarimetry of the Interstellar Medium* (*Astronomical Society of the Pacific Conference Series* vol 97) ed W G Roberge & D C B Whittet pp 118–+
[3] Piirola V, Berdyugin A, Mikkola S and Coyne G V 2005 *ApJ* **632** 576–589
[4] Pereyra A and Magalhães A M 2007 *ApJ* **662** 1014–1023 (*Preprint* `arXiv:astro-ph/0702550`)
[5] Piirola V, Vornanen T, Berdyugin A and Coyne G V S J 2008 *ApJ* **684** 558–568 (*Preprint* `0805.4289`)
[6] Wiktorowicz S J and Matthews K 2008 *PASP* **120** 1282–1297 (*Preprint* `0810.5561`)
[7] Bailey J, Lucas P W and Hough J H 2010 *MNRAS* **405** 2570–2578 (*Preprint* `1003.1753`)
[8] Frisch P C, Andersson B G, Berdyugin A, Piirola V, DeMajistre R, Funsten H O, Magalhaes A M, Seriacopi D B, McComas D J, Schwadron N A, Slavin J D and Wiktorowicz S J 2012 *ApJ* **760** 106
[9] Piirola V, Berdyugin A and Berdyugina S 2014 *Society of Photo-Optical Instrumentation Engineers (SPIE) Conference Series* (*Society of Photo-Optical Instrumentation Engineers (SPIE) Conference Series* vol 9147) p 8
[10] Berdyugin A, Piirola V and Teerikorpi P 2014 *A&A* **561** A24
[11] Wiktorowicz S J, Nofi L A, Jontof-Hutter D, Kopparla P, Laughlin G P, Hermis N, Yung Y L and Swain M R 2015 *ArXiv e-prints* (*Preprint* `1507.03588`)
[12] Bailey J, Kedziora-Chudczer L, Cotton D V, Bott K, Hough J H and Lucas P W 2015 *MNRAS* **449** 3064–3073 (*Preprint* `1503.02236`)
[13] Wiktorowicz S J and Nofi L A 2015 *ApJ* **800** L1 (*Preprint* `1412.6117`)
[14] Frisch P C, Berdyugin A, Piirola V, Magalhaes A M, Seriacopi D B, Wiktorowicz S J, Andersson B, Funsten H O, McComas D J, Schwadron N A, Slavin J D, Hanson A J and Fu C W 2015 *ApJ* **814** 112 (*Preprint* `ArXiv:1510.04679`)
[15] Frisch P C, Andersson B G, Berdyugin A, Piirola V, Funsten H O, Magalhaes A M, Seriacopi D B, McComas D J, Schwadron N A, Slavin J D and Wiktorowicz S J 2015 *ApJ* **805** 60 (*Preprint* `1503.00353`)
[16] Marshall J P, Cotton D V, Bott K, Ertel S, Kennedy G M, Wyatt M C, del Burgo C, Absil O, Bailey J and Kedziora-Chudczer L 2016 *ApJ* **825** 124 (*Preprint* `1604.08286`)
[17] Cotton D V, Bailey J, Kedziora-Chudczer L, Bott K, Lucas P W, Hough J H and Marshall J P 2016 *MNRAS* **455** 1607–1628 (*Preprint* `1509.07221`)
[18] Marshall J P, Cotton D V, Bott K, Ertel S, Kennedy G M, Wyatt M C, del Burgo C, Absil O, Bailey J and Kedziora-Chudczer L 2016 *ApJ* **825** 124 (*Preprint* `1604.08286`)
[19] Bailey J, Cotton D V and Kedziora-Chudczer L 2017 *MNRAS* **465** 1601–1607 (*Preprint* `1611.01596`)
[20] Cotton D V, Marshall J P, Bailey J, Kedziora-Chudczer L, Bott K, Marsden S C and Carter B D 2017 *MNRAS* **467** 873–897 (*Preprint* `1701.02890`)
[21] Cotton D V, Bailey J, Howarth I D, Bott K, Kedziora-Chudczer L, Lucas P W and Hough J H 2017 *Nature Astronomy* **1** 690–696
[22] Frisch P C, Berdyugin A, Piirola V, Cole A, Hill K, Harlingten C, Magalhaes A M, Seriacopi D B, Ferrari T, Ribeiro N L, Wiktorowicz S J, Cotton D V, Bailey J, Kedziora-Chudczer L, Marshall J P, Bott K, Santos F, Heiles C, McComas D J, Funsten H O, Schwadron N A, Livadiotis G and Redfield S 2018 *ApJ, submitted*

[23] McComas D J, Allegrini F, Bochsler P, Bzowski M, Christian E R, Crew G B, DeMajistre R, Fahr H, Fichtner H, Frisch P C, Funsten H O, Fuselier S A, Gloeckler G, Gruntman M, Heerikhuisen J, Izmodenov V, Janzen P, Knappenberger P, Krimigis S, Kucharek H, Lee M, Livadiotis G, Livi S, MacDowall R J, Mitchell D, Möbius E, Moore T, Pogorelov N V, Reisenfeld D, Roelof E, Saul L, Schwadron N A, Valek P W, Vanderspek R, Wurz P and Zank G P 2009 *Science* **326** 959–

[24] Schwadron N A, Bzowski M, Crew G B, Gruntman M, Fahr H, Fichtner H, Frisch P C, Funsten H O, Fuselier S, Heerikhuisen J, Izmodenov V, Kucharek H, Lee M, Livadiotis G, McComas D J, Moebius E, Moore T, Mukherjee J, Pogorelov N V, Prested C, Reisenfeld D, Roelof E and Zank G P 2009 *Science* **326** 966–

[25] Zirnstein E J, Heerikhuisen J, Funsten H O, Livadiotis G, McComas D J and Pogorelov N V 2016 *ApJ* **818** L18

[26] Kubiak M A, Swaczyna P, Bzowski M, Sokół J M, Fuselier S A, Galli A, Heirtzler D, Kucharek H, Leonard T W, McComas D J, Möbius E, Park J, Schwadron N A and Wurz P 2016 *ApJS* **223** 25 (*Preprint* 1603.01741)

[27] Bzowski M, Kubiak M A, Czechowski A and Grygorczuk J 2017 *ApJ* **845** 15 (*Preprint* 1707.02193)

[28] Schwadron N A and McComas D J 2017 *ApJ* **135** 135

[29] Redfield S and Linsky J L 2008 *ApJ* **673** 283–314

[30] Schwadron N A, Moebius E, Leonard T, Fuselier S A, McComas D J, Heirtzler D, Kucharek H, Rahmanifard F, Bzowski M, Kubiak M A, Sokol J, Swaczyna P and Frisch P 2015 *ApJS* **220** 25

[31] Wolleben M 2007 *ApJ* **664** 349–356 (*Preprint* arXiv:0704.0276)

[32] Heiles C 1998 *IAU Colloq. 166: The Local Bubble and Beyond* (*Lecture Notes in Physics, Berlin Springer Verlag* vol 506) ed Breitschwerdt D, Freyberg M J and Truemper J pp 229–238

[33] Frisch P C, Redfield S and Slavin J 2011 *ARA&A* **49**

[34] Frisch P and Dwarkadas V V 2018 *ArXiv e-prints* (*Preprint* 1801.06223)

[35] Schwadron N A, Adams F C, Dingus B, Funsten H O, Desiati P, McComas D J and Frisch P C 2012 *in preparation* **00** 00

[36] Slavin J D and Frisch P C 2008 *A&A* **491** 53–68

[37] Wood B E, Müller H R and Witte M 2015 *ApJ* **801** 62 (*Preprint* 1501.02725)

[38] Bzowski M, Swaczyna P, Kubiak M A, Sokół J M, Fuselier S A, Galli A, Heirtzler D, Kucharek H, Leonard T W, McComas D J, Möbius E, Schwadron N A and Wurz P 2015 *ApJS* **220** 28 (*Preprint* 1510.04835)

[39] Herschel W E F R S 1783 *Phil. Trans. R. Soc. Lond.* **73** 247–283

[40] Frisch P C, Dorschner J M, Geiss J, Greenberg J M, Grün E, Landgraf M, Hoppe P, Jones A P, Krätschmer W, Linde T J, Morfill G E, Reach W, Slavin J D, Svestka J, Witt A N and Zank G P 1999 *ApJ* **525** 492–516

[41] Kimura H, Mann I and Jessberger E K 2003 *ApJ* **582** 846–858

[42] Jeans S J 1948 *The Stars in Their Courses* (University Press Cambridge)

[43] Hebrard G, Mallouris C, Ferlet R, Koester D, Lemoine M, Vidal-Madjar A and York D 1999 *A&A* **350** 643–658

[44] Gry C and Jenkins E B 2001 *A&A* **367** 617–628

[45] Grün E and Landgraf M 2000 *J. Geophys. Res.* **105** 10291–10298

[46] Altobelli N, Postberg F and Fiege K e a 2016 *Science* **352** A130

[47] Ellison D C, Drury L O and Meyer J 1997 *ApJ* **487** 197

[48] Clayton G C and et al 1992 *ApJ* **385** L53–L57

[49] Czechowski A and Mann I 2003 *A&A* **410** 165–173

[50] Czechowski A and Mann I 2003 *J. Geophys. Res.* **108** 13

[51] Slavin J D, Frisch P C, Müller H R, Heerikhuisen J, Pogorelov N V, Reach W T and Zank G 2012 *ApJ* **760** 46 (*Preprint* 1210.1127)

[52] Sterken V J, Strub P, Krüger H, von Steiger R and Frisch P 2015 *ApJ* **812** 141

[53] Slavin J D, Frisch P C, Müller H R, Heerikhuisen J, Pogorelov N V, Reach W T and Zank G 2012 *ApJ* **760** 46

[54] Burlaga L F, Florinski V and Ness N F 2018 *ApJ* **854** 20

[55] Voshchinnikov N V, Henning T, Prokopjeva M S and Das H K 2012 *A&A* **541** A52 (*Preprint* 1203.3677)

[56] Frisch P C, Andersson B, Berdyugin A, Funsten H O, Magalhaes M, McComas D J, Piirola V, Schwadron N A, Slavin J D and Wiktorowicz S J 2010 *ApJ* **724** 1473–1479

[57] Kubiak M A, Bzowski M, Sokół J M, Swaczyna P, Grzedzielski S, Alexashov D B, Izmodenov V V, Moebius E, Leonard T, Fuselier S A, Wurz P and McComas D J 2014 *ArXiv e-prints* (*Preprint* 1405.7742)

[58] Funsten H O, DeMajistre R, Frisch P C, Heerikhuijsen J, Higdon D M, Janzen P, Larsen B, Livadiotis G, McComas D J, Möbius E, Reese C, Reisenfeld D B, Schwadron N A and Zirnstein E J 2013 *ApJ* **776** 30

[59] Kurth W S and Gurnett D A 2003 *J. Geophys. Res.* **108** 2–16

[60] Mitchell J J, Cairns I H and Robinson P A 2004 *Journal of Geophysical Research (Space Physics)* **109** 6108

[61] Gurnett D A, Kurth W S, Burlaga L F and Ness N F 2013 *Science* **341** 1489–1492
[62] Krueger H, Strub P, Sterken V J and Grün E 2015 *ApJ* **812** 140
[63] Landgraf M, Baggaley W J, Grün E, Krüger H and Linkert G 2000 *J. Geophys. Res.* **105** 10343–10352
[64] Mathis J S, Rumpl W and Nordsieck K H 1977 *ApJ* **217** 425–433
[65] Landgraf M 2000 *J. Geophys. Res.* **105** 10303–10316
[66] Ma Q, Matthews L S, Land V and Hyde T W 2013 *ApJ* **763** 77 (*Preprint* 1210.0459)
[67] Mann I and Czechowski A 2004 *AIP Conf. Proc. 719: Physics of the Outer Heliosphere* pp 53–58
[68] Grün E, Gustafson B, Mann I, Baguhl M, Morfill G E, Staubach P, Taylor A and Zook H A 1994 *A&A* **286** 915–924
[69] Baguhl M, Grün E, Hamilton D P, Linkert G, Riemann R, Staubach P and Zook H A 1995 *Space Sci. Rev.* **72** 471–476
[70] Mann I 2010 *ARA&A* **48** 173–203
[71] Burlaga L F and Ness N F 2016 *ApJ* **829** 134

IOP Conf. Series: Journal of Physics: Conf. Series **1100** (2018) 012012 doi:10.1088/1742-6596/1100/1/012012

A database of small-scale magnetic flux ropes in the solar wind from Wind spacecraft measurements

Qiang Hu, Jinlei Zheng, and Yu Chen

Department of Space Science and CSPAR, The University of Alabama in Huntsville, Huntsville, AL 35805, USA

E-mail: qiang.hu@uah.edu

Abstract. We present an extensive study of identifying small-scale magnetic flux ropes with duration between 9 and 361 minutes from Wind spacecraft in-situ measurements. The approach is based on the well-established Grad-Shafranov (GS) reconstruction method applicable to two-dimensional (2D) magnetohydrostatic structures. An automated computer algorithm is developed to sift through the time-series data covering two solar cycles, from 1996 to 2016. A large number of flux rope events is identified and an online database is built to accommodate the outcome from this detection algorithm. We provide detailed descriptions of the key steps of the GS-based detection algorithm and the general features of the online database. Selected preliminary statistical analyses are shown, which yield the following main results. (1) The occurrence of small-scale flux ropes has strong solar cycle dependency. (2) The small-scale magnetic flux ropes in the ecliptic plane tend to align along the Parker spiral direction. (3) Both the duration and scale size distributions of the small-scale magnetic flux ropes obey power laws. We show the similarities and discrepancies in their properties as compared with their large-scale counterparts, the magnetic clouds. We also discuss the implications on their origination mechanisms.

1. Introduction

Magnetic flux ropes are a fundamental type of magnetic field structures occurring in space and laboratory plasmas. In space environment, one prominent type is the relatively large-scale flux ropes, i.e., the so-called magnetic clouds (MCs), which has a clear origin from the Sun. They correspond to the magnetic field structures embedded in coronal mass ejections (CMEs) and possess a winding magnetic field line configuration with a prolonged duration up to days when detected by in-situ spacecraft at a heliocentric distance 1 AU from the Sun. They have been extensively studied in the past decades. On the other hand, the relatively small-scale flux ropes, of duration down to a few minutes or less, just begin to gain renewed interest in the heliospheric community. Although previous attempts had been made, a comprehensive study especially by using the most advanced approaches to the most modern spacecraft observations is still lacking. Whether or not these small-scale flux ropes form a continuum population from their large-scale counterparts, i.e., MCs, has yet to be determined from a more extensive event database that did not exist before. Then the fundamental question regarding their properties and origination can be addressed.

Cartwright and Moldwin [1] carried out an earlier and representative study of identifying and characterizing the small-scale flux ropes in the solar wind with duration ≥ 10 minutes, similar to

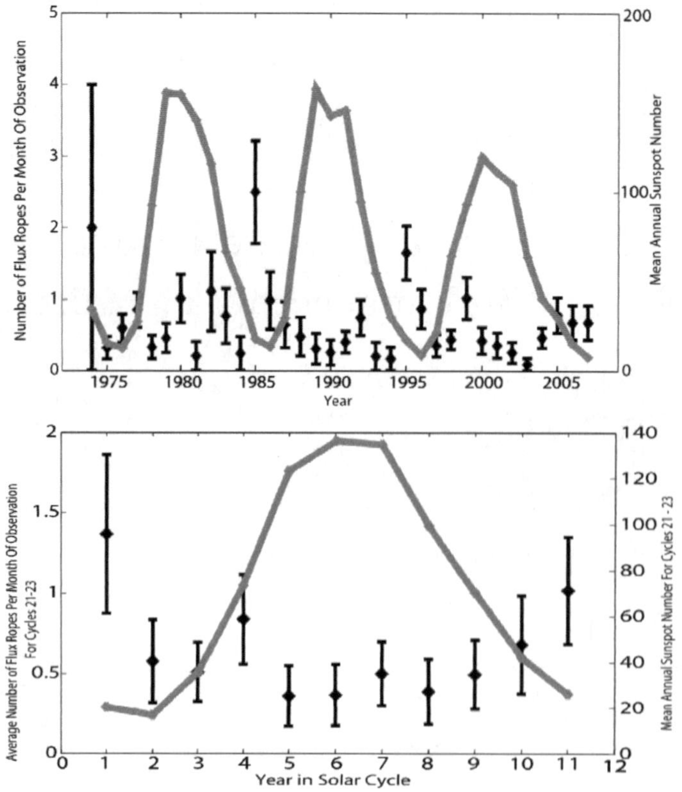

Figure 1. The upper panel is small-scale flux rope occurrence counts per month of observation for 1974 $\sim$ 2006. The lower panel is the monthly occurrence counts averaged over solar cycles 21 to 23 (adapted from [1]).

the range of duration we examine. They surveyed small-scale flux ropes between the heliocentric distances 0.3 and 5.5 AU, from 1974 to 2007. They found that the occurrence rate of small-scale magnetic flux ropes has a negative solar cycle dependency. More events tend to occur during solar minimum rather than solar maximum (see Figure 1; however, be cautioned about the fairly low event counts). We show, however, in Figure 2, based on our database, the event occurrence rate that correlates with the sunspot numbers in solar cycles 23 and 24. Cartwright and Moldwin's database included small-scale flux ropes from 0.3 to 5 AU. Our database only includes the ones at 1 AU. However due to relatively simplistic approach, their event count is only on the order of $\sim$1 per month, which is a common feature of a handful of such similar studies.

Feng et al. [2, 3] investigated the small to intermediate size magnetic flux ropes that had durations mostly beyond the range we examine. They suggested that the small-scale flux ropes are interplanetary manifestations of small CMEs, originating from weak solar eruptions and forming in the solar corona, just like magnetic clouds. Feng et al. [3] found a positive correlation between the occurrence rate of small- and intermediate-scale magnetic flux ropes and the occurrence rate of magnetic clouds from 1995 to 2005. Therefore, they called these small- and intermediate-scale magnetic flux ropes as small magnetic clouds (SMCs). However, the occurrence trend of SMCs shown by Feng et al. [3] (Figure 4 in their paper) is not consistent with the trend of sunspot numbers. As pointed out by Cartwright and Moldwin [1], Feng et al. [3] did not exclude the Alfvénic waves in their database. Another caveat weakening their conclusion is that the total number of events in their database is also very small.

Gopalswamy et al. [4] investigated the correlation between the annual number of magnetic clouds events and the frontside halo CMEs, and compared both with the sunspot numbers. They found that the main trend of magnetic cloud occurrence counts is consistent with the trend of sunspot numbers. It is because that the magnetic clouds originate from the Sun, as is widely accepted. If both magnetic clouds and small-scale flux ropes originate from solar eruptions, they may share the same dependency with solar activity. However, this fact is still not a sufficient condition such that the small-scale flux ropes originate from the Sun. On one hand, there may be two populations of small-scale flux ropes that have different origins and different solar cycle dependencies, i.e., one population has solar cycle dependency but the other does not. When they are mixed, the solar cycle dependency may still appear. On the other hand, even if all small-scale flux rope events have solar cycle dependency as magnetic clouds do, we still cannot conclude that they originate from the Sun, since magnetic clouds have clear solar eruption correspondences but small-scale flux ropes do not. There may be other local plasma dynamic processes far away from the Sun that could generate the small-scale flux ropes and are also modulated by the solar activity cycle.

In addition, all these previous studies used relatively simple approaches, mostly based on single-point time-series analysis. At times, simple and mostly one-dimensional flux rope models were employed to fit the in-situ data [1]. Hu and Sonnerup [5, 6] developed a new approach to determine the flux rope axial orientation base on the Grad-Shafranov (GS) equation, describing quasi-2D magnetohydrostatic equilibrium suitable for examining cylindrical magnetic flux rope configurations based on in-situ spacecraft data [7]. Specifically, the analysis relies on the theoretical requirement that the transverse pressure P_t, the sum of the plasma pressure and the axial magnetic pressure, *versus* the 2D flux function, A, be single-valued and double-folded through a flux rope interval, given a proper determination of a cylindrical geometry (z being the cylindrical axis). An optimal z-axis orientation is found via an optimization procedure to minimize a residue representing the degree of satisfaction for $P_t(A)$ being single-valued. We refer readers to Chen et al. [this volume] and [7] for detailed descriptions of the Grad-Shafranov (GS) equation and the corresponding procedures for the GS-based reconstruction technique. In short, a trial-and-error process was devised to search for the optimal z axis orientation in the whole space. An appropriately spaced search grid is constructed on the upper-half hemisphere of a unit sphere in a spherical coordinate, in which each grid point represents one trial z-axis orientation (represented by the polar angle θ and the azimuthal angle ϕ). All grid points are traversed to find the z-axis orientation with the minimum fitting residue of $P_t(A)$, which is taken as the optimal flux rope axial orientation. In our flux rope detecting algorithm, we extend the usage of Hu and Sonnerup's approach. This approach is not only used for determining the optimal z axis of a flux rope, but also used for checking flux rope candidate. Since in the case of a cylindrical flux rope, an optimal z-axis with a reasonable small fitting residue will surely be found. Conversely, small fitting residue of P_t *versus* A is taken as a main criterion in identifying flux rope candidates, together with additional criteria to be presented in the following sections.

We would like to stress that what distinguishes our analysis from all previous studies is that we have far more number of events, about hundreds per month as opposed to about 1 per month, on average. Therefore, our analysis results will be based on much better statistics with a significantly large number of events.

2. Automated Detection Algorithm Based on the GS Method

In the present study, we detect flux ropes with duration from about 10 minutes to 360 minutes through an automated computer routine based on the GS method. We split this task into multiple iterations. These iterations, corresponding to the ranges of duration of flux ropes to be identified, are: 10-15 minutes, 15-20 minutes, 20-25 minutes, ..., and 355-360 minutes. Each iteration identifies flux rope candidates with the duration falling in the time range specified. For example, when we run the 10-15 minutes iteration, we set a sliding window width to 15 minutes, and the lower limit of flux rope duration to 10 minutes. With this setting, the program checks the data segment in a sliding 15-minute window, and if the double folded part of $P_t(A)$ is less than 10 minutes, the program will discard it. When this iteration is done, the flux ropes with durations longer than 10 minutes and less than 15 minutes will be discovered and recorded. After all the given data is scanned by the 15-minute window, we go to the next iteration to find flux ropes with other lengths. In practice, we extend the boundary of each iteration by 1 minute to make smooth conjunction between two adjacent iterations. Then the iterations become: 9-16 minutes, 14-21 minutes, 19-26 minutes, ..., and 354-361 minutes.

Our automated detection algorithm is based on the fact that the magnetic flux function $A(x, y)$ is a field line invariant, and the transverse pressure P_t is a single-valued function of A, based on the GS equation. As a spacecraft passes through the cross section of a magnetic flux rope with closed transverse field lines, it firstly crosses some transverse magnetic field lines in its first-half path toward the center, then it crosses exactly the same set of transverse field lines, but in reverse order, in its second-half path. Therefore, along the spacecraft path, the measured magnetic flux function A associated with the field lines traversed twice by the spacecraft shows a double-folded pattern, or contains a turn point where an extremum is reached. Since the transverse pressure P_t is a single-valued function of A, the two branches of the data points along the first and second halves of the spacecraft path for P_t *versus* A should coincide as well. Conversely, given a specific time interval of interest, the transverse pressure P_t and the flux function A are calculated from the in-situ spacecraft data. Then we check whether the P_t *versus* A curve has double-folded feature and how good the overlapping is. Later we will define the selection criteria. If the criteria are satisfied, the structure under checking is considered as a flux rope candidate.

In general, the ground rules we play by are: (1) as many as possible flux rope candidates are to be identified by an exhaustive sifting process through the time-series data, and (2) a single flux rope is to be identified for each interval/event.

A fixed width sliding window is used to select data segment in each iteration. The window width defines the maximum duration of the flux rope to be detected during this iteration. We also define a lower limit of the flux rope duration. The flux ropes only with its length between lower and upper limits will be processed during this iteration. Later we are going to run multiple iterations with different window widths to detect flux ropes with different sizes. Specifying the lower limit will avoid the duplication among the windows with different widths. For the time series magnetic field data within a given window, to make the P_t *versus* A curve double folded, the A array must have one and only one inflection point (or turn point), defined as the place where the magnetic field component $B_{y'}$ changes sign (see Equation 1). As discussed above, since the transverse pressure P_t is a single-valued function of A, the only way to make P_t *versus* A curve double folded is that the A array has to fold onto itself so that the P_t *versus* A curve has two branches. If the A array has more than one inflection point, the corresponding P_t *versus* A curve will be multiple folded, which does not meet the ground rule (2) for a single flux rope configuration. When such a situation occurs, the window may contain more than one flux rope structures. We just need to narrow down the window size to make it contain only one single flux rope structure. On the other hand, a narrow window cannot detect the flux ropes with the lengths longer than it. Therefore, to detect flux ropes with different lengths, we have adopted the strategy of running multiple windows with different widths.

The core procedure for each iteration consists of the following two major steps. A frame of reference co-moving with the structures has to be chosen beforehand, usually as the deHoffmann-Teller (HT) frame or the frame with the constant average solar wind velocity. Both can be used robustly in this analysis of fairly uniform solar wind flows, at least within each data segment.

- **Step 1.** As the sliding detection window moves forward, we calculate the A array along the projected spacecraft path (at $y' = 0$) by

$$A(x', 0) = \int_0^{x'} -B_{y'}(\xi, 0) d\xi. \tag{1}$$

Apparently the inflection point corresponds to the point along the spacecraft path where the field component $B_{y'}$ changes sign and a point at which the extreme value in A is reached. If we find the calculated A array within the window is monotonic or has more than one inflection points, we will do nothing but simply move the window forward by one data point. A monotonic A array contains no extrema and can never lead to a double-folded P_t *versus* A curve. An A array with more than one inflection points indicates that the current window may contain multiple flux rope structures. For the former situation, there is no further action needed to be done, and for the latter, a smaller size detection window in another run will take care of it. Once the A array with only one inflection point is discovered, the A array will be split into two branches at the inflection point. Then the two branches will be trimmed to have the same A values at the boundary. Note that the trimming is not according to the number of data points, but is according to the A value, because a flux rope should have the same A value at its boundary. After trimming, the two branches may not have the same length, but must have the same or similar boundary A values. After getting two branches of A values with the same boundary, we can calculate P_t values corresponding to each A value. With both P_t and A values obtained along the spacecraft path, we get the two branches of the P_t *versus* A curve.

- **Step 2.** The next step is to examine how well the two branches of the P_t *versus* A curve overlap. Before doing this, another check process can be performed to reduce the further workload. As a physical nature of a flux rope, the total transverse pressure P_t must reach its maximum in the flux rope center. Reflected in the two branches of the P_t *versus* A curve, the turning point (corresponding to the inflection point in the A array) must be on the top. We need to remove the cases with turning points not on top. Taking into account the measurement error and small fluctuations, we introduce the tolerance. With tolerance, we require the P_t value at the turning point be in top 15% of all P_t values. If the data segment would survive the check after undergoing all the aforementioned procedures, we are ready to obtain two more metrics as defined below to check the double-folding quality,

$$R_{dif} = [\frac{1}{2N}\sum_{i=1}^{N}((P_t)_i^{1st} - (P_t)_i^{2nd})^2]^{\frac{1}{2}}/|\max(P_t) - \min(P_t)|, \tag{2}$$

and

$$R_{fit} = [\frac{1}{L}\sum_{i=1}^{L}(P_t(x_i, 0) - P_t(A(x_i, 0)))^2]^{\frac{1}{2}}/|\max(P_t) - \min(P_t)|. \tag{3}$$

We determined that the conditions $R_{dif} \leq 0.12$ and $R_{fit} \leq 0.14$ could guarantee good flux rope quality while keeping as many candidates as possible.

Equation (2) is modified from Equation (5) in [6], and Equation (3) is taken from [8], subject to adjusted normalization factors. The quantity R_{dif} represents the point-wise difference between two branches, in which both $(P_t)_i^{1st}$ and $(P_t)_i^{2nd}$ are calculated from observational data. Only R_{dif} alone is not sufficient to decide if a segment of data is a good flux rope candidate or not, because a small R_{dif} can only guarantee the good double-folding of the two branches of P_t *versus* A curves, no matter what the shape of the folded curve is. A reliable threshold for R_{dif} is hard to set for acceptable flux rope candidates. To help with this, we obtain an additional fitting residue by using a 3rd order polynomial to fit the data points of P_t *versus* A. This fitting ignores the time sequence of the data points and merges two branches into one. Its fitting residue is defined in Equation (3) and denoted as R_{fit}, where $P_t(x_i, 0)$ is calculated from measured data and $P_t(A(x_i, 0))$ is calculated from the fitting function. In the flux rope searching process, we use only R_{dif} to look for optimal z-axis orientation. For a given data segment, the R_{dif} for each trial z-axis orientation is calculated, then the z-axis orientation with the minimum R_{dif} is taken as the optimal axial orientation. With the determined optimal axial orientation, if both R_{dif} and R_{fit} satisfy our criteria, this data segment will be labeled as a flux rope candidate. The threshold values of R_{dif} and R_{fit} are selected based on examining thousands of data segments with double-folding features in $P_t(A)$, and are given in Table 1.

When a sliding window process is finished for one iteration, we will get a record list of identified flux rope intervals. However, this record list has many overlapped records or intervals. Each group of such overlapping intervals is called a cluster. We clean up such a cluster from the record list in order to obtain uniquely identified flux rope events. We usually pick the interval with the smallest R_{dif} values and discard the others to eliminate overlapping. We have considered possible improvements to avoid the overlapping. One possible method is to move the entire sliding window out of the time range of a flux rope once it is detected. However, we cannot guarantee that the one we picked is the best one in a flux rope cluster. Eventually we decided to slide the detection window continuously to guarantee the detection of the maximum number of flux ropes.

The last check step is the standard Walén test to rule out possible Alfvénic structures or waves [9]. An Alfvén wave or structure may show the similar magnetic field profile in the GSE coordinate system, for instance. The scale size of small flux ropes is comparable with that of Alfvén wave structures. We have to do further test to remove Alfvén waves from our flux rope database. The Walén slope is defined as the slope of the linear regression line between the remaining flow velocity $\mathbf{v}'$ in the co-moving reference frame and the local Alfvén velocity, component-wise. We remove the records whose absolute values of Walén test slope are greater than or equal to 0.3 (indicating significant remaining flows in the reference frame). Up to this point, we have finished all major steps to build the flux rope database. At last, we have the option to apply one more criterion. Because the average magnitude of magnetic field in the ambient solar wind is about 5 nT, we remove the flux rope records whose average magnitude of magnetic field is less than 5 nT.

In summary, Table 1 lists the set of metrics and criteria we use in the identification of small-scale flux rope events from the in-situ Wind solar wind observations. The duration $9 \sim 361$ minutes covers the range of most small scale flux ropes, and this range is easily to be extended to intermediate or large size flux ropes, e.g., $6 \sim 12$ hours, to bridge the gap between small-scale flux ropes and MCs. The condition $\bar{B} \geq 5$ nT excludes small fluctuations in solar wind. The metrics R_{dif} and R_{fit} guarantee the good quality of small-scale flux ropes based on the GS method. And the last but a critical condition, Walén test slope ≤ 0.3, removes the Alfvénic structures or waves.

Table 1. Small-scale Magnetic Flux Rope Detection Metrics and Criteria

Duration	$\bar{B}$	R_{dif}	R_{fit}	Walén test slope
$9 \sim 361$ (minutes)	≥ 5 (nT)	≤ 0.12	≤ 0.14	≤ 0.3

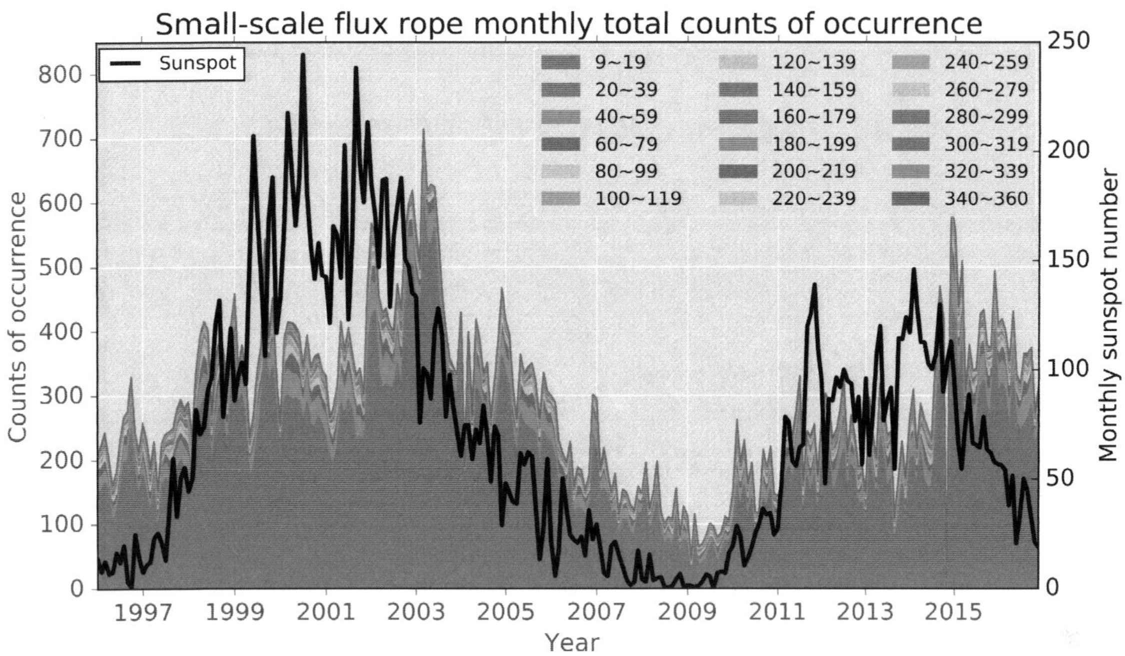

Figure 2. The monthly counts of flux rope events occurrence (left axis) and the monthly sunspot numbers (black curve; right axis) during 1996-2016. Colors represent counts of events with different durations as indicated.

3. Small-scale Magnetic Flux Rope Database from Wind Spacecraft Measurements

We apply the flux rope detection algorithm based on the GS reconstruction technique to the Wind spacecraft measurements during 1996-2016, covering nearly two solar cycles. We successfully detected a large number of small-scale magnetic flux ropes with more general configurations, including non-force-free and non-axisymmetric configurations. There are a total number of 74,241 small-scale magnetic flux ropes detected, with an average number more than 3,500 per year. This database provides sufficient number of samples for researcher to carry out statistical analysis, correlate with other structures, and examine some special cases in detail.

Figure 2 shows the monthly occurrence counts of small-scale flux ropes from 1996 to 2016, covering solar cycles 23 (from May 1996 to January 2008), and 24 (which began in December 2008, reached its maximum in April 2014, and may have ended in early 2017). The different colors represent different durations of small-scale flux ropes (from 9 minutes to 360 minutes), and the thick black line is the corresponding monthly sunspot number. The events of smaller durations generally have greater rates of occurrence. Clearly the total counts including all events of variable durations follow the monthly sunspot numbers, hinting at solar-cycle dependency of these events. Note that the occurrence counts also vary cycle by cycle. From the sunspot numbers we can see that the solar activity in cycle 23 is more intense than that of cycle 24, and accordingly, the overall small-scale flux rope occurrence counts in cycle 23 is larger than those of cycle 24, approximatelly proportional to sunspot numbers. The peaks of occurrence counts tend to appear in the declining phase of each solar cycle.

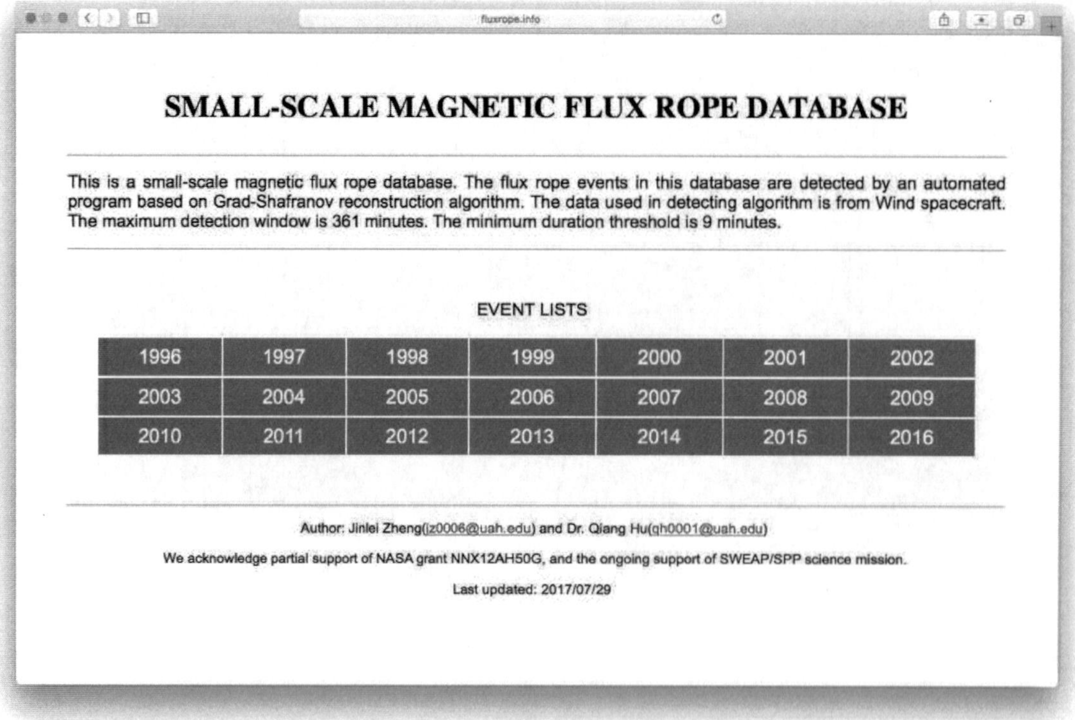

Figure 3. The home page of the small-scale magnetic flux rope database website.

3.1. On-line Database

We have established a website (*https://fluxrope.info*) to host this database online. We make the database open to public and keep it up to date. More flux rope cases with longer duration and at higher latitude locations will be added in near future. Figure 3 is a screenshot of the home page of the small-scale magnetic flux rope database website. When clicking on any year on the "EVENT LISTS" table, the annual event list page will show up, which is presented in Figure 4 for year 1996. The event list page lists every detected flux rope event in one year in chronological order. For each flux rope record in each row, some basic characteristics are listed such as time range in UT, duration in minutes, fitting residue (R_{fit}), average magnetic field strength, maximum magnetic field strength, average plasma β, average proton β_p, average solar wind speed, average proton temperature, and flux rope axial orientation. More information on magnetic field and plasma profiles is stored in database and can be put online in new versions of the website. These quantities make it very convenient for us and other researchers to further apply more selection criteria to pick desirable subset of events for different purposes.

The time range for each record is a clickable hyperlink which will navigate to the detailed flux rope information page for each particular event. Figure 5 shows the upper part of an event page and Figure 6 shows the lower part of the event page. In Figure 5, the upper panel in the first column is the plot of P_t *versus* A, from which one can easily judge the "double-folding" quality of $P_t(A)$, a necessary condition for being a flux rope. The lower panel in the first column is the plot of Walén test with the slope and correlation coefficient shown on the plot. From these two quantities one can know the magnitude of the remaining flows for each velocity component in the frame of reference moving with the flux rope structure. Generally speaking, a spread of

IOP Conf. Series: Journal of Physics: Conf. Series **1100** (2018) 012012 doi:10.1088/1742-6596/1100/1/012012

fluxrope.info/1996/year1996.html

HOME

Small-scale Flux Rope Events in 1996

No.	Time Range	Duration	Residue	<B> (nT)	B_{max} (nT)	$\langle\beta\rangle, \langle\beta_p\rangle$	<Vsw> (km/s)	<Tp> (10^6K)	Orientation
1	1996/01/01 02:01 ~ 1996/01/01 02:56	56	0.1017	5.25	6.04	nan, 1.01	388	0.10	θ=60, φ=340
2	1996/01/01 03:27 ~ 1996/01/01 03:59	33	0.1284	5.47	6.00	nan, 0.65	388	0.09	θ=20, φ=220
3	1996/01/01 05:07 ~ 1996/01/01 05:45	39	0.1127	5.37	6.12	nan, 0.85	387	0.09	θ=50, φ=340
4	1996/01/01 05:47 ~ 1996/01/01 06:56	70	0.1336	5.42	6.25	nan, 0.91	389	0.08	θ=50, φ=280
5	1996/01/01 08:51 ~ 1996/01/01 09:37	47	0.1293	5.07	6.02	nan, 1.42	407	0.08	θ=70, φ=0
6	1996/01/01 09:40 ~ 1996/01/01 09:51	12	0.0812	5.17	5.79	nan, 1.05	409	0.07	θ=10, φ=40
7	1996/01/01 10:34 ~ 1996/01/01 10:53	20	0.1320	6.70	6.94	nan, 0.23	414	0.04	θ=50, φ=80
8	1996/01/01 11:39 ~ 1996/01/01 11:52	14	0.1098	5.60	5.89	nan, 0.58	411	0.06	θ=60, φ=80
9	1996/01/01 11:54 ~ 1996/01/01 12:02	9	0.1040	5.58	6.39	nan, 0.64	409	0.07	θ=70, φ=280
10	1996/01/01 12:09 ~ 1996/01/01 12:35	27	0.0903	5.74	6.16	nan, 0.61	407	0.07	θ=80, φ=240
11	1996/01/02 14:24 ~ 1996/01/02 14:42	19	0.0939	5.90	6.17	nan, 0.49	377	0.07	θ=10, φ=260
12	1996/01/02 17:58 ~ 1996/01/02 18:06	9	0.0564	7.73	9.16	nan, 0.55	386	0.07	θ=80, φ=320
13	1996/01/02 18:15 ~ 1996/01/02 18:23	9	0.1045	6.97	7.74	nan, 0.95	396	0.09	θ=60, φ=280
14	1996/01/02 18:24 ~ 1996/01/02 20:24	121	0.1343	9.49	11.70	nan, 0.37	425	0.09	θ=70, φ=280
15	1996/01/02 20:44 ~ 1996/01/02 20:57	14	0.0699	8.80	9.99	nan, 0.70	480	0.15	θ=50, φ=120
16	1996/01/02 20:59 ~ 1996/01/02 21:07	9	0.0733	5.55	6.50	nan, 3.56	491	0.28	θ=20, φ=320
17	1996/01/02 21:07 ~ 1996/01/02 21:17	11	0.0571	6.03	7.33	nan, 2.65	491	0.26	θ=60, φ=20
18	1996/01/02 21:48 ~ 1996/01/02 22:09	22	0.1045	6.78	7.53	nan, 1.22	459	0.21	θ=80, φ=160
19	1996/01/02 22:25 ~ 1996/01/02 22:33	9	0.0830	6.32	6.74	nan, 1.48	476	0.23	θ=70, φ=40
20	1996/01/03 00:39 ~ 1996/01/03 00:53	15	0.1056	5.72	6.37	nan, 1.53	489	0.19	θ=80, φ=100
21	1996/01/03 02:50 ~ 1996/01/03 02:59	10	0.1296	6.28	6.66	nan, 1.19	466	0.19	θ=60, φ=120
22	1996/01/03 03:27 ~ 1996/01/03 04:41	75	0.1149	6.82	7.99	nan, 0.77	456	0.15	θ=20, φ=100
23	1996/01/03 04:41 ~ 1996/01/03 04:50	10	0.1152	7.67	7.98	nan, 0.34	458	0.10	θ=10, φ=160
24	1996/01/03 04:51 ~ 1996/01/03 05:32	42	0.1253	7.70	8.06	nan, 0.33	460	0.09	θ=30, φ=20

Figure 4. The annual event list page of the small-scale magnetic flux rope database website.

the data points horizontally indicates satisfaction of the detection criterion based on the Walén test. In the second and the third columns, the upper panels are the hodograms of the flux rope magnetic field components in the Minimum Variance Analysis (MVA) principal-axes frame and the lower panels are the hodograms of the flux rope magnetic field components in the flux rope reconstruction frame. The movement of the end points of magnetic field vectors is displayed in these hodograms. Generally a smooth rotation in one or two magnetic field components, typical of a flux rope configuration, may be visualized in these plots. The other panels in Figure 5 and Figure 6 are time series data of magnetic field and plasma parameters, including the pitch angle distribution of suprathermal electrons, the temperature of protons and electrons, and plasma β, when available.

This website provides a large number of small-scale magnetic flux rope events and lots of essential information, which can benefit the relevant studies on small-scale magnetic flux ropes in the solar wind. In the next subsection, we show some selected statistical results obtained from the current small-scale flux rope database based on the Wind spacecraft measurements.

3.2. Selected Statistical Properties

In this subsection, we are going to present some statistical properties of the flux ropes in our database. First of all, after analyzing the solar wind speed distribution from our flux rope database, in the following analysis, we split the entire database into two subsets according to the corresponding average solar wind speed either greater or less than 400 km/s. Note that this value is close to the mode and median of the speed distribution (not shown), 402 km/s and 435

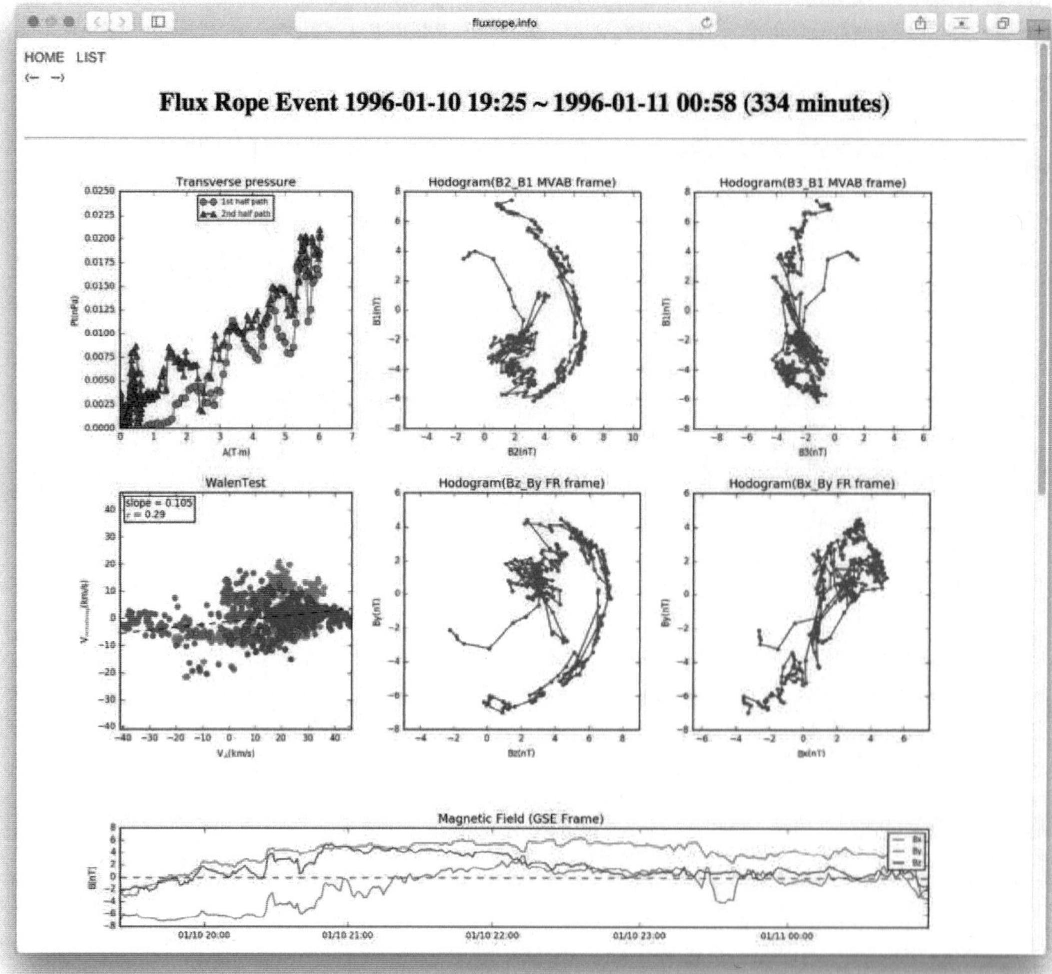

Figure 5. Detailed information for one small-scale magnetic flux rope record: event page part 1.

km/s, respectively.

Figure 7 (a) and (b) show the small-scale flux rope axial orientations in the GSE spherical coordinates based on the last column of the list of even properties. Figure 7 (a) is the polar angle histogram which is binned by 10°, and Figure 7 (b) is the azimuthal angle histogram which is binned by 20°. These two plots indicate that the small-scale flux ropes have preferential axial orientations. From Figure 7 (a), one can see that most small-scale flux ropes have large polar angels, i.e., most of them tend to lie on the ecliptic plane. Figure 7 (b) shows two peaks located at bin 120° $\sim$ 140° and the bin 300° $\sim$ 320°. In fact, these two bins represent two parallel but opposite directions in the GSE XY plane, so they differ by about 180°. Either one of these directions happens to be the tangential direction of the Parker spiral at 1 AU (corresponding to $\phi \approx 135°$ or 315°). This indicates that the projection of the flux rope axis tends to align with the Parker spiral on the ecliptic plane. The red and blue bars represent the events occurring

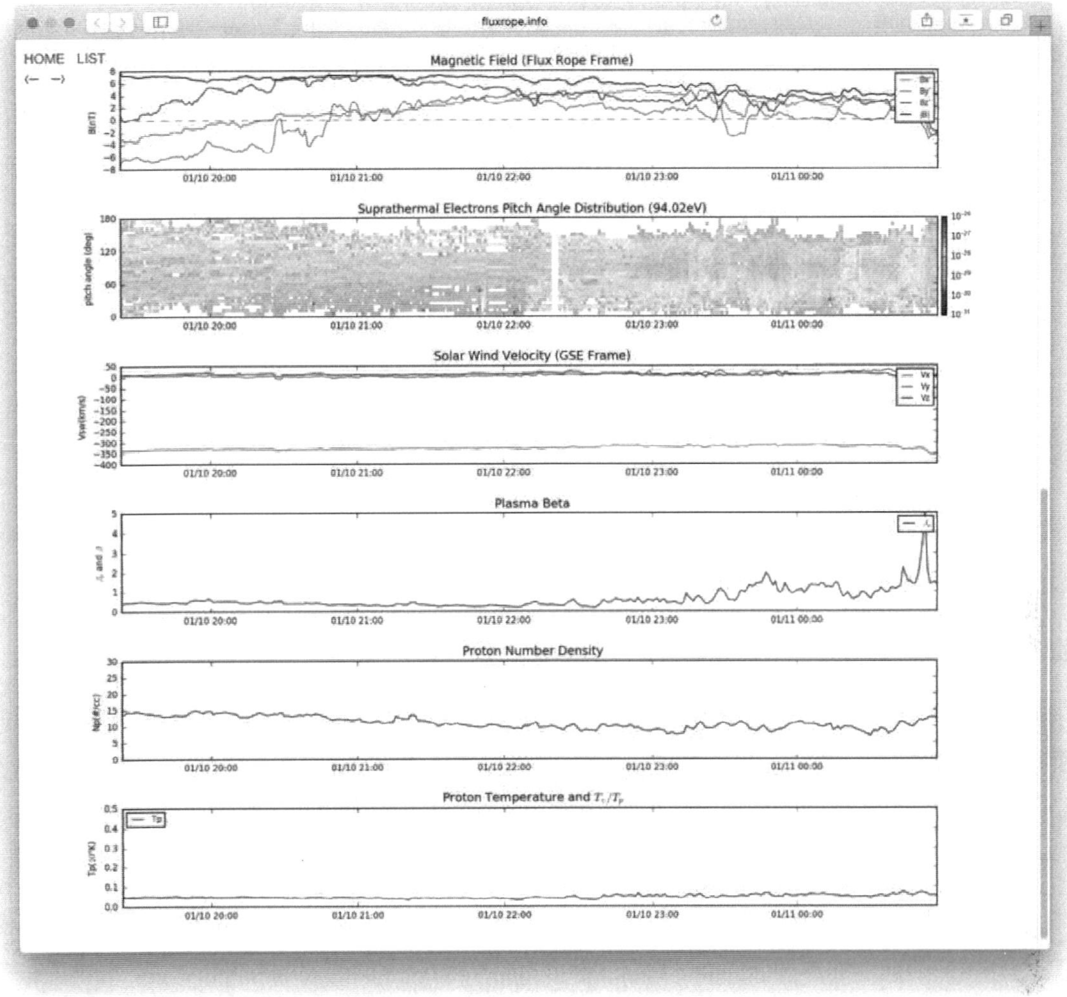

Figure 6. Detailed information for one small-scale magnetic flux rope record: event page part 2.

under different solar wind speed conditions (blue: $\overline{V}_{sw}$ < 400 km/s; red: $\overline{V}_{sw}$ >= 400 km/s). Figure 7 (a) and (b) indicate that the small-scale flux ropes have similar orientation preferences in both fast and slow solar wind on the ecliptic at 1 AU.

Figure 8 (a) and (b) show the durations and scale size distribution of small-scale flux ropes in our database. The data points in black in each plot represent the histogram of the entire event set, and the points in blue or red represent the histogram of the the subsets for slow ($\overline{V}_{sw}$ <400 km/s) and fast ($\overline{V}_{sw}$ >=400 km/s) solar wind speed, respectively. In Figure 8 (a), the shapes of these three curves are close to straight lines on the log-log scale except for the high tails to the right. After examining the generation process of our flux rope database, we find that the high tail is due to cutoff effect of a finite range of window widths. In the present database, the longest duration range is 354∼361 minutes, which means that the flux ropes within this duration range will not be merged into longer flux ropes, since there is no longer durations allowed beyond this

IOP Conf. Series: Journal of Physics: Conf. Series **1100** (2018) 012012 doi:10.1088/1742-6596/1100/1/012012

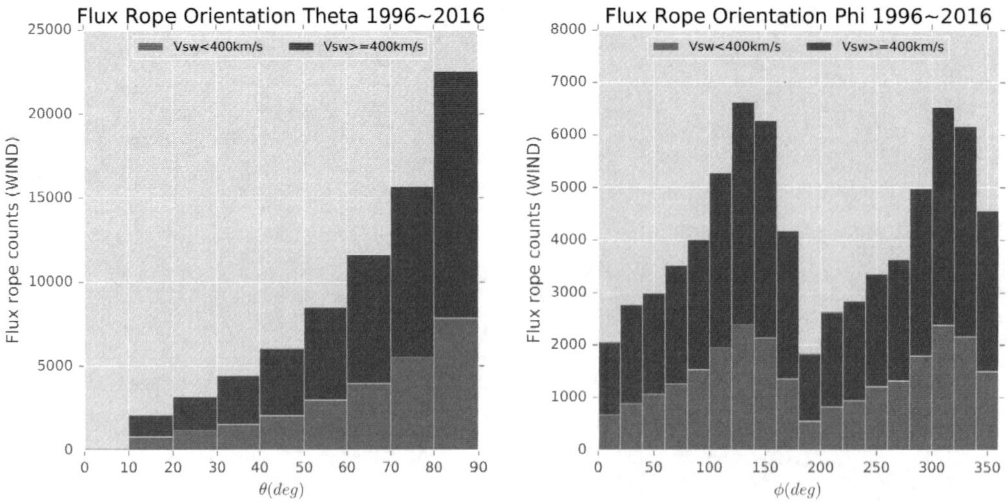

(a) (b)

Figure 7. (a) Flux rope axial orientation: polar angle θ histogram (limited to between 0° and 90°). (b) Flux rope axial orientation: azimuthal angle ϕ histogram. This angle is measured from the positive GSE-X axis toward the projection of the flux rope axis onto the ecliptic plane.

range, thus the cutoff. Actually, the flux ropes in duration ranges shorter than 354~361 minutes may also be enclosed by longer flux ropes with durations beyond 354~361 minutes. This cutoff leads to the enhanced fluctuation and a high tail near the end to the right. To assure this explanation, we manually shifted the cutoff boundary to shorter durations, the high tail also shifted accordingly. In Figure 8 (b), there are low tails at both ends in each curve, which are also caused by the same cutoff effect of finite duration ranges. Due to the boundary cutoff in duration ranges, when convert the duration to scale size, there is a lack of events near the lower and upper scale size boundaries. We exclude these abnormal sections of the data points in the subsequent analyses.

Except for the high tails due to the cutoff effect, Figure 8 (a) shows a linear relation between the flux rope occurrence counts and the duration under logarithmic scales, which indicates a power law distribution of flux rope durations. The color coded straight lines are the corresponding fitted power law functions as denoted. The flux ropes under the slow solar wind condition obey the power law with a power index ~ -1.74, while the ones in fast solar wind obey the power law with power index ~ -2.06. One can see that the red curve ($\overline{V}_{sw}$ >=400 km/s) has larger absolute slope value than the blue curve ($\overline{V}_{sw}$ <400 km/s). The intersection of the blue and red curves is located at about 80 minutes. To the left side of the intersection point, there are more flux rope events in fast solar wind, while to the right, there are more flux rope events with low solar wind speed. This indicates that the longer duration flux ropes (duration>100 minutes) tend to occur under the slow solar wind speed.

Figure 8 (b) also exhibits approximately linear relations (in logarithmic scales) between flux rope scale size and occurrence counts, excluding the low ends due to the cutoff effect. This indicates a power law distribution of flux rope scale sizes, properly calculated taking into account the axial orientations. At a glance, the absolute slope of the red curve ($\overline{V}_{sw}$ >=400 km/s) is greater than that of the blue curve ($\overline{V}_{sw}$ <400 km/s), which is consistent with Figure 8 (a). In other words, the power law slope of the red curve is softer than the blue curve. Comparing with Figure 8 (a), one can see that the flux ropes ranging from 9~361 minutes have the approximately

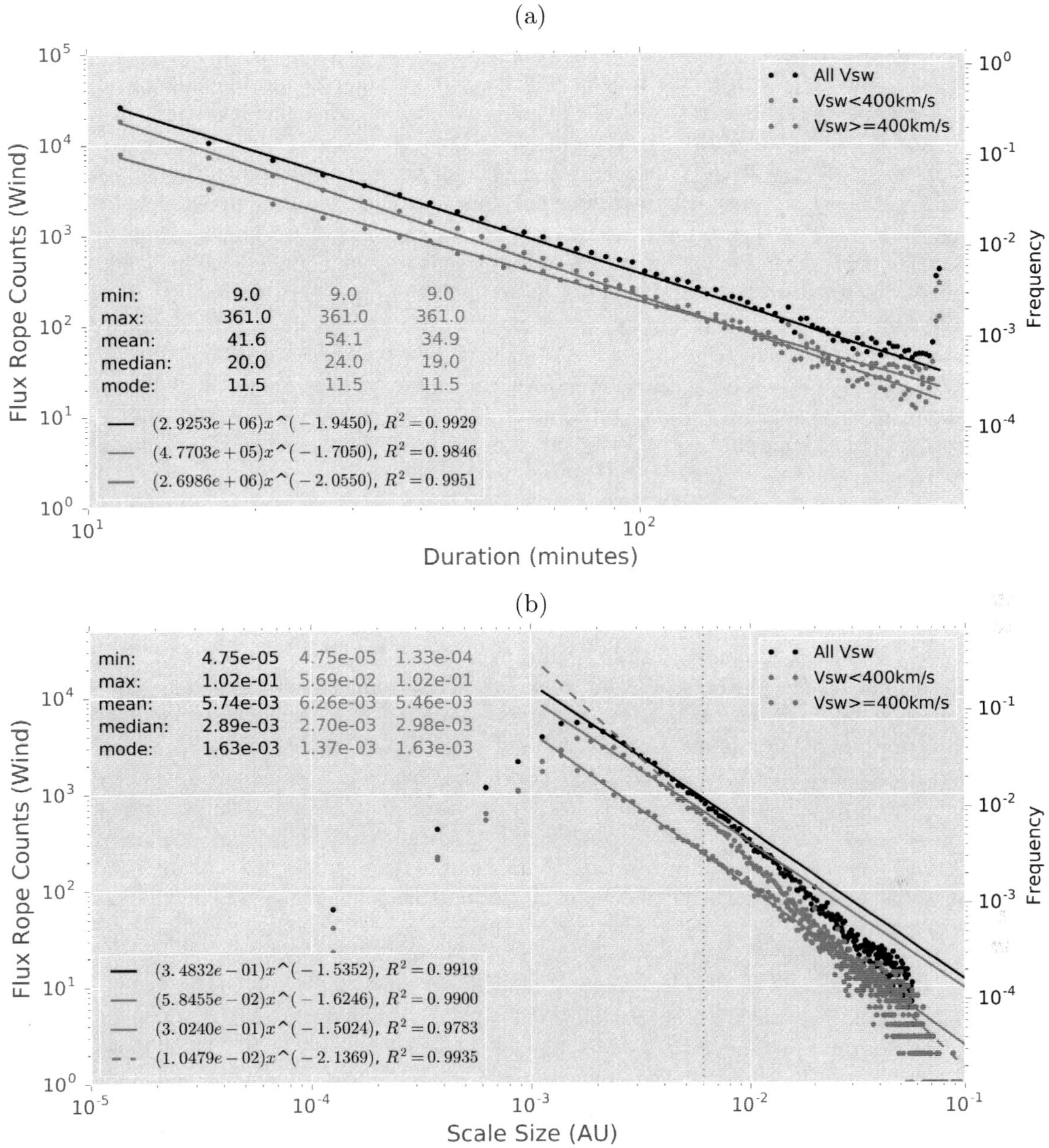

Figure 8. (a) The histogram of small-scale flux rope duration plotted in logarithmic scales. The time range is from 9 minutes to 361 minutes, with 5 minutes bin size. Note that the high tails to the right ends of the curves are due to duration boundary cutoff effect (see explanations in text). (b) The histogram of small-scale flux rope cross section scale sizes plotted in logarithmic scales, with 0.00025 AU bin size. The flux rope scale size is calculated from the flux rope duration taking into account the axial orientation. Again, the low points to the left and right ends of the curves are due to boundary cutoff effect (see explanations in text). For both (a) and (b), the basic statistical moments and linear regression parameters to each curve are listed, respectively.

corresponding scale-size range of 0.002∼0.05 AU (excluding the lower end). When we use power law functions to fit the data, we find that the data points in blue color are fitted very well by a single power-law function with power index ~ -1.64, but for the red and black data points, the tails show noticeable deviations from a single fitted lines. Apparently, the deviations in the black data points are due to the deviations in the red ones. We use two power functions with different power indices to fit the red data points, and find that in the scale size range $0.001 \sim 0.01$ AU, the red data points are well fitted by a power law function with a power index ~ -1.69, while in the range $0.01 \sim 0.05$ AU, the red data points are well fitted by a power law function with a power index ~ -2.27. The breakpoint of the two power laws is at ~ 0.006 AU.

It is interesting to note that both Figure 8 (a) and (b) obey the power law distribution. Because the flux ropes can take any axial orientations, it is not guaranteed that a longer duration always corresponds to a larger scale size. However, in Figure 8 (a) and (b), since both of these two quantities obey the power law, they may have a simple linear relationship. The power law distributions are fairly common in nature. these analysis results are relatively new, concerning small-scale flux ropes. We will speculate on these new results in what follows from the perspective of self-organized criticality theory.

The self-organized criticality (SOC) theory was first proposed by Bak et al. [10], and then applied by Lu et al. [11, 12] to solar physics to explain the power law distributions of flare occurrence rate over flare energy, peak flux, and duration. This model is usually referred to as the avalanche model and has been widely used to explain the statistical characteristics of hard X-ray (HXR) flares [12, 13, 14, 15]. The avalanche model predicts a power law distribution for the total energy, the peak luminosity, and the duration of individual events. Li et al. [16] studied the solar flares and CMEs during the solar cycle 23. They found that the solar flare duration distribution obeys a power law with a power index ~ -2.55. In Figure 8 (a), we also produce a power law, implying that the occurrence of small-scale flux ropes may be explained by SOC theory. Note that the absolute values of the power indices from our fitted functions are generally smaller than those in Li et al.'s [16] result, since solar flares or CMEs and small-scale flux rope probably involve two different kinds of processes. Although they share the similar statistical characteristics in terms of the power law distributions in certain properties, complying with the SOC theory, the underling physical mechanism responsible for generating such behavior still cannot be revealed.

Figure 9 (a) is the histogram of average proton temperature within flux ropes plotted in logarithmic scales. We can see that the peaks of blue and red curves are separated, corresponding to different modes. The peak of the red curve ($\overline{V}_{sw}$ >=400 km/s) is near $\overline{T}_p \approx 0.1$ (10^6 K), while the peak of the blue curve ($\overline{V}_{sw}$ <400 km/s) is near $\overline{T}_p \approx 0.03$ (10^6 K). Therefore, the small-scale flux ropes under low speed solar wind ($\overline{V}_{sw}$ <400 km/s) tend to have low proton temperature, while the ones under medium and high speed solar wind ($\overline{V}_{sw}$ >=400 km/s) tend to have high proton temperature. Note that the black curve and the red curve are overlapping beyond $\overline{T}_p = 0.2$ (10^6K), which means that the small-scale flux ropes with proton temperature greater than 0.2 (10^6 K) occur dominantly under medium and high speed solar wind conditions. Figure 9 (b) is the histogram of average proton β_p (excluding the electron temperature T_e) within flux ropes plotted in log-log scales. When the electron temperature T_e measurements are included in calculating plasma pressure, the plasma β is further increased with the mean and median values from the distributions slightly above 1.

The low proton temperature T_p as well as low β_p is a key characteristic of magnetic clouds. However, for small-scale magnetic flux ropes, the T_p varies case by case and the plasma β can become significant. Figure 10 is the 2-D histogram of flux rope T_p *versus* scale size. The triangle shape distribution stretching down to the right in Figure 10 indicates that the flux ropes with larger scale size (≥ 0.02 AU) usually have lower T_p (≤ 0.1 10^6 K). Given that the large-scale magnetic flux ropes (MCs) have low T_p, this seems to be a smooth transition from the small-scale

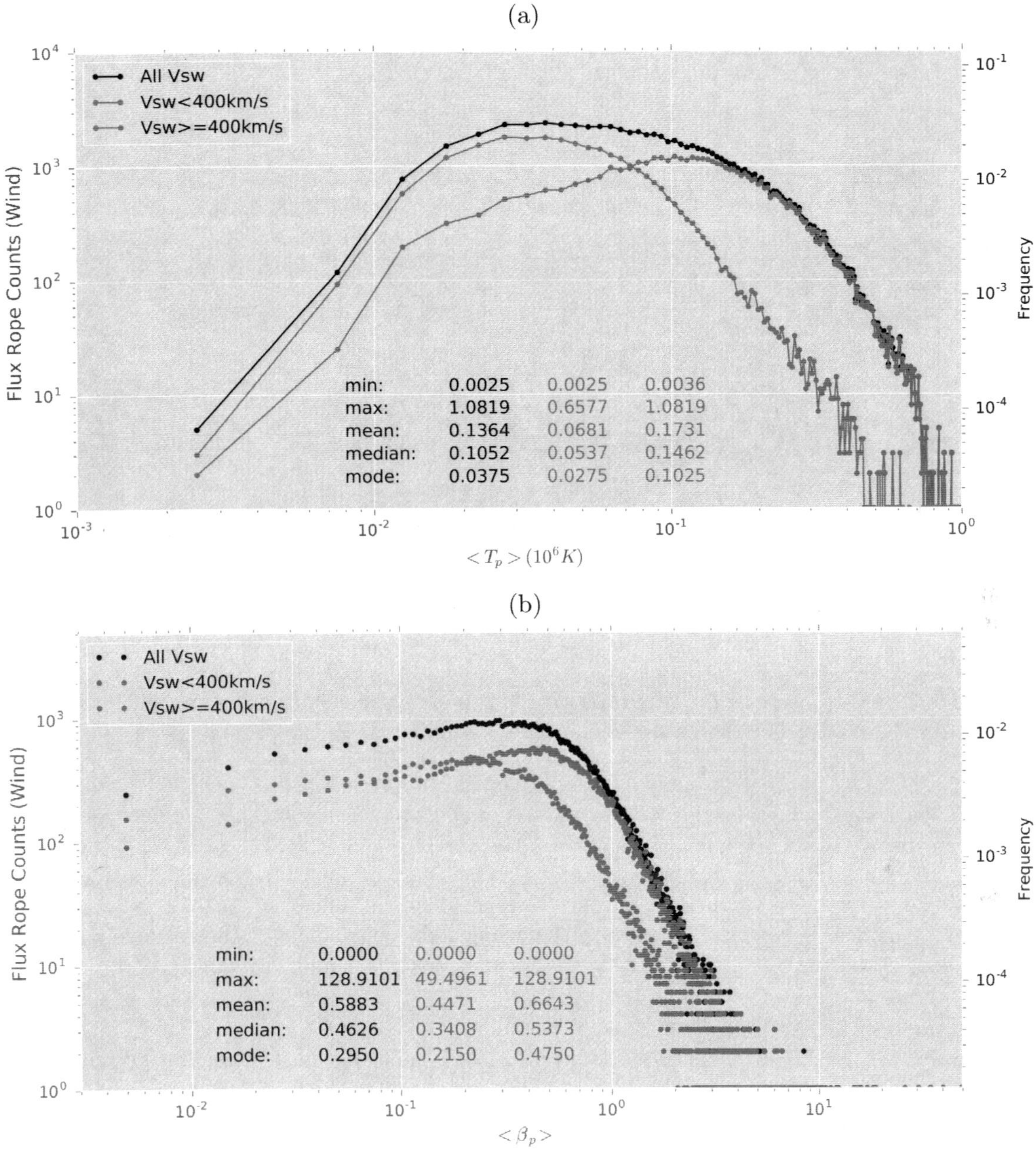

Figure 9. (a) The histogram of average proton temperature within flux ropes in logarithmic scales, with 0.005 (10^6 K) bin size. (b) The histogram of average proton β_p within flux ropes in logarithmic scales, with 0.01 bin size. For both (a) and (b), the blue curve represents the flux rope events under solar wind speed $\overline{V}_{sw}$ <400 km/s, and the red curve with solar wind speed $\overline{V}_{sw}$ >=400 km/s. The black curve represents the entire data set.

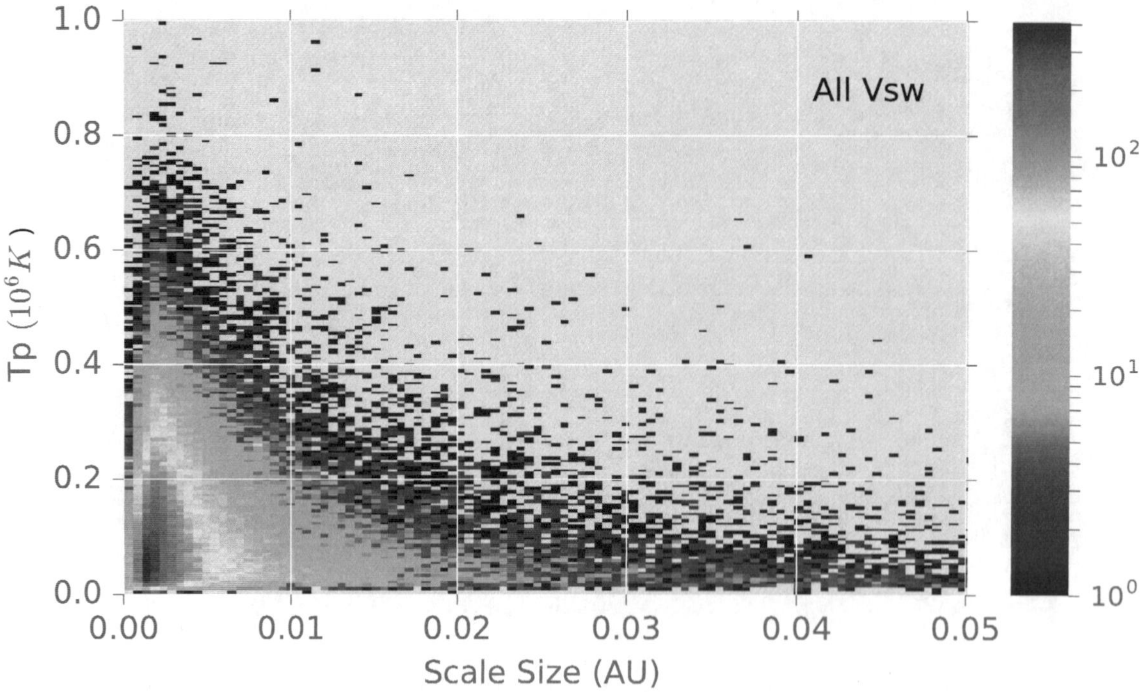

Figure 10. The 2-D histogram of flux rope proton temperature *versus* scale size. The bin grids are 200×200. The color bar represents the small-scale flux rope event counts.

magnetic flux ropes to their larger counterparts. For the smaller size small-scale flux ropes, the range of T_p spreads widely. Figure 10 shows that the range of T_p of the flux ropes with scale sizes less than 0.05 AU spreads from near 0 to 0.8 (10^6 K). When looking into the distribution under different solar wind speed conditions, we find that most of the relatively larger size small-scale flux ropes ($\geq$ 0.02 AU) appear in the slow solar wind with low T_p. As for the relatively smaller size small-scale flux ropes ($\leq$ 0.01 AU), they appear in both fast and slow solar wind.

4. Summary and Discussion

In summary, we have built a small-scale magnetic flux rope database from the Wind in-situ spacecraft measurements in the solar wind, using an automated algorithm based on the GS reconstruction method. We have found, on average, about 3,500 events per year, for the past twenty years. The event duration is between 9 and 361 minutes, and the scale sizes are on the order of $\sim 10^{-3}$ - 10^{-2} AU. Preliminary statistical analysis yields the following results.

(1) The occurrence of small-scale flux ropes has strong solar cycle dependency, which has the same trend as the occurrence of their large-scale counterparts, the magnetic clouds. (2) The small-scale magnetic flux ropes in the ecliptic plane tend to align along the Parker spiral, indicating that they belong to the general population of "flux tubes" [17]. (3) In low speed ($<$ 400 km/s) solar wind, the flux ropes tend to have lower proton temperature and higher proton number density, while in high speed ($>$ 400 km/s) solar wind, they tend to have higher proton temperature and lower proton number density. (4) Both the duration and scale size distributions of the small-scale magnetic flux ropes obey power laws.

We reported in [18] that the wall-to-wall time distribution obeys double power laws with the break point at 60 minutes (corresponding to the correlation length in solar wind turbulence), which is consistent with the waiting time distributions of magnetohydrodynamic (MHD) turbulence simulations and the related observations. We also performed case studies on the small-scale magnetic flux ropes downstream of interplanetary shocks, and found that the peaks of enhanced ions flux correspond to the merging edges of two adjacent flux ropes, indicating that the merging flux ropes are able to energize particles with appropriate energy bands [19].

Table 2. Observational evidence in support of two competing views on the origin of small-scale flux ropes.

Solar Origin	Local Origin
Flux rope configuration similar to MCs	—
—	Properties similar to 2D MHD turbulence [18]
—	Plasma properties different from MCs

Table 2 lists, tentatively, the observational evidence obtained from our analysis, in support of the two competing views on the origin of small-scale flux ropes. Each column contains the results that have been considered well established and exclusive for the corresponding origination mechanism. This is nowhere near to be complete and we expect the table to be updated when further analyses are to be carried out, especially when the new discoveries are expected to be returned by the upcoming Parker Solar Probe and Solar Orbiter missions. Additionally, based on the present analysis results, the event occurrence rate, the axial orientations, and the power-law distributions in duration and scale sizes seem to be arguably supporting both hypotheses, at least not in direct contradiction to either view. For example, Borovsky [17] analyzed the orientations of 65,860 flux tubes from 1998 to 2004 identified by searching for their boundaries, usually corresponding to discontinuities from the ACE spacecraft in-situ measurements. He found that the axial orientations are mostly aligned with the nominal Parker spiral direction at 1 AU. He therefore advocates the view that these flux tubes are distinct and still rooted on the Sun. They form the "spaghetti-like" structure of the solar wind, connecting back to the source [20]. On the other hand, such axial orientations do not contradict the view that these structures are generated from solar wind turbulence [21, 18] which has a prominent 2D component and the associated (perpendicular) guide field along the Parker spiral direction.

The on-going effort is to extend the range of flux rope duration and scale sizes to larger values including the automated identification of intermediate and large size flux ropes, i.e., interplanetary CMEs. The goal is to find out whether there exists a continuous distribution in the flux rope sizes that can extend into the regime of well-identified ICMEs which have an apparent solar origin. If there were two distinct populations of different origin, then there likely exists an overlapping range for flux rope duration of a few hours. This will need to be examined with additional observations such as elemental compositions and charge states of the plasma.

Acknowledgments

We are grateful to our colleagues at SPA/CSPAR, UAH, Drs. Laxman Adhikari, Jakobus le Roux, Gang Li, Gary Webb, Gary Zank, and Lingling Zhao for on-going collaborations. We also thank Dr. Olga Khabarova for stimulating discussions. We acknowledge NASA grants NNX12AH50G, NNX14AF41G, NNX17AB85G, subawards NRL N00173-14-1-G006 and SAO SV4-84017, and NSF grant AGS-1650854 for support. Special thanks also go to the SCOSTEP/VarSITI program for support of the development and maintenance of the on-line small-scale flux rope database.

References

[1] Cartwright M L and Moldwin M B 2010 *Journal of Geophysical Research (Space Physics)* **115** A08102 ISSN 2156-2202 a08102 URL http://dx.doi.org/10.1029/2009JA014271
[2] Feng H Q, Wu D J and Chao J K 2007 *Journal of Geophysical Research: Space Physics* **112** ISSN 2156-2202 a02102 URL http://dx.doi.org/10.1029/2006JA011962
[3] Feng H Q, Wu D J, Lin C C, Chao J K, Lee L C and Lyu L H 2008 *Journal of Geophysical Research: Space Physics* **113** A12105 ISSN 2156-2202 a12105 URL http://dx.doi.org/10.1029/2008JA013103
[4] Gopalswamy N, Yashiro S, Xie H, Akiyama S and Mkel P 2015 *Journal of Geophysical Research: Space Physics* **120** 9221–9245 ISSN 2169-9402 2015JA021446 URL http://dx.doi.org/10.1002/2015JA021446
[5] Hu Q and Sonnerup B U Ö 2001 *Geophysical Research Letters* **28** 467–470
[6] Hu Q and Sonnerup B U O 2002 *Journal of Geophysical Research: Space Physics* **107** SSH 10–1–SSH 10–15 ISSN 2156-2202 URL http://dx.doi.org/10.1029/2001JA000293
[7] Hu Q 2017 *Sci. China Earth Sciences* **60** 1466–1494
[8] Hu Q, Smith C W, Ness N F and Skoug R M 2004 *Journal of Geophysical Research: Space Physics* **109** ISSN 2156-2202 a03102 URL http://dx.doi.org/10.1029/2003JA010101
[9] Paschmann G and Sonnerup B U O 2008 *ISSI Scientific Reports Series* **8** 65–74
[10] Bak P, Tang C and Wiesenfeld K 1987 *Phys. Rev. Lett.* **59**(4) 381–384 URL https://link.aps.org/doi/10.1103/PhysRevLett.59.381
[11] Lu E T and Hamilton R J 1991 *The Astrophysical Journal Letters* **380** L89–L92
[12] Lu E, Hamilton R, Mctiernan J and Bromund K 1993 *Astrophysical Journal* **412** 841–852 ISSN 0004-637X
[13] Vlahos L, Georgoulis M, Kluiving R and Paschos P 1995 *Astronomy and Astrophysics* **299** 897
[14] Georgoulis M K and Vlahos L 1998 *Astronomy and Astrophysics* **336** 721–734
[15] MacKinnon A L and MacPherson K P 1997 *Astronomy and Astrophysics* **326** 1228–1234
[16] Li C, Yang J, Li C, Tian Y, Yang Y and Huang Q 2016 *Scientia Sinica Physica, Mechanica and Astronomica* **46** 29501 (pages 0) URL http://phys.scichina.com:8083/sciG/EN/abstract/article_519739.shtml
[17] Borovsky J E 2008 *Journal of Geophysical Research (Space Physics)* **113** A08110
[18] Zheng J and Hu Q 2018 *Astrophys. J. Lett.* **852** L23 (*Preprint* 1801.01771)
[19] Zheng J, Hu Q, Chen Y and le Roux J 2017 *Journal of Physics Conference Series* (*Journal of Physics Conference Series* vol 900) p 012024
[20] Borovsky J E 2010 *Journal of Geophysical Research: Space Physics* **115** ISSN 2156-2202 a09101 URL http://dx.doi.org/10.1029/2009JA015040
[21] Zank G P, Adhikari L, Hunana P, Shiota D, Bruno R and Telloni D 2017 *The Astrophysical Journal* **835** 147 URL http://stacks.iop.org/0004-637X/835/i=2/a=147

IOP Conf. Series: Journal of Physics: Conf. Series **1100** (2018) 012013 doi:10.1088/1742-6596/1100/1/012013

Voyager Evidence for Large-Scale Structure in Plasma and Energetic Particle Behavior Beyond the Termination Shock

D S Intriligator[1], W D Miller[1], J Intriligator[1,2], and W Webber[3]
[1]Carmel Research Center, Inc., Santa Monica, CA
[2]Tufts University, Medford, MA
[3]New Mexico State University, Las Cruces, NM

E-mail: devriei@aol.com

Abstract. Voyager 1 crossed the termination shock at the end of 2004, more than a full solar cycle ago. Similarly, Voyager 2 crossed it and entered the heliosheath at the end of August, 2007, which was before the beginning of SC 24. Voyager 1 crossed the heliopause in August, 2012, when solar activity was high. Thus, at each spacecraft changing readings over time in the heliosheath could be due either to the increasing solar distance or to changing solar activity. Comparing energetic particle observations by the two spacecraft at corresponding distances shows cases for which distance is more important. In particular, at both spacecraft the spectral index of protons in the anomalous cosmic ray energy range shows corresponding variations with distance, despite the differing times and the great separation between the spacecraft, and so do ratios of electron count rates in differential energy bands. These comparisons provide insight into the expectation that Voyager 2 will soon cross the heliopause. Consideration of the large-scale spatial structure of moving flare-generated disturbances also provides insight into particle behaviour associated with the plasma wave events observed by Voyager 1 since it crossed the heliopause. We discuss a variety of observations of these phenomena from the heliosheath and the local interstellar medium and suggest a possibility for future simulations.

1. Introduction

As Voyagers 1 and 2 (V1 and V2) continue ever farther from the Sun, their measurements of space environment conditions not observable from the Earth pose continuing challenges of coherently understanding them. The heliographic inertial latitude of the trajectory of V1 is about 34 degrees north, while the latitude of the trajectory of V2 is about 30 degrees south, and they are also separated by nearly 45 degrees in longitude. Thus, the distance between the two spacecraft is of the same order of magnitude as their respective distances from the Sun, although at a solar distance of nearly 142 AU, V1 is now significantly farther from the Sun than V2 is at more than 117 AU.

For a number of years, we analyzed understanding various physical phenomena associated with the Voyager data. We followed the Halloween 2003 events from the Sun to Earth, Ulysses, Cassini, and the Voyagers [1-3]. We also followed a number of other specific events from the Sun to the Voyagers e.g., July 2012 [4]. Without the Ulysses data these events are more difficult to follow across the heliosphere. V1 has been in the local interstellar medium (LISM) for more than 5 years, but the V1 plasma instrument (PLS) has not worked since 1980. Since then the only available estimates of the plasma environment at V1 have come from when they were inferred indirectly using data from the Low Energy Charged Particle (LECP) instrument (Decker [5]). Recently we have used the V1 Cosmic Ray System (CRS) (Stone [6]) for similar inferences. However, the V2 PLS is still working well enough that it is expected to be able to make direct observations of the LISM plasma after the heliopause (HP) is crossed, so there is great interest in when this will happen. V1 crossed the HP in August, 2012 (Webber and McDonald; Stone, et al. [7]), near 121.7 AU from the Sun. If the HP near V2's trajectory is at the same distance, then the HP crossing is still more than a year away. Webber and Intriligator [8] implied that V2 may see the HP in the South at a radial distance of ~10 AU closer than that observed at V1. However, assuming that the LISM particle environment near V2's trajectory is similar to that near V1's trajectory implies that comparing the histories of particle data in the heliosheath on V1 and V2 may provide insights into the approach to the HP. Evidently, these comparisons also provide insight into conditions in the heliosheath. On the other hand, data from V1

IOP Conf. Series: Journal of Physics: Conf. Series **1100** (2018) 012013 doi:10.1088/1742-6596/1100/1/012013

since the HP crossing show that the behavior of particles in the LISM is more complex than previously expected, because the strongest interplanetary disturbances generated by solar flares propagate all the way through the heliosheath and out into the LISM, where they generate plasma wave events as well as perturbing particle propagation. This complex behavior needs to be studied further.

Thus, the results we present here go substantially beyond our recent studies [9,10] connecting major flares of SC 24 with events in the heliosheath and LISM. We describe evidence that the electron spectrum observed by the V1 CRS changed very little at the HP, despite the relatively large change in count rates, and that the electron spectrum observed by the V2 CRS is now closely similar to the V1 spectrum near and beyond the HP, implying that the crossing may be near. We also compare these data with CRS proton data in the anomalous cosmic ray (ACR) energy range and discuss how these analyses might be affected by particle effects associated with plasma wave events generated by interplanetary disturbances generated by solar flares as they propagate out into the LISM.

2. Differential Electron Energy Bands

As described in more detail at the Voyager CRS website [11], the telescopes of the CRS consist of stacks of solid-state detectors, mounted so that particles can enter either end of each stack, with a guard detector surrounding each stack and coincidence logic incorporated into the data system. This logic identifies simultaneous energy deposition events in various combinations of detectors and thereby distinguishes counts caused by electrons from counts caused by positive ions. Among positive ions, it discriminates between protons, alpha particles, and heavier nuclei, as long as the guard detector is not triggered. Triggering this detector indicates that a particle came through the side of the telescope instead of from the desired field of view; thus readings from other detectors would be an incomplete or mistaken characterization of the particle, and are therefore rejected. The data system also reports for each accepted event how much energy was deposited in each detector, but the rates page of the website [11] only lists count rates for various types of events, including not only coincidence events but also count rates in individual detectors, without including energy information.

The CRS has two high-energy telescopes, designated HET-I and HET-II, which are identical except for being mounted with their fields of view in different directions. These telescopes have detector coincidences that count both ions and electrons, and in particular there are coincidence combinations that identify electrons that stop in detector C4 of the HET: electrons with higher energies that penetrate C4 but stop in C3; and electrons with still higher energies that penetrate both C4 and C3, but stop in C2. Figures 1(a), 1(b), and 1(c) show the time series of the count rates for the respective detectors of HET-I on V1 from the beginning of 2004 to the end of 2017. (The corresponding plots for HET-II are so nearly identical that they are not shown.)

Although these histories look similar during most of this period of 14 years, with the HP crossing emphatically evident in all of them, the C4 plot in Figure 1(a) shows a small peak around the time of the TS crossing at the end of 2004 that has little if any counterpart in Figures 1(b) and 1(c). Thus, for a clearer comparison the C4/C3 and C3/C2 ratios were calculated and are shown in Figures 1(d) and 1(e). Figure 1(d) evidently shows not only a prominent brief peak in the C4/C3 ratio around the TS crossing but also a longer subsequent period of ratios (around 8) that then gradually decline to around 4 (indicating that this portion of the spectrum became harder) while Figure 1(e) shows very little evidence of a peak near the TS crossing, but a gradual rise in the C3/C2 ratio from around 2.0, with a great amount of scatter, to between 2.5 and 3.0 with less scatter. However, despite the sharp rises in the individual count rates at the HP crossing, neither ratio plot shows a visible change then, suggesting that the electrons observed in these energy bands in the heliosheath as the spacecraft was approaching the HP may have diffused inward from the LISM.

Thus, Figures 2(a) through 2(e) show the corresponding time series for HET-I of the V2 CRS, which is the only high-energy telescope on V2 on which these coincidence counters still function. Each part of Figure 2 is plotted on the same scale as the corresponding part of Figure 1 to aid comparisons.

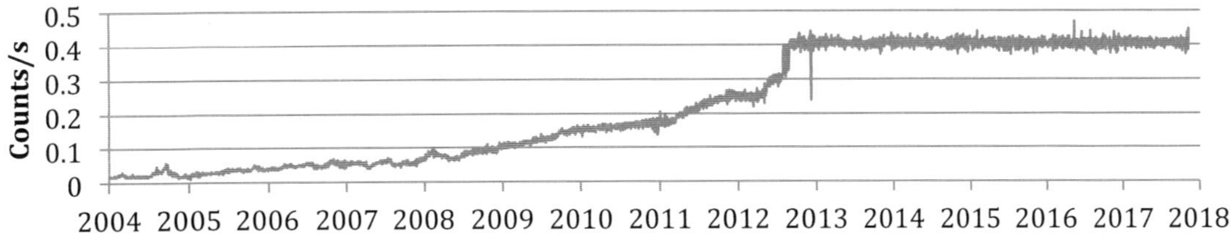

Figure 1(a). Count rate of electrons stopping in detector C4 of HET-I of the V1 CRS.

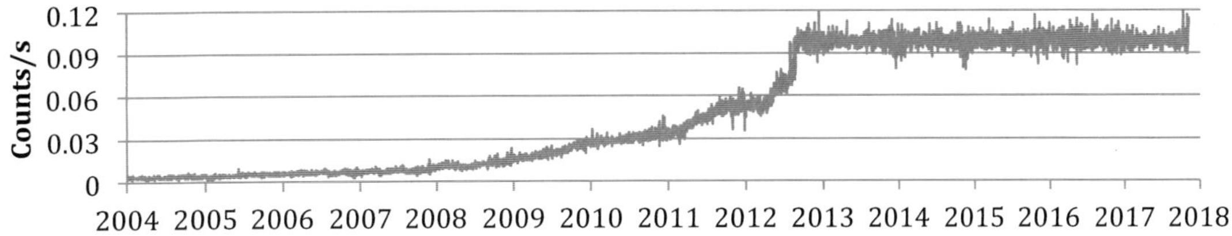

Figure 1(b). Count rate of electrons stopping in detector C3 of HET-I of the V1 CRS.

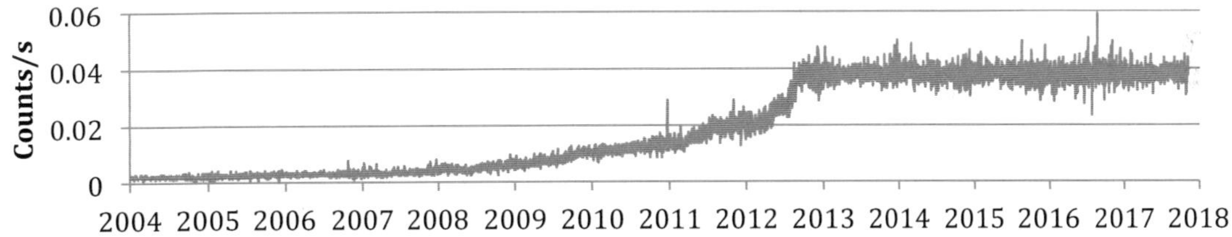

Figure 1(c). Count rate of electrons stopping in detector C2 of HET-I of the V1 CRS.

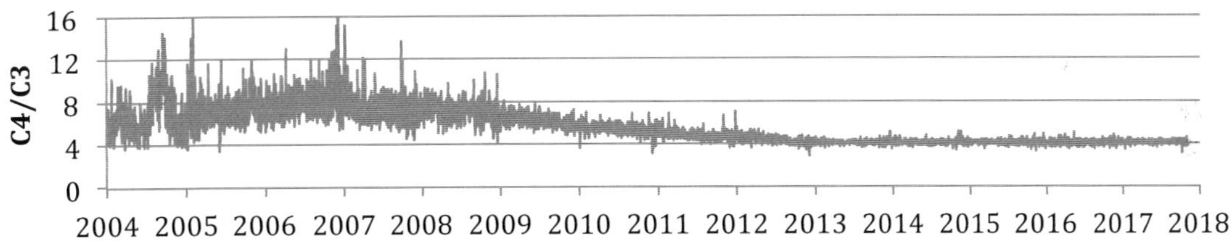

Figure 1(d). C4/C3 count rate ratio for HET-I of the V1 CRS.

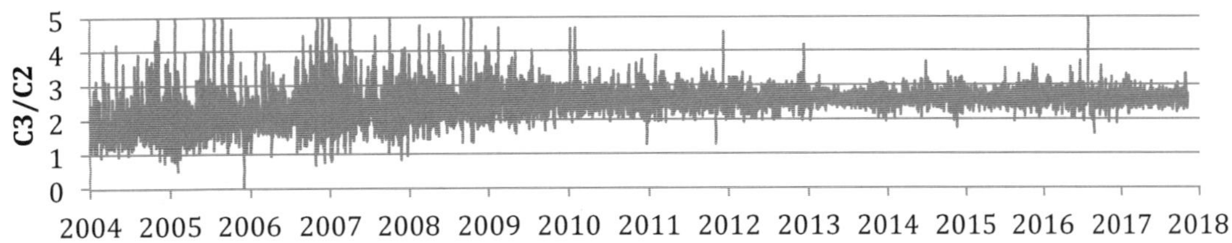

Figure 1(e). C3/C2 count rate ratio for HET-I of the V1 CRS.

Comparing Figure 1 with Figure 2 shows that as of the most recent data in late 2017 the V2 count rates in Figures 2(a), 2(b), and 2(c) were approximately the same as the corresponding V1 count rates in late 2011 in Figures 1(a), 1(b), and 1(c). Furthermore, the V2 ratios in Figures 2(d) and 3(e) show the same trends as the V1 ratios in Figures 1(d) and 1(e) (C4/C3 downward toward 4 and C3/C2 upward toward 2.5) and by late 2017 were approximately the same as the corresponding V1 ratios were in late 2011. This may be more significant than the correspondence of the count rates, because using the ratios compensates for the possibility that the count rates at the HP may be different for V2.

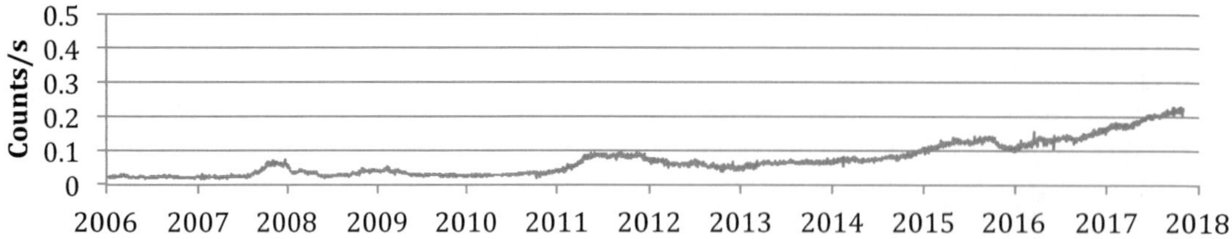

Figure 2(a). Count rate of electrons stopping in detector C4 of HET-I of the V2 CRS.

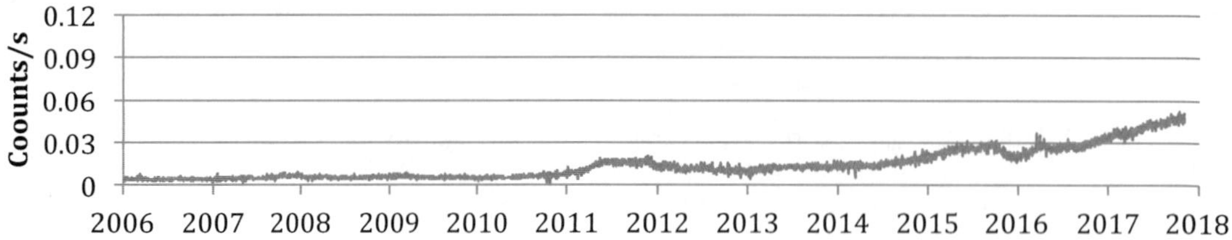

Figure 2(b). Count rate of electrons stopping in detector C3 of HET-I of the V2 CRS.

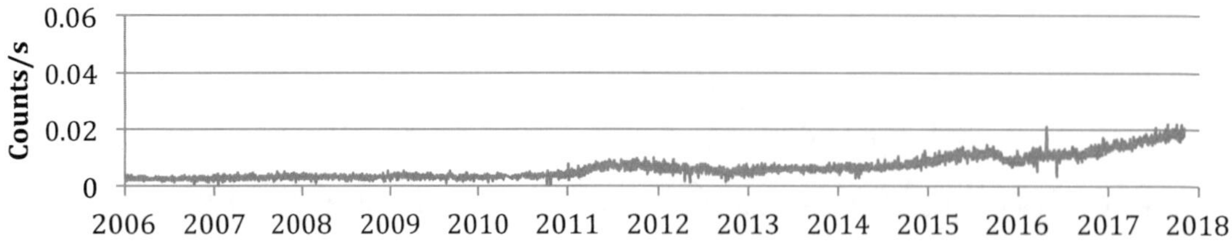

Figure 2(c). Count rate of electrons stopping in detector C2 of HET-I of the V2 CRS.

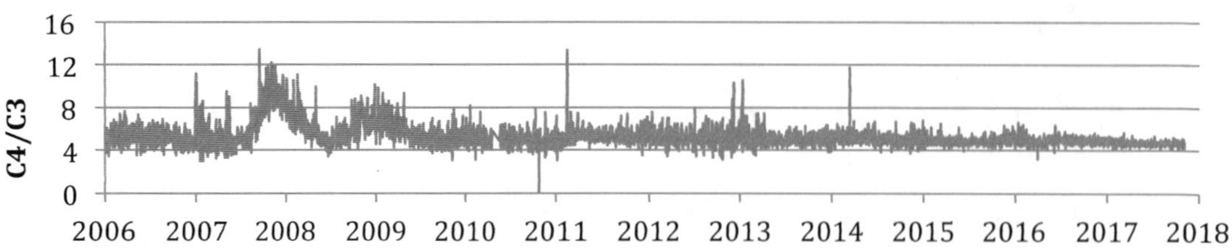

Figure 2(d). C4/C3 count rate ratio for HET-I of the V2 CRS.

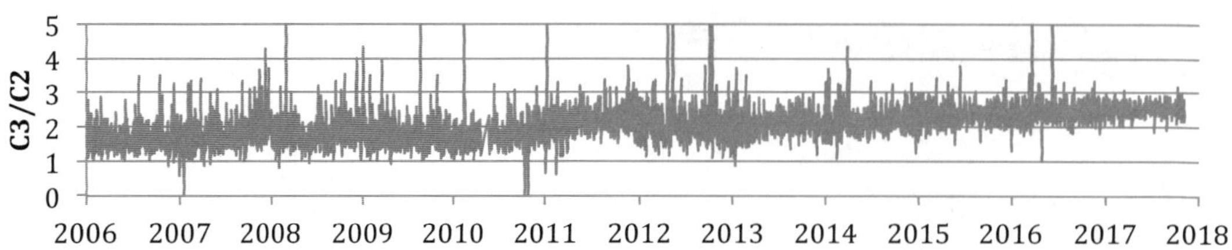

Figure 2(e). C3/C2 count rate ratio for HET-I of the V2 CRS.

This could happen if, for example, Opher, et al. [12] are correct that beyond the TS the fluctuations in the heliospheric current sheet (HCS) are compressed by the deceleration of the solar wind until portions of the HCS reconnect to form a disordered structure of magnetic islands and bubbles. This reconnection would accelerate anomalous cosmic rays and energetic electrons and the magnetically disordered plasma would interact with the LISM to produce a north-south asymmetry between the electron intensities at V1 and V2, with energetic electrons being locally trapped in the islands and bubbles. They performed a high-resolution 3-D MHD simulation predicting that the LISM field is stronger on the south side of the nose of the heliosphere and would therefore have higher pressure. This would cause most of the islands and bubbles to be carried to the northern hemisphere. Then the TS in the south would be closer to the Sun than in the north, consistent with the Voyager TS crossings at 94 AU in the north and 84 AU in the south [8]. Having fewer magnetic islands and bubbles in the south would allow easier diffusion of energetic electrons in this region, so that when V2 reaches the HP it may not see the steep rise in electron count rates observed at V1, as shown in Figures 1(a) - 1(c). Another possibility is that the LISM ahead of the heliosphere is like the simulations of Pogorelov, et al. and of Borovikov and Pogorelov [13], who predict a lumpy and rippled HP due to the Rayleigh-Taylor and Kelvin-Helmholtz instabilities, resulting in irregular compression of the LISM near the HP.

However, it remains to be seen whether the results of the simulation in [12] are consistent with the work in [14]. As we noted last year, it combined Voyager LECP observations and energetic neutral atom (ENA) data from Cassini to estimate the thickness of the heliosheath in the direction of V2. This work updated estimates originally made for V1 and V2 in 2010 in [15], in which the thickness for V1 was estimated at around 30 AU and for V2 was estimated as at least 40 AU. As the actual thickness observed by V1 was about 27.7 AU and V2 has now travelled nearly 34 AU through the heliosheath without reaching the HP (though it may be close), the ENA estimates appear to have been sufficiently reliable thus far that they may deserve additional comparisons with estimates made by other methods.

3. Proton Spectral Indices in the ACR Energy Range in the Heliosheath

As the ratio plots in Figures 1 and 2 imply that electron energy spectra in the heliosheath change with distance, this naturally raises the question of corresponding changes in positive ion energy spectra. The V1 and V2 pages of the COHO website include 6-hour and daily averages of CRS proton fluxes in 8 adjacent energy bands from 3.4 to 56.0 MeV. Example plots of the logarithms of these fluxes versus the logarithms of the energies of the midpoints of the energy bands showed approximately straight lines with substantial changes of slope over time in the heliosheath for both spacecraft, with the spectra soon after the TS crossing being substantially softer than those of particles observed later.

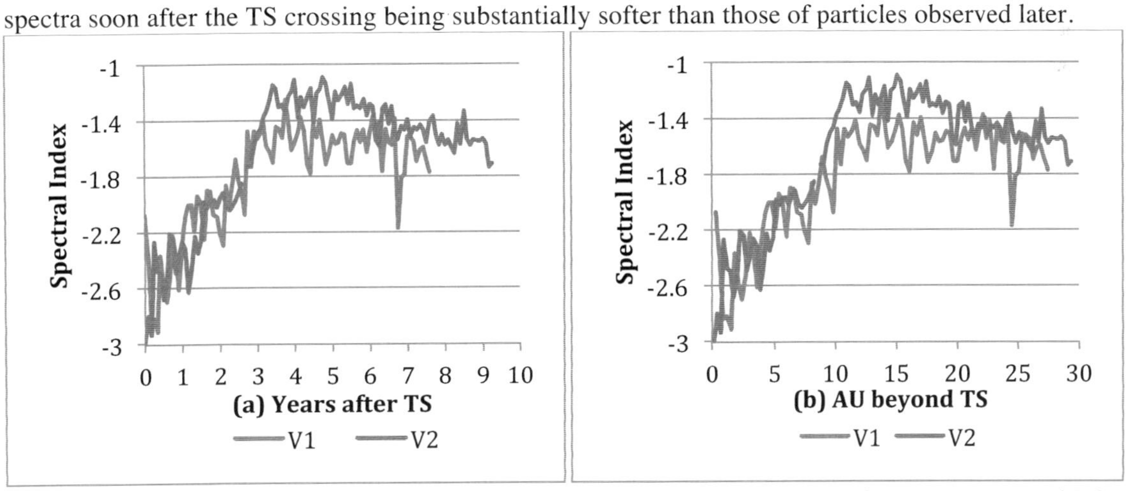

Figure 3. Spectral indices of monthly samples of CRS protons from 3.4 to 56.0 MeV at V1 and V2 as functions of (a) time after crossing the TS and (b) distance beyond the TS.

For a more comprehensive view of this behavior, a monthly sample spectrum from the middle day of each month was taken from the daily averages of these CRS data and their spectral indices were calculated as the slopes of the spectra as they would appear on a log-log plot. These results in Figure 3(a) as functions of the time from each V1 and V2 TS crossing show a strong similarity in the time histories of the V1 and V2 data spectral index values even though the times at V2 are about 2 years and 9 months after the corresponding times at V1. This suggested that distance from the TS was the primary influence on the spectral index values. Figure 3(b) shows the spectral indices as functions of distance from the TS crossing. As V1 was traveling about 1.6 km/s faster than V2 during this period, in the distance plot the V1 line is stretched out farther than in the time plot. This shows that V2 has traveled only a little farther in the heliosheath in 10 years than V1 traveled in 7 years and 9 months.

This dependence on distance from the TS appears consistent with the blunt termination shock model of McComas and Schwadron [16], since as each spacecraft moves into the heliosheath it first traverses field lines that are connected to the TS at increasingly greater distances along the flank of the heliosphere and consequently encounters particles that have received more acceleration. This effect reaches a maximum at about 10 AU beyond the TS, and after that the connection to the acceleration region weakens (though not ending as abruptly as depicted by McComas and Schwadron in [17]). The higher peak spectral index observed at V2 may be related to the fact that the V2 trajectory is at a heliographic inertial longitude of about 217 degrees, while the V1 trajectory is at about 174 degrees, so that V2 is closer in longitude to the acceleration region at 300 or more degrees hypothesized in [17].

4. Particle Behavior Associated with Plasma Wave Events

Another perspective on the particle environment in the LISM (in addition to the simulations discussed in Section 2) and how it may relate to the inward-diffusing particles observed by V2 is provided by the particle behavior associated with the plasma wave events that have been observed in the LISM at around 3 kHz by the V1 Plasma Wave System (PWS). In [10] we discussed how the 3 kHz events that had been observed by V1 up to that time were probably associated with dynamic pressure increases observed by the plasma instrument (PLS) on V2 and with periods of major solar flares, so that solar sources were identified for the merged interaction regions (MIRs) generating the shocks that were discussed by Gurnett et al. [18] as the sources of the PWS events.

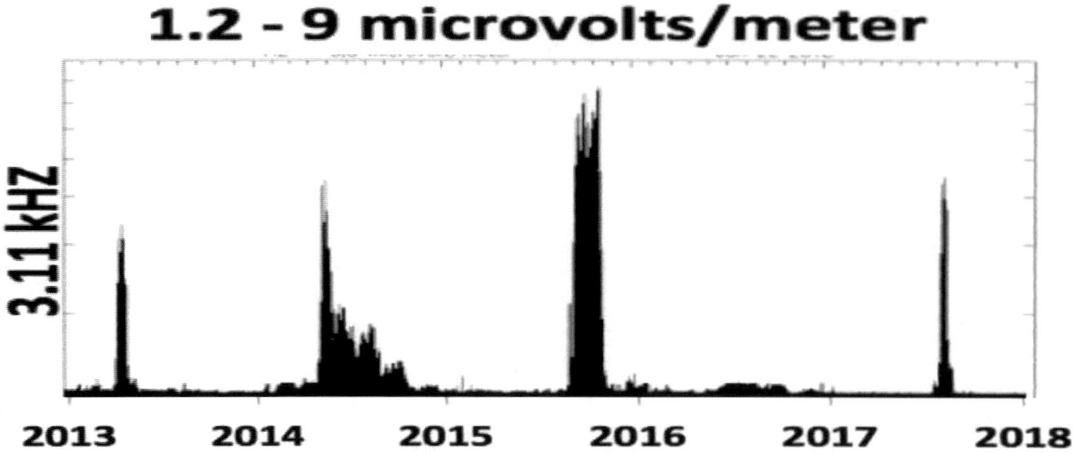

Figure 4. V1 PWS 3.11 kHz channel peak signal in μvolts/m, 2013 through 2017 [19]. The peaks here result from intense electron plasma oscillation events after V1 entered the LISM. The V-shaped dips in the CRS data (Figure 5) extend before and after the PWS events above.

At the time [18] was written only three PWS events had been observed: one in late 2012, one in April and May of 2013, and one that lasted through most of 2014, but the most intense oscillations were in May. Since then, as shown in Figure 4 (which starts at the beginning of 2013, so it omits the event in late 2012; from the University of Iowa website [19]), there have been two more: the strongest one of all in late 2015 and a weaker and much briefer one in August, 2017. As described by Gurnett et al., the data from the PWS broadband analyzer show that the peak frequencies of the plasma wave events have successively increased slightly, indicating slight increases in the plasma density as the spacecraft has travelled more deeply into the LISM, but they all have been within the frequency range of the 11th channel of the 16-channel analyzer of the PWS. This channel is centered on 3.11 kHz, as indicated in the figure and caption, but actually covers frequencies from about 2.4 to about 4.2 kHz.

Gurnett et al. also discuss the variations of particle count rates associated with the PWS events. The angular resolution of the LECP observations shows that the count rate of the galactic cosmic ray particles gyrating perpendicular to the magnetic field (i.e., with pitch angles near 90 degrees with respect to the field) varies significantly around the times of the PWS events, while the particles moving parallel to the field (i.e., with pitch angles near 0 and 180 degrees) show very little variation. They show in [18] that corresponding variations also appear in the count rate of the guard detector of HET-I in the CRS, although since this detector is omnidirectional it counts particles with all pitch angles, so that the changing count rates of the particles with pitch angles near 90 degrees are superimposed on the nearly constant count rates of the particles with other pitch angles, and hence the percentage changes of the guard count rate are small. Thus, these are more complex phenomena.

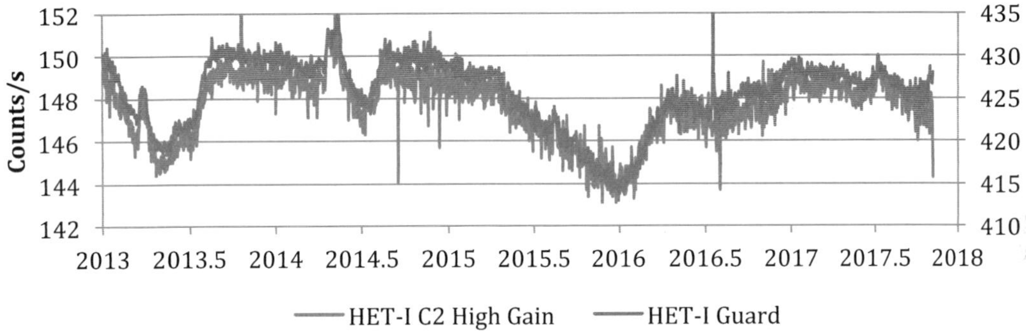

Figure 5. V1 HET-I Detector C2 count rate with no coincidence analysis (left scale) and guard count rate (right scale). The V-shaped dips in Figure 5 correspond to the PWS peaks in Figure 4.

The rates page [11] of the CRS website includes not only count rates for various detector coincidence combinations used in Section 2 but also the rates for individual detectors in the telescopes of the CRS. These individual count rates are unconstrained by coincidence conditions, so they are omnidirectional like the guard count rates, and as exemplified by Figure 5, they show variations associated with the PWS peaks in the same way as the guard count rates. However, Figure 5 is also an example of the smaller count rates from the detectors in the telescope stacks, which are smaller than the cylindrical guard detectors surrounding the telescopes.

In Figure 5 the count rate variations can be seen to have followed the same general pattern for each of the PWS enhancements in Figure 4: a slow decline lasting weeks or months, interrupted by a brief peak, which approximately coincides with the most intense days of the PWS peak, followed by a return to the decline and an eventual bottom, and then a recovery. **Thus in Figure 5, except for each brief peak, the longer-term variation of each count rate resembles a letter V-shaped dip**. However, the pattern was not repeated exactly, since the peaks early in 2013 and in 2014 for the first two events were prominent and the accompanying V-shaped depressions were relatively brief. In contrast, for the third event the V-shaped depression was much deeper and longer-lasting, with a peak barely visible in the figure just after 2015.5. Similarly, for the fourth event the peak was only a little larger than for the third event, and the much shallower depression reached its minimum in the middle of October, about three weeks before the end of the data on November 6.

Gurnett et al. note in the caption to their Figure 3 that the count rate for the guard detector of HET-I is mainly due to protons with energies >20 MeV, but the count rates for the HET detectors provide information about the effects on particles over a wider range of energies. This is because each detector is connected to an amplifier from which the outputs are digitized for the data system, so that each detector has a threshold for the minimum energy of the particles that it detects. Furthermore, in each observation cycle the control system switches the amplifiers between low-gain and high-gain modes.

Thus, plotting the count rates for the HET detectors in each gain mode and taking the ratio between the highest count rate in the peak of the 2014 event and the lowest count rate at the bottom of the depression of the 2015 event provided quantitative measures of the degrees of modulation in a number of energy ranges. Similar degrees of modulation are associated with the PWS events for particles with energies from hundreds of keV to several GeV, so that the most plausible interpretation is that the modulation over all these energies is as described by Gurnett et al., occurring primarily for the particles with pitch angles near 90 degrees, with much less effect on other particles.

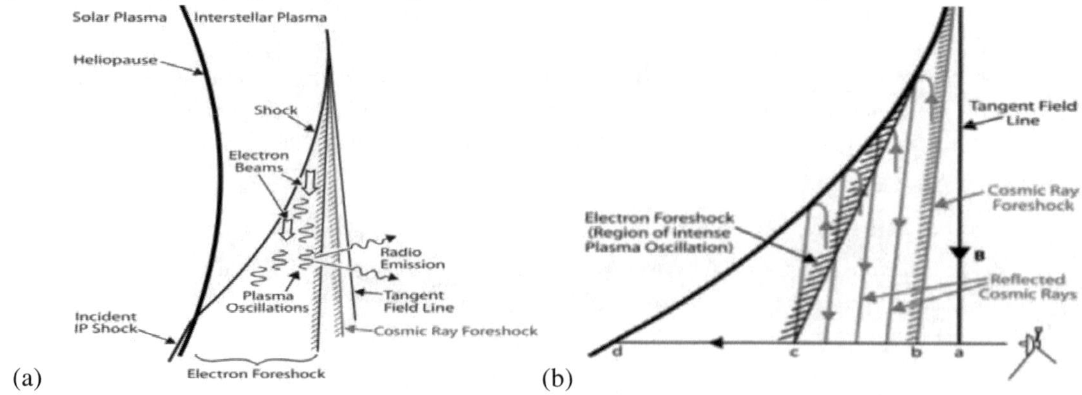

Figure 6. (a) Schematic depiction of the mechanism of plasma wave generation by a beam in the electron foreshock. (b) Schematic depiction of shock motion past the spacecraft and particle peak formation by cosmic ray reflection. (from Figures 7 and 8(a) of [18])

Using an analogy with the bow shock in the solar wind as it collides with the terrestrial magnetosphere, [18] explains the brief peaks in the particle count rates near the times of peak plasma wave intensity. They concluded that this analogy did not provide an explanation for the longer-term V-shaped depressions. They questioned whether the depressions seen near the peaks for the plasma wave events actually were associated with the MIRs for these events, since the depression for the 2013 event started ahead of where the tangent field line could have been, and likewise the depression for the 2012 event began soon after the spacecraft crossed the HP, which also seemed too early to be associated with the shock for that event. They suggest that the depressions are "due to a pre-existing shallow expansion of magnetic field lines draped around the nose of the HP, which via conservation of the first adiabatic invariant would act to decrease the cosmic ray intensities, especially near 90° pitch angles."

Since the particularly broad and deep depression associated with the 2015 event lasted from late winter or early spring of 2015 to early spring of 2016 (so it began about the time the paper was submitted), while V1's solar distance increased from about 130.8 to about 134.5 AU, far beyond the HP crossing at about 121.7 AU, and then the depression for the 2017 event lasted from early 2017 past November 6, from about 138 to over 140 AU, this explanation is no longer plausible. However, the reference to the effect of conservation of the first adiabatic invariant leads to an interpretation consistent with the observations of the more recent plasma wave events as well as the earlier ones.

Figure 6(a) shows Figure 7 of [18] and Figure 6(b) reproduces Figure 8(a) of [18], showing their model of how a shock emerging from the heliosheath generates a cosmic ray foreshock that follows the tangent line to the shock, as shown in Figure 6(a), and an electron foreshock following the cosmic ray foreshock, in which an electron beam generates both the electron plasma oscillations that provide the most intense signals detected by the PWS and also radio signals that propagate ahead of the shock, as were observed in the 2014 event. Figure 6(b) shows how cosmic rays reflected from the shock explain the peak in the count rates near the time when the electron plasma oscillations are observed.

We suggest that the key to understand the V-shaped depressions in the count rates, as shown in Figure 5, is to recognize that the diagrams in Figure 6 represent slices out of the overall 3-dimensional configuration of the moving MIR, representing effects on field lines that are connected to the spacecraft or are close enough to it that the spacecraft is within the gyroradii of many cosmic ray particles orbiting such field lines with pitch angles near 90 degrees. Figure 7(a) schematically depicts a side view of successive stages of the advance of a shock, taking into account the facts that V1 is at a heliographic inertial latitude of about 34 degrees north, while all the flares that we identified in [10] as likely sources of the MIRs producing the plasma wave events were closer to the solar equator in the northern hemisphere, or were in the southern hemisphere.

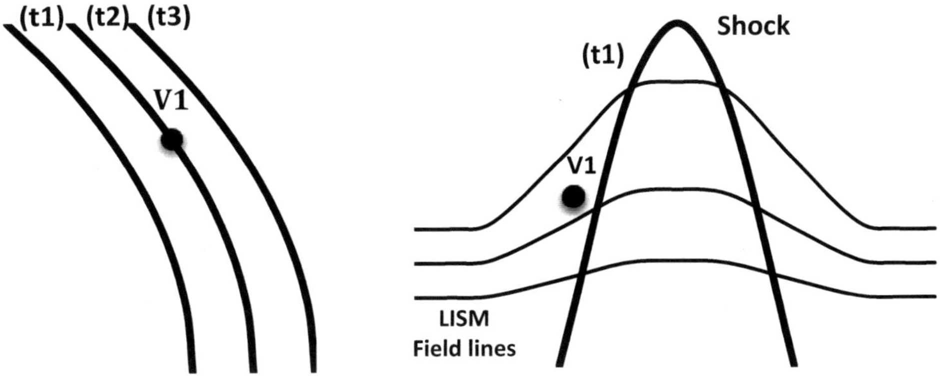

(a) (b)

Figure 7. (a) Schematic depiction of shock profiles at successive times. Horizontal direction is radial and vertical direction is latitude. (b) Schematic depiction of the shock effect on the LISM field at time t_1. Horizontal direction is longitude and vertical direction is radial.

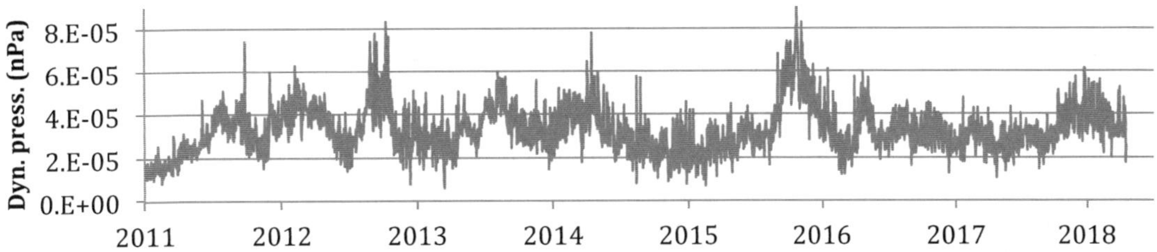

Figure 8. History of PLS plasma dynamic pressures at V2 in nanopascals, from the website in [20].

Thus, at time t_1 the shock has not yet reached V1, but a portion of the shock at lower latitudes has already exceeded the radial distance of V1. At this time the effects depicted in Figure 6 are occurring on field lines too far south of the spacecraft for it to observe them. Time t_2 is the time when the shock reaches the spacecraft and it observes the effects in Figure 6. Then at time t_3 the shock has passed V1 and the effects depicted in Figure 6 are occurring on field lines too far north of the spacecraft for it to observe them and it is also inside the MIR.

However, although at time t_1 the shock has not yet reached the spacecraft, the portion that is already south of the spacecraft would be expected to be deforming the LISM field lines over a wide region. This is schematically depicted in Figure 7(b), showing that V1 would be in a region where the field is being weakened by being stretched by the portion of the shock that is south of the spacecraft, with a corresponding reduction of cosmic ray intensity by conservation of the first adiabatic invariant.

5. Plasma Pressure and Future Prospects

Finally, we consider the hints of possible future developments provided by the recent plasma pressure history at V2. Figure 8 updates Figure 1 from [10], using the file of most recent V2 PLS plasma parameter daily averages from the website in [20]. In addition to the peaks from the middle of 2011 to early 2016 discussed by Richardson, et al. [20], a new pressure peak comparable in magnitude to several of the previous ones has been observed since late 2017. Since magnetometer readings for these recent years are not currently available, it is not known yet if this peak is a MIR like all but the first

IOP Conf. Series: Journal of Physics: Conf. Series **1100** (2018) 012013 doi:10.1088/1742-6596/1100/1/012013

two discussed by Richardson, et al., but if so then it may eventually produce another plasma wave event. In any case, activity evidently continues in the outer heliosphere despite declining solar activity.

6. Concluding Remarks

V2 appears to be approaching the HP and may reach it in a few more weeks or months. However, the HP and the LISM just beyond it can be seen from the V1 and V2 observations to be dynamic plasma and particle environments subject to effects propagating from regions not directly observed.

The modeling efforts reported in [12] and [13] give contrasting results, but neither used estimates of actual solar wind emission at a source surface a few solar radii from the Sun, as were used by our modeling [10]. They assumed typical values much farther from the Sun. However, the recent more realistic results of Washimi et al. [21] modified a steady-state model with OMNI data from 1 AU.

We believe comprehensive and consistent understanding of the Voyager observations from great distances in recent years will only come from matching these observations with MHD simulations that include pickup ions and neutral hydrogen and follow the outward propagation of the solar wind actually emitted from the Sun during these times. However, considering the iterative mesh refinements used by Opher et al [12] and by Pogorelov and his collaborators [13], and the lengthy computations of Washimi et al. [21], we recognize the great computational challenges of accomplishing the needed simulations. We encourage others to join us in seeking ways to make them feasible.

7. References

[1] Intriligator D S, et al., 2005, *J. Geophys. Res.*, **110**, A09S10
[2] Detman T R, et al., 2011, *J. Geophys. Res.*, **116**, A03105
[3] Intriligator D S, et al., 2012, *J. Geophys. Res.*, **117**, A06104
[4] Intriligator D S, et al., 2015, *J. of Phys. Conf. Ser.* **577**, 012013
[5] Decker R B, et al., 2010, *AIP Conf. Proc.*, **1302**, 51
[6] Stone, E C, et al., 1977, *Space Sci. Rev.*, **21**, 355
[7] Webber W R, and McDonald F B, 2013, *Geophys. Res. Lett.*, **40,** 1665; Stone, E C et al., 2013, *Science*, **341**, 150
[8] Webber W R, and Intriligator D S, 2011, *J. Geophys. Res.*, **116**, A06105
[9] Intriligator D S, et al., 2016, *15th AIAC, "The Science of Ed Stone Celebrating his 80th Birthday:" Journal of Physics Conference Series,* **761**, 012013
[10] Intriligator D S, et al., 2017, *16th AIAC, J. of Phys. Conf. Ser.*, **900,** 105
[11] CRS homepage: https://voyager.gsfc.nasa.gov; rates: https://voyager.gsfc.nasa.gov/rates.html
[12] Opher M, et al, 2011, *Astrophys. J.*, **734**, 71
[13] Pogorelov N, et al., 2017, *Astrophys. J.*, **845**, 9; Borovikov S, et al., 2014, *Astrophys. J. Lett.*, **783**, L16
[14] Roelof E, et al., 2012, *AIP Conf. Proc.*, **1436**, 239; Krimigis S, et al., AGU Fall Mtg, **SH23A-04,** 2016
[15] Krimigis S M, et al., 2010, 9th AIAC, *AIP Conf. Proc.*, **1302**, 79
[16] McComas D J, and Schwadron N A, 2006, *Geophys. Res. Lett.*, **33**, L04102
[17] McComas D J, and Schwadron N A, 2012, *Astrophys. J.*, **758**, 19
[18] Gurnett D A, et al, 2015, *Astrophys. J.*, **809**, 121
[19] PWS Gurnett and Kurth, http://www-pw.physics.uiowa.edu/das/voyager-pws-sa-list
[20] Richardson J D, et al., 2017, *Astrophys. J.*, **834,** 190; http://web.mit.edu/afs/athena/org/s/space/www/voyager/voyager_data/voyager_data.html
[21] Washimi H, et al., 2017, *Astrophys. J. Lett.*, **846,** L9

Acknowledgments

We gratefully thank the Voyager projects for the vast information we have received. We acknowledge the support of Carmel Research Center, Inc. We gratefully obtained the: CRS data from [11]; PWS data from Drs. Gurnett and Kurth, and the University of Iowa website [19]; and PLS data from [20].

IOP Conf. Series: Journal of Physics: Conf. Series **1100** (2018) 012014 doi:10.1088/1742-6596/1100/1/012014

Flat Proton Spectra in Large Solar Energetic Particle Events

D Lario[1], L Berger[2], L B Wilson III[3], R B Decker[1], D K Haggerty[1], E C Roelof[1], R F Wimmer-Schweingruber[2] and J Giacalone[4]

[1]The Johns Hopkins University. Applied Physics Laboratory
[2]Institut für Experimentelle und Angewandte Physik (IEAP). Christian-Albrechts-Universität zu Kiel
[3]NASA, Goddard Space Flight Center
[4]Lunar and Planetary Laboratory. University of Arizona

Corresponding author's e-mail: david.lario@jhuapl.edu

Abstract. We present solar energetic particle events observed at 1 AU from the Sun for which the proton energy spectra at energies between ~50 keV to ~1 MeV flatten during a period of at least ~12 hours prior to the passage of the associated interplanetary shock. The flattening of the proton energy spectra occurs when the source of the particles (presumably the traveling interplanetary shock) is still downwind from the spacecraft and particle intensities are still continuously increasing. The arrival of the shock at the spacecraft is then characterized by a steepening of the spectra, where low-energy proton intensities show a more pronounced enhancement than the high-energy proton intensities. We discuss the mechanisms that may result in this flattening of the spectra in terms of current models presented in the literature.

1. Introduction

The time evolution of the solar energetic particle (SEP) intensities measured in interplanetary (IP) space depends on both the acceleration processes responsible for the energization of the particles and the transport processes undergone by the particles during their travel from their sources to the observing spacecraft. Both the acceleration and the transport processes depend on a large number of factors, among them the energy achieved by the particle, which controls its ability to escape from the acceleration region and regulates the effects that the ambient medium has on its transport before reaching a spacecraft. The evolution of the kinetic energy spectra of the particles measured at a given heliospheric location during a SEP event provides information of the acceleration mechanisms undergone by the particles once the processes of particle transport and particle release from their acceleration region have been properly deconvolved.

Several functional forms have been adopted to describe the particle energy spectra integrated over the duration of a SEP event [e.g., 1, 2]. However, in order to determine the time history of the acceleration processes, an analysis of the precise temporal evolution of the energy spectra throughout the SEP event is required. This type of analysis shows how the energy spectra evolve from the onset of the event, when the most energetic particles arrive at the spacecraft, to later times when the energy spectra typically straighten into a power-law shape at low energies mostly in association with the arrival of IP shocks [e.g., 3, 4, 5]. Related to the processes of particle acceleration at IP shocks, the study of the energy spectra around the passage of shocks has revealed a broad range of energy spectra and diverse time evolutions that depend, among other factors, on the properties of the shocks [e.g., 6, 7, 8, 9]. However, the shock parameters alone do not unequivocally determine the characteristics of the energetic particle enhancements observed in association with the passage of the shocks [10].

Here we pay attention to a recurrently observed peculiarity in intense SEP events associated with the passage of IP shocks. Long time intervals of at least ~12 hours are observed before the passage of the shocks showing flat proton energy spectra (in the energy range between ~50 keV and ~1 MeV) that only steepen when the shock approaches the spacecraft when low-energy proton intensities become more abundant than higher-energy intensities. We corroborate that those periods are commonly observed by a diversity of energetic particle instruments on board spacecraft located at ~1 AU from the Sun. Therefore, unless all particle instruments have similar functional problems, this phenomenon results from an actual physical mechanism that culminates in the measurement of equal proton intensities at different energies. Since those periods occur when the traveling shock is still far from the spacecraft, we believe that the mechanism responsible for their observation is predominantly related to the processes of particle escape from the vicinity of the shock and their subsequent transport to the spacecraft, rather than the shock acceleration process itself.

2. Observations

The Low-Energy Magnetic Spectrometer (LEMS120) telescope of the Electron Proton and Alpha Monitor (EPAM) on the Advanced Composition Explorer (ACE) measures ions between ~47 keV and 4.8 MeV in eight differential energy channels [11]. The magnet or open (O) semi-conductor detector telescope (SST-O) of the Three-Dimensional Plasma and Energetic Particle Investigation (3DP) on board the Wind spacecraft measures ions from a few tens of keV up to 11 MeV [12, 13]. Neither ACE/EPAM/LEMS120 nor Wind/3DP/SST distinguish among protons, alpha particles and other ion species. We will assume that intensities measured by both instruments are dominated by protons and the contribution of alpha particles and heavy ions is small compared to that of protons. The Energetic Particle Sensor (EPS) on board the Geostationary Operational Environmental Satellites (GOES) detects protons in eight energy intervals ranging from 0.8-4.2 MeV up to 200-500 MeV in the case of GOES-11 [14, 15]. The combination of the Low Energy Detector (LED) and High Energy Detector (HED) of the Energetic and Relativistic Nuclei and Electron (ERNE) experiment on board the Solar and Heliospheric Observatory (SOHO) provides proton intensities in the energy range from ~1 MeV to hundreds of MeV [16]. This assortment of energetic particle instruments allows us to build continuous proton energy spectra covering a broad range of energies starting just above ~50 keV.

Figure 1 shows four intense SEP events where long-time intervals with flat energy spectra were observed. Each panel shows, from top to bottom, (i) proton intensities measured by (a-b) ACE/EPAM/LEMS120 and GOES-11/EPS and (c-d) ACE/EPAM/LEMS120 and SOHO/ERNE; (ii) the proton solar wind speed measured by the Solar Wind Electron Proton Alpha Monitor (SWEPAM) on board ACE [17]; and (iii) the magnetic field magnitude as observed by the Magnetic Field Experiment on board ACE [18] (when gaps of ACE/SWEPAM data exist we overplot, using red dots, proton solar wind speed data from the Solar Wind Experiment (SWE) on board the Wind spacecraft [19] shifted to the time of the shock passage at ACE). Table 1 summarizes the main properties of the solar flares and plane-of-sky CME speeds associated with the origin of the SEP events included in this study as well as the arrival times of the associated IP shocks. The black vertical arrows in Figure 1 indicate the time of the parent solar flares and the solid vertical lines the passage of IP shocks by the ACE spacecraft. The dashed vertical lines in panels (a) and (c) indicate the passage of the leading edge of the interplanetary counterpart of coronal mass ejections (ICMEs) as identified by Richardson and Cane [20].

The tilted gray rectangles in Figure 1 indicate the time intervals when flat low-energy proton spectra were observed. These periods are easily identifiable as those time-intervals when intensities at different energies overlap. We have indicated in panels (b) and (d) with dotted traces those periods when the low-energy channels of ACE/EPAM/LEMS120 showed contamination due to high-energy particles penetrating in the instrument that occurs mainly at the onset of the SEP events. In these two events (Figures 1b and 1d), the flat energy spectra period was observed when the contribution of these high-energy particles in the respective channels had already diminished.

For the events in panels (a) and (b), the flat spectra were observed over the energy range from ~68 keV up to ~2-3 MeV starting several hours before the arrival of the shocks. Considering the average transit speed of the shock to travel from the Sun to 1 AU (last column in Table 1), at the time when these flat spectra started being observed, the shocks were located at ~0.5 and ~0.6 AU from the spacecraft for the events (a) and (b), respectively. Whereas in panel (a) flat spectra were observed almost up to the shock arrival, in panel (b) the intensities at different energies separated out as the shock approached the spacecraft resulting in steeper energy spectra at the time of the shock passage.

For the event in panel (c), the flat-spectra period started ~17 hours before the shock passage, i.e. when the shock was still ~0.4 AU away from spacecraft (assuming that it propagated at the averaged transit speed quoted in Table 1). The flat spectra continued until the shock arrival even when low-energy intensities were still increasing. The energy range over which these flat spectra were observed extended from ~50 keV up to ~0.5 MeV.

For the event in panel (d), the flat spectra developed about ~17 hours after the onset of the event when low-energy particles gradually began to arrive at the spacecraft. However, the low-energy intensities started to separate about ~17 hours before the shock arrival (when the shock was ~0.3 AU away from the spacecraft considering that the shock propagated at the average transit speed quoted in Table 1).

One may wonder whether the observation of these flat-spectra periods is due to an instrumental effect. Figure 2 shows the same set of events shown in Figure 1 but using Wind/3DP/SST-O data. The time-intensity profiles, the intensities at the different energy channels, and the periods with flat-spectra were similar as those observed by ACE/EPAM/LEMS120. Either, both instruments had a similar malfunction or the development of the flat-spectra periods responds to a real physical mechanism.

Table 1. Main properties of the solar flares and shocks associated with the selected SEP events

Event	Flare SXR onset time yy/doy/hh:mm	Flare Class	Flare Site	CME (km/s)[a]	Shock time passage doy/hh:mm	Average shock transit speed (km/s)
Figs. 1(a)-2(a)	05/133/16:13	M8.0/2B	N12E12	1689	135/02:11	1223
Figs. 1(b)-2(b)	03/306/17:03	X8.3/2B	S14W56	2598	308/05:59	1125
Figs. 1(c)-2(c)	01/116/12:11	M7.8/2B	N17W31	1006	118/04:31	1030
Figs. 1(d)-2(d)	00/256/11:31	M1.0/2N	S17W09	1550	259/03:59	645
Fig. 3(a)	00/313/22:42	M7.4/1N	N10W77	1738	315/06:19	1314
Fig. 3(b)	78/266/09:47	/2B	N35W50	-	268/07:18*	913
Fig. 3(c)	98/273/13:08	M2.8/2N	N23W81	-	275/06:53	995
Fig. 3(d)	74/262/22:53	/2B	N08W63	-	264/12:45*	1097
Fig. 4(a)	12/147/~20:35	-	N16W122	1966	149/02:48	1375
Fig. 4(b)	11/080/02:15	-	N16W130	1341	081/18:05	1043
Fig. 5	12/067/00:02	X5.4/3B	N17E27	2684	068/10:45	1197
	12/067/01:05	X1.3/SF	N22E12	1825		

[a] CME plane-of-sky speed as reported in the catalog of CMEs observed by SOHO at cdaw.gsfc.nasa.gov/CME_list/

*Time of a Sudden Storm Commencement used as a proxy for the time of IP shock passage

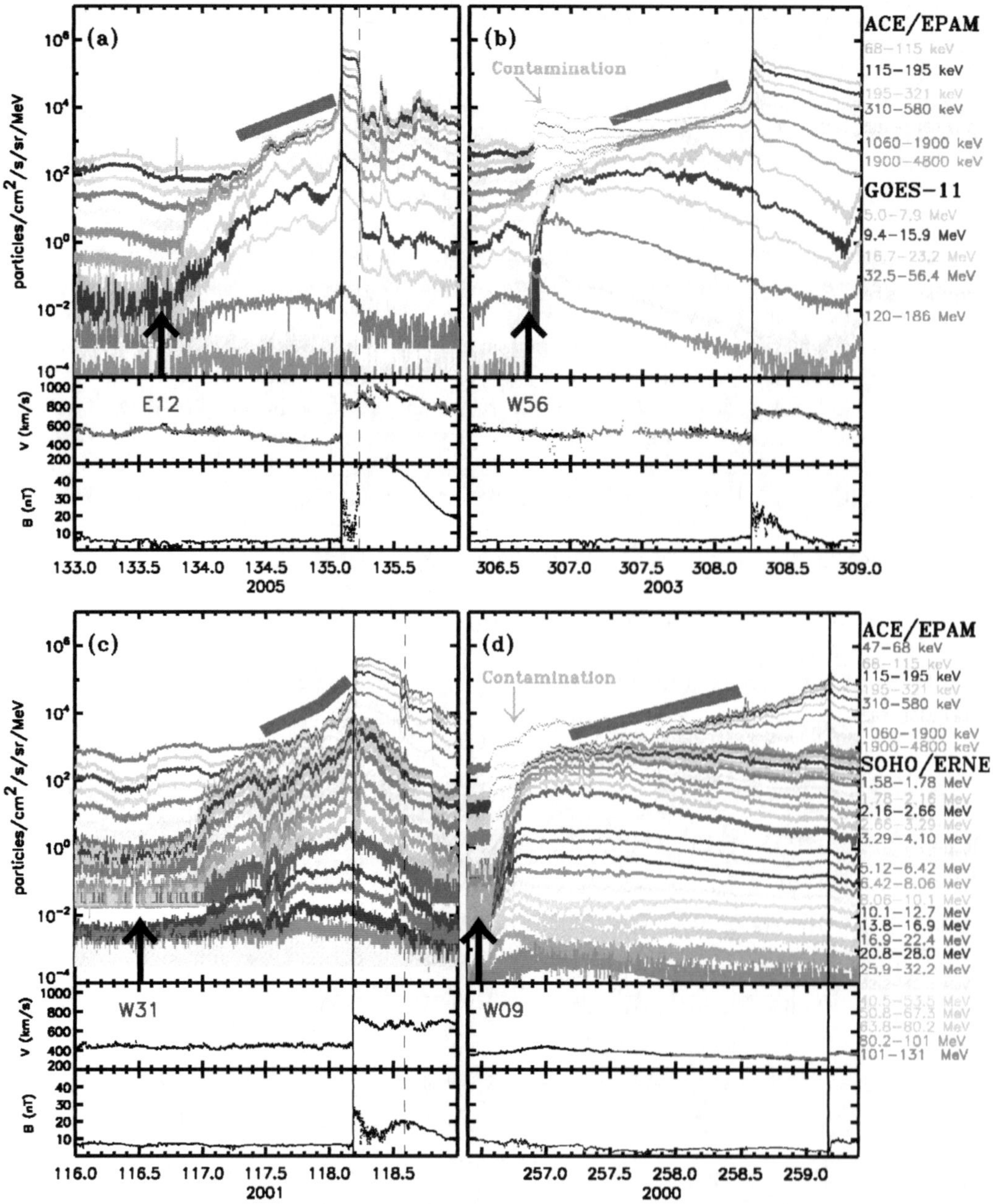

Figure 1. One-minute averages of the spin-averaged ion intensities measured by ACE/ EPAM/LEMS120 and of the proton intensities measured by GOES-11-EPS (panels a and b) and by SOHO/ERNE (panels c and d), together with the proton solar wind speed measured by ACE/SWEPAM and the magnetic field magnitude measured by the magnetic field experiment on board ACE. The black arrows indicate the occurrence of the solar eruption that generated the event at the longitude indicated below. The vertical solid lines indicate the passage of IP shocks, and the dashed vertical lines the leading edge of ICMEs. The tilted gray rectangles indicate the periods when flat spectra at low energies (<~1 MeV) were observed. The noisy 47-68 keV ion channel from ACE/EPAM/LEMS120 during the events in panels (a) and (b) has not been represented. The red dots in the solar wind panels indicate Wind/SWE data shifted to the time of the shock at ACE.

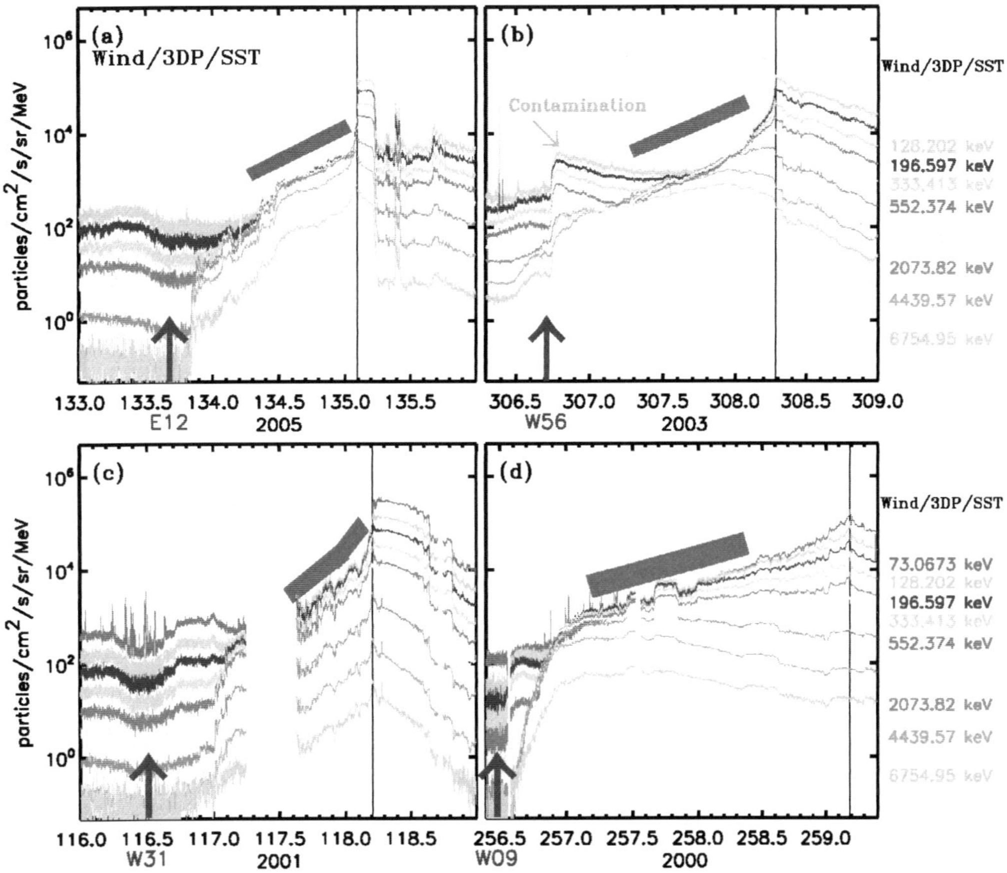

Figure 2. Wind/3DP/SST-O spin-averaged ion intensities for the same events shown in Figure 1.

We have also performed a search for SEP events showing flat-spectra periods in historical data records. The Charged Particle Measurement Experiment (CPME) on board the Interplanetary Monitoring Platform (IMP-8) measured protons in ten differential channels over the energy range 0.29-145 MeV [21]. The stack of detectors and the on-board logic allowed a distinction between proton, alpha, and Z≥3 heavy ions. Alpha particle intensities were used to correct some of the proton channels that were also sensitive to alpha particles [21]. Figure 3 shows some of the SEP events selected within the IMP-8/CPME data set for which long-lasting flat-spectra periods were observed. The properties of the parent solar flares for these SEP events are listed in Table 1. The IMP-8 SEP events showing periods with flat spectra were also associated with the passage of IP shocks (in the column of Table 1 with the times of the shock passage, we have indicated with an asterisk those times when a Sudden Storm Commencement was used as a proxy for an IP shock passage instead of an actual shock observed in magnetic field and solar wind plasma data). Since IMP-8/CPME data do not reach the low proton energies measured by ACE/EPAM/LEMS120 and Wind/3DP/SST-O, the effects of the shock passage on the intensity-time profiles are not so accentuated as for the events shown in Figures 1 and 2. The flat spectra in the events in Figures 3(a) and 3(b) extended up to ~4.6 MeV, whereas for the events in Figures 3(c) and 3(d) they extended up to ~2 MeV, with intensities still increasing. Several hours before the shock arrival, proton intensities at different energies separated resulting in energy spectra that steepened.

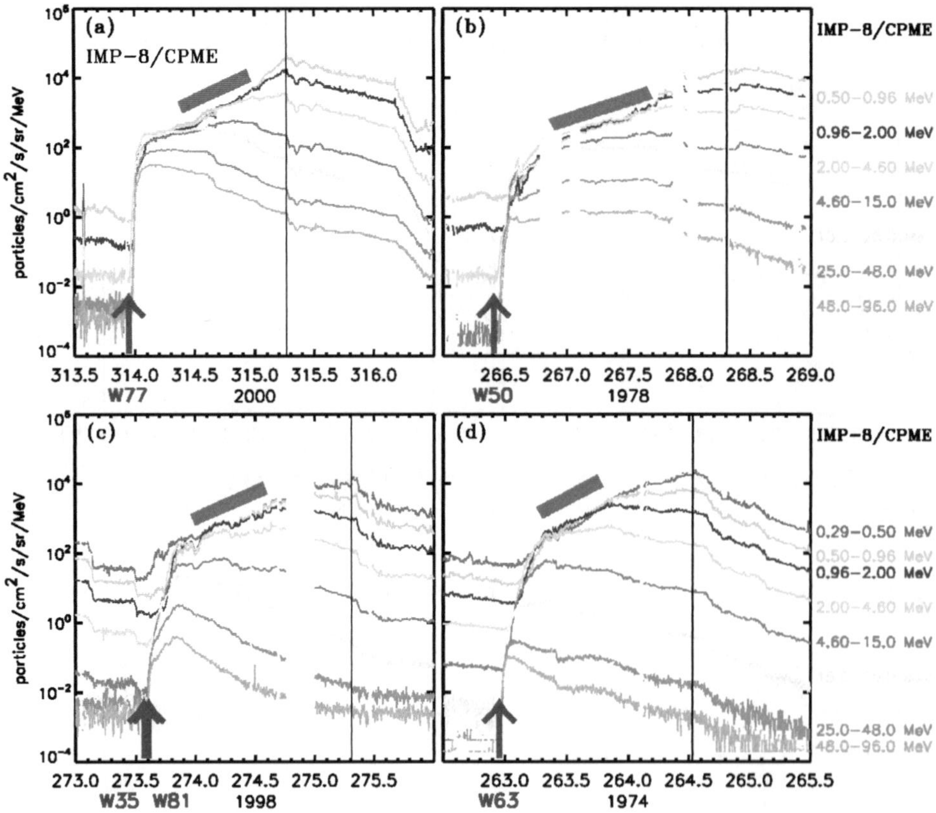

Figure 3. 5.5-minute averages of proton intensities measured by IMP-8/CPME in four selected events. The solid vertical lines indicate the passage of the interplanetary shocks and the purple arrows the occurrence of the parent solar eruption as described in Table 1. Note that for the event in panel (c) two solar eruptions may be responsible for the SEP event and the IP shock as described in [22].

More recent data sets also show some SEP events with flat-spectra periods. Figure 4 shows two SEP events observed by the suite of energetic particle instruments of the In-situ Measurements of Particles And CME Transients (IMPACT) on board the spacecraft A of the Solar TErrestrial RElations Observatory (STEREO-A) [23]. IMPACT consists of a series of telescopes; in particular, the Solar Electron and Proton Telescope (SEPT) that measures ions from ~60 to 6500 keV in 32 energy bins without distinguishing protons from alpha particles or other heavy ions [24], the Low-Energy Telescope (LET) that measures protons from 1.8 to ~15 MeV [25] and the High-Energy Telescope (HET) that measures protons from 13 to ~100 MeV [26]. It is important to mention that, similarly to ACE/EPAM/LEMS120 and Wind/3DP/SST-O, STEREO/SEPT ion channels might have some contamination by electrons and high-energy particles penetrating into the ion telescope apertures, especially at the onset of intense SEP events (as indicated in Figure 4). Moreover, during periods of reduced counts rates at low energies (such as in a flat-spectra period) but with significant ion fluxes above 2.2 MeV, limitations in the SEPT electronics produce extra counts in specific low-energy bins. These bins have not been included in Figure 4. Such caveats in the STEREO/SEPT data are described in detail in [27].

Figure 4 shows that, after the periods with possible contamination in the low-energy channels early in the events, flat-spectra developed prior to the arrival of the shocks. For the event in Figure 4(a), the flat energy spectra extended up to ~2 MeV with intensities still increasing. For the event in Figure 4(b), a first period with flat-spectra starting on day ~80.5 and extending up to ~2 MeV was observed, when the lower energy channels were still contaminated by penetrating high-energy particles. Just prior to the arrival of the shock in Figure 4(b), another flat-spectra interval was observed extending up to ~1 MeV with almost flat intensity-time profiles.

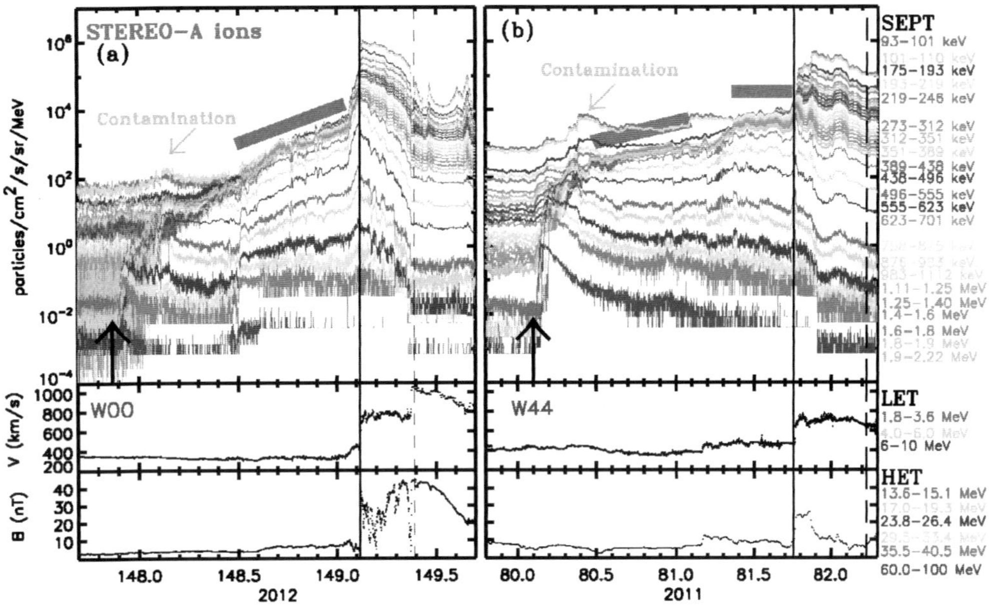

Figure 4. One-minute averages of, from top to bottom, ion intensities averaged over the four telescopes of STEREO-A/SEPT and proton intensities as measured by STEREO-A/LET and STEREO-A/HET, solar wind proton speed and magnetic field as measured by the Plasma and Suprathermal Ion Composition (PLASTIC) investigation [28] and the Magnetic Field Experiment on board STEREO-A [29]. The black arrows indicate the occurrence of the solar eruption that generated the event at the indicated longitude (longitude given with respect to STEREO-A, see [30]). The vertical solid lines indicate the passage of IP shocks, and the dashed vertical lines the leading edge of ICMEs. The gray tilted rectangles indicate the time intervals where the energy spectra at low energies flatten.

3. Discussion

The examples shown in the previous section illustrate some of the SEP events where long-lasting periods (>12 h) with flat proton energy spectra (mostly at energies below 1 or 2 MeV) are observed. These SEP events tend to be intense and associated with the passage of interplanetary shocks. The flat-spectra periods are usually observed prior to the arrival of the shocks when low-energy proton intensities are still increasing. The arrival of the shock is usually characterized by a subsequent intensity increase, where low-energy proton intensities show a more pronounced enhancement than the high-energy proton intensities, resulting in a steepening of the energy spectra.

The continuous injection and acceleration of particles by the approaching traveling IP shock seems to be a common characteristic of these events. Hybrid simulations of particle acceleration at shocks, including processes produced by magnetic field fluctuations excited by the shock-accelerated particles propagating in the upstream region of the shocks, predict similar intensities of protons at different energies at a certain distance upstream of the shock [e.g., 31]. At small upstream distances from the shock, the lower the energy of the particles the more intense the event becomes, resulting in a softer energy spectrum (see Figure 1e in [31]). Whereas these simulations are consistent with the observations, the limited simulation domain of these models does not allow us to analyze how far from the shock and up to what energies these flat spectra can extend.

As a matter of fact, the events analyzed in Section 2 show that the observation of these flat-spectra periods starts when the shocks are still located at large distances downwind from the spacecraft and continues for a long-lasting interval. It is well-known that the observation of shock-accelerated particles by spacecraft located at 1 AU depends not only on the processes of particle acceleration at the approaching shock, but also on the particle release from the vicinity of the shock and the subsequent particle transport from the shock up to the spacecraft. Particle acceleration may start as soon as the shock forms close to the corona [32] and continues as the shock expands into the IP medium and passes the spacecraft at 1 AU [33]. The conditions of particle acceleration at the shock throughout all this time may vary, not only because shock properties change as the shock propagates but also because the spacecraft establishes magnetic connection with different regions of the shock where shock parameters and particle acceleration conditions vary as the shock expands toward the spacecraft [34]. Therefore, the flattening of these spectra intervals must be more a consequence of the processes of particle escape from the shock vicinity and particle transport toward the spacecraft rather than the varying processes of particle acceleration at the shock.

Several models have been developed over the years to couple the propagation of IP shocks as a mobile source of shock-accelerated particles and the transport of SEPs in IP space [e.g., 35-41, and references therein]. Some of these models have been developed in earlier efforts that modeled simultaneous observations of SEP events by multiple spacecraft (i.e. Helios, ISEE-3 and IMP-8) [e.g., 42, 34]. In particular, the models by Li et al. [38] and Hu et al. [39] included particle acceleration processes via the diffusive shock acceleration mechanism in a series of shells formed around the shock. These models predict hard energy spectra early in the SEP events as observed at 1 AU from the Sun that with time extend to low energies (as slower particles reach the spacecraft) and soften as the shock approaches 1 AU and becomes less and less efficient at accelerating particles to higher energies (cf., Figures 8-11 in [38] and Figures 9-10 in [39]). These models consider that, once particles escape from the shock vicinity where they are accelerated, their propagation in the upstream unperturbed solar wind is described by a focused-diffusion transport where the strength of the pitch-angle scattering processes is taken from a 2D-slab turbulence model conveniently scaled with radial distance and particle energy. The predicted energy spectrum integrated over time-intervals of different durations shows that early in the SEP events it may flatten with energy but quickly steepens after a few hours (e.g., ~6 hours after the shock initiation time in Figures 9-10 in [39]). The evolution of the energy spectra in such models depends on the combination of the particle acceleration processes at the region magnetically connected with the observer, the escape of particles from the shock complex, and the processes of particle transport in IP space.

The flat-spectra periods shown in Section 2 are instantaneously observed (as opposite to integrated over time) and start not just after the onset of the SEP event but when the shocks are already located at some distance from the Sun (for example, at ~0.3-0.6 AU from the Sun for the events in Figures 1 and 2). That the flat-spectra periods are observed in intense SEP events (all of them associated with fast CMEs, see Table 1), are of long duration (>12 h), and occur while particle intensities are still increasing, suggest that the intensity of the event and the way particles propagate from a strong shock able to inject a large number of particles toward the spacecraft play essential roles in the development of these flat spectra.

The model of Ng et al. [41] coupled the evolution of the shock accelerated particles with that of Alfvén waves in the solar wind (where the wave growth depended on the momentum gradients of the particle populations and shock acceleration was represented by a continuous injection of particles with a prescribed temporal evolution and energy source spectra). This model predicts proton energy spectra that flatten at some energy below ~10 MeV and that can be observed only after a certain time has passed since the onset of the SEP event (e.g., Figures 6 and 9 in [41]). In this model, high-energy protons streaming outward from the shock amplify the pre-existing ambient Alfvén waves, in such a way that the slower low-energy protons, as well as energetic minor ions, find themselves traveling through resonant Alfvén waves previously amplified by the higher velocity protons. The process results in a self-regulatory mechanism where the waves amplified by the higher-energy protons control the transport of the lower-energy protons. The harder the spectra of the injected shock-accelerated particles the faster the growth of the amplified waves, and hence the stronger the effect on the lower energy particle transport. The result is a retardation of the outward SEP transport by the moving region of amplified waves, generating regions of flat intensity spatial profiles behind this region but steep profiles in and ahead of this region. The measurement of a flat-energy spectrum by a distant spacecraft depends on how this region of amplified waves propagates, how the particle acceleration at the shock evolves, and how the waves are continuously amplified by the higher energy protons. The model, however, predicts flat-spectra intervals of short duration (e.g., Figure 8 in [41]) that contrast with the long-lasting periods shown in Section 2.

Le Roux and Webb [43] developed a time-dependent focused transport model to investigate particle acceleration at fast IP shocks via first-order Fermi mechanism with a prescribed Alfvén wave spectrum upstream and across the shock to scatter the particles. A minimum particle speed is required for the particles traveling along the magnetic field to stay ahead of the traveling shock, which depends on the shock speed V_{sh}, the angle between the upstream magnetic field and the shock normal θ_{Bn}, and the particle speed along the magnetic field $v\mu$ (where μ is the particle pitch-angle and v is the particle speed). Whereas particles with $v\mu<(V_{sh}-V_{sw})sec(\theta_{Bn})$ cannot escape from the shock vicinity and participate in the acceleration processes, particles with $v\mu>(V_{sh}-V_{sw})sec(\theta_{Bn})$ are capable of moving forward along the magnetic field (where V_{sw} is the upstream solar wind speed). The result is that the energy spectra observed at some distance from the shock show (i) a deficit of particles at the energies corresponding to the shock speed, (ii) a flattening at higher energies corresponding to particle speeds and pitch-angles μ such that their escape from the shock vicinity is impeded, and (iii) a decrease at even higher intensities (e.g., Figure 3a in [43]). The arrival of the shock is then characterized by a steepening of the energy spectra coinciding with the arrival of the particles confined in the shock vicinity. Note that the temporal evolution of V_{sh} and θ_{Bn}, as well as the varying conditions for particle scattering close to the shock during the transit of the shock from the Sun to 1 AU lead to different conditions for particle escape from the shock vicinity. Therefore, this mechanism needs to include this variability to be consistent with the long-lasting flat spectra observed in multiple events.

According to these models, both the capability of the particles to escape from the shock vicinity and the interactions undergone by the particles during their transport to the spacecraft are at the root of the observation of these flat-spectra periods. The amplification of waves by high-energy particles seems to play an essential role in the transport of the lower energy particles [41]. Most of the SEP events exhibiting flat-energy spectra shown in Section 2 are intense and associated with solar eruptions capable of generating >100 MeV protons. However, for the SEP events shown in Figures 1(c) and 4(a), the contribution of high-energy particles was minimal. In most cases, the flat-spectra periods were observed when the high-energy intensities had already declined suggesting that, if they were the cause of the amplification of the resonating Alfvén waves, those were generated earlier in the event and did not decay throughout the event. We note that the presence of these amplified waves did not limit the intensities measured at 1 AU, since in most of the cases particle intensities continued to increase even when flat energy spectra were observed.

The elevated intensities and the presence of high-energy protons do not seem to be conditions necessary for the observation of flat-spectra at low energies. For example, Figure 5 shows an intense SEP event measured by ACE/EPAM/LEMS120 and GOES-13 with elevated intensities (even higher than those of the events shown in Section 2) and where even ~400 MeV protons were observed. The origin of the shock in this event may be related to two consecutive solar eruptions [44], and the elevated proton intensities might have resulted from the passage of a pre-existing IP shock early on day 67 [45]. It is possible that the pre-conditioning of the ambient medium may inhibit the amplification of Alfvén waves [46] and hence prevent the formation of SEP flat spectra. Therefore, the medium where the particles propagate plays also a role in the observation of flat-spectra intervals.

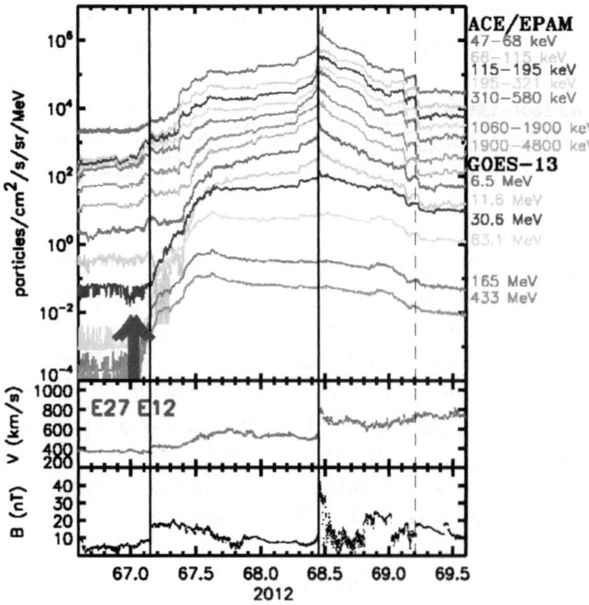

Figure 5. Five-minute averages of spin-averaged ion intensities measured by ACE/EPAM/LEMS120 and of proton intensities as measured by GOES-13-SEM during an intense SEP event in March 2012, together with the proton solar wind speed as measured by Wind/SWE and the magnetic field magnitude as measured by the magnetic field experiment on board ACE. The purple arrows indicate the occurrence of the solar eruptions that generated the event at the longitude indicated below. The vertical solid lines indicate the passage of interplanetary shocks, and the dashed vertical line the leading edge of an ICME.

4. Conclusions

The goal of this paper is to highlight the peculiarity of long-lasting periods with flat energy spectra observed in intense SEP events associated with the passage of IP shocks, especially at low-energies, from about ~50 keV up to about ~1 MeV. The processes by which shock-accelerated particles propagate from the shock toward the observer are likely the dominant causes for the formation of these flat-spectra periods, where the amplification of Alfvén waves supposedly plays a substantial role. A straightforward question to ask concerns the extension of these spectra to energies below the lower thresholds of the energetic particle instruments, i.e. whether these spectra continue flat to energies below ~50 keV, or their intensities rise with decreasing energy up to the more abundant thermal particle intensities. Extended measurements of proton distributions from solar wind thermal through suprathermal energies for this type of events will be analyzed in future works.

Acknowledgments

All data used in this paper can be downloaded from www.srl.caltech.edu/ACE/ASC/ sprg.ssl.berkeley.edu/wind3dp/data/wi/3dp/ , sd-www.jhuapl.edu/IMP/ , srl.utu.fi/erne_data/, satdat.ngdc.noaa.gov/sem/goes/data/avg/ , www-ssc.igpp.ucla.edu/ssc/stereo/ www2.physik.uni-kiel.de/stereo/ , www.srl.caltech.edu/STEREO/ . We acknowledge all the teams of the science instruments for making their data used in this paper available. This work is supported by NASA-HGI grant NNX16AF73G and NASA/LWS grant NNX15AD03G.

References

[1] Mewaldt R A, Looper M D, Cohen C M S, Haggerty D K, Labrador A W, Leske R A, Mason G M, Mazur J E and von Rosenvinge T T 2012 *Space Sci. Rev.* **171** 97
[2] Desai M I, Mason G M, Dayeh M A, Ebert R W, McComas D J, Li G, Cohen C M S, Mewaldt R A, Schwadron N A and Smith C W 2016 *Astrophys. J.* **816** 68
[3] Doran D J and Dalla S 2016 *Solar Phys.* **291** 2071
[4] Laurenza M, Consolini G, Storini M and Damiani A 2012 *Astrophys. Space Sci. Trans.* **8** 19
[5] Laurenza M, Consolini G, Storini M and Damiani A 2015 *J. Phys. Conf. Ser.* **632** 012066
[6] van Nes P, Reinhard R, Sanderson T, Wenzel K-P and Zwickl R 1984 *J. Geophys. Res.* **89** 2122
[7] Ho G C, Lario D, Decker R B, Smith C W and Hu Q 2008 *AIP Conf. Proc.* **1039** 184
[8] Laurenza M, Consolini G, Storini M, Pallocchia G and Damiani A 2016 *J. Phys. Conf. Ser.* **767**

012015
[9] Pallocchia G, Laurenza M, Consolini G 2017 *Astrophys. J.* **837** 158
[10] Lario D, Hu Q, Ho G C, Decker R B, Roelof E C and Smith C W 2005 Proc. Solar Wind 11/ SOHO 16, Connecting Sun and Heliosphere (Eds. B Fleck et al.), ESA SP **592** 81
[11] Gold R E, Krimigis S M, Hawkins S E, Haggerty D K, Lohr D A, Fiore E, Armstrong T P, Holland G and Lanzerotti L J 1998 *Space. Sci. Rev.* **86** 541
[12] Lin R P, Anderson K A, Ashford S, Carlson C et al. 1995 *Space. Sci. Rev.* **71** 125
[13] Wilson III L B, Cattell C A, Kellog P J, Goetz K et al. 2010 *J. Geophys. Res.* **115** A12104
[14] Onsager T G, Grubb R, Kunches J, Matheson L, Speich D, Zwickl R and Sauer H 1996 *Proc. SPIE (Ed. E R Washwell)* **2812** 281
[15] Sandberg I, Jiggens P, Heynderickx and Daglis I A 2014 *Geophys. Res. Lett.* **41** 4435
[16] Torsti J, Valtonen E, Lumme M, Peltonen P, Eronen T, et al. 1995 *Solar Phys.* **162** 505
[17] McComas D J, Bame S J, Baker P, Feldman W C, Phillips J L, Riley P and Griffee J W 1998 *Space. Sci. Rev.* **86** 563
[18] Smith C W, L'Heureux J, Ness N F, Acuna M H, Burlaga L et al. 1998 *Space. Sci. Rev.* **86** 613
[19] Ogilvie K W, Chornay D J, Fritzenreiter R J, Hunsaker F., et al. 1995 *Space. Sci. Rev.* **71** 55
[20] Richardson I G and Cane H V 2017 www.srl.caltech.edu/ACE/ASC/DATA/level3/icmetable2
[21] Sarris E T, Krimigis S M and Armstrong T P 1976 *J. Geophys. Res.* **81** 2341
[22] Lario D, Marsden R G, Sanderson T R, Maksimovic M et al. 2000 *J. Geophys. Res.* **105** 18251
[23] Luhmann J G, Curtis D W, Scroeder P, et al. 2008 *Space. Sci. Rev.* **136** 117
[24] Muller-Mellin R, Bottcher S, Falenski J, Rode E, et al. 2008 *Space. Sci. Rev.* **136** 363
[25] Mewaldt R A, Cohen C M S, Cook W R, Cummings A C et al. 2008 *Space. Sci. Rev.* **136** 285
[26] von Rosenvinge T T, Reames D V, Baker R, Hawk J, et al. 2008 *Space. Sci. Rev.* **136** 391
[27] Gomez-Herrero R 2013 "STEREO/SEPT Level 2 Science Data Format Specifications and Caveats" in www2.physik.uni-kiel.de/stereo/data/sept/level2/SEPT_L2_description.pdf
[28] Galvin A B, Kistler L M, Popecki M A, Farrugia C J et al. 2008 *Space. Sci. Rev.* **136** 437
[29] Acuna M H, Curtis D, Scheifele J L, Russell C T, Schroeder P, Szabo A and Luhmann J G 2008 *Space. Sci. Rev.* **136** 203
[30] Lario D, Aran A, Gomez-Herrero R, Dresing N, Heber B, Ho G C, Decker R B and Roelof E C 2013 *Astrophys. J.* **767** 41
[31] Giacalone J 2004 *Astrophys. J.* **609** 452
[32] Lario D, Kwon R Y, Riley P and Raouafi N E 2017 *Astrophys. J.* **847** 103
[33] Cane H V, Reames D V and von Rosenvinge T T 1988 *J. Geophys. Res.* **93** 95558
[34] Lario D, Sanahuja B and Heras A M 1998 *Astrophys. J.* **509** 415
[35] Pomoell J, Aran A, Jacobs C, Rodriguez-Gasen R, Poedts S and Sanahuja B 2015 *J. Space Weather and Space Climate* **5** A12, 10.1051/swsc/2015015
[36] Luhmann J G, Mays M L, Odstrcil D, Li Y, Bain H, Lee C O, Galvin A B, Mewaldt R A, Cohen C M S, Leske R A, Larson D and Futaana Y 2017 *Space Weather* **15** 934
[37] Schwadron N, Gorby M, Torok T, Downs C, et al. 2014 *Space Weather* **12** 323
[38] Li G, Zank G P and Rice W K M 2003 *J. Geophys. Res.* **108** 1082, 10.1029/2002JA009666
[39] Hu J, Li G, Ao X, Zank G P and Verkhoglyadova O 2017 *J. Geophys. Res.* **122** 10938
[40] Boronikov D, Sokolov I V, Roussev I I, Taktakishvili A and Gombosi T I 2018, 2018arxiv180607030B
[41] Ng C K, Reames D V and Tylka A J 2003 *Astrophys. J.* **591** 461
[42] Heras A M, Sanahuja B, Lario D, Smith Z, Detman T and Dryer M 1995 *Astrophys. J.* **445** 497
[43] Le Roux J A and Webb G 2012 *Astrophys. J.* **746** 104
[44] Lario D, Ho G C, Roelof E C, Anderson B J and Korth H 2013 *J. Geophys. Res.* **118** 63
[45] Lario D and Karelitz A 2014 *J. Geophys. Res.* **119** 4185
[46] Lario D, Aran A and Decker R B 2008 *Space Weather* **6** S12001, 10.1029/2008SW000403

IOP Conf. Series: Journal of Physics: Conf. Series **1100** (2018) 012015 doi:10.1088/1742-6596/1100/1/012015

Investigation of different small-scale flux-rope acceleration scenarios for energetic particles in the solar wind near Earth

J A le Roux[1,2], G P Zank[1,2] and O V Khabarova[3]

[1] Department of Space Science, University of Alabama in Huntsville (UAH), 320 Sparkman Drive, AL 35805, USA
[2]Center of Space Plasma Science and Aeronomic Research (CSPAR), University of Alabama in Huntsville (UAH), 320 Sparkman Drive, AL 35805, USA
[3]Heliophysics Laboratory, Pushkov Institute for Terrestrial Magnetism, Ionosphere, and Radio Wave Propagation (IZMIRAN), Moscow, Russia

jar0013@uah.edu

Abstract. Our previous kinetic transport theoretical development for energetic particle acceleration by and large-scale transport through solar wind regions with numerous dynamic small-scale flux ropes in the strong guide/background field limit is further analyzed and extended. The basic flux-rope acceleration mechanisms and the issue of compressibility are further clarified by applying concepts such as magnetic curvature and shear flow to these structures. A set of new coupled focused-transport-MHD turbulence equations is presented for modeling coherent and stochastic energetic particle acceleration by small-scale flux ropes self-consistently. Furthermore, test particle coherent and stochastic acceleration rates are compared for the different flux-rope acceleration mechanisms, and stochastic acceleration and pitch-angle scattering rates for flux ropes and Alfvén waves are compared, for energetic protons at Earth.

1. Introduction

From recent observations near 1 AU we learned: (i) Solar wind regions near primary current sheets (the heliospheric current sheet or current sheets associated with interplanetary coronal mass ejections behind traveling shocks and corotating interaction regions) are filled with contracting and merging small-scale flux ropes with cross sections belonging to the turbulence inertial range. These structures are generated when primary current sheets undergo turbulent magnetic reconnection [1,2]. (ii) Enhanced energetic particle fluxes up to MeV energies correlate well with these flux-rope regions [1,2,3]. (iii) Dynamic small-scale flux ropes are especially efficient accelerators during strong compression [1,2]. (iv) An unprecedented number of small-scale flux ropes were identified at 1 AU using the Grad-Shafranov reconstruction approach [4]. The latter result is consistent with the common occurrence of dynamic small-scale flux ropes in the low-latitude solar wind near 1 AU as a natural development of local MHD turbulence in a high conductivity plasma with a strong guide/background field and a plasma β of order 1 or less [5]. In this limit, the observed formation of energetic particle power-law spectra at 1 AU can be interpreted as a consequence of flux-rope dynamics, such as

IOP Conf. Series: Journal of Physics: Conf. Series **1100** (2018) 012015 doi:10.1088/1742-6596/1100/1/012015

merging, primarily occurring in the 2D plane perpendicular to a strong guide/background field [6]. Following up on the promising idea of efficient acceleration when energetic particles interact with numerous contracting and merging small-scale flux ropes [7,8], Zank et al. and le Roux et al. developed comprehensive focused transport theories of energetic particle acceleration in a turbulent magnetized plasma containing dynamic small-scale flux ropes in the strong guide field limit [9,10].

Here, the basic small-scale flux rope acceleration mechanisms are further investigated using concepts such as magnetic curvature and the shear-flow tensor. The difference between the incompressible and compressible limits of flux-rope dynamics for acceleration, and the close link between guiding center kinetic and focused transport theory are addressed anew. An extended theory is presented for modeling self-consistent energetic particle acceleration by small-scale flux ropes. This involves coupling of the focused transport equation to a new MHD equation for total magnetic island energy density advective transport in the non-uniform solar wind medium, derived from a recent version of nearly incompressible MHD theory for solar wind turbulence [5]. The latter equation includes magnetic island damping rates derived from assuming conservation of total energy in the exchange of energy between energetic particles and magnetic islands during coherent and stochastic acceleration. Finally, for energetic protons at 1 AU, coherent and stochastic flux-rope acceleration rates for the different flux-rope acceleration mechanisms are discussed, and stochastic acceleration and pitch-angle scattering rates for flux ropes and Alfvén waves are compared.

2. The small-scale flux rope acceleration mechanisms

2.1 *A Guiding center kinetic theory perspective*

Small-scale flux ropes detected near Earth have cross sections of L_l~0.01-0.001 AU [1,2]. Near 1 AU, suprathermal protons, e.g., have gyro-radii $r_g << L_l$ for a wide range of energies that easily include MeV energies. Thus, standard guiding center kinetic theory, which is restricted to gyro-radii much less than scale of the electromagnetic field in the plasma, is well suited for modeling energetic particle transport through and acceleration in solar wind regions near Earth with numerous contracting and merging small-scale flux rope structures. In guiding center kinetic theory, the gyro-phase-averaged rate of change in kinetic energy for energetic charged particles can be expressed in different ways:

$$\begin{aligned}\left\langle \frac{dK}{dt} \right\rangle_\phi &\approx \left[q\underline{E} \bullet \left(v_\parallel \underline{b} + \frac{mv_\parallel^2}{qB} \underline{b} \times \underline{\kappa} \right) \right]_\parallel + \left[M \frac{\partial B}{\partial t} + q\underline{E} \bullet \left(\frac{M}{q} \frac{\underline{B} \times \underline{\nabla} B}{B^2} + \frac{M}{q} (\underline{\nabla} \times \underline{b})_\parallel \underline{b} \right) \right]_\perp \\ &\approx \left[qE_{REC\parallel} v\mu + mv^2\mu^2 (\underline{V}_E \bullet \underline{\kappa}) \right]_\parallel + \left[(1-\mu^2) M' \left(\frac{\partial B}{\partial t} + (\underline{V}_E \bullet \underline{\nabla}) B \right) + (1-\mu^2) B \frac{dM'}{dt} \right]_\perp \qquad (1) \\ &\approx \left[qE_{REC\parallel} v\mu + mv^2\mu^2 (\underline{V}_E \bullet \underline{\kappa}) \right]_\parallel + \left[-mv^2 \frac{1}{2} (1-\mu^2) \left[(\underline{V}_E \bullet \underline{\kappa}) + (\underline{\nabla} \bullet \underline{V}_E) \right] \right]_\perp \end{aligned}$$

where q is the net particle charge, $v_\parallel$ is the parallel guiding velocity component, $\underline{b}$ is the unit vector along the flux-rope magnetic field, $\underline{\kappa} = (\underline{b} \bullet \underline{\nabla})\underline{b}$ is the curvature of the flux-rope magnetic field, M is the magnetic moment of a gyrating particle, $\underline{E} \bullet \underline{b} = E_{REC\parallel}$ is the parallel reconnection electric field component associated with merging flux-rope pairs, μ is the cosine of the particle pitch angle, $\underline{V}_E$ is the electric field drift (plasma drift) velocity (velocity at which the curved flux-rope magnetic field is contracting or merging), and M' is the magnetic moment for $\mu = 0$. Equation (1), 1st line, states the basic flux-rope acceleration mechanisms grouped in terms of parallel kinetic energy changes (terms in 1st square bracket) and perpendicular kinetic energy changes (terms in 2nd square bracket). The mechanisms are: (1) Parallel guiding center motion acceleration by the parallel reconnection electric field component $E_{REC\parallel}$ generated in reconnection regions between merging neighbouring flux ropes (1st term in 1st square bracket), (2) curvature drift acceleration by the motional electric field induced by contracting or merging small-scale flux ropes (2nd term in the 1st square bracket), (3) betatron acceleration due time variations in the flux-rope magnetic field strength (1st term in the 2nd square bracket), (4) grad-B drift acceleration by the motional electric field induced by contraction and merging of flux ropes (2nd term in the 2nd square bracket) and, (5) parallel drift acceleration by $E_{REC||}$ (the last term in 2nd square bracket). Direct comparison of the terms in the 1st line of (1) with the corresponding terms in the 2nd line reveals that one can express the curvature drift acceleration term in terms of $\underline{V}_E \bullet \underline{\kappa}$ (advection of curved flux-rope magnetic field at plasma drift velocity), and the grad-B drift acceleration term in terms of $(\underline{V}_E \bullet \underline{\nabla})B$ (advection of the perpendicular gradient in flux-rope field strength at the plasma drift velocity). This version of the grad-B drift acceleration term can be combined with the betatron acceleration term into a generalized betatron acceleration expression $M'dB/dt = M'(\partial B/\partial t + (\underline{V}_E \bullet \underline{\nabla})B)$ (1st 2 terms in 2nd square bracket in line 2) [11]. Comparison of the last terms in line 1 and line 2 indicates that approximate conservation of magnetic moment requires a small $E_{REC||}$-value. Furthermore, comparing the terms in the 2nd square bracket in line 2 with the corresponding terms in line 3 reveals that the generalized betatron acceleration expression in line 2 can be related to a combination of the $\underline{V}_E \bullet \underline{\kappa}$ and the $\underline{\nabla} \bullet \underline{V}_E$- terms, assuming approximate magnetic moment conservation (a small $E_{REC||}$-value). Thus, generalized betatron acceleration is determined by a competition between incompressible flux-rope contraction or merging ($\underline{V}_E \bullet \underline{\kappa} > 0$) and compressible contraction or merging ($\underline{\nabla} \bullet \underline{V}_E < 0$). However, such a competition does not appear in the curvature drift acceleration term that depends only on $\underline{V}_E \bullet \underline{\kappa}$. To investigate this issue further it is useful to introduce the relationships

$$\underline{V}_E \bullet \underline{\kappa} = -\underline{b} \bullet (\underline{b} \bullet \underline{\nabla})\underline{V}_E = -\left[b_i b_j \sigma_{ij} + \frac{1}{3}(\underline{\nabla} \bullet \underline{V}_E)\delta_{ij} \right] \tag{2}$$

where we relate the magnetic curvature advection term $\underline{V}_E \bullet \underline{\kappa}$ to the parallel shear-flow term $\underline{b} \bullet (\underline{b} \bullet \underline{\nabla})\underline{V}_E$ which in turn is expressed in terms of the shear-flow tensor σ_{ij}

$$\sigma_{ij} = \frac{1}{2}\left[\frac{\partial V_{Ei}}{\partial x_j} + \frac{\partial V_{Ej}}{\partial x_i} - \frac{2}{3}(\underline{\nabla} \bullet \underline{V}_E)\delta_{ij} \right] \tag{3}$$

using the Cauchy Stokes theorem [12]. Upon inserting (2) in the bottom line of equation (1), we get

$$\left\langle \frac{dK}{dt} \right\rangle_\phi \approx qE_{REC||} v\mu - \frac{1}{3}mv^2(\underline{\nabla} \bullet \underline{V}_E) - \frac{1}{2}mv^2(3\mu^2 - 1)b_i b_j \sigma_{ij} \tag{4}$$

where the term containing $\underline{\nabla} \bullet \underline{V}_E$ can be recognized as the standard Parker cosmic-ray transport equation term for the combination of curvature drift, grad-B drift, betatron and parallel drift acceleration that collectively becomes plasma drift compression acceleration acting on the isotropic part of the particle distribution $f_0(p)$ [13,14]. The last term in (4) can be interpreted likewise as combining parallel shear flow tensor acceleration associated with curvature drift acceleration $(-mv^2\mu^2 b_i b_j \sigma_{ij})$ with parallel shear flow tensor acceleration associated with unified grad-B drift, betatron and parallel drift acceleration $(+mv^2 1/2(1-\mu^2)b_i b_j \sigma_{ij})$ that collectively becomes parallel shear-flow tensor acceleration acting on the anisotropic part of the particle distribution related to the 2nd moment of the particle distribution $f_2(p)$. The interpretation in terms of particle anisotropy moments is made assuming a Legendre moment expansion for the energetic particle distribution that is nearly isotropic so that $f(\underline{p}) = f_0(p) + 3\mu f_1(p) + 5/2(3\mu^2 - 1)f_2(p)$, where $f_1(p)$ is the 1st moment and $f_2(p)$ is the 2nd moment in the anisotropic part of the energetic particle distribution, and averaging over all μ-values. Upon decomposing the shear flow tensor according to $\sigma_{ij} = \sigma_{ij}^{sh} - 1/3(\underline{\nabla} \bullet \underline{V}_E)\delta_{ij}$, where $b_i b_j \sigma_{ij}^{sh} = 1/2(b_i b_j \partial V_{Ei}/\partial x_j + b_j b_i \partial V_{Ej}/\partial x_i) = \underline{b} \bullet (\underline{b} \bullet \underline{\nabla})\underline{V}_E$, in (4), the result is

$$\left\langle \frac{dK}{dt} \right\rangle_\phi \approx qE_{REC\|} v\mu - \frac{1}{3} mv^2 (\underline{\nabla} \bullet \underline{V}_E) + \frac{1}{3} mv^2 \frac{1}{2} (3\mu^2 - 1)(\underline{\nabla} \bullet \underline{V}_E) - \frac{1}{2} mv^2 (3\mu^2 - 1) \underline{b} \bullet (\underline{b} \bullet \underline{\nabla}) \underline{V}_E \quad (5)$$

In addition to the Parker transport compression term, there is a new compression term (3rd term in (5)). Just like the Parker compression term, this extra term also combines compression acceleration linked to curvature drift acceleration $\left(+1/3mv^2\mu^2(\underline{\nabla} \bullet \underline{V}_E)\right)$ with compression acceleration associated with unified betatron, grad-B drift, and parallel drift acceleration $\left(-1/3mv^2 1/2(1-\mu^2)(\underline{\nabla} \bullet \underline{V}_E)\right)$ to form collectively a plasma drift compression acceleration term. Different from the Parker transport compression term that acts on the isotropic part of the particle distribution $f_0(p)$, the new compression term acts on the anisotropic part of the particle distribution connected with $f_2(p)$. The last term in (5) combines parallel shear flow acceleration associated with curvature drift acceleration $\left(-mv^2\mu^2\underline{b} \bullet (\underline{b} \bullet \underline{\nabla})\underline{V}_E\right)$ with parallel shear flow acceleration linked to unified betatron, grad-B drift, and parallel drift acceleration $\left(+mv^2 1/2(1-\mu^2)\underline{b} \bullet (\underline{b} \bullet \underline{\nabla})\underline{V}_E\right)$ to form collectively a shear-flow acceleration term (reduced shear flow tensor without the compression term) that acts only on the anisotropic part of the energetic particle distribution related to $f_2(p)$. Upon combining the two $\underline{\nabla} \bullet \underline{V}_E$-terms in (5), and doing the substitution $-\underline{b} \bullet (\underline{b} \bullet \underline{\nabla})\underline{V}_E = \underline{V}_E \bullet \underline{\kappa}$,

$$\left\langle \frac{dK}{dt} \right\rangle_\phi \approx qE_{REC\|} v\mu - \frac{1}{2} mv^2 (1-\mu^2)(\underline{\nabla} \bullet \underline{V}_E) + \frac{1}{2} mv^2 (3\mu^2 - 1)(\underline{V}_E \bullet \underline{\kappa}) \quad (6)$$

whereby the last line of equation (1) is recovered. In the process we gained new insight in the $\underline{\nabla} \bullet \underline{V}_E$-term in equation (1), having determined that it can be interpreted as a combination of curvature drift acceleration with unified betatron, grad-B drift, and parallel drift acceleration acting collectively as plasma drift compression acceleration on both the isotropic and the anisotropic part of the energetic particle distribution.The $\underline{V}_E \bullet \underline{\kappa}$-term, on the other hand, can be viewed as a combination of curvature drift acceleration with unified betatron, grad-B drift, and parallel drift acceleration acting collectively as plasma drift shear flow acceleration only on the anisotropic part of the distribution.

2.2 *Compressible versus incompressible flux-rope drift and betatron acceleration*

Consider first the limit of incompressible flux-rope contraction or merging (in the strong guide field limit we interpret this to mean magnetic island area conservation during contraction or merging in the 2D plane perpendicular to the guide/background magnetic field) when $0 < -\underline{\nabla} \bullet \underline{V}_E \ll \underline{V}_E \bullet \underline{\kappa} = -\underline{b} \bullet (\underline{b} \bullet \underline{\nabla})\underline{V}_E > 0$. Then, from equations (1), (5), and (6) we have

$$\begin{aligned} \left\langle \frac{dK}{dt} \right\rangle_\phi &\approx \left[qE_{REC\|} v\mu + mv^2\mu^2 (\underline{V}_E \bullet \underline{\kappa})\right]_\| + \left[(1-\mu^2)M'\left(\frac{\partial B}{\partial t} + (\underline{V}_E \bullet \underline{\nabla})B\right) + (1-\mu^2)B\frac{dM'}{dt}\right]_\perp \\ &= \left[qE_{REC\|} v\mu + mv^2\mu^2 (\underline{V}_E \bullet \underline{\kappa})\right]_\| + \left[-mv^2\frac{1}{2}(1-\mu^2)(\underline{V}_E \bullet \underline{\kappa})\right]_\perp \\ &= \left[qE_{REC\|} v\mu - mv^2\mu^2 \underline{b} \bullet (\underline{b} \bullet \underline{\nabla})\underline{V}_E\right]_\| + \left[mv^2\frac{1}{2}(1-\mu^2)\underline{b} \bullet (\underline{b} \bullet \underline{\nabla})\underline{V}_E\right]_\perp \end{aligned} \quad (7)$$

For magnetic island contraction or merging ($\underline{V}_E \bullet \underline{\kappa} > 0$) it is clear that curvature drift acceleration must result in parallel kinetic energy gain (2nd term in 1st square bracket in (7)). However, for generalized betatron acceleration (compare 1st 2 terms in 2nd square bracket in line 1 with the term in 2nd square bracket in lines 2 and 3) the result is perpendicular kinetic energy loss, assuming approximate conservation of M (neglecting the last term in the 2nd square bracket in line 1). it means that $M'dB/dt = M'(\partial B/\partial t + (\underline{V}_E \bullet \underline{\nabla})B) < 0$. In this way we can relate the perpendicular energy loss to a decreasing magnetic field strength inside flux-rope structures following the plasma drift $\underline{V}_E$ during incompressible flux-rope contraction or merging. If in the generalized betratron expression grad-*B* drift acceleration is negligible compared to betatron acceleration [15], perpendicular kinetic energy loss is caused by the time variation in the field strength rather than the spatial variation so that energy loss is predominantly associated with the standard betatron acceleration term $M'(\partial B/\partial t)$. Note that incompressible contraction or merging is associated with a negative parallel shear flow component in flux ropes ($\underline{V}_E \bullet \underline{\kappa} = -\underline{b}\bullet(\underline{b}\bullet\underline{\nabla})\underline{V}_E > 0$).

In conclusion, energetic particle acceleration in incompressible contracting and merging small-scale flux ropes is related to parallel plasma drift shear flow acceleration involving a competition between parallel kinetic energy gain from curvature drift acceleration, and perpendicular kinetic energy loss predominantly from betatron acceleration. On average, the net acceleration from the combination of the two acceleration processes only involves the anisotropic part $f_2(p)$ of the particle distribution, assuming an expansion of the distribution to the 2nd moment.

When flux-rope contraction and merging occurs in the compressible limit (area reduction during contraction or merging in 2D magnetic island plane perpendicular to the guide magnetic field) so that $0< -\underline{\nabla}\bullet\underline{V}_E >> \underline{V}_E\bullet\underline{\kappa} = -\underline{b}\bullet(\underline{b}\bullet\underline{\nabla})\underline{V}_E > 0$, we find from equations (1), (5), and (6) that

$$\left\langle \frac{dK}{dt} \right\rangle_\phi \approx \left[qE_{REC\|}v\mu + mv^2\mu^2(\underline{V}_E \bullet \underline{\kappa}) \right]_\| + \left[(1-\mu^2)M'\left(\frac{\partial B}{\partial t} + (\underline{V}_E \bullet \underline{\nabla})B \right) + (1-\mu^2)B\frac{dM'}{dt} \right]_\perp$$
$$\approx \left[qE_{REC\|}v\mu - mv^2\mu^2\frac{1}{3}(\underline{\nabla}\bullet\underline{V}_E) \right]_\| + \left[-mv^2\frac{1}{2}(1-\mu^2)\frac{2}{3}(\underline{\nabla}\bullet\underline{V}_E) \right]_\perp \quad (8)$$

where for simplicity compression terms acting on $f_2(p)$ has been neglected in favour of those acting on $f_0(p)$ (nearly isotropic particle distribution). During contraction and merging in the compressible limit $\underline{\nabla}\bullet\underline{V}_E < 0$ and, although relatively small, $\underline{V}_E\bullet\underline{\kappa} > 0$. Thus, the 2nd term in the 1st square bracket in both lines 1 and 2 of (8) suggests that curvature drift acceleration will contribute to parallel kinetic energy gain. By comparing the generalized betatron expression (1st two terms in the 2nd square bracket in line 1) with the term in the 2nd square bracket in line 2, it follows that the generalized betatron acceleration term will result in perpendicular kinetic energy gain ($\underline{\nabla}\bullet\underline{V}_E < 0$), assuming $dM'/dt \approx 0$. Therefore, $M'dB/dt = M'(\partial B/\partial t + (\underline{V}_E\bullet\underline{\nabla})B) > 0$, relating the perpendicular energy gain to an increasing magnetic field strength with time following the plasma drift flow in flux ropes contracting and merging in the compressible limit . If grad-*B* drift acceleration is negligible compared to betatron acceleration in the generalized betatron expression, as mentioned above, the perpendicular kinetic energy gain is associated with the standard betatron acceleration term $M'(\partial B/\partial t)$ indicating an increasing flux-rope field strength in time rather than spatially.

To summarize, in contrast to finding curvature drift energy gain and betatron energy loss associated with a negative parallel component of flux-rope shear flow during incompressible contraction or merging, we find both curvature drift and betatron energy gain associated with flux-rope flow during compressible flux-rope contraction or merging [9,10]. Furthermore, on average, net energy gain for the two acceleration processes in the incompressible limit is linked only to the anisotropic part of the energetic particle distribution, whereas net energy gain for the same two processes in the compressible limit involves both the isotropic and anisotropic part of the distribution. For a further discussion of the role of shear flow and compression in small-scale flux rope acceleration, see [16].

Consider finally the 1st term in equation (1) which is associated with parallel guiding center motion acceleration by the parallel reconnection electric field formed at the interface of merging small-scale flux ropes. In this case, the result is parallel kinetic energy gain or loss depending on whether the guiding center motion is in the direction of or in the opposite direction of the reconnection electric field force. Therefore it makes sense that, averaged over all μ-values, net acceleration only occur for the anisotropic part of the distribution (zero net acceleration for an isotropic distribution). In this sense the acceleration has the same characteristics as parallel shear flow acceleration, but the difference is that net acceleration in the latter case only depends on $f_2(p)$ whereas acceleration by the parallel reconnection electric field involves $f_1(p)$.

2.3 The focused transport theory connection

We model the induced electric field in contracting and merging flux ropes as $\underline{E}_I = -\underline{U}_I \times \underline{B}$, where $\underline{U}_I$ is the flux-rope plasma flow velocity. Then the contraction/merging velocity in flux ropes $\underline{V}_E = \underline{U}_I = \underline{U}_\perp$. Making this substitution in equations (5) and (6) reveals the close connection between standard guiding center kinetic theory and focused transport kinetic theory that we use to model particle acceleration by dynamic small-scale flux ropes because, according to focused transport theory,

$$\frac{1}{p}\left\langle\frac{dp}{dt}\right\rangle_\phi^I = \mu\left(\frac{q\underline{E}_{REC}}{p} - \frac{1}{v}\frac{d\underline{U}_I}{dt}\right)\bullet\underline{b} - \frac{1}{2}\left(1-\mu^2\right)\left(\underline{\nabla}\bullet\underline{U}_I\right) - \frac{1}{2}\left(3\mu^2-1\right)\underline{b}\bullet\left(\underline{b}\bullet\underline{\nabla}\right)\underline{U}_I \qquad (9)$$

The only difference between equations (5) and (6), and (9) is the presence of an additional acceleration term referring to parallel guiding center motion acceleration by the non-inertial force associated with the parallel component of the acceleration of the flux-rope flow $d\underline{U}_I/dt\bullet\underline{b}$ ($d/dt = \partial/\partial t + \underline{U}\bullet\underline{\nabla}$).

In (9), the reconnection electric field in merging (reconnecting) flux ropes is modeled assuming that flux rope dynamics occur mainly in a 2D plane containing the magnetic island (twist) component $\underline{B}_I$ and flow $\underline{U}_I$ of the flux rope perpendicular to the guide field (axial) component $\underline{B}_0$ when the guide field is strong [6]. Near Earth it appears that assuming $B_I/B_0 \ll 1$ is reasonable [17], and furthermore there is evidence that the flux-rope guide field is aligned with the solar wind spiral magnetic field [18]. Since our focus is on large-scale transport of energetic particles through multiple flux ropes with cross sections on turbulence inertial range scales, we model the reconnection electric field on macroscopic (MHD) scales as $\underline{E}_{REC} = -\underline{U}_I \times \underline{B}_I \parallel \underline{B}_0 \approx \underline{B}_{TOT} = \underline{B}_0 + \underline{B}_I$. Therefore, $\underline{E}_{REC} \approx \underline{E}_{REC\parallel}$. In the strong guide field limit, the electric field induced by flux-rope contraction and merging is mainly in the 2D plane perpendicular to $\underline{B}_0$, because $\underline{E}_I \approx -\underline{U}_I \times\underline{B}_0 \perp \underline{B}_0$. Therefore, curvature and grad-B drift acceleration occur largely in the 2D plane, while parallel guiding center motion acceleration by the parallel reconnection electric field is mainly restricted to the guide field direction. In the strong guide field limit, the magnetic field unit vector $\underline{b}$ can be decomposed as $\underline{b} \approx \underline{b}_0 + \underline{B}_I/B_0$, where $\underline{b}_0$ is the unit vector along the guide/background field. Accordingly, the flux-rope acceleration mechanisms in quasi-2D flux ropes in focused transport theory are classified as follows:

$$\frac{1}{p}\left\langle\frac{dp}{dt}\right\rangle_\phi^I = \mu\left(v_{REC}^I + v_{ACC}^I\right) + \frac{1}{2}\left(3\mu^2-1\right)v_{INC}^I + \frac{1}{2}\left(1-\mu^2\right)v_{COM}^I \qquad (10)$$

where

$$\begin{aligned}
v_{REC}^I &= \frac{q\underline{E}_{REC}\bullet\underline{b}}{p} \approx -\frac{q}{p}\underline{U}_I\times\underline{B}_I\bullet\underline{b}_0 \\
v_{ACC}^I &= -\frac{1}{v}\frac{d\underline{U}_I}{dt}\bullet\underline{b} \\
v_{INC}^I &= -\underline{b}\bullet\left(\underline{b}\bullet\underline{\nabla}\right)\underline{U}_I \approx -\frac{\underline{B}_I}{B_0}\bullet\left(\frac{\underline{B}_I}{B_0}\bullet\underline{\nabla}\right)\underline{U}_I \\
v_{COM}^I &= -\left(\underline{\nabla}\bullet\underline{U}_I\right)
\end{aligned} \qquad (11)$$

In (11) we list the approximate relative momentum rates of change (without the μ-dependence) for the different flux-rope acceleration cases: (i) v_{REC} refers to parallel guiding center motion acceleration by the parallel reconnection electric field force generated in merging flux ropes structures, (ii) v_{ACC} denotes parallel guiding center motion acceleration by the parallel non-inertial force associated with the acceleration of the flux-rope flow, (iii) v_{INC} indicates combined curvature and generalized betatron acceleration (parallel shear-flow acceleration) when flux ropes contract and merge in the incompressible limit ($\underline{U}_I \bullet \underline{\kappa} >> |\underline{\nabla} \bullet \underline{U}_I|$), (iv) and v_{COM} represents curvature drift and generalized betatron acceleration (compression acceleration) for flux ropes contracting and merging in the compressible limit ($|\underline{\nabla} \bullet \underline{U}_I| >> \underline{U}_I \bullet \underline{\kappa}$). The expressions in (11) also serve as approximate expressions for the rate of change of particle pitch angle (related to $\langle d\mu/dt \rangle_\phi$ when including the μ-dependence) induced by the different flux-rope acceleration cases. To this, focused transport theory adds the pitch-angle rate of change generated by the magnetic mirroring force acting on energetic particles in flux ropes which we express as $v_{REF} = v(\underline{\nabla} \bullet \underline{b}) \approx v(\underline{\nabla} \bullet \underline{B}_I / B_0)$.

3. The coupled focused-transport-MHD equations for self-consistent energetic particle acceleration by dynamic small-scale flux ropes

Using perturbation analysis involving the decomposition of the different momentum/pitch-angle variation rates as $v_i^I = \langle v_i^I \rangle + \delta v_i^I$, where $\langle v_i^I \rangle$ is the mean rate and δv_i^I a random fluctuating rate, a focused transport equation for energetic particle interaction with numerous small-scale flux ropes was derived that models both coherent energetic particle acceleration in response to mean, and stochastic (2nd order Fermi) acceleration in response to statistical fluctuations in flux-rope dynamic properties [10]. The current version of the equation includes additional transport terms related to v_{ACC} with additional and more detailed transport coefficient expressions than before. Furthermore, based on nearly incompressible MHD theory for low-frequency solar wind turbulence [5], a new equation was derived for the transport of the energy of the quasi-2D magnetic island component of small-scale flux ropes in a non-uniform solar wind medium. Coupling between the two equations was established by deriving damping coefficients for magnetic island energy on the basis of total energy conservation in the exchange of energy between energetic particles and magnetic islands and including them in the magnetic island transport equation. The basic structure of the coupled equations is as follows:

$$\left(\frac{df}{dt}\right)_{SW} = -\left\langle \nabla \bullet \left(\left\langle \frac{d\underline{x}}{dt} \right\rangle_\varphi^I (\varepsilon_I) f \right) \right\rangle - \frac{1}{p^2}\frac{\partial}{\partial p}\left(p^2 \left\langle \frac{dp}{dt} \right\rangle_\phi^I (\varepsilon_I) f \right) - \frac{\partial}{\partial \mu}\left(\left\langle \frac{d\mu}{dt} \right\rangle_\phi^I (\varepsilon_I) f \right)$$
$$+ \frac{\partial}{\partial \mu}\left(D_{\mu\mu}^I(\varepsilon_I)\frac{\partial f}{\partial \mu} + D_{\mu p}^I(\varepsilon_I)\frac{\partial f}{\partial p} \right) + \frac{1}{p^2}\frac{\partial}{\partial p}\left(p^2 \left[D_{p\mu}^I(\varepsilon_I)\frac{\partial f}{\partial \mu} + D_{pp}^I(\varepsilon_I)\frac{\partial f}{\partial p} \right] \right) \quad (12)$$
$$\left(\frac{d\varepsilon_I}{dt}\right)_{SW} = \gamma_I^{coh}(f)\varepsilon_I + \gamma_I^{stoch}(f)\varepsilon_I$$

where the energetic particle distribution is $f(\underline{x}, p, \mu, t)$. The top equation in (12) is an extended focused transport equation for the propagation of energetic particles through and acceleration by numerous dynamic small-scale flux ropes in the non-uniform solar wind medium. On the left hand side of this equation $(df/dt)_{SW}$ represents the standard focused transport equation for energetic particle transport in the non-uniform solar wind flow and magnetic field [19]. On the right hand side are additional terms for modeling the interaction of energetic particles with dynamic small-scale flux ropes. This includes

$$\left\langle \frac{dp}{dt} \right\rangle_\phi^I (\varepsilon_I) = p\left[\frac{1}{2}(1-\mu^2)\langle v_{COM}^I \rangle(\varepsilon_I) + \frac{1}{2}(3\mu^2 - 1)\langle v_{INC}^I \rangle(\varepsilon_I) \right]$$
$$+ p\left[\mu\left(\langle v_{REC}^I \rangle(\varepsilon_I) + \langle v_{ACC}^I \rangle(\varepsilon_I) \right) \right] \quad (13)$$

which is the average, coherent energetic particle momentum rate of change in response to mean flux-rope properties for all the flux-rope acceleration cases, and

$$D_{pp}^{I}(\varepsilon_I) = p^2\left\langle\left[\frac{1}{2}(1-\mu^2)\delta v_{COM}^{I}(\varepsilon_I) + \frac{1}{2}(3\mu^2-1)\delta v_{INC}^{I}(\varepsilon_I) + \mu\left(\delta v_{REC}^{I}(\varepsilon_I) + \delta v_{ACC}^{I}(\varepsilon_I)\right)\right]^2\right\rangle\tau_{dec} \quad (14)$$

is the momentum diffusion coefficient associated with the variance in the momentum rate of change for all the flux-rope acceleration mechanisms (stochastic acceleration) in response to fluctuations in flux-rope dynamic properties [20], and τ_{dec} is the energetic particle decorrelation time, the time scale on which propagating particles see decorrelated magnetic island properties. The expression for this time scale depends on the assumed model for particle propagation for which two limits, the quasi-linear and the non-linear transport limits, can be specified in our theory depending on the strength of the island magnetic field [10,21]. In the quasi-linear limit, decorrelation for energetic particles occurs through undisturbed guiding center motion, whereas in the non-linear transport limit decorrelation is realized in terms of assumed diffusive guiding center motion predominantly in the guide/background field direction (scattering on smaller magnetic islands produced in a forward cascade). Viewed on large scales, both transport limits manifest as parallel diffusion of energetic particles across numerous flux-rope structures mainly along the guide field, found to be an important element for more efficient acceleration in 3D simulations of particle acceleration by dynamic flux ropes [22]. In the presence of strong energetic particle scattering, as might occur in strong turbulence conditions behind heliospheric shocks, coherent particle acceleration by flux ropes becomes stochastic, resulting in additional momentum diffusion coefficient expressions for 2nd order Fermi acceleration that can be derived by taking the diffusion approximation limit (not shown) [8,9,10].

In equation (12), the bottom equation models the transport of the total energy density ε_I (kinetic plus magnetic) of the magnetic island component of small-scale flux ropes. On the left hand side of this transport equation $(d\varepsilon_I/dt)_{SW}$ is given by

$$\frac{\partial\varepsilon_I}{\partial t} + \nabla\bullet\left(\varepsilon_I\underline{U}_0\right) + \frac{1}{2}\left[1+(4a-1)\sigma_D^I\right]\left(\nabla\bullet\underline{U}_0\right)\varepsilon_I$$
$$-\frac{1}{4}\left[\left(1+\sigma_C^I\right)^{3/2} - \left(1+\sigma_C^I\right)^{1/2}\sigma_D^I + \left(1-\sigma_C^I\right)^{3/2} - \left(1-\sigma_C^I\right)^{1/2}\sigma_D^I\right]\left(\frac{2^{1/2}}{\rho_0^{3/2}}(n\bullet\nabla)\rho_0\right)\varepsilon_I^{3/2} \quad (15)$$
$$+\frac{\left(1-\left(\sigma_C^I\right)^2\right)^{1/2}}{L_I}\left[\left(1+\sigma_C^I\right)^{1/2} + \left(1-\sigma_C^I\right)^{1/2}\right]\left(\frac{2}{\rho_0}\right)^{1/2}\varepsilon_I^{3/2}$$

where $\underline{U}_0$ is the background solar wind flow velocity, a = 1/2 indicates that, statistically, quasi-2D magnetic island energy is distributed axisymmetrically around $\underline{B}_0$, σ_C^I is the normalized cross helicity and σ_D^I is the normalized residual energy associated with magnetic island turbulence [5], ρ_0 is the background solar wind density, and $\underline{n}$ is a unit vector pointing in an arbitrary direction along magnetic island turbulence in the 2D plane perpendicular to $\underline{B}_0$. Equation (15), which models the transport of ε_I in the non-uniform background solar wind medium, was derived from the quasi-2D magnetic island turbulence equations in Elsässer variables in nearly incompressible MHD theory for turbulence in a solar wind with plasma $\beta \sim 1$, thus making it suitable for application in the supersonic solar wind flow near Earth [5]. An interesting aspect of this equation is that in the advection term (2nd term in (15)), $\underline{U}_0$ is present, but not the Alfvén speed, indicating that magnetic island structures are advected with the solar wind. The 3rd and 4th terms in equation (15) describe how large-scale solar wind flow compression and a large-scale density gradient in the solar wind, respectively, can enhance the energy density of magnetic islands structures. The last term models how magnetic island energy density is reduced by a forward cascade of energy during non-linear interactions of magnetic islands. On the right hand side in the bottom equation of (12), $\gamma_I^{coh}(f)$ represents the damping rate of magnetic island energy density when energetic particles experience coherent acceleration by small-scale flux ropes, and $\gamma_I^{stoch}(f)$ models the damping rate of magnetic island energy density when energetic particle are stochastically accelerated by small-scale flux ropes. The expressions of these damping coefficients are given by

$$\begin{aligned}\gamma_I^{coh}(f) &= -\frac{1}{2\pi}\frac{1}{\varepsilon_I}\int_{-1}^{1} d\mu \int_0^{\infty} dp p^2 v \left\langle \frac{dp}{dt} \right\rangle_{\phi}^{I} (\varepsilon_I) f \\ \gamma_I^{stoch}(f) &= \frac{1}{2\pi}\frac{1}{\varepsilon_I}\int_{-1}^{1} d\mu \int_0^{\infty} dp p^2 v \left(D_{p\mu}^I(\varepsilon_I)\frac{\partial f}{\partial \mu} + D_{pp}^I(\varepsilon_I)\frac{\partial f}{\partial p} \right)\end{aligned} \tag{16}$$

These expressions, derived assuming total energy conservation in the energy exchange between energetic particles and small-scale flux ropes, enable us to model energetic particle acceleration by numerous small-scale flux ropes in a self-consistent fashion. This extension of the theory was motivated by test particle solutions for energetic particle acceleration by small-scale flux ropes that generated power-law spectra with high particle pressure [9,10]. It is interesting to note that whereas $\gamma_I^{stoch}(f)$ depends on the pitch-angle and momentum gradients of the energetic particle distribution as we are accustomed to in quasi-linear kinetic theories, $\gamma_I^{coh}(f)$ depends only on the particle distribution function itself. The pitch-angle gradient $\partial f/\partial \mu$ can potentially also result in growth in magnetic island energy density.

4. Comparing coherent flux-rope acceleration mechanisms at Earth

Based on our focused transport theory for energetic particle acceleration discussed above, we derived expressions for the average coherent momentum gain rate ratios for the different flux-rope acceleration cases listed in equation (11). The expressions for energetic ions are

$$\begin{aligned}\frac{\langle dp/dt\rangle_{\phi,\mu}^{INC}}{\langle dp/dt\rangle_{\phi,\mu}^{COM}} &\approx \frac{\langle v_{INC}^I\rangle}{\langle v_{COM}^I\rangle}\frac{3f_2(p)}{f_0(p)} = \frac{\sigma_{INC}^I}{\sigma_{COM}^I}\frac{\langle \delta B_I^2\rangle}{B_0^2}\frac{3f_2(p)}{f_0(p)} \\ \frac{\langle dp/dt\rangle_{\phi,\mu}^{REC}}{\langle dp/dt\rangle_{\phi,\mu}^{COM}} &\approx \frac{\langle v_{REC}^I\rangle}{\langle v_{COM}^I\rangle}\frac{3f_1(p)}{f_0(p)} = \frac{\sigma_{INC}^I}{\sigma_{COM}^I}\frac{Z}{A}\frac{L_I}{d_i}\frac{V_{A0}}{v}\frac{\langle \delta B_I^2\rangle^{1/2}}{B_0}\frac{3f_1(p)}{f_0(p)} \\ \frac{\langle dp/dt\rangle_{\phi,\mu}^{REC}}{\langle dp/dt\rangle_{\phi,\mu}^{INC}} &\approx \frac{\langle v_{REC}^I\rangle}{\langle v_{INC}^I\rangle}\frac{f_1(p)}{f_2(p)} = \frac{\sigma_{INC}^I}{\sigma_{INC}^I}\frac{Z}{A}\frac{L_I}{d_i}\frac{V_{A0}}{v}\frac{B_0}{\langle \delta B_I^2\rangle^{1/2}}\frac{f_1(p)}{f_2(p)}\end{aligned} \tag{17}$$

The derivation of these expressions includes (i) the weighting factor of a nearly isotropic particle distribution, modeled by expanding it up to the 2nd moment in terms of a Legendre polynomial expansion with respect to μ, as discussed above, and, (ii) averaging the expressions over all μ-values. In (17), $(\langle \delta B_I^2\rangle/B_0^2$ is the ratio of the average magnetic field energy density of the magnetic island component over the magnetic field energy density of the guide/background field component of small-scale flux ropes, V_{A0} is the Alfvén speed of the background solar wind, L_I is the flux-rope cross section, d_i is the ion inertial scale length, Z/A is the ratio of the ion atomic number over the mass number, and $\sigma_i^I, i = INC, COM, REC, ACC$, represents control parameters for the efficiency of coherent acceleration for the different flux-rope acceleration cases.

Consider the ratio of the coherent acceleration rate of combined curvature and generalized betatron acceleration (unified betatron and grad-B drift acceleration) in contracting and merging small-scale flux ropes operating in the incompressible limit $\langle dp/dt\rangle_{\phi,\mu}^{INC}$ (shear-flow acceleration) to the coherent acceleration rate for the same acceleration mechanisms in the compressible limit $\langle dp/dt\rangle_{\phi,\mu}^{COM}$ (adiabatic compression acceleration) in the 1st line of (14). Assuming $\sigma_{INC}^{I}/\sigma_{INC}^{I}\approx 1$, and that at 1 AU, small-scale flux ropes with cross sections in the inertial range are in the strong guide field limit ($\langle\delta B_I^2\rangle/B_0^2\approx 0.1$ [17]), we find that $\langle dp/dt\rangle_{\phi,\mu}^{INC}/\langle dp/dt\rangle_{\phi,\mu}^{COM}\approx 0.1<<1$ if we assume that the expression holds for a significant energetic particle anisotropy $3f_2(p)/f_0(p)\approx 1$. However, one would expect near-isotropic energetic particle distributions to exist in the enhanced turbulence conditions behind heliospheric shocks, for example. For a small anisotropy, $\langle dp/dt\rangle_{\phi,\mu}^{INC}/\langle dp/dt\rangle_{\phi,\mu}^{COM}<<0.1$. In the extreme limit of a purely isotropic particle distribution ($f_2(p)=0$), $\langle dp/dt\rangle_{\phi,\mu}^{INC}=0$. This points to a key difference alluded to above that coherent shear-flow acceleration by incompressible flux ropes only yields net acceleration when the particle distribution is anisotropic whereas coherent flux-rope compression acceleration produces net acceleration for both isotropic and anisotropic particle distributions. In the limit of a strictly isotropic distribution, no net shear-flow acceleration associated with incompressible flux-rope contraction and merging indicates that the probability for parallel kinetic energy gain from curvature drift acceleration equals the probability for perpendicular kinetic energy loss from generalized betatron acceleration [8,9,10].

In conclusion, in the test particle limit, coherent energetic ion acceleration at Earth involving combined curvature and generalized betatron acceleration in quasi-2D contracting and merging small-scale flux ropes operating in the compressible limit is estimated to be much more efficient compared to when the same acceleration mechanisms occur in small-scale flux ropes acting in the incompressible limit. Equivalently, compression acceleration is predicted to be much more efficient than shear-flow acceleration. This conclusion holds as long as the anisotropic part of the particle distribution does not strongly dominate the isotropic part of the distribution. According to (17), what is required for shear-flow acceleration in incompressible flux ropes to rival compression acceleration in compressible flux ropes would be to maintain a particle anisotropy on the level $3f_2(p)/f_0(p)\approx 1$, combined with a weaker guide field so that $\langle\delta B_I^2\rangle/B_0^2\approx 1$. There is evidence from kinetic particle simulations that, for a guide field of approximately this strength, the energetic particle anisotropy can be sufficiently large so that shear-flow and compression acceleration reach a comparable level of efficiency [16].

Based on the above analysis, one could ask whether there is any reason to expect flux-rope dynamics in the solar wind to be in the compressible limit when the strong guide field limit applies. Although it appears that small-scale flux ropes tend to contract predominantly incompressibly in discussions of particle simulations with a significant guide field [16], and also is thought of as intrinsically incompressible in its manifestation as the quasi-2D turbulence component of coherent structures in nearly incompressible MHD theory of solar wind turbulence [5], there is observational evidence to the contrary. For example, when primary current sheets associated with interplanetary coronal mass ejections (ICMEs) interact with the heliospheric current sheet, the current sheets are disturbed and several small-scale flux rope structures may be formed when turbulent magnetic reconnection occur in these structures. The flux ropes, being trapped between the converging heliospheric current sheet and the primary current sheets of ICMEs, experience compression which may lead to efficient particle acceleration by compressing flux ropes [1,2]. Furthermore, in nearly incompressible MHD theory of quasi-2D magnetic island turbulence, incompressible flux ropes can become compressible under the influence of large-scale density and flow velocity gradients gradients in the non-uniform solar wind [5]. For example, the flow compression across heliospheric shocks might result in the emission of compressible small-scale flux ropes [23]. Closer to the Sun, Guidoni et al. [24] discusses the possibility of strong plasma compression during magnetic island contraction for islands propagating sunward during a solar flare event.

IOP Conf. Series: Journal of Physics: Conf. Series **1100** (2018) 012015 doi:10.1088/1742-6596/1100/1/012015

Next, we analyze the ratio of the coherent parallel guiding center motion acceleration rate due to the mean parallel reconnection electric field in merging small-scale flux ropes $\langle dp/dt\rangle_{\phi,\mu}^{REC}$ to the coherent acceleration rate of combined curvature and generalized betatron acceleration in small-scale flux ropes contracting and merging in the compressible limit $\langle dp/dt\rangle_{\phi,\mu}^{COM}$ (2nd line of (17)). Overall, $\langle dp/dt\rangle_{\phi,\mu}^{REC}/\langle dp/dt\rangle_{\phi,\mu}^{COM} >> 1$ at Earth for energetic protons, even though we assume the strong guide field limit $\langle \delta B_I^2\rangle/B_0^2 \approx 0.1$at Earth, because of the large ratio of L_I/d_i (maximum values for $L_I \approx 0.01$ AU and $d_i \approx 6\times10^{-7}$ AU). The domination of $\langle dp/dt\rangle_{\phi,\mu}^{REC}$ is reduced significantly at higher particle speeds because $\langle dp/dt\rangle_{\phi,\mu}^{REC}/\langle dp/dt\rangle_{\phi,\mu}^{COM} \propto V_{A0}/v$, but not sufficiently to affect the domination of $\langle dp/dt\rangle_{\phi,\mu}^{REC}$. However, if we extend the scope of the analysis by considering all flux-rope cross sections belonging to the turbulence inertial range (~0.01 AU < L_I < ~6 ×10^{-7} AU at Earth), we find that $\langle dp/dt\rangle_{\phi,\mu}^{COM}/\langle dp/dt\rangle_{\phi,\mu}^{REC} > 1$ for energetic protons above ~1 keV when L_I < ~10^{-4} AU even if $3f_1(p)/f_0(p) \approx 1$. Domination by $\langle dp/dt\rangle_{\phi,\mu}^{COM}$ can be achieved for suprathermal protons for all cross sections in the inertial range if we specify a sufficiently smaller particle anisotropy $3f_1(p)/f_0(p) \approx 0.01$. Assuming a purely isotropic particle distribution ($f_1(p) = 0$), there is no net particle acceleration by the mean parallel reconnection electric field $\left(\langle dp/dt\rangle_{\phi,\mu}^{REC} = 0\right)$ because there is an equal probability for particle motion along and opposite to the mean parallel reconnection electric field force as discussed above. In summary, in the strong guide field limit applicable at 1 AU, coherent parallel guiding center motion acceleration by the mean parallel reconnection electric field due to flux-rope merging tend to be more efficient than combined curvature drift and generalized betatron acceleration in compressible flux ropes for the largest flux-rope cross sections in the inertial range, but compression acceleration by flux ropes can dominate for all cross sections in the inertial range if the energetic particle anisotropy is sufficiently small.

Consider the ratio $\langle dp/dt\rangle_{\phi,\mu}^{REC}/\langle dp/dt\rangle_{\phi,\mu}^{INC}$ (3rd line of (17)). We find that $\langle dp/dt\rangle_{\phi,\mu}^{REC}/\langle dp/dt\rangle_{\phi,\mu}^{INC} > 1$ for suprathermal protons at Earth for all flux-rope cross sections belonging to the inertial range when applying the strong guide field limit and limiting the particle anisotropy to $f_1(p)/f_2(p) \approx 1$. Domination by the parallel reconnection electric field is further strengthened for a small energetic particle anisotropy $f_2p)/f_1(p)$ << 1. A way for $\langle dp/dt\rangle_{\phi,\mu}^{INC}$ to rival the efficiency of $(\langle dp/dt\rangle_{\phi,\mu}^{REC} = 0)$ for at least the smallest flux-rope cross sections in the inertial range would be to maintain a sufficiently strong particle anisotropy $f_1(p)/f_2(p) \approx 1$ combined with a weaker guide field so that($\langle \delta B_I^2\rangle/B_0^2 \approx 1$ which, qualitatively, is in agreement with kinetic simulation results [16].

Coherent energetic particle parallel guiding center motion momentum gain by the mean non-inertial force associated with the parallel acceleration of the flux-rope flow $\langle dp/dt\rangle_{\phi,\mu}^{ACC}$ (expression not shown) appears to be less efficient than momentum gain from both parallel guiding center motion momentum gain by the mean parallel reconnection electric field and combined curvature drift and generalized betatron acceleration in compressible flux ropes, and this is even more so in the case of a nearly isotropic particle distribution. Also in this case, there is no net coherent acceleration when the energetic particle distribution is strictly isotropic.

Based on our estimates for coherent energetic proton acceleration in response to mean flux-rope dynamic properties, which were made in the strong guide field limit and for flux-rope cross sections in the inertial range at 1 AU, we conclude that the two most efficient acceleration scenarios involve

IOP Conf. Series: Journal of Physics: Conf. Series **1100** (2018) 012015 doi:10.1088/1742-6596/1100/1/012015

combined curvature drift and generalized betatron acceleration in contracting and merging flux ropes in the compressible limit, and parallel guiding center motion acceleration by the parallel reconnection electric field of merging flux ropes. The latter tends to dominate for the largest magnetic island cross sections in the inertial range when the anisotropy in the particle distribution not too small.

5. Stochastic acceleration by small-scale flux ropes

A main factor that contributes to differences between the efficiency of coherent acceleration in response to mean flux-rope properties, and the efficiency of stochastic acceleration due to fluctuations in flux-rope properties, is the anisotropy in the energetic particle distribution. In the case of stochastic acceleration, both the isotropic and anisotropic part of the distribution function plays a role in all four acceleration scenarios. As discussed above, net coherent acceleration only occurs for the anisotropic part of the particle distribution for most acceleration cases. The exception is acceleration by flux-rope flow compression which also yields net acceleration when acting on the isotropic part of the distribution. We find for stochastic acceleration that all four acceleration cases contribute to particle acceleration when the particle anisotropy is strictly zero, as long as we take the quasi-linear transport limit of our theory. In the non-linear transport limit the acceleration expressions becomes undetermined in this limit. When comparing ratios of coherent acceleration rates with ratios of stochastic acceleration rates, the results are similar qualitatively for the most part. Quantitative differences in the acceleration ratios are most noticeable when near-isotropic energetic particle distributions are assumed, which is a disadvantage for most coherent acceleration scenarios.

6. Comparing stochastic acceleration by small-scale flux-ropes and Alfvén waves

We find that stochastic acceleration involving parallel guiding center motion acceleration of suprathermal protons at 1 AU in response to fluctuations in the parallel reconnection electric field of merging small-scale flux ropes to be the only flux-rope acceleration scenario in the quasi-linear spatial transport limit of our theory that is more effective than stochastic acceleration by parallel propagating Alfvén waves in standard quasi-linear theory [25]. However, stochastic acceleration by active small-scale flux ropes in the non-linear transport regime of our kinetic transport theory is significantly more efficient when compared to the quasi-linear limit of our theory. Consequently, in the non-linear limit also combined stochastic curvature drift and generalized betatron acceleration in response to fluctuations in the properties of small-scale flux ropes contracting and merging in the compressible limit, is more effective than stochastic acceleration by Alfvén waves for a wide range of suprathermal proton kinetic energies above ~ 10 keV. The enhanced acceleration efficiency can be attributed to the fact that, in the non-linear transport regime of our theory, energetic particles are modeled to have diffusively distorted guiding center trajectories inside flux ropes (scattering on smaller-scale flux ropes) when traversing these structures in the background/guide field direction. Thus, energetic particles spend more time being accelerated in each active flux rope (quasi-trapped) compared to the quasi-linear regime. In the quasi-linear regime, particles traverse flux ropes in the guide field direction more rapidly because of the assumption of undisturbed guiding center motion, leaving less time for acceleration in each flux rope.

7. Pitch-angle scattering by small-scale flux-ropes and Alfvén waves

In our current focused transport approach, the variance in the magnetic mirroring force present in small-scale flux ropes is predicted to play an important role in energetic particle pitch-angle scattering in solar wind conditions at 1 AU. This is related to our finding that energetic particle pitch-angle scattering by small-scale flux ropes in the quasi-linear spatial transport limit of our focused transport theory approach is more efficient compared to previous quasi-linear kinetic theories for particle interaction with 2D turbulence, where particle scattering is determined by the variance in the magnetic Lorentz force associated with 2D turbulence. Another difference is that in earlier approaches the particle decorrelation time is determined by a competition between the time scales for gyromotion around $\underline{B}_0$ and the turbulence dynamic time scale [e.g., 26]. Because the turbulence was treated as strictly 2D, the option exercised in our current theory of particles experiencing decorrelated flux-rope turbulence while propagating along the guide field was not considered. In addition, we find that energetic proton pitch-angle scattering by small-scale flux ropes at 1 AU should be more efficient than pitch-angle scattering by Alfvén waves, provided that the non-linear spatial transport limit of our theory is applicable, but less efficient than pitch-angle scattering by Alfvén waves when we take the quasi-linear limit of our theory. This raises the question which limit of our theory applies best to solar wind conditions near 1 AU. Fits to observed intensity time profiles of solar energetic events at 1 AU using focused transport theory suggest that the energetic ion parallel mean free path $\lambda_{\parallel}$ can vary widely between ~2×10^{-2} - 1 AU during quiet solar wind conditions in the absence of interplanetary shocks [27]. Nonetheless, it appears that $\lambda_{\parallel} > L_I$ because the maximum small-scale flux-rope cross section at 1 AU is $L_I \approx 0.01$ AU, indicating scatter-free transport of energetic particles through these structures so that the quasi-linear limit of our theory is more appropriate in quiet solar wind conditions. However, one would expect that the values of $\lambda_{\parallel}$ in the enhanced turbulence levels behind traveling shocks should be significantly smaller, providing potential conditions for the application of the non-linear transport limit of our theory.

8. Summary

It was discussed how the basic small-scale flux rope drift and betatron acceleration mechanisms present in guiding center kinetic theory relate to flux-rope plasma drift contraction or merging in both the incompressible and incompressible limits. It was shown that drift and betatron acceleration in flux-ropes undergoing incompressible contraction and merging can be interpreted in terms of shear flow acceleration linked to a negative parallel component of shear flow inside flux ropes. Drift and betatron acceleration in flux-ropes contracting and merging in the compressible limit can be viewed as compression acceleration in flux ropes with a relatively negligible magnetic curvature or parallel component of shear flow. We found that when the drift and betatron acceleration mechanisms in guiding center kinetic theory are expressed in term of plasma drift compression and shear flow acceleration, a close link exists between the flux-rope acceleration mechanisms of guiding center kinetic theory and those in focused transport theory which we use to model energetic particle acceleration by small-scale flux ropes. Focused transport theory includes an additional acceleration mechanism associated with parallel guiding center motion acceleration by the parallel component of the acceleration of the flux-rope flow (non-inertial force). Furthermore, it was discussed that, only in the case of compression acceleration, net energetic particle acceleration occurs for both the isotropic and anisotropic parts of the energetic particle distribution. For all other flux-rope acceleration scenarios (shear-flow acceleration during incompressible contraction and merging, parallel guiding center motion acceleration by the parallel reconnection electric field in merging flux ropes and the parallel component of the acceleration of the flux-rope flow (non-inertial force)), net acceleration occurs only for the anisotropic part of the distribution.

In an extension of previous work, we presented the outline of a statistical transport theory for self-consistent energetic particle acceleration by and large-scale transport through numerous contracting and merging quasi-2D small-scale flux ropes. The theory applies to flux rope cross-sections belonging to the inertial range of turbulence and is valid for small-scale flux ropes in the strong guide/background field limit. It was assumed that flux-rope dynamics are occurring essentially in the 2D plane of the magnetic island (twist) component perpendicular to the guide field (axial) component of these structures. The extended theory, as presented, consisted of two coupled equations: (i) a focused transport equation which divide all flux-rope acceleration mechanisms into coherent acceleration in response to mean magnetic island dynamic quantities, and stochastic (2nd order Fermi) acceleration due to statistical fluctuations in these quantities and, (ii) a new equation for the transport of total energy density (kinetic and magnetic) of magnetic islands advected with the non-uniform solar wind flow based on a recent nearly incompressible MHD theory for solar wind turbulence. Coupling between the two equations were established with the inclusion of newly derived magnetic island damping coefficients in the magnetic island transport equation derived on the basis of conservation of total energy in the energy exchange between energetic particles and magnetic islands for both coherent and stochastic acceleration.

Test particle, coherent energetic proton acceleration rates at Earth were compared for different flux-rope acceleration cases in the strong guide field limit for flux-rope cross sections in the inertial range. We found the two most efficient acceleration scenarios to be combined curvature drift and generalized betatron acceleration in contracting and merging flux ropes in the compressible limit (compression acceleration), and parallel guiding center motion acceleration by the parallel reconnection electric field in merging flux ropes. The latter tended to dominate for the largest magnetic island cross sections in the inertial range provided that the anisotropy in the particle distribution not too small.

Stochastic flux-rope acceleration rates were also investigated. A key difference between the two types of acceleration is that both the isotropic and anisotropic parts of the distribution function contribute to net acceleration for all acceleration cases during stochastic acceleration. Coherent acceleration is solely determined by the anisotropic part of the particle distribution for all the acceleration cases except acceleration by flux-rope flow compression which also yields net acceleration from the isotropic part of the distribution. Comparison of stochastic acceleration for small-scale flux ropes and parallel-propagating Alfvén waves (standard quasi-linear theory) revealed that parallel guiding center motion acceleration by the parallel reconnection electric field in merging flux ropes was more efficient than Alfvén waves in the quasi-linear transport limit of our theory. In the non-linear transport regime of our theory, combined stochastic curvature drift and generalized betatron acceleration in flux ropes contracting and merging in the compressible limit (compression acceleration) can also be more efficient than stochastic acceleration by Alfvén waves for energetic protons at 1 AU.

Finally, our theory suggested that the variance in the magnetic mirroring force present in small-scale flux ropes plays a potential important role in energetic particle pitch-angle scattering in solar wind conditions at 1 AU. Analysis of pitch-angle scattering showed that energetic proton pitch-angle scattering by small-scale flux ropes should be more efficient than pitch-angle scattering by Alfvén waves, provided that the non-linear spatial transport limit of our theory is applicable, but less efficient than pitch-angle scattering by Alfvén waves when we take the quasi-linear limit of our theory.

Acknowledgments

le Roux and Zank acknowledge support from NASA grant NNX15AI65G, and NSF-DOE grant PHY-1707247.

References

[1] Khabarova O V, Zank G P, Li G, le Roux J A, Webb G M, Dosch A and Malandraki O E 2015 *Astrophys. J.* **808** 181
[2] Khabarova O V, Zank G P, Li G, Malandraki O E, le Roux J A and Webb G M 2016 *Astrophys. J.* **827** 122
[3] Khabarova O V, Zank G P 2017 *Astrophys. J.* **843** 4
[4] Zheng J, Hu Q 2018 *Astrophys. J.* **852** L23
[5] Zank G, Adhikari L, Hunana P, Shiota D, Bruno R and Telloni D 2017 *Astrophys, J.* **835** 147
[6] Dmitruk P, Matthaeus W H and Seenu N 2004 *Astrophys. J.* **617** 667
[7] Drake J F, Swisdak M, Che H and Shay M A 2006 *Nature* **443** 553
[8] Drake J, Swisdak M and Fermo R 2013 *Astrophys. J.* **763** L5
[9] Zank G P, le Roux J A, Webb G M, Dosch A and Khabarova O V 2014 *Astrophys. J.* **797** 28
[10] le Roux J A, Zank G P, Webb G M and Khabarova O V 2015 *Astrophys J.* **801** 112
[11] Dahlin J T, Drake J F and Swisdak M 2016 *Phys. Plasmas* **23** 120704
[12] le Roux J A, Webb G M 2012 *Astrophys. J.* **764** 104
[13] Kóta J 1977 *Proc. 15th Cosmic Ray Conf.*, Povdiv, Bulgaria, MG-4
[14] Webb G M, Martinic M J and Moraal H 1981, *Proc. 17th Int. Cosmic Ray Conf.*, Paris, France, **10** 109

[15] Pritchett P 2008 *Phys. Plasmas* **15** 102105
[16] Li X, Guo F, Li H and Birn J 2018 *Astrophys. J.* **855** 80
[17] Smith C W, Isenberg P A, Matthaeus W H and Richardson J D 2016 *Astrophys. J.* **638** 508[18]
[18] Zheng J and Hu Q 2018 *Astrophys. Lett.* **852** L23
[19] Isenberg P A 1987 *J. Geophys. Res.* **92** 1067
[20] Bian N H and Kontar, E P 2013 *Phys. Rev. Lett.* **110** 151101
[21] le Roux J A, Zank G P, Webb G M and Khabarova O V 2016 *Astrophys. J.* **827** 47
[22] Dahlin J T, Drake J. F. and Swisdak M 2017 *Physics of Plasmas* **24** 092110
[23] Karimabadi H, Roytershteyn V, Vu H X, Omelchenko Y A, Scudder J, Daughton W, Dimmock A, Nykyri K, Wan M, Sibeck D, Tatineni M, Majumdar A, Loring B and Geveci B 2014 *Phys. Plasmas* **21** 062308
[24] Guidoni S E, DeVore C R, Karpen J T, Lynch B J 2016 *Astrophys. J.* **820** 60
[25] Schlickeiser R 1989 *Astrophys. J.* **336** 243
[26] le Roux J A and Webb G M 2007 *Astrophys. J.* **667** 930
[27] Dröge W 2005 *Adv. Space Res.* **35** 532

IOP Conf. Series: Journal of Physics: Conf. Series **1100** (2018) 012016 doi:10.1088/1742-6596/1100/1/012016

The Local Interstellar Medium: Comparison with IBEX Results, Sightlines to Exoplanet Host Stars, and Trajectories of the Voyager Spacecraft

Jeffrey L Linsky[1], **Seth Redfield**[2], **Eric Edelman**[3], **Julia Zachary**[2] **and Brian Wood**[4]

[1] JILA, University of Colorado and NIST, Boulder CO, USA
[2] Wesleyan University, Middletown CT, USA
[3] Embry-Riddle Aeronautical University, Prescott AZ, USA
[4] Naval Research Laboratory, Washington DC, USA

E-mail: [1] jlinsky@jila.colorado.edu

Abstract. We report on our recent studies of the local interstellar medium (LISM) using high-resolution ultraviolet spectra of nearby stars with the *Space Telescope Imaging Spectrograph (STIS)* instrument on the *Hubble Space Telescope (HST)*. Our objective is to measure the physical properties, kinematics, and morphology of the LISM near the Sun. After describing our techniques, we call attention to the different direction and speed of the gas in the Local Interstellar Cloud (LIC) surrounding the Sun compared to the infall vector of neutral helium entering the heliosphere as measured by the *Interstellar Boundary Explorer (IBEX)* and *Ulysses* spacecraft. We then suggest possible explanations for the differences. We describe the cloud structure and gas properties along the lines of sight to 3 nearby exoplanet host stars. Finally, we determine the cloud structure and properties of the interstellar gas in the directions that the *Voyager* spacecraft are heading. In the trajectory of *Voyager 2*, there is no absorption from the LIC or the very close G cloud suggesting that the Sun may lie very close to the outer edge of the LIC or perhaps even outside of the LIC.

1. Inferring kinematic and morphological properties of the LISM gas

For the last 16 years, we have been acquiring a large spectroscopic database of interstellar gas absorption properties within 100 pc of the Sun, the region of space that we call the local interstellar medium (LISM). The observations with several ultraviolet spectrographs on the *Hubble Space Telescope (HST)* consist of stellar emission line spectra with narrow absorption lines produced by neutral and ionized gas in the lines of sight to these stars. We have now acquired spectra of more than 200 stars that are distributed across the sky sampling nearly all Galactic coordinates (see Figure 1). The *HST* spectra were obtained mostly with the *Space Telescope Imaging Spectrograph (STIS)* instrument with resolutions of 3.0–6.5 km/s, but we also use some ground based spectra of the Ca II H and K lines. The *STIS* spectra include interstellar absorption lines of H I (121.6 nm Lyman-α), O I (130.2, 130.5, 130.6 nm), C II (133.4, 133.5 nm), Mg II (279.5, 280.3 nm), and Fe II (258.6, 260.0 nm) superimposed on emission lines formed in stellar chromospheres.

The high-resolution spectra of these lines allow us to measure the radial velocities (v_r) and line widths (b) of each interstellar velocity component detected against a stellar emission line or

IOP Conf. Series: Journal of Physics: Conf. Series **1100** (2018) 012016 doi:10.1088/1742-6596/1100/1/012016

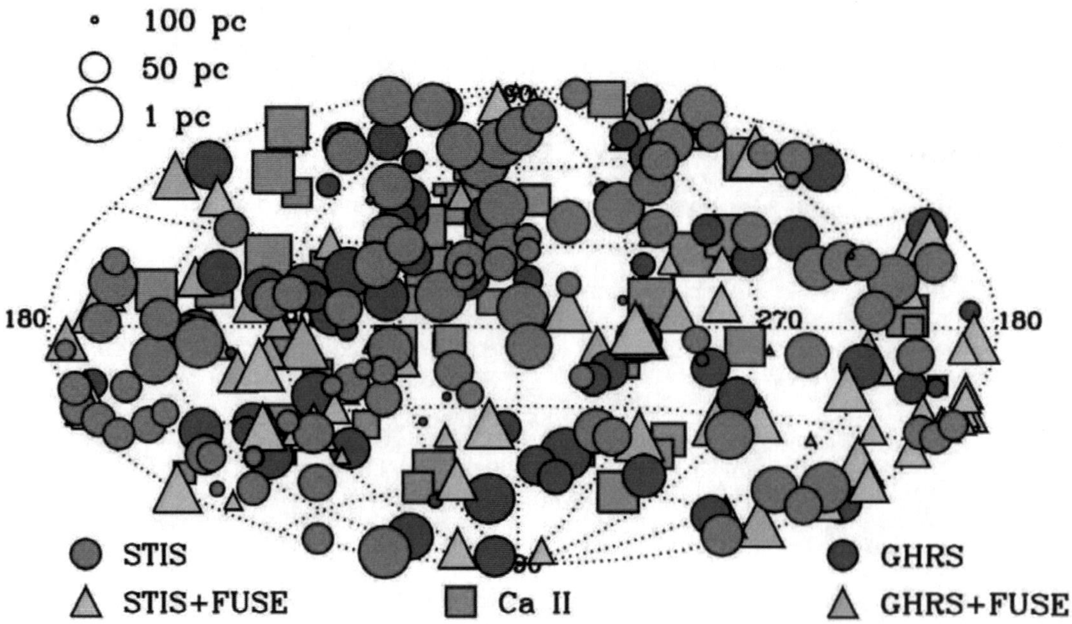

Figure 1. The locations in Galactic coordinates [longitude (l) and latitude (b)] of 160 stellar targets analyzed by RL08 are located within 100 pc of the Sun. Observations with the *STIS* and *Goddard High Resolution Spectrograph (GHRS)* instruments on *HST*, the *Far Ultraviolet Spectrograph Explorer (FUSE)*, and ground based Ca II H and K lines have identified 270 LISM absorption components. The symbol sizes identify distances to the targets. Figure from Redfield (2009).

continuum. For every sightline, we detect at least one velocity component and often two to six components even for stars located within 5 pc. Our detection of multiple velocity components along the same short sightline indicates that the LISM gas has velocity structure with scales of order one pc or smaller. With this large data set, Redfield & Linsky (2008) (hereafter RL08) developed a model of the partially ionized LISM gas consisting of many discrete components, which we call clouds, for which we assume that the gas moves coherently with the same velocity vector and physical properties (temperature and turbulence). While this model is simple, it fits the data quite well (Redfield & Linsky 2015), but we intend to test this model when there is sufficient data to see whether the clouds have inhomogeneous properties, as proposed by Gry & Jenkins (2014), or may have nonthermal velocity distributions.

With spectra for 160 sightlines, RL08 identified 15 interstellar clouds located within 15 pc of the Sun and derived temperatures and turbulence values for many of the velocity components identified in each sightline. The technique for inferring these results is described schematically in Figure 2. While the observed profile of a single ion in a single sightline provides only v_r and the column density of that ion $N_{\rm ion}$, the observation of multiple ions in the same sightline provides a wealth of information. Comparison of the line widths (b) of different ions or neutrals with different atomic masses (m) allows one to determine the temperature (T) and turbulent velocity (ξ) of the gas (see Figure 3). The best leverage for this measurement comes from the comparison of the widths of lines of D I (m=2) with Fe II (m=26). The difference in the column density of element X relative to hydrogen, log (X/H), compared to the solar ratio is the depletion of the element that probably results from atoms being absorbed onto the surfaces of dust grains and thus removed from the interstellar gas phase. Accurate depletions require measurements

IOP Conf. Series: Journal of Physics: Conf. Series **1100** (2018) 012016 doi:10.1088/1742-6596/1100/1/012016

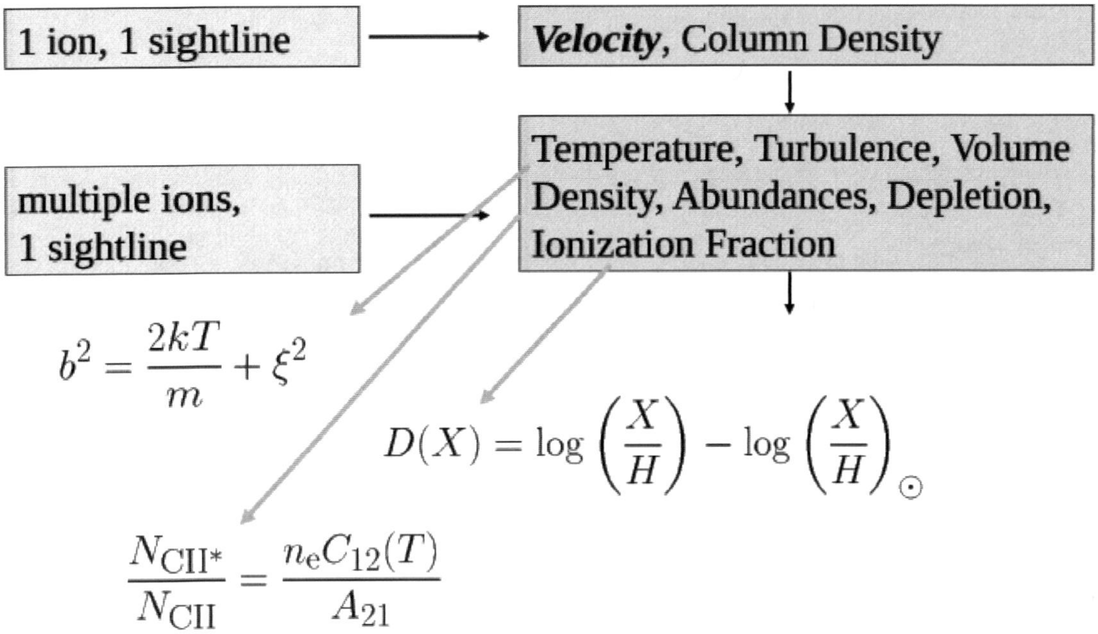

Figure 2. Observational diagnostics — the interstellar gas properties that can be obtained from high-resolution spectra of one ion or multiple ions in a single sightline showing the equations used to infer temperature, turbulence, element depletion, and electron density.

of the important ionization stages of an element in order to obtain the total abundance of the element in the LISM gas. Finally, there are line ratios that are sensitive to electron density because the radiative de-excitation of one of the lines is very small such that de-excitation from the upper level is by electron collisions (Redfield & Falcon 2008). The ratio of Mg I/Mg II is also sensitive to electron density (Gry & Jenkins 2014). An upper limit to the neutral hydrogen number density can be obtained from the hydrogen column density and sightline distance or from theoretical models (e.g., Slavin & Frisch 2008, Frisch et al. 2011).

The morphological structure of interstellar clouds requires a determination of the velocity vector for each cloud from the radial velocity data. We started this iterative process by asking how accurately a single velocity vector fits all of the radial velocity data. The differences of individual radial velocities relative to the projected radial velocities computed from the best fit vector are very large: the ratios of these differences divided by the velocity measurement errors are more than 10σ away from the predicted values for 45% of the data points (Redfield 2009). The velocity structure of the LISM is therefore very clumpy. We then deleted data points one by one starting with the most discrepant data until the remaining discrepancies were consistent with the velocity measurement errors (about 1.5 km s^{-1} for the *STIS* data). The velocity vector for these 79 sightlines define a two-dimensional region in space roughly centered opposite from the Galactic Center and covering about 45% of the sky. This cloud called the Local Interstellar Cloud (LIC) has a mean temperature $T = 7500 \pm 1300$ K, mean turbulence $\xi = 1.62 \pm 0.75$ km s^{-1}, and logarithmic depletions of Fe and Mg (-1.12 ± 0.10 and -0.97 ± 0.23, respectively). A similar analysis of the data for the remaining sightlines led to the identification of 14 other clouds. The nearest clouds to the Sun are the G cloud in the direction of α Cen (1.3 pc) toward the Galactic center, and the Blue cloud in the direction of Sirius (2.6 pc). The G cloud covering about 20% of the sky has a mean temperature $T = 5500 \pm 400$ K, while the Blue cloud covering about 6%

IOP Conf. Series: Journal of Physics: Conf. Series **1100** (2018) 012016 doi:10.1088/1742-6596/1100/1/012016

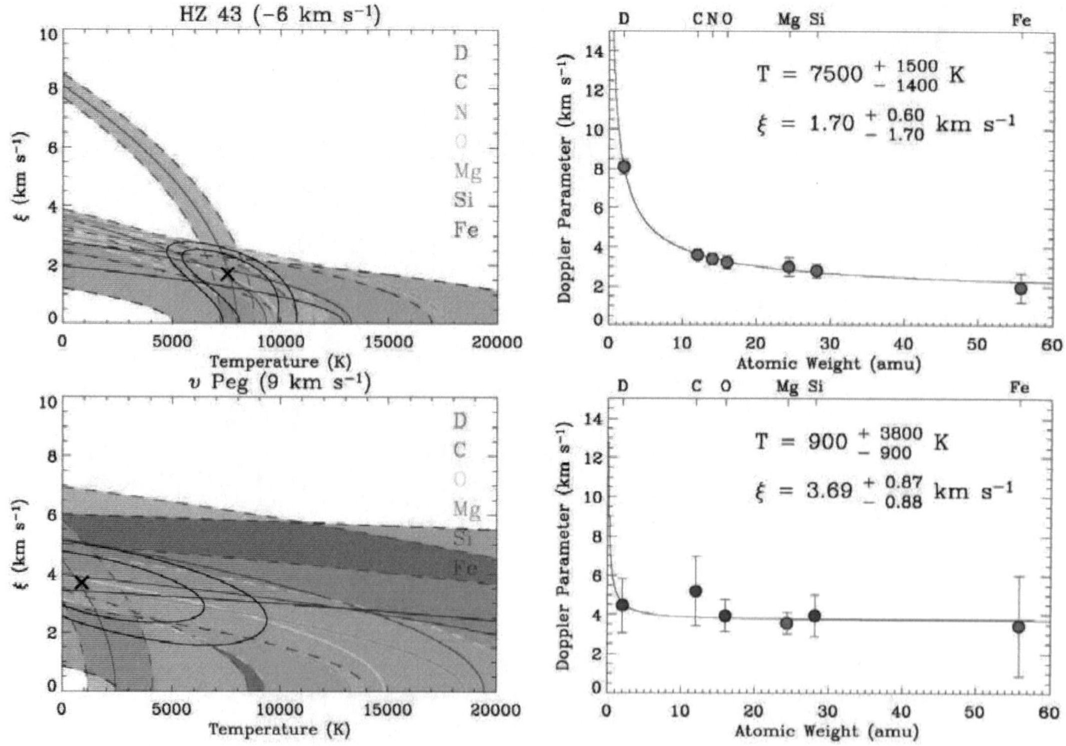

Figure 3. Determination of temperature (T) and turbulence (ξ) from the widths of spectral lines of many elements with different atomic weights. *Left panels:* Plots of the range of parameter space (ξ and T) consistent with the measured widths (solid lines) and their measurement errors (dashed lines) for different elements and ions. The symbol **X** shows the best agreement and the solid lines around **X** are 1σ and 2σ uncertainties in the best fit parameters. *Right panels:* Plots of the Doppler width parameters (b) for each element and the least squares best fit for T and ξ. For HZ 43 (top panels) deuterium is thermally broadened and the heavier elements are increasingly broadened by turbulence with increasing mass. For the sightline to υ Peg, the inferred temperature is low and thermal broadening is very small for all elements. Figure from Redfield & Linsky (2004).

of the sky may have even cooler gas ($T = 3900 \pm 2300$ K). Figure 4 shows the location of the four closest clouds in Galactic coordinates. These clouds are partially ionized with hydrogen about 26% ionized if the LIC is representative (Slavin & Frisch 2008).

After this 15 cloud model was constructed, we obtained observations of 34 stars randomly selected from a much larger list in an *HST* SNAP program. Malamut et al. (2014) found that every radial velocity predicted by the 15 velocity vectors is consistent with the observed radial velocities of these stars, thereby confirming the 15-cloud model. A calculator for the radial and tangential velocities predicted by each cloud's velocity vector for all directions is available at http://sredfield.web.wesleyan.edu. Figure 5 shows the complete 15 cloud model.

An important question is whether the Sun lies inside or outside of the LIC, because the properties of the LISM gas immediately outside of the heliosphere set the outer boundary condition for heliosphere models. Since the LIC covers about 45% of the sky, the Sun could lie either just inside or just outside of the LIC. The absence of interstellar absorption at the

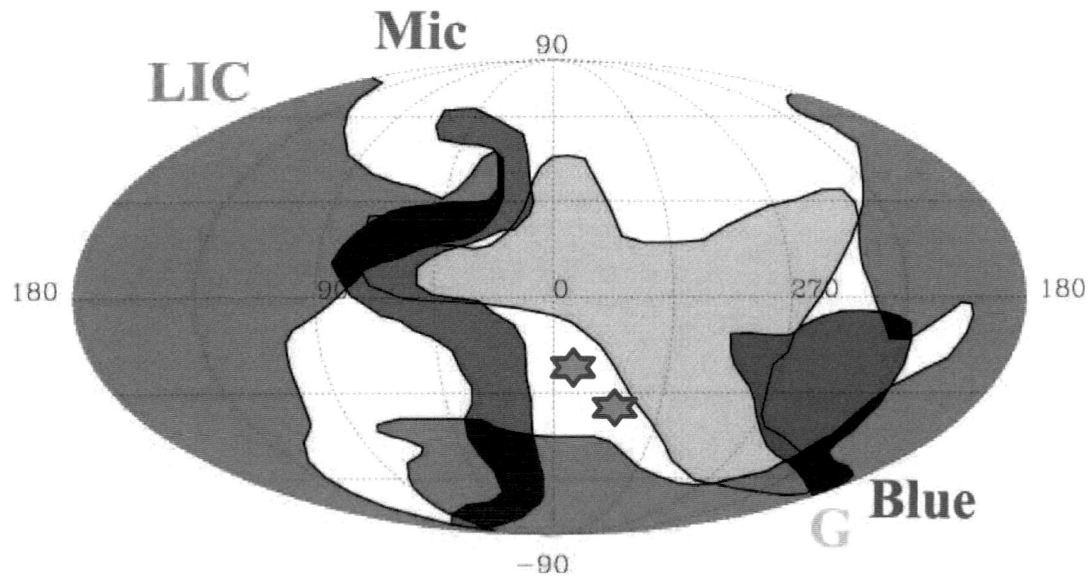

Figure 4. Location in Galactic coordinates of four nearby interstellar clouds: LIC, G, Blue, and Mic. The star symbols indicate the location of GJ 780 (lower symbol) and GJ 754 (upper symbol). Both stars are in directions outside of the four closest interstellar clouds. Original figure from RL08.

velocity predicted by the LIC vector in the direction of α Cen suggests that the Sun lies outside of the LIC or barely inside with insufficient optical depth to detect absorption by Mg II or other lines. If the Sun is located inside of the LIC, then the Mg II absorption upper limits and the Sun's motion with respect to the LIC indicate that the Sun will leave the LIC in less than 3000 years (Redfield & Linsky 2000), perhaps very soon.

2. The flow vectors of the Local Interstellar Cloud and the neutral helium flowing into the heliosphere do not agree and possible explanations for the disagreement

The flow of interstellar helium atoms into the heliosphere is the best test of the inflow vector of interstellar gas because charge exchange and ionization by solar extreme ultraviolet radiation should have only small effects on helium atoms and thus may not substantially change their inflow direction and speed until very close to the Sun. Wood et al. (2015) reanalyzed the Ulysses spacecraft observations of neutral helium, and the *Interstellar Boundary Explorer (IBEX)* observations of helium have been analyzed by Möbius et al. (2015), Leonard et al. (2015), and Bzowski et al. (2015). Table 1 summarizes *Ulysses* and *IBEX* results and compares them to the predicted inflow properties of gas in the LIC and nearby G clouds (RL08).

The speed and direction of the interstellar helium gas flowing into the heliosphere do not agree with the mean properties of the LIC and G clouds, although the LIC cloud properties are closer to the helium inflow values than the G cloud. The absence of agreement could be explained in several ways:

Uncertain LIC mean properties: RL08 obtained the mean kinematic and temperature parameters for the LIC from the analysis of data for 160 sightlines. Since then we have

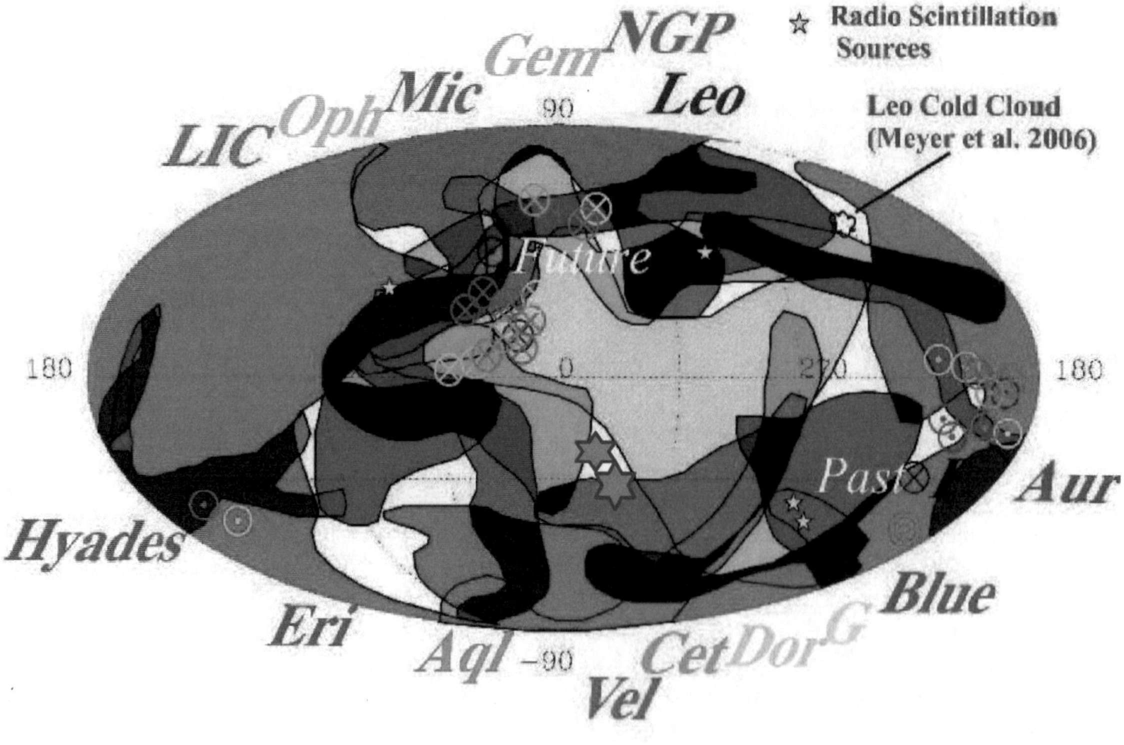

Figure 5. Location in Galactic coordinates of the 15 interstellar clouds identified by RL08. The star symbols indicate the location of GJ 780 (lower symbol) and GJ 754 (upper symbol). Neither of the two sightlines shows absorption by the LIC or G clouds or spatial agreement with these clouds, but both show absorption likely produced by the Vel cloud. The ⊗ symbols indicate the upwind directions and the ⊙ symbols indicate the downwind directions of the cloud velocity vectors. Original figure from RL08.

Table 1. Comparison of heliospheric and astronomical measurements of inflowing LISM neutrals

Source	v_r (km s^{-1})	T (K)	λ (ecliptic)	β (ecliptic)	Reference
IBEX	26	8710^{+440}_{-680}	75.75 ± 0.92	-5.29 ± 0.05	Möbius et al. (2015)
IBEX	27.0 ± 1.3		74.5 ± 1.7	-5.2 ± 0.3	Leonard et al. (2015)
IBEX	25.8 ± 0.4	7440 ± 260	75.8 ± 0.5	-5.16 ± 0.10	Bzowski et al. (2015)
Ulysses	26.08 ± 0.21	7260 ± 270	75.54 ± 0.19	-5.44 ± 0.24	Wood et al. (2015)
LIC cloud	23.84 ± 0.90	7500 ± 1300	78.55 ± 3.4	-7.19 ± 3.3	RL08
G cloud	29.6 ± 1.1	5500 ± 400	71.11 ± 1.9	-8.52 ± 3.6	RL08

increased the number of LIC sightlines by about 25%. The resulting change in the LIC parameters is very small with $\Delta v_r = 0.1$ km s^{-1}, $\Delta\lambda = +0.1$, and $\Delta\beta = +0.2$. As a result, we do not anticipate a major change in the mean LIC parameters with future analyses of larger data sets.

Turbulence near the heliosphere: The radial velocity dispersion of the LIC gas, which we have called turbulence, is ± 1.6 km s^{-1}. If the gas speed in the LIC near the heliosphere is 1.6 km s^{-1} larger than the mean speed for the LIC, then $v_r = 23.84 + 1.6 = 25.46$ km s^{-1},

IOP Conf. Series: Journal of Physics: Conf. Series **1100** (2018) 012016 doi:10.1088/1742-6596/1100/1/012016

which is consistent with the *IBEX* and *Ulysses* measurements of the speed of inflowing helium atoms. The large uncertainty in the flow direction of the LIC gas is not inconsistent with the flow directions measured by *IBEX* and *Ulysses*, and the LIC gas temperature is in good agreement with their results. Thus a higher speed of the LIC gas close to the heliosphere is a viable explanation for the discrepancy as suggested by Bzowski et al. (2015).

Helium Warm Breeze: The derived speed and direction of the inflowing helium atoms observed by *IBEX* requires two components: the main component and a secondary component now called the "Helium Warm Breeze" discovered by Kubiak et al. (2014). According to Bzowski et al. (2015), the Warm Breeze component needed to be included when fitting the *IBEX* helium data is substantial (4–10% of the main component flux) but highly uncertain, leading to possible systematic errors in the inflow speed and direction of the main component. They argued that there may be an as yet unaccounted for additional component in the helium flow. It is important to decrease the uncertainty in the helium flow parameters though more observations and more sophisticated data analysis. Simulations presented by Bzowski et al. (2017) show that the Warm Breeze can be explained by charge exchange reactions between neutral He and He^+ in the outer heliosphere that is distorted by the interstellar magnetic field.

Non-Maxwellian velocity distribution: Since densities in the LISM clouds are very low (about 0.2 cm^{-3}), the velocity distribution of atoms and ions in the LISM gas may differ substantially from a Maxwellian distribution. Space plasmas, such as those seen by *Voyager* beyond the termination shock, often have power-law tails that can be fitted with kappa functions (e.g., Zirnstein & McComas 2015). Whether the electron and proton energy distributions in partially ionized LISM clouds should be characterized by kappa functions is an open question.

Helium inflow parameters are fitted inside of heliosphere: The *IBEX* observations are fitted with helium inflow parameters extrapolated to a distance of 150 AU. However, the peak neutral hydrogen density in the solar hydrogen wall occurs near 300 AU and "pristine" interstellar densities occur beyond 600 AU in the plasma model of Zank et al. (2013). Thus the comparison of the helium inflow parameters and the LIC velocity vector refer to different locations. Just as the inflow speed of neutral hydrogen is decelerated by charge-exchange interactions, especially in the heliopause and hydrogen wall regions, neutral helium atoms may experience a much smaller but still significant deceleration by charge exchange or other collisional interactions between 150 and 600 AU.

Helium inflow parameters are tightly correlated: Bzowski et al. (2015) showed that T, v_r, and λ are tighly correlated in the analysis of *IBEX* observations of inflowing helium. If for some reason the inflow speed of helium is 24 km s^{-1} rather than 25.8 km s^{-1}, then the tight correlations among parameters require that $T = 6300$ K, $\lambda = 78.5$, and $\beta = -5$. These parameters would then all be consistent with the LIC parameters. This calculation highlights the need to better understand the Helium Warm Breeze and its effects on the determination of the helium inflow parameters from the *IBEX* observations. However, the analysis of the *Ulysses* data from all parts of its orbit does not involve any tight correlations among parameters and may, therefore, provide a more precise determination of the He flow vector.

At this point in time, we can likely rule out only the first possible explanation for the disagreement in flow parameters. The other possible explanations require further study or observations. It is important to understand the reason or reasons for the present disagreement, in particular whether the disagreement results from errors in the analysis of *IBEX* and *Ulysses* data or from departures from the mean flow of LIC gas near the heliosphere.

3. Properties of LISM gas along the sightlines to three exoplanet host stars

One objective of our study of LISM gas is to identify the interstellar gas properties along sightlines to nearby planet-hosting stars. This information is important for the analysis of Lyman-α and other stellar emission lines with interstellar absorption components during planetary transits in order to separate absorption by an exoplanets's atmosphere from interstellar absorption. We are also looking for evidence of stellar winds that can be measured from Lyman-α absorption in the star's astrosphere. Edelman et al. (2018) will be submitting a paper soon analyzing *STIS* spectra of the Mg II, Fe II, and Lyman-α lines for three planet-hosting stars: HD 192310 (K2 V), HD 9826 (F9 V), and HD 206860 (G0 V). We identify absorption by the LIC and two other clouds in the sightlines to each star, but none of the stars shows astrospheric absorption in the Lyman-α line. We also list the clouds in the sightlines to all planet-hosting stars within 20 pc of the Sun based either on measurements or from the kinematic calculator described above.

4. Properties of the LISM gas along the trajectories of the Voyager spacecraft

On April 16, 2018 *Voyager 1* was at a distance of 141.84 AU from the Sun past the heliopause and heading toward the edge of the heliosphere at a speed of 3.6 AU/year. At this time *Voyager 2* was at 117.27 AU in the heliosheath and in the process of leaving the heliosphere at a speed of 3.26 AU/year. What will be the interstellar environments that these spacecraft will encounter after leaving the heliosphere? Zachary et al. (2018) have investigated this question by analyzing *STIS* spectra of four stars located within 10 pc — two stars (GJ 678.1A and GJ 686) in the general direction that *Voyager 1* is headed and two (GJ 780 and GJ 754) in the direction that *Voyager 2* is headed (see Figure 6). The present positions of the *Voyager* spacecraft, coordinates (Galactic and ecliptic) of the four stars and the angular distances from the *Voyager* sightlines ($\Delta\theta$) are listed in Table 2. Both GJ 678.1A and GJ 754 show absorption predicted by the LIC velocity vector and second components consistent with other clouds. We, therefore, predict that *Voyager 1* will first pass through the peak of the hydrogen wall at 300 AU in 44 years and then in about 127 years enter the pristine LIC at 600 AU. These locations are from the plasma model of Zank et al. (2013). The RL08 LISM model predicts that *Voyager 1* will leave the LIC in about 1,900 years. It may then enter the Oph, NGP, or Mic clouds.

Table 2. Distance and Directions to the *Voyager* Spacecraft and Stars

Spacecraft or Star	Distance	l (Galactic)	b (Galactic)	λ (Ecliptic)	β (Ecliptic)	$\Delta\theta$ (degrees)
Voyager 1	141.84 AU	33.31	+27.15	256.17	+35.20	
GJ 678.1A	9.98 pc	28.57	+20.54	261.84	+28.76	8.1
GJ686	8.09 pc	42.24	+24.30	263.20	+41.89	9.0
Voyager 2	117.27 AU	340.03	−32.73	290.63	−36.36	
GJ 780	6.11 pc	329.77	−32.42	287.87	−44.70	9.2
GJ 754	5.92 pc	352.36	−23.90	285.50	−23.18	13.1

The trajectory of *Voyager 2* is very different from *Voyager 1*. The sightlines of GJ 780 and GJ 754 do not show absorption at the predicted LIC or G cloud velocities. Instead, both show absorption by the Vel cloud and other clouds (Dor, Mic, and Aql). The absence of absorption at the predicted LIC velocity indicates that the Sun is located either barely inside of the LIC

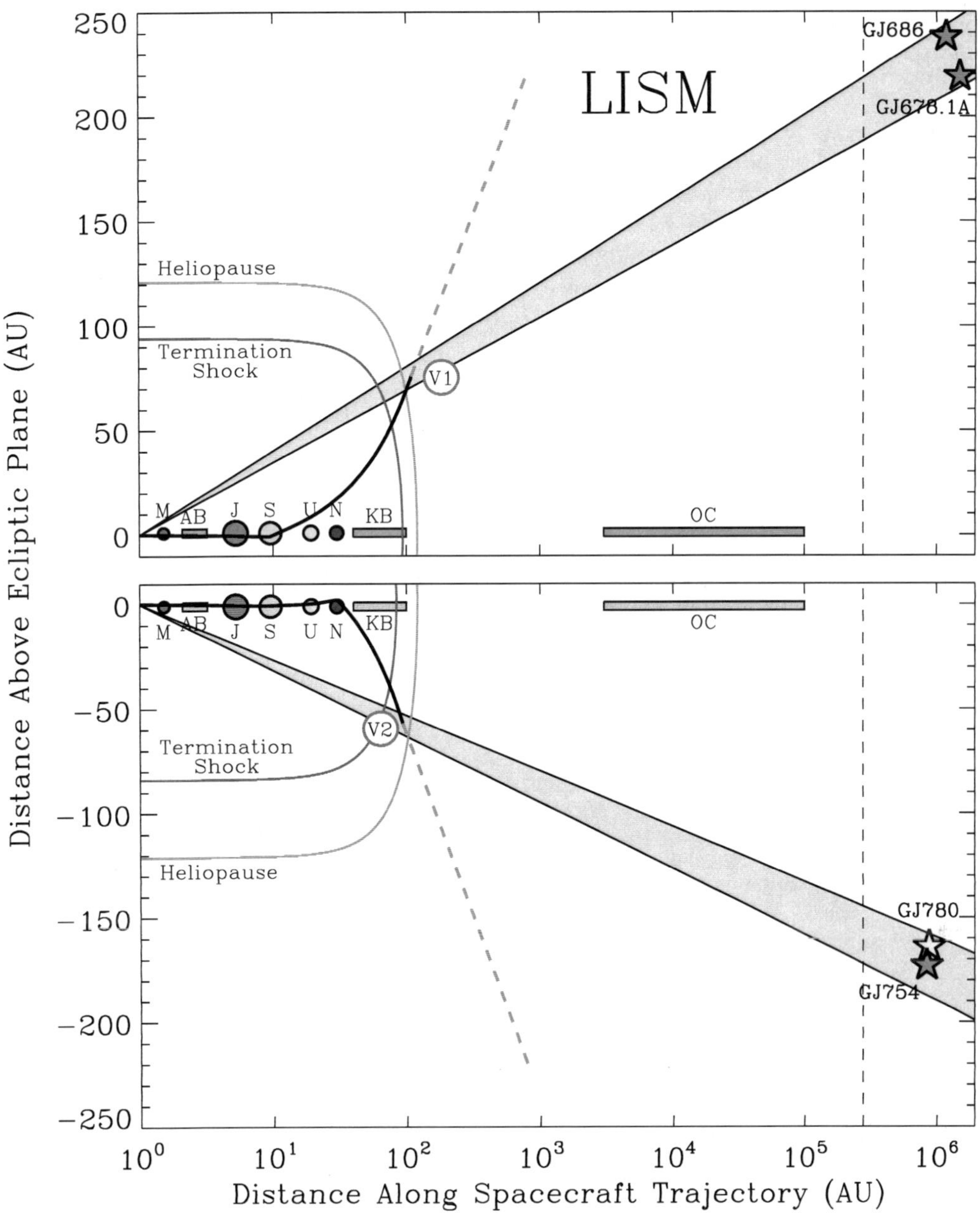

Figure 6. Trajectories of the *Voyager 1 (V1)* and *Voyager 2 (V2)* spacecraft. Solid lines indicate their trajectories through the solar system, the termination shock, and the heliopause (for *V1*) and dashed lines indicate future paths above and below the ecliptic plane. Solid lines enclose the LISM that the Voyagers will encounter. The vertical dashed line is for the distance of 1 pc. KB refers to the location of the Kuiper belt and OC refers to the location of the Oort cloud.

with only an upper limit to the column density of LIC gas in the *Voyager 2* sightline or perhaps just outside of the LIC. Figure 4 shows that GJ 780 and GJ 754 are located away from LIC and G clouds but near or perhaps inside of the Vel or other clouds (see Figure 5). We predict that *Voyager 2* will enter the hydrogen wall peak in about 56 years and enter the pristine LISM in about 148 years. Which LISM cloud it will enter, however, is uncertain. It may traverse a thin layer of the LIC for less than 2000 years or directly enter the Vel or another cloud.

5. Conclusions

The outer edge of the heliosphere can be studied from an external perspective by analyzing interstellar absorption line measurements and from an internal perspective by analyzing observations of inflowing helium atoms obtained by the *IBEX* and *Ulysses* satellites. At present these two techniques do not provide consistent results for the gas flowing into the heliosphere. Whether the different inferred properties of the inflowing gas result from (a) the LIC having inhomogeneous kinematics, with the gas flowing into the heliosphere moving slightly faster than the mean LIC flow, or (b) from inaccurate knowledge of the Helium Warm Breeze or another component of the helium flow is a challenge that must be resolved. It is essential that both techniques be studied in detail until we have a robust understanding of the interface between the outer heliosphere and the inner LISM.

6. Acknowledgments

Our study of the LISM based on *HST* observations is supported by grants to Wesleyan University and the University of Colorado at Boulder by the Space Telescope Science Instutute, which is operated by the Association of Universities for Research in Astronomy, Inc., for NASA under contract NAS 5-26555. This paper is based on data from our SNAP programs (HST-GO-11568 and HST-GO-13332), our astrospheres around exoplanet host stars program (HST-GO-12475), and our Voyager sight line program (HST-GO-13658). JLL thanks the organizers of the 17th Annual International Astrophysics Conference for the invitation to present this paper.

7. References

Bzowski M, Kubiak M A, Czechowski A and Grygorczuk J 2017 *Astrophys. J.* **845** 15
Bzowski M, Swaczyna P, Kubiak M A, Sokól J M, Fuselier S A, Galli A, Heirtzler D, Kucharek H, Leonard T W, et al 2015 *Astrophys. J. Suppl. Ser.* **220** 28
Edelman E, Redfield S, Linsky J L, Wood B E and Müller H 2018 to be submitted to *Astrophys. J.*
Frisch P C, Redfield S and Slavin J D 2011 *Ann. Rev. Astron. Astrophys.* **49** 237
Gry C and Jenkins E B 2014 *Astron. Astrophys.* **567** A58
Kubiak M A, Bzowski, M, Sokól, J M, Swaczyna P, Grzedzielski S, Alexashov D B, Izmodenov V V, Möbius E Leonard T, Fuselier S A, Wurz P and McComas D J 2014 *Astrophys.J. Suppl. Ser.* **213** 29
Leonard T W, Möbius E, Bzowski M, Fuselier S A, Heirtzler D, Kubiak M A, Kucharek H, Lee M A, McComas D J, Schwadron N A and Wurz P 2015 *Astrophys. J.* **804** 42
Malamut C, Redfield S, Linsky J L, Wood B E and Ayres T R 2014 *Astrophys. J.* **787** 75
Möbius E, Bzowski M, Frisch P C, Fuselier S A, Heirtzler D, Kubiak M A, Kucharek H, Lee M A, Leonard T, McComas D J et al. 2015 *Astrophys. J. Suppl. Ser.* **220** 24
Redfield S 2009 *Space Sci. Rev.* **143** 323
Redfield S and Falcon R E 2008 *Astrophys. J.* **683** 207
Redfield S and Linsky J L 2000 *Astrophys. J.* **534** 825
Redfield S and Linsky J L 2004 *Astrophys. J.* **613** 1004
Redfield S and Linsky J L 2008 *Astrophys. J.* **673** 283 [RL08]
Redfield S and Linsky J L 2015 *Astrophys. J.* **812** 125
Slavin J D and Frisch, P C 2008 *Astron. Astrophys.* **491** 53
Wood, B E, Müller H R and Witte, M 2015 *Astrophys. J.* **801** 62
Zachary J, Redfield S, Linsky J L and Wood B E 2008 submitted to *Astrophys. J.*
Zank G P, Heerikhuisen J., Wood B E, Pogorelov N V, Zirnstein E and McComas D J 2013 *Astrophys. J.* **763** 20
Zirnstein E J and McComas D J 2015 *Astrophys. J.* **815** 31

IOP Conf. Series: Journal of Physics: Conf. Series **1100** (2018) 012017 doi:10.1088/1742-6596/1100/1/012017

Kappa distributions: Thermodynamic origin and Generation in space plasmas

George Livadiotis

Southwest Research Institute, USA
glivadiotis@swri.edu

Abstract. The paper reviews the new developments on the thermodynamic origin of kappa distributions and their generation in space plasmas. We present the correct formulation of kappa distributions that is consistent with thermodynamics. Then, we present the characterization of mechanisms that generate kappa distributions. There are three fundamental plasma conditions for the kappa distributions to be generated in space plasmas. Several applications of kappa distributions in space plasmas are presented and discussed. Finally, the thermodynamic validity of kappa distributions is presented. According to the zeroth law of thermodynamics, the most generalized form of particle distribution assigned with a temperature, is given by the kappa distributions, where temperature and kappa are two independent parameters spanning the 2-D abstract space of thermodynamics.

1. Introduction

In the last ~60 years, numerous analyses have established the theory of kappa distributions and provided a plethora of different applications in space and astrophysical plasmas exhibiting non-Maxwellian behavior (c.f., Figure 1 in [1], Chapter 1).

Classical collisional particle systems residing in thermal equilibrium have their particle velocity/energy distribution function stabilized into a Maxwell-Boltzmann distribution. On the contrary, space and astrophysical plasmas are collisionless particle systems residing in stationary states characterized by the so-called kappa distribution function. Kappa distributions have become increasingly widespread across the physics of space plasma processes, describing particles in the heliosphere, from the solar wind and planetary magnetospheres to the heliosheath and beyond, the interstellar and intergalactic plasmas (Tables 1 &2).

Kappa distributions were introduced to describe magnetospheric electron data, by Olbert and its Ph.D. students and colleagues [2-4]; Binsack (1966) [2] was the first to publish the usage of kappa distributions, but he acknowledged that the kappa function was actually "introduced by Prof. Olbert of MIT in his studies of IMP-1."

Single types of kappa distributions are usually sufficient to model the space plasma populations (e.g., [5-7]); however, more complicated models of kappa distributions superposition have been employed to describe rare features (e.g., anisotropy) (e.g., [8-10]). A kappa distribution is primarily formulated to describe a Hamiltonian, i.e., the particle kinetic and potential energy ([11,12]; [1], Chapter 3). In several cases, though, the potential energy of a particle is small compared to its kinetic energy and can often be ignored, and thus, the system's statistical description is reduced to the kappa distribution of the particle velocities.

Table 1. Examples of Applications of Kappa Distributions in Space and Astrophysical Plasmas

Space/Astrophysical Plasma	References
Solar Atmosphere	[13]
Solar Spectra	[14,15]
Solar Corona	[16-20]
Solar Flares	[21-24]
Solar Radio Bursts	[25,26]
Solar Wind	[8,10,27-44]
Solar Energetic Particles	[45,46]
Corotating Interaction Regions	[47]
Planetary Magnetospheres	(See Table 2)
Cometary Magnetospheres	[48]
Inner Heliosheath	[6,36,49-58]
HII Regions	[60]
Planetary Nebulae	[61,62]
Supernovae Magnetospheres	[63]
Cosmic Rays	[64,65]
Cosmology Scales	[66]

(Note: See also [1], Chapters 10-17)

Table 2. Examples of Applications of Kappa Distributions in Planetary Magnetospheric Plasmas

Magnetospheric Plasma	References
Terrestrial Magnetosphere	
Magnetosheath	[4,67,68]
Magnetopause	[69]
Magnetotail	[70]
Ring Current	[71]
Plasma Sheet	[72-74]
Magnetospheric Substorms	[24,75,76]
Aurora	[77]
Magnetospheres of Giant Planets	
Jovian	[78,79]
Saturnian	[5,80-82]
Uranian	[83]
Neptunian	[84]
Magnetospheres of Planetary Moons	
Io	[85]
Enceladus	[86]

The purpose of this paper is to review the new developments on the thermodynamic origin of kappa distributions and their generation in space plasmas. In Section 2, we present the formulation of kappa distributions that is consistent with thermodynamics. In Section 3, we present what the observations and measurements of kappa distributions have revealed regarding the trends between the kappa index and other plasma parameters. These trends can be used as indications to detect the mechanisms generating the observed kappa distributions. In section 4, we briefly picture the theory of mechanisms responsible for generating kappa distributions in space plasmas, and discuss three conditions that affect the generation of these distributions. In Section 5, we examine the thermodynamic validity of kappa distributions. The zeroth law of thermodynamics can be used to derive the most generalized form of particle velocity/energy distributions assigned with a temperature, where temperature and kappa are two independent parameters spanning the 2-D abstract space of thermodynamics. Finally, in Section 6, we summarize the main conclusions of the paper.

2. Formulation of Kappa distributions

A breakthrough in the field came with the connection of kappa distributions with the solid background of non-extensive statistical mechanics. Understanding the statistical origin of kappa distributions was a cornerstone of further theoretical developments and applications. The origin of the kappa distribution in non-extensive statistical mechanics has already been examined by several authors [87-90].

Historically, the first evidence of a connection between the theory of kappa distributions and the framework of non-extensive statistical mechanics was mentioned by Treumann (1997) [87] in the appendix of his investigation regarding the superdiffusion near the magnetopause. A detailed derivation of the kappa distribution through non-extensive statistical mechanics was referred to in the analysis of Milovanov & Zelenyi (2000) [88] and Leubner (2002) [89]. They showed the connection of kappa distributions with the canonical probability distribution that maximizes the Tsallis entropy under the constraints of canonical ensemble (e.g., [91]).

Livadiotis & McComas [10,34,90) completed the connection of kappa distributions with the modern non-extensive statistical mechanics using the concept of physical temperature that unifies the thermodynamic and kinetic definitions of temperature (see also [7]; [1], Chapter 1). The developed kappa distribution formulation is the only one that includes correctly the notion of temperature:

$$P(\varepsilon) \propto \left[1 + \frac{1}{\kappa} \cdot \frac{\varepsilon - U}{k_B T}\right]^{-\kappa-1} , \quad (1)$$

where $U = \langle \varepsilon \rangle$ stands for the mean kinetic energy.

Thereafter, the connection of kappa distributions with non-extensive statistical mechanics revealed the robust physical meaning of temperature, thermal pressure, and other thermodynamic parameters. The properties of kappa distributions and the proven tools of non-extensive statistical mechanics have been successfully applied to a variety of space plasmas throughout the heliosphere. These analyses led to the determination of the thermodynamic variables of these plasmas, as well as, the understanding of the underpinning plasma processes.

The standard 3-dimensional kappa distribution function of particle velocities is given by

$$P(\vec{u};\theta;\kappa) = \pi^{-\frac{3}{2}} \cdot \mathrm{A}(\kappa) \cdot \theta^{-3} \cdot \left[1 + \frac{1}{\kappa - \frac{3}{2}} \cdot \frac{(\vec{u} - \vec{u}_b)^2}{\theta^2}\right]^{-\kappa-1} , \quad \mathrm{A}(\kappa) \equiv (\kappa - \tfrac{3}{2})^{-3/2} \cdot \frac{\Gamma(\kappa+1)}{\Gamma(\kappa - \frac{1}{2})} , \quad (2)$$

which is parametrized by the 3-dimensional kappa index κ (in general depends on the dimensionality), and the thermal speed $\theta = \sqrt{2k_B T / m}$; the latter is defined in order to have no dependence on the kappa index but only on the temperature (it can be seen as the temperature in speed units). The bulk velocity $\vec{u}_b$ gives the average velocity of the particles, that is, the flow velocity.

The classical thermal equilibrium corresponds to $\kappa \to \infty$, where we have $\mathrm{A}(\kappa) \to 1$ and $[1 + x / (\kappa - \frac{3}{2})]^{-\kappa-1} \to e^{-x}$, thus, the distribution recovers the Maxwellian formulation:

$$P(\vec{u};\theta;\kappa \to \infty) = \pi^{-\frac{3}{2}} \cdot \theta^{-3} \cdot \exp\left\{-\frac{(\vec{u} - \vec{u}_b)^2}{\theta^2}\right\} . \quad (3)$$

As mentioned above, the kappa index κ depends on the degrees of freedom with the linear relation:

$$\kappa(f) = \kappa_0 + \tfrac{1}{2} f , \; f = d \cdot N , \quad (4)$$

where f are the total kinetic degrees of freedom from all the N particles of d degrees each. The kappa index, κ_0, called "invariant kappa index", arranges the different stationary states along the kappa index, independently of the dimensionality, the degrees of freedom, or the number of particles (e.g., [7,54]).

Then, using the invariant kappa index we can formulate the d-dimensional kappa distribution:

$$P(\vec{u};\theta;\kappa) = \pi^{-\frac{d}{2}} \cdot \mathrm{A}_d(\kappa_0) \cdot \theta^{-d} \cdot \left[1 + \frac{1}{\kappa_0} \cdot \frac{(\vec{u} - \vec{u}_\mathrm{b})^2}{\theta^2}\right]^{-\kappa_0 - 1 - \frac{d}{2}} , \quad \mathrm{A}_d(\kappa_0) \equiv \kappa_0^{-\frac{d}{2}} \cdot \frac{\Gamma(\kappa_0 + 1 + \frac{d}{2})}{\Gamma(\kappa_0 + 1)} . \quad (5)$$

For example, if using a 3D kappa distribution to describe a plasma population we find $\kappa \approx 3$, while using a 1D kappa distribution we find $\kappa \approx 2$, then there should be no conflict between those two results; both lead to the same (invariant) kappa index: $\kappa(3) = \kappa_0 + \frac{3}{2} \approx 3$ and $\kappa(1) = \kappa_0 + \frac{1}{2} \approx 2$ lead to $\kappa_0 \approx \frac{3}{2}$.

3. Observations

Recently, we have shown that the values of the kappa index (the parameter that labels and governs kappa distributions) are related to other solar wind plasma parameters (e.g., density n, temperature T) [7]. These trends reveal relationships that vary with location, time, and the type of space plasmas. In Figure 1, we illustrate the representative values of density n, temperature T, and kappa index κ of ~40 different space plasmas. We also plot the quantity M=1/(κ-0.5), which provides a measure of how far the system resides from thermal equilibrium [34,35]. (a) The parameters are plotted in a (n,T) diagram, using a color-map to illustrate the values of κ and M. Panel (b) shows that M is negatively correlated (or, κ is positively correlated) with T and n; more precisely, M is linearly decreasing as $M \approx 2.27 - 0.19 \cdot \log(nT^{0.6})$.

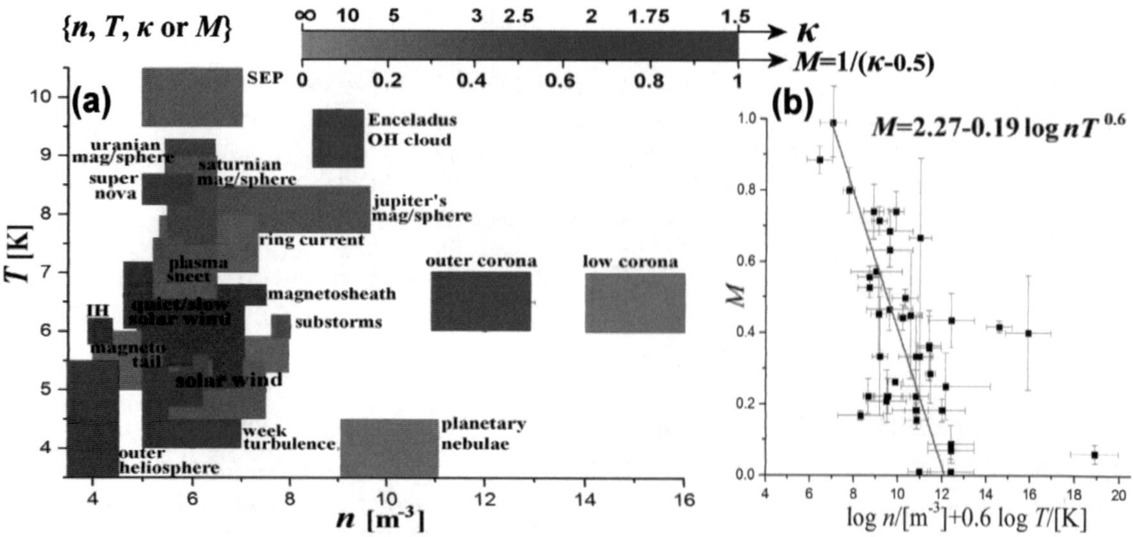

Figure 1. (a) Color-map of the 3D kappa indices of various space plasmas with different values of density n and temperature T. The kappa index κ or the measure M=1/(κ-1/2) spans the whole interval from $\kappa \to \infty$ or M=0 (thermal equilibrium) to $\kappa \to 1.5$ or M=1 (Anti-equilibrium, i.e., the furthest state from thermal equilibrium). (b) The measure M is negatively correlated (or, the kappa index is positively correlated) with the temperature and density. (Adapted from [7])

Figure 1 shows that, on average, the kappa index is positively correlated with temperature and density. Similar cases where the kappa index is positively correlated with temperature have also been observed in individual space plasmas, e.g., the magnetosheath [68] or the inner heliosheath [6,56,57], as shown in Figure 2.

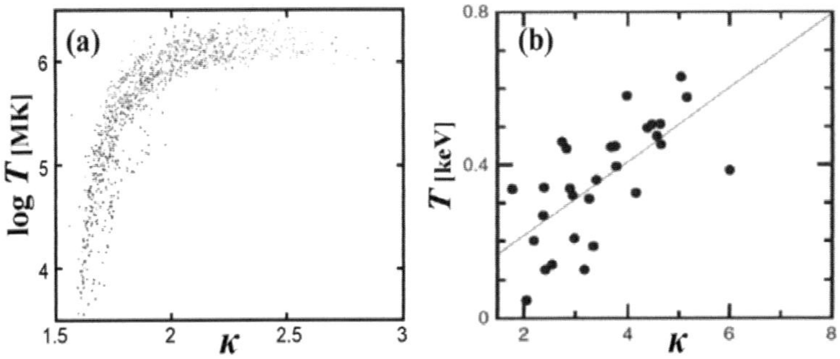

Figure 2. Examples of empirical relations between temperature and kappa index characterizing (a) inner heliosheath and (b) magnetosheath. (Adapted from [6] and [68])

The kappa index is mostly positively correlated with temperature, while temperature and density can be either correlated or anti-correlated, depending on the polytropic index (if it is $a>1$ or $a<1$, respectively). A polytropic relation connects the values of thermal variables (such as, density n, temperature T, thermal pressure P, the kappa index, κ, etc.) along a certain streamline of the plasma flow. A polytrope is a certain thermodynamic process characterized by such a relation. Typically, this is a power-law between two thermodynamic variables,

$$[P(\vec{r})/P(\vec{r}_*)]=[n(\vec{r})/n(\vec{r}_*)]^a \text{ , or } [n(\vec{r})/n(\vec{r}_*)]=[T(\vec{r})/T(\vec{r}_*)]^\nu \text{ , with } \nu\equiv 1/(a-1) \text{ ,} \quad (6)$$

where the exponent a denotes the typical polytropic index; ν is another way of expressing the polytropic index, which corresponds to the effective degrees of freedom.

While the most of space plasmas exhibit positive correlations between density and temperature, there are several plasmas with negative correlations, consistent with constant or quasi-constant thermal pressure, were found in heliosheath [6,38] and planetary magnetosheaths (e.g., the low latitude boundary layer at the terrestrial magnetosheath [92]; in the terrestrial central plasma sheet [93,94]; and in the Jovian magnetosheath, [95]).

Table 3. Examples of Polytropic Indices Measurements

Space Plasma	Datasets	Resulting a	References
Bow Shock	ISEE-1	2	[96]
Bow Shock	ISEE-1	1.6-1.7	[97]
Venus bow Shock	Pioneer Venus Orbiter	1.85	[98]
Plasma Sheet	AMPTE/IRM	1.67 ± 0.5, 1.4	[99]
Plasma Sheet	Cluster	-0.15, 0-1	[93,94]
Magnetosheath	ISEE-1 & -2	~0	[92]
Jovian Mag/sheath	New Horizons	~0	[95]
Solar Wind	Helios-1	1.46, 1.58	[100]
Solar Wind	Pioneer Venus Orbiter	5/3, 2	[101]
Solar Wind	Wind	0.5-2.5	[102]
Solar Wind	OMNI	1.8 ± 2.4	[103]
Solar Wind	Wind	~1.66	[104]
Solar Wind	Wind	~1.15, ~1.65	[43]
Magnetic Clouds	IMP-8	0.5, 0.6	[105]
CMEs	Ulysses/ CMEs	0.73, 0.78	[106]
Inner Heliosheath	IBEX	-0.04 ± 0.07	[38]
Inner Heliosheath	IBEX	~0	[107]

Aside from the temperature and density, the kappa index was also found to be connected with other plasma parameters, e.g., polytropic index (e.g., [43,108]; [1], Chapter 5), thermal pressure (e.g., [6]), potential energy (e.g., [43,77]), and magnetic field (e.g., [44]). The latter has significant interplay in the generation of kappa distributions in space plasmas, where the kappa index is found to decrease as the magnetic field strength increases (e.g., [68,109,110]).

The observed trends between the kappa index and other space plasma parameters can be caused by mechanisms responsible for generating kappa distributions.

4. Generation of kappa distributions

There are various mechanisms mentioned in the literature that can generate kappa distributions, and thus, produce relationships among the kappa index and other plasma parameters. Some examples are studied in the book of kappa distributions [1]: Connection with polytropes ([43,107]; [1], Chapter 5); strong Debye shielding and magnetic coupling ([44]; [1], Chapter 5); superstatistics ([111,112]; [1], Chapter 6); presence of weak turbulence ([40]; [1], Chapter 8); effect of pickup ions especially in the inner heliosheath ([37]; [1], Chapter 10); turbulence related to diffusion inverse proportional to velocity ([22]; [1], Chapter 13); pump acceleration mechanism ([113]; [1], Chapter 15); and, effect of shock waves ([114]; [1], Chapter 16).

Systems with no correlations between their particles are stabilized into the classical thermal equilibrium – the state where the particle distribution function of velocities/energies is described by a Maxwell-Boltzmann distribution. Systems with particle correlations, however, reside in stationary states out of this classical thermal equilibrium, described by kappa distributions. In fact, the kappa index is inversely proportional to the correlation coefficient ρ between the energies of any two particles, $\kappa=(3/2)/\rho$ [115,54,12]. The largest kappa index is infinity, corresponding to the system residing at thermal equilibrium, where the particles are characterized by zero correlation ($\rho=0$). The smallest kappa index is 3/2 (for 3D distributions), corresponding to the furthest state from thermal equilibrium, called anti-equilibrium, where the particles have the highest correlation ($\rho=1$). All other cases reside between these extremes, with kappa indices between 3/2 and ∞, i.e., with correlation ρ between 0 and 1 (Figure 3). Therefore, local correlations among the velocities or energies of particles are interwoven with kappa distributions.

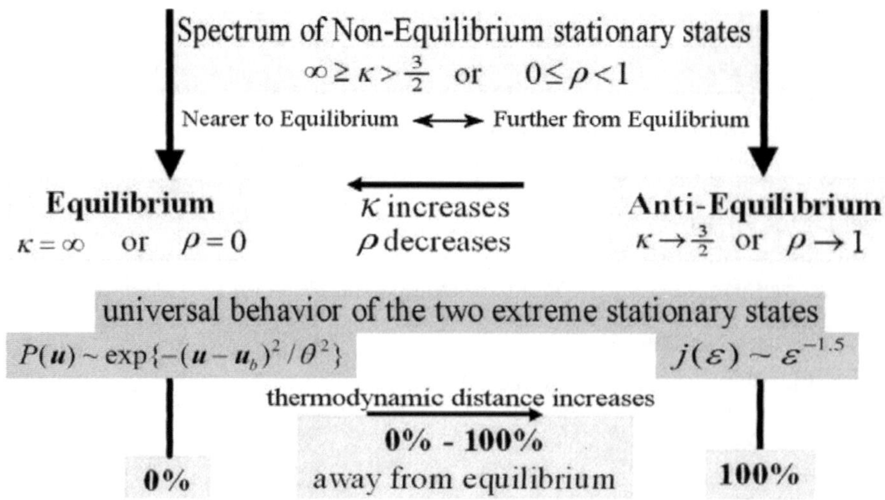

Figure 3. Arrangement of stationary states depending on the kappa index κ or the correlation ρ between particle energies. The two extreme stationary states correspond to the classical thermal equilibrium ($\kappa\rightarrow\infty$) and the recently studied state of anti-equilibrium ($\kappa\rightarrow0$). (Taken from [10])

A mechanism that generates kappa distributions involves plasma processes that induce local correlations among particles. Livadiotis et al. (2018) [44] identified three fundamental conditions that can generate kappa distributions in space plasmas, noted as: (i) Debye shielding, (ii) magnetic field coupling, and (iii) temperature fluctuations, each one prevailing in different scales of solar wind plasma and magnetic field properties (Figure 4):

(i) *Debye shielding*: Debye shielding produces a natural ordering of the particle correlations in plasmas (e.g., [116]). Inside a Debye sphere, particles are highly correlated with each other and act together through their electromagnetic interactions. In contrast, at distances greater than a Debye length λ_D, particles are largely uncorrelated owing to the Debye shielding of the closer particles. Therefore, each Debye sphere represents a cluster of correlated particles (ions and electrons), which is essentially uncorrelated to the more distant particles and their Debye spheres. Weakly coupled plasmas have small coupling parameter Γ, or equivalently, a large number of particles within their Debye spheres, N_D (Debye number), because of the relation $\Gamma \cong 0.1 \cdot N_D^{-2/3}$. Local correlations among particles within the Debye spheres can persist for a substantial length of time without being destroyed by collisions. The larger the Debye number N_D, the more effective the collective behavior in plasma, which inclines the system even further from thermal equilibrium; thus, the smaller the kappa index becomes. Hence, N_D is negatively correlated with κ.

(ii) *Magnetic field binding*: The magnetic field's long-range interactions induce correlations among particles, turning the system away from thermal equilibrium, i.e., the kappa index decreases. Unfortunately, we do not know the exact functional relationship between B and κ, but we expect an inverse correspondence (negative correlation): the largest the magnetic field, the more intense the induced correlations (strongest "binding"), and the lower the kappa index; thus, B is negatively correlated with κ. While the magnetic energy "binds" the particles together, the thermal energy, that is a measure of collisions, disorders the particles and destroys their correlations, thus competes with the magnetic binding, leading to a positive correlation between T and κ. Then, the densities of magnetic and thermal energy have an opposite effect on the kappa index, that is, negative vs. positive correlation, respectively. The competition between magnetic and thermal pressure can be represented by the plasma beta β (the ratio of thermal over magnetic pressure). As β increases, the magnetic over thermal pressure decreases, and thus, long-range interactions due to the magnetic field become weaker and the kappa index increases. Hence, β is positively correlated with κ.

(iii) *Temperature fluctuations*: Complex driven nonequilibrium systems, such as space plasmas exhibit temperature fluctuations on a large spatio-temporal scale compared to local relaxation times. There is a relatively fast dynamics, given by the local particle distributions and their statistical mechanics, and also a slow one, given by a spatio-temporally inhomogeneous parameter that varies globally with the environment. The two effects produce a superposition of two statistics, called "superstatistics" [111]. According to the concept of superstatistics, local fluctuations of temperature shift the system into stationary states described by kappa distributions, where the stabilized kappa index is linearly related to the inverse of the variance of temperature fluctuations.

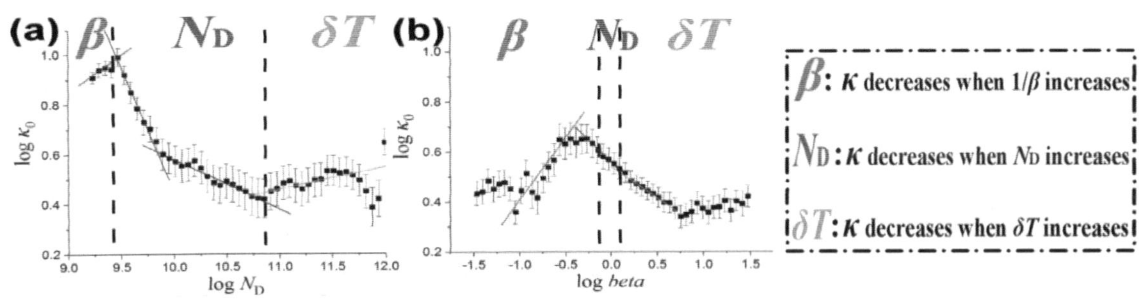

Figure 4. We observe three regions with different coupling and dynamics, corresponding to three processes that can generate kappa distributions: magnetic field coupling, Debye shielding, and

temperature fluctuations, respectively. The three regions are noted by the parameters beta plasma β, number of particles in a Debye sphere N_D, and standard deviation of temperature fluctuations δT. (Taken from [44])

The general scheme that characterizes these mechanisms is composed of the following two parts: (1) Generation of local correlations among particles, and (2) "Thermalization", that is, the re-stabilization of the particle system into a non-equilibrium stationary state assigned by a temperature, and thus, described by kappa distributions (Figure 5).

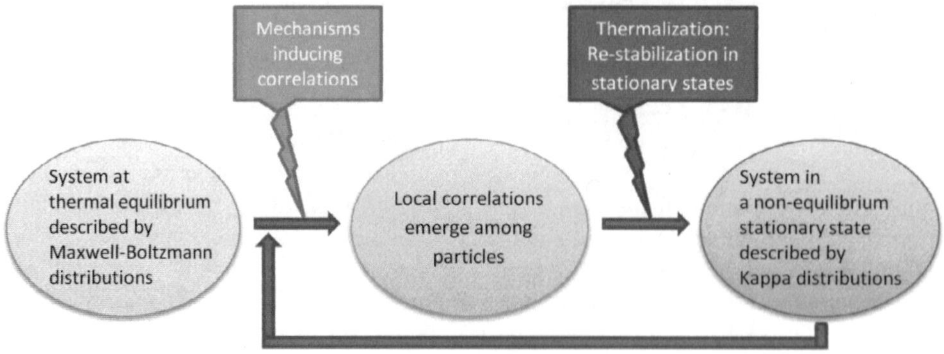

Figure 5. General scheme of mechanisms generating kappa distributions: Correlation and Thermalization. The input can be systems described either by Maxwell-Boltzmann or kappa distributions. (Taken from [44])

While there are several mechanisms capable of generating kappa distributions in space plasmas, none of these mechanisms can explain why the kappa distributions are allowed by the laws of thermodynamics. Fortunately, it has been shown that this is the case; namely, the zeroth law of thermodynamics is consistent with kappa distributions [117].

5. Thermodynamics of kappa distributions

One of the most frequent misinterpretations of temperature is that it is assumed to be dependent on the kappa index, instead of being a fundamental independent parameter.

The temperature is a well-understood concept in physics when it comes to particle systems at thermal equilibrium. This is based on the equivalence of the two fundamental definitions of temperature, that is (i) the kinetic definition of Maxwell (1866) [118] and (ii) the thermodynamic definition of Clausius (1862) [119]. The same equivalence is shown to exist for particle systems in stationary states out of the classical thermal equilibrium, where the particle distribution functions of velocities/energies cannot be described by the Maxwell-Boltzmann behavior. Indeed, when the system resides in any arbitrary stationary state assigned with a temperature, then its particle distribution function can be generally described by kappa distributions and combinations thereof.

According to the zeroth law of thermodynamics, if two thermodynamic systems are each in thermal equilibrium with a third, then they are in thermal equilibrium with each other. This is equivalent to have two systems in thermal equilibrium to each other. Indeed, two systems are in thermal equilibrium [120], if (a) they are linked by a wall permeable only to energy transfer (heat), and (b) they do not change over time. "Thermalization" is the characterization of a system of particles residing in any stationary state assigned by a temperature; in such a state, the particle velocities or energies can be stabilized into a Maxwell-Boltzmann distribution, or a kappa distribution, in general.

The mean particle kinetic energy is given by $U \equiv \frac{1}{2} d\, k_B T$. Then, the kinetic definition is given by $T_{\rm Kin} \equiv [2/(d\, k_B)] \cdot U$. On the other hand, the thermodynamic definition of Clausius is given by the notion of "physical temperature", $T_{\mathrm{phys}} \equiv (\partial S / \partial U)^{-1} \cdot [1 - \frac{1}{\kappa} \cdot S / k_{\mathrm{B}}]$, which is consistent with the zeroth law of thermodynamics; (S is the entropy and U is the internal mean energy of the system). The physical temperature T_{phys} is obtained in accordance with the generalized zeroth law (e.g., [117,121,122), while serves also the role of the kinetic definition of temperature within the framework of non-extensive Statistical Mechanics [90]. Therefore, all the advantages of a kinetically defined temperature, in contrast to other configurational definitions [123], can be ascribed to T_{phys}. Several inconsistencies concerning the kinetic definition in regards to the zeroth law of thermodynamics [124,125] are fully recovered, since the origin of T_{phys} establishes the generalized zeroth law. (See also: [7,10,34,54,90,116]; [1], Chapter 1).

The outcome of mixing two plasmas that reside in two different stationary states (different kappa indices) is similar to the mixing of classical gases, which obey simple calorimetry rules; e.g., if we mix two plasmas with different temperatures and densities, the result will be a mass-weighted average temperature of the combined plasma (Figure 6). This was resolved by demonstrating the equivalence of the two basic temperature definitions, (1) kinetic and (2) thermodynamic, for kappa distributions of particles out of equilibrium [90,34].

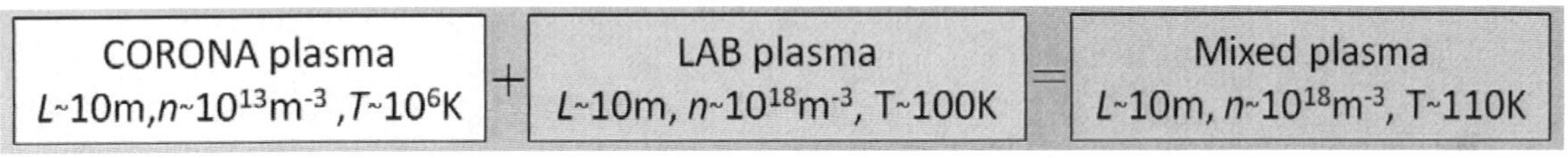

Figure 6. Space and laboratory plasmas can be understood under the same unique thermodynamics. The temperature of solar corona can be calculated using the kinetic definition, while the temperature of a laboratory plasma can be measured using simply a thermometer (that makes use of the thermodynamic definition). Thus, the resulting temperature follows from simple calorimetry rules. (Taken from [7])

Having understood the independence of temperature on other thermodynamic parameters, we now need to interpret correctly the observations of correlations between temperature and the kappa index. A similar example can be retrieved from plasmas at classical thermal equilibrium: The plasma particles are described by Maxwell distributions, dependent only on temperature T and density n. While these are two independent parameters, along the streamlines of a plasma flow, we frequently observe various relationships between n and T, typically, described by a polytropic relation, $T=A\cdot n^{a-1}$. However, each streamline has different constant A, while the polytropic index a characterizes the specific streamline ([103,104,126,127]; [1], Chapter 5).

There are other descriptions of generalized polytropic relations, $T=f(n)$ [107]. Generalized polytropic relations may also include the kappa index for space plasmas out of thermal equilibrium, i.e., $T=f(n,\kappa)$. In fact, very often the dependence on the density is ignored, dealing only with the more interesting relation of κ-T. For example, see the positively correlated kappa index and temperature values in the inner heliosheath [6], in the magnetospheres of Earth [69], Jupiter [78], and Saturn [5].

The observed relationships between the temperature and the kappa index by no means imply any universal dependence between those two thermodynamic parameters, in the same way that the observed relations between the temperature and the density do not imply any universal dependence between them. Care should be shown to distinguish the local relations that depend on certain plasma streamlines or the whole plasma, and not misinterpret these with any universal behavior that indicates to the foundations of physical laws.

Therefore, the stationary state at the classical thermal equilibrium is neither unique nor has any special properties that can be identified by measuring the internal energy U. All stationary states can equivalently describe system's internal energy, leading to the *Relativity Principle for Statistical Mechanics*: "A particle system is identified with the same internal energy from any of the stationary states that can attain" ([1], Chapter 1).

6. Conclusions

In this paper, we presented new developments on the thermodynamic origin of kappa distributions and their generation in space plasmas. We presented the formulation of kappa distributions that is consistent with thermodynamics, as well as a brief synopsis of the observations and measurements of kappa distributions. We discussed the observed trends between the kappa index and other plasma parameters, their origin and importance, as they are indications of the mechanisms generating kappa distributions. We then pictured the theory of mechanisms generating kappa distributions in space plasmas, and discussed three conditions that affect the generation of these distributions. We finally showed the thermodynamic validity of kappa distributions: The zeroth law of thermodynamics can be used to derive the most generalized form of particle distributions which are assigned with a temperature, where temperature and kappa are two independent parameters spanning the 2-D abstract space of thermodynamics.

In summary, we presented the following topics:

- The physical meaning of the temperature for stationary states out of thermal equilibrium was established, showing the equivalence of the thermodynamic and kinetic definitions of temperature. The concept of temperature was exclusively known for systems residing at the classical thermal equilibrium – the stationary state described by a Maxwellian distribution of velocities and the BG statistical mechanics.
- Therefore, the stationary state at the classical thermal equilibrium is neither unique nor has any special properties that can be identified by measuring the internal energy U. All stationary states can equivalently describe system's internal energy, leading to the *Relativity Principle for Statistical Mechanics*: "A particle system is identified with the same internal energy from any of the stationary states that can attain".
- Resolution of misinterpretations of temperature. The temperature is a fundamental independent parameter, and should not be considered dependent on the kappa index.
- The outcome of mixing two plasmas that reside in two different stationary states (different kappa indices) is similar to the mixing of classical gases, which obey simple calorimetry rules.
- The physical meaning of the kappa index is interwoven with the statistical correlation between the energy of the particles. The kappa index classifies the different stationary states independently of the dimensionality, the degrees of freedom, or the number of particles.
- The kappa index depends on the particle degrees of freedom f, with $\kappa(f) = \kappa_0 + \frac{1}{2}f$. The invariant kappa index, κ_0, classifies the different stationary states independently of the dimensionality, the degrees of freedom, or the number of particles, and allows the construction of multi-dimensional kappa distributions.
- The general scheme that characterizes these mechanisms is composed of the following two parts: (1) Generation of local correlations among particles, and (2) "Thermalization", that is, the re-stabilization of the particle system into a non-equilibrium stationary state assigned by a temperature, and thus, described by kappa distributions.
- There are three fundamental plasma conditions that need to be fulfilled for the kappa distributions to be generated in space plasmas, noted as: (i) Debye shielding, (ii) magnetic field coupling, and (iii) temperature fluctuations, each one prevailing in different scales of solar wind plasma and magnetic field properties.

References

[1] Livadiotis, G. 2017 *Kappa distribution: Theory & Applications in plasmas* Netherlands, UK, US: Elsevier.
[2] Binsack, J. H. 1966 Plasma studies with the IMP-2 satellite, Ph.D. Thesis, MIT.
[3] Olbert, S. 1968 Summary of experimental results from M.I.T. detector on IMP-1, in Physics of the Magnetosphere, R. L. Carovillano, J. F. McClay, and H. R. Radoski (Eds), New York: Springer, 641.
[4] Vasyliũnas, V. M. 1968 A survey of low-energy electrons in the evening sector of the

magnetosphere with OGO 1 and OGO 3. *J. Geophys. Res.* **73**, 2839–2884.

[5] Dialynas, K., S. M. Krimigis, D. G. Mitchell, D. C. Hamilton, N. Krupp, and P. C. Brandt 2009 Energetic ion spectral characteristics in the Saturnian magnetosphere using Cassini/MIMI measurements. *J. Geophys. Res.* **114**, A01212.

[6] Livadiotis, G., McComas, D.J, Dayeh, M.A., Funsten, H.O. and Schwadron, N.A. 2011 First sky map of the inner heliosheath temperature using IBEX spectra. *Astrophys. J.* **734**, 1.

[7] Livadiotis, G. 2015 Statistical background and properties of kappa distributions in space plasmas. *J. Geophys. Res.* **120**, 1607–1619.

[8] Štverák, S., Maksimovic M., Travnicek P. M., Marsch E., Fazakerley A. N., and Scime E. E. 2009 Radial evolution of nonthermal electron populations in the low-latitude solar wind: Helios, Cluster, and Ulysses Observations. *J. Geophys. Res.* **114**, A05104.

[9] Lazar, M., V. Pierrard, R. Schlickeiser and S. Poedts 2012 Modeling space plasma dynamics with anisotropic Kappa distributions, *Astrophys. Space Sci.* **33**, 97-107.

[10] Livadiotis, G. and McComas, D.J. 2013 Understanding kappa distributions: A toolbox for space science and astrophysics. *Space Sci. Rev.* **75**, 183–214.

[11] Livadiotis, G. 2015 Kappa distribution in the presence of a potential energy. *J. Geophys. Res.* **120**, 880.

[12] Livadiotis, G. 2015 Kappa and *q* indices: Dependence on the degrees of freedom. *Entropy* **17**, 2062.

[13] Fleishman, G. D. and Kuznetsov, A. A. 2014 Theory of gyroresonance and free-free emissions from non-Maxwellian quasi-steady-state electron distributions. *Astrophys. J.* 781, 77.

[14] Dzifčáková, E., and J. Dudík 2013 H to Zn ionization equilibrium for the non-Maxwellian electron κ-distributions: Updated calculations, *Astrophys. J. Suppl. Ser.* **206**, 6.

[15] Dzifčáková, E., J. Dudík, P. Kotrč, F. Fárník and A. Zemanová 2015 KAPPA: A package for synthesis of optically thin spectra for the non-Maxwellian κ-distributions based on the Chianti database. *Astrophys. J. Suppl. Ser.* **217**, 14.

[16] Owocki, S.P., and J.D. Scudder 1983 The effect of a non-Maxwellian electron distribution on oxygen and iron ionization balances in the solar corona. *Astrophys. J.* **270**, 758-768.

[17] Vocks, C., G. Mann, and G. Rausche 2008 Formation of suprathermal electron distributions in the quiet solar corona. *Astron. Astrophys.* **480**, 527–536.

[18] Kašparová, J., and Karlický, M. 2009 Kappa distribution and hard X-ray emission of solar flares. *Astron. Astrophys.* **497**, L13-L16.

[19] Lee, E., D. R. Williams, and G. Lapenta 2013 Spectroscopic indication of suprathermal ions in the solar corona, arXiv:1305.2939v.

[20] Cranmer, S. R. 2014 Suprathermal electrons in the solar corona: Can nonlocal transport explain heliospheric charge states? *Astrophys. J. Lett.* **791**, L31.

[21] Oka, M., Ishikawa, S., Saint-Hilaire, P., Krucker, S., and Lin, R.P. 2013 Kappa distribution model for hard X-ray coronal sources of solar flares. *Astrophys. J.* **764**, 6.

[22] Bian, N., G.A. Emslie, D.J. Stackhouse, and E.P. Kontar 2014 The formation of a kappa-distribution accelerated electron populations in solar flares. *Astrophys. J.* **796**, 142.

[23] Jeffrey, N. L. S., Fletcher, L., and Labrosse, N. 2016 First evidence of non-Gaussian solar flare EUV spectral line profiles and accelerated non-thermal ion motion. *Astron. Astrophys.* **590**, A99.

[24] Livadiotis, G., and McComas, D. J. 2013 Evidence of large-scale quantization in space plasmas. *Entropy* **15**, 1118–1134.

[25] Li, B., and Cairns, I.H. 2013 Type III radio bursts in coronal plasmas with kappa particle distributions. *Astrophys. J.* 763, L34.

[26] Li, B., and Cairns, I.H. 2014 Fundamental Emission of Type III Bursts produced in non-Maxwellian coronal plasmas with kappa-distributed background particles. *Sol. Phys.* **289**, 951-976.

[27] Collier, M. R., D. C. Hamilton, G. Gloeckler, P. Bochsler, and R. B. Sheldon (1996), Neon-20, oxygen-16, and helium-4 densities, temperatures, and suprathermal tails in the solar wind determined with WIND/MASS. *Geophys. Res. Lett.* **23**, 1191–1194.

[28] Maksimovic, M., V. Pierrard and J. Lemaire 1997 A kinetic model of the solar wind with Kappa distributions in the corona. *Astron. Astrophys.* **324**, 725-734.

[29] Maksimovic M., Zouganelis, I., Chaufray, J.-Y., Issautier, K., Scime, E. E., Littleton, J. E., Marsch, E., McComas, D. J., Salem, C., Lin, R. P., Elliott, H. 2005 Radial evolution of the electron

distribution functions in the fast solar wind between 0.3 and 1.5 AU. *J. Geophys. Res.* **110**, A09104.

[30] Pierrard, V., Maksimovic, M. and Lemaire, J. 1999 Electron velocity distribution function from the solar wind to the corona. *J. Geophys. Res.* **104**, 17021-17032.

[31] Mann, G., A. Warmuth, and H. Aurass 2009 Generation of highly energetic electrons at reconnection outflow shocks during solar flares. *Astron. Astrophys.* **494**, 669-675.

[32] Marsch, E. 2006 Kinetic physics of the solar corona and solar wind, *Living Rev. Sol. Phys.* **3**, 1.

[33] Zouganelis, I. 2008 Measuring suprathermal electron parameters in space plasmas: Implementation of the quasi-thermal noise spectroscopy with kappa distributions using in situ Ulysses/URAP radio measurements in the solar wind. *J. Geophys. Res.* **113**, A08111.

[34] Livadiotis, G. and McComas, D.J. 2010 Exploring transitions of space plasmas out of equilibrium. *Astrophys. J.* **714**, 971.

[35] Livadiotis, G. and McComas, D.J. 2010 Measure of the departure of the q-metastable stationary states from equilibrium. *Phys. Scr.* **82**, 035003.

[36] Livadiotis, G., and McComas, D. J. 2010 Non-equilibrium stationary states in the Heliosphere: The influence of pick-up ions. *AIP Conf. Proc.*, **1302**, 70-76.

[37] Livadiotis, G. and McComas, D.J. 2011 The influence of pick-up ions on space plasma distributions. *Astrophys. J.* **738**, 64.

[38] Livadiotis, G., and McComas, D. J. 2013 Fitting method based on correlation maximization: Applications in Astrophysics. *J. Geophys. Res.* **118**, 2863-2875.

[39] Livadiotis, G. and McComas, D.J. 2013 Near-equilibrium heliosphere – Far-equilibrium heliosheath. *AIP Conf. Proc.* **1539**, 344-350.

[40] Yoon, P. H. 2014 Electron kappa distribution and quasi-thermal noise. *J. Geophys. Res.* **119**, 7074.

[41] Pierrard, V. and Pieters, M. 2015 Coronal heating and solar wind acceleration for electrons, protons, and minor ions, obtained from kinetic models based on kappa distributions. *J. Geophys. Res.* **119**, 9441.

[42] Pavlos, G.P., Malandraki, O.E., Pavlos, E.G., Iliopoulos, A.C., and Karakatsanis, L.P. 2016 Non-extensive statistical analysis of magnetic field during the March 2012 ICME event using a multi-spacecraft approach. *Physica A* **464**, 149–181.

[43] Livadiotis, G. 2018 Using kappa distributions to identify the potential energy. *J. Geophys. Res.* **123**, 1050–1060.

[44] Livadiotis, G., Desai, M.I., and Wilson III, L.B. 2018 Generation of kappa distributions in solar wind at 1 AU. *Astrophys. J.*, **853**, 142.

[45] Xiao, F., Shen, C., Wang, Y., Zheng, H., and Whang, S. 2008 Energetic electron distributions fitted with a kappa-type function at geosynchronous orbit. *J. Geophys. Res.* **113**, A05203.

[46] Laming, J. M., J. D. Moses, Y.-K. Ko, C. K. Ng, C. E. Rakowski, and A. J. Tylka 2013 On the remote detection of suprathermal ions in the solar corona and their role as seeds for solar energetic particle production. *Astrophys. J.* **770**, 73.

[47] Chotoo, K., et al. 2000 The suprathermal seed population for corotating interaction region ions at 1AU deduced from composition and spectra of H+, He++, and He+ observed by Wind. *J. Geophys. Res.* **105**, 23107–23122.

[48] Broiles, T.W., Livadiotis, G., Burch, J.L., Chae, K., Clark, G., Cravens, T.E., Davidson, R., Eriksson, A., Frahm, R.A., Fuselier, S.A., Goldstein, J, Goldstein, R, Henri, P, Madanian, H, Mandt, KE, Mokashi, P., Pollock, C., Rahmati, A., Samara, M., Schwartz, S.J. 2016 Characterizing cometary electrons with kappa distributions. *J. Geophys. Res.* **121**, 7407-7422.

[49] Decker, R. B. and Krimigis, S. M. 2003 Voyager observations of low-energy ions during solar cycle 23. *Adv. Space Res.* **32**, 597–602.

[50] Decker, R. B., et al. 2005 Voyager 1 in the foreshock, termination shock, and heliosheath. *Science* **309**, 2020–2024.

[51] Heerikhuisen, J., Pogorelov, N. V., Florinski, V., Zank, G. P. and le Roux, J. A. 2008 The effects of a k-distribution in the heliosheath on the global heliosphere and ENA flux at 1 AU. *Astrophys. J.* **682**, 679-689.

[52] Heerikhuisen, J., Zirnstein, E., Pogorelov, N. 2015 κ-distributed protons in the solar wind and their

charge-exchange coupling to energetic hydrogen. *J. Geophys. Res.* **120**, 1516–1525.

[53] Zank, G. P.; J. Heerikhuisen, N. V. Pogorelov, R. Burrows, D. J. McComas 2010 Microstructure of the heliospheric termination shock: implications for energetic neutral atom observations. *Astrophys. J.* **708**, 1092.

[54] Livadiotis, G. and McComas, D.J. 2011 Invariant kappa distribution in space plasmas out of equilibrium. *Astrophys. J.* **741**, 88.

[55] Livadiotis, G. 2016 Curie law for systems described by kappa distributions. *EPL* **113**, 10003.

[56] Livadiotis, G., McComas, D. J., Randol, B., Möbius, E., Dayeh, M. A., Frisch, P. C., Funsten, H. O., Schwadron, N. A. and Zank, G. P. 2012 Pick-up ion distributions and their influence on ENA spectral curvature. *Astrophys. J.* **751**, 64.

[57] Livadiotis, G., McComas, D. J., Schwadron, N. A., Funsten, H. O. and Fuselier, S. A. 2013 Pressure of the proton plasma in the inner heliosheath. *Astrophys. J.* **762**, 134.

[58] Fuselier, S. A., Allegrini, F., Bzowski, M., Dayeh, M. A., Desai, M., Funsten, H. O., Galli, A., Heirtzler, D., Janzen, P., Kubiak, M. A., Kucharek, H., Lewis, W., Livadiotis, G., McComas, D. J., Möbius, E., Petrinec, S. M., Quinn, M., Schwadron, N., Sokół, J. M., Trattner, K. J., Wood, B. E., Wurz, P. 2014 Low energy neutral atoms from the heliosheath. *Astrophys. J.*, **784**, 89.

[59] Zirnstein, E. J. and McComas, D. J. 2015 Using kappa functions to characterize outer heliosphere proton distributions in the presence of charge-exchange. *Astrophys. J.* **815**, 31.

[60] Nicholls, D. C., Dopita, M. A., Sutherland, R. S. 2012 Resolving the Electron Temperature Discrepancies in H II Regions and Planetary Nebulae: κ-distributed Electrons. *Astrophys. J.* **752**, 148.

[61] Nicholls, D. C., Dopita, M. A., Sutherland, R. S., Kewley, L. J., Palay, E. 2013 Measuring nebular temperatures: the effect of new collision strengths with equilibrium and κ-distributed electron energies. *Astrophys. J. Supp.* **207**, 21.

[62] Zhang, Y., Liu, X.-W. and Zhang B. 2014 H-I free-bound emission of planetary nebulae with large abundance discrepancies: Two-component models vs. κ-distributed electrons. *Astrophys. J.* **780**, 93.

[63] Raymond, J. C., Winkler, P. F., Blair, W. P., Lee, J.-J., and Park, S. 2010 Non-Maxwellian Hα profiles in Tycho's supernova remnant, *Astrophys. J.* **712**, 901.

[64] Le Roux, J. A., Webb, G. M., Shalchi, A., Zank, G. P. 2010 A generalized nonlinear guiding center theory for the collisionless anomalous perpendicular diffusion of cosmic rays. *Astrophys. J.* **716**, 671-692.

[65] Yalcin, G.C., and Beck, C. 2018 Generalized statistical mechanics of cosmic rays: Application to positron-electron spectral indices. *Sci. Rep.* **8**, 1764.

[66] Hou, S. Q., He, J. J., Parikh, A., Kahl, D., Bertulani, C. A., Kajino, T., Mathews, G. J., and Zhao, G. 2017 Non-extensive statistics to the cosmological lithium problem. *Astrophys. J.* **834**, 165.

[67] Formisano, V., Moreno, G., Palmiotto, F., and Hedgecock, P. C. 1973 Solar Wind Interaction with the Earth's Magnetic Field 1. Magnetosheath, *J. Geophys. Res.* **78**, 3714-3730.

[68] Ogasawara, K., Angelopoulos, V., Dayeh, M.A., Fuselier, S.A., Livadiotis, G., McComas, D.J., McFadden, J.P. 2013 Characterizing the dayside magnetosheath using ENAs: IBEX and THEMIS observations. *J. Geophys. Res.* **118**, 3126-3137.

[69] Ogasawara, K., Dayeh, M.A., Funsten, H.O., Fuselier, S.A., Livadiotis, G., and McComas, D.J. 2015 Interplanetary magnetic field dependence of the suprathermal energetic neutral atoms originated in subsolar magnetopause. *J. Geophys. Res.* **120**, 964-972.

[70] Grabbe, C. 2000 Generation of broadband electrostatic waves in Earth's magnetotail. *Phys. Rev. Lett.* **84**, 3614.

[71] Pisarenko, N. F., Budnik, E. Yu., Ermolaev, Yu. I., Kirpichev, I. P., Lutsenko, V. N. Morozova, E. I., and Antonova, E. E. 2002 The ion differential spectra in outer boundary of the ring current: November 17, 1995 case study. *J. Atm. Solar-Terr. Phys.* **64**, 573–583.

[72] Christon, S. P. 1987 A comparison of the Mercury and earth magnetospheres: electron measurements and substorm time scales. *Icarus* **71**, 448-471.

[73] Wang, C.-P., L. R. Lyons, M. W. Chen, R. A. Wolf, and F. R. Toffoletto 2003 Modeling the inner plasma sheet protons and magnetic field under enhanced convection. *J. Geophys. Res.* **108**, 1074.

[74] Kletzing, C. A., J. D. Scudder, E. E. Dors, and C. Curto 2003 Auroral source region: Plasma

properties of the high latitude plasma sheet. *J. Geophys. Res.* **108**, 1360.

[75] Zaharia, S., Cheng, C. Z., and Johnson, J. R. 2000 Particle transport and energization associated with disturbed magnetospheric events. *J. Geophys. Res.* **105**, 18741–18752.

[76] Hapgood, M., C. Perry, J. Davies and M. Denton 2011 The role of suprathermal particle measurements in CrossScale studies of collisionless plasma processes, *Planet. Space Sci.* **59**, 618–629.

[77] Ogasawara, K., Livadiotis, G., Grubbs, G.A., Jahn, J.-M., Michell, R., Samara, M., Sharber, J. R., and Winningham, J. D. 2017 Properties of suprathermal electrons associated with discrete auroral arcs, *Geophys. Res. Lett.*, **44**, 3475–3484.

[78] Collier, M.R., and D.C. Hamilton 1995 The relationship between kappa and temperature in the energetic ion spectra at Jupiter. *Geophys. Res. Lett.* **22**, 303-306.

[79] Mauk, B. H., D. G. Mitchell, R. W. McEntire, C. P. Paranicas, E. C. Roelof, D. J. Williams, S. M. Krimigis, and A. Lagg 2004 Energetic ion characteristics and neutral gas interactions in Jupiter's magnetosphere. *J. Geophys. Res.* **109**, A09S12.

[80] Schippers et al. 2008 Multi-instrument analysis of electron populations in Saturn's magnetosphere. *J. Geophys. Res.*, **113**, A07208.

[81] Livi, R., Goldstein, J., Burch, J. L., Crary, F., Rymer, A. M., Mitchell, D. G., and Persoon, A. M. 2014 Multi-instrument analysis of plasma parameters in Saturn's equatorial, inner magnetosphere using corrections for spacecraft potential and penetrating background radiation. *J. Geophys. Res.* **119**, 3683.

[82] Carbary, J. F., M. Kane, B. H. Mauk and S. M. Krimigis 2014 Using the kappa function to investigate hot plasma in the magnetospheres of the giant planets. *J. Geophys. Res.* **119**, 8426–8447.

[83] Mauk, B. H., S. M. Krimigis, E. P. Keath, A. F. Cheng, T. P. Armstrong, L. J. Lanzerotti, G. Gloeckler, and D. C. Hamilton 1987 The hot plasma and radiation environment of the Uranian magnetosphere. *J. Geophys. Res.* **92**, 15283.

[84] Krimigis, S. M., Armstrong, T.P., Axford, W.I., Bostrom, C.O., Cheng, A.F., Gloeckler, G., Hamilton, D.C., Keath, E.P., Lanzerotti, L.J., Mauk, B.H., Van Allen, J.A. 1989 Hot plasma and energetic particles in Neptune's magnetosphere. *Science* **246**, 1483-1489.

[85] Moncuquet, M., Bagenal, F., and Meyer-Vernet, N. 2002 Latitudinal structure of the outer Io plasma torus. *J. Geophys. Res.* **108**, 1260.

[86] Jurac, S., McGrath, M. A., Johnson, R. E., Richardson, J. D., Vasyliunas, V. M., and Eviatar, A. 2002 Saturn: Search for a missing water source. *Geophys. Res. Lett.* **29**, 2172.

[87] Treumann, R A. 1997 Theory of superdiffusion for the magnetopause. *Geophys. Res. Lett.* **24**, 1727-1730.

[88] Milovanov A. V., and L. M. Zelenyi 2000 Functional background of the Tsallis entropy: "coarse-grained" systems and "kappa" distribution functions, *Nonlin. Processes Geophys.* **7**, 211–221.

[89] Leubner, M. P. 2002 A nonextensive entropy approach to kappa distributions. *Astrophys. Space Sci.* **282**, 573-579.

[90] Livadiotis, G. and McComas, D.J. 2009 Beyond kappa distributions: Exploiting Tsallis statistical mechanics in space plasmas. *J. Geophys. Res.* **114**, A11105.

[91] Tsallis, C. 2009 Introduction to Nonextensive Statistical Mechanics, New York: Springer.

[92] Sckopke, N., Paschmann, G., Haerendel, G., Sonnerup, B. U. Oe., Bame, S. J., Forbes, T. G., Hones, E. W., Jr., Russell, C. T. 1981 Structure of the low-latitude boundary layer. *J. Geophys. Res.* **86**, 2099-2110.

[93] Pang, X. X., Cao, J. B., Liu, W., Ma, Y., Lu, H., Yang, J., Li, L., Liu, X., Wang, J., Wang, T., and Yu, J. 2015 Polytropic index of central plasma sheet ions based on MHD Bernoulli integral. *J. Geophys. Res.* **120**, 4736-4747.

[94] Pang, X. X., Cao, J. B., Liu, W. L., Ma, Y., Lu, H., Yang, J., Li, L., Liu, X., Wang, J., Wang, T., and Yu, J. 2015 Case study of small scale polytropic index in the central plasma sheet. *Sci. China Earth Sci.* **58**, 1993-2001.

[95] Nicolaou, G., McComas, D. J., Bagenal, F., Elliott, H. A., and Wilson, R. J. 2015 Plasma properties in the deep Jovian magnetotail. *Plan. Space Sci.* **119**, 222-232.

[96] Zhuang, H. C., and Russell, C. T. 1981 An analytic treatment of the structure of the bow shock and

magnetosheath. *J. Geophys. Res.* **86**, 2191-2205.

[97] Winterhalter, D., Kivelson, M. G., Walker, R. J., and Russell, C. T. 1984 The MHD Rankine-Hugoniot jump conditions and the terrestrial bow shock: A statistical comparison. *Adv. Space Res.* **4**, 287-292.

[98] Tatrallyay, M., Russell, C. T., Luhmann, J. G., Barnes, A., and Mihalov, J. G. 1984 On the proper Mach number and ratio of specific heats for modeling the Venus bow shock. *J. Geophys. Res.* **89**, 7381-7392.

[99] Baumjohann, W. and Paschmann, G. 1989 Determination of the polytropic index in the plasma sheet. *Geophys. Res. Lett.* **16**, 295-298.

[100] Totten, T. L., Freeman, J. W., and Arya, S. 1995 An empirical determination of the polytropic index for the free-streaming solar wind using Helios 1 data. *J. Geophys. Res.* **100**, 13-17.

[101] Newbury, J. A., Russell, C. T., and Lindsay, G. M. 1997 Solar wind polytropic index in the vicinity of stream interactions. *Geophys. Res. Lett.* **24**, 1431-1434.

[102] Kartalev, M., Dryer, M., Grigorov, K., and Stoimenova, E. 2006 Solar wind polytropic index estimates based on single spacecraft plasma and interplanetary magnetic field measurements. *Geophys. Res. Lett.* **111**, A10107.

[103] Nicolaou, G., Livadiotis, G., and Moussas, X. 2014 Long term variability of the polytropic Index of solar wind protons at ~1AU. *Sol. Phys.* **289**, 1371-1378.

[104] Livadiotis, G., and Desai, M.I. 2016 Plasma-field coupling at small length scales in solar wind near 1au. *Astrophys. J.*, **829**, 88.

[105] Osherovich, V. A., Farrugia, C. J., and Burlaga, L. F. 1993 Dynamics of aging magnetic clouds. *Adv. Space Res.* **13**, 57-62.

[106] Hammond, C. M., Phillips, J. L., Crawford, G. K., Balogh, A. 1996 The relationship between electron density and temperature inside coronal mass ejections. *AIP Conf. Proc.* **382**, 558-561.

[107] Livadiotis, G. 2016 Superposition of polytropes in the inner heliosheath. *Astrophys. J. Suppl. Ser.*, **223**, 13.

[108] Meyer-Vernet, N., Moncuquot, M., and Hoang, S. 1995, Temperature inversion in the Io plasma torus, *Icarus* **116**, 202-213.

[109] Tao, X., and Lu, Q. 2014 Formation of electron kappa distributions due to interactions with parallel propagating whistler waves. *Phys. Plasmas* **21**, 022901.

[110] Stepanova, M., and Antonova, E. E. 2015 Role of turbulent transport in the evolution of the κ distribution functions in the plasma sheet. *J. Geophys. Res.* **120**, 3702–3714.

[111] Beck, C. and Cohen, E.G.D. 2003 Superstatistics. *Physica A* **322**, 267–275.

[112] Schwadron, N. A., Dayeh, M. A., Desai, M., Fahr, H., Jokipii, J. R., and Lee, M. A. 2010, Superposition of stochastic processes and the resulting particle distributions. *Astrophys. J.* **713**, 1386–1392.

[113] Fisk, L. A., and G. Gloeckler 2014 The case for a common spectrum of particles accelerated in the heliosphere: Observations and theory. *J. Geophys. Res.* **119**, 8733.

[114] Zank, G. P., Li, G., Florinski, V., Hu, Q., Lario, D., and Smith, C. W. 2006 Particle acceleration at perpendicular shock waves: Model and observations. *J. Geophys. Res.* **111**, A06108.

[115] Abe, S. 1999 Correlation induced by Tsallis' nonextensivity. *Physica A* **269**, 403-409.

[116] Livadiotis, G. and McComas, D.J. 2014 Electrostatic shielding in plasmas and the physical meaning of the Debye length. *J. Plasma Phys.* **80**, 341-378.

[117] Abe, S. 2001 General pseudoadditivity of composable entropy prescribed by the existence of equilibrium. *Phys. Rev. E* **63**, 061105.

[118] Maxwell, J. C. 1866 On the Dynamical Theory of Gases. *Phil. Mag.* **32**, 390-393.

[119] Clausius, R.J.E. 1862 Sixth Memoir: On the Application of the Theorem of the Equivalence of Transformations to Interior Work, 215.

[120] Bailyn, M. 1994 *A Survey of Thermodynamics*, New York: American Institute of Physics Press.

[121] Wang, Q.A., Nivanen, L., Méhauté, A.L., and Pezeril, M. 2002 On the generalized entropy pseudoadditivity for complex systems. *J. Phys. A* **35**, 7003–7007.

[122] Toral, R. 2003 On the definition of physical temperature and pressure for nonextensive thermostatistics. *Physica A* **317**, 209–212.

[123] Hoover, W. G., and Hoover, C. G. 2008 Nonequilibrium temperature and thermometry in heat-conducting models. *Phys. Rev. E*, **77**, 041104.
[124] Baranyai, A. 2000 Numerical temperature measurement in far from equilibrium model systems. *Phys. Rev. E* **61**, R3306–R3309.
[125] Baranyai, A. 2000 Temperature of nonequilibrium steady-state systems. *Phys. Rev. E* **62**, 5989–5997.
[126] Nicolaou, G. and Livadiotis, G. 2016 Misestimation of temperature when applying Maxwellian distributions to space plasmas described by kappa distributions. *Astrophys. Space Sci.* **361**, 359.
[127] Nicolaou, G., and Livadiotis, G. 2017 Modeling the plasma flow in the inner heliosheath with a spatially varying compression ratio. *Astrophys. J.* **838**, 7.

Acknowledgments
The work was supported in part by the project NNX17AB74G of NASA's HGI Program.

The Evolution of Interplanetary Shocks Propagating into the Very Local Interstellar Medium

Parisa Mostafavi[1,2] and Gary P. Zank[1,2]

[1] Department of Space Science, University of Alabama in Huntsville, Huntsville, AL 35899, USA

[2] Center for Space Plasma and Aeronomic Research (CSPAR), University of Alabama in Huntsville, Huntsville, AL 35899, USA

E-mail: psm0011@uah.edu

Abstract. Voyager 1 has made in situ measurements of the very local interstellar medium (VLISM) since August 2012 and its magnetometer and plasma wave instrument have detected several VLISM shock waves. Interplanetary shocks propagate through the supersonic solar wind and then through the inner heliosheath after colliding with the heliospheric termination shock (HTS). Interplanetary shock waves are transmitted partially across the heliopause (HP) into the VLISM and partially reflected back into the inner heliosheath. Previous studies showed that the in situ VLISM shocks observed by Voyager 1 were very weak and remarkably broad and had properties different than shocks inside the heliosphere [1, 2]. We model the first VLISM shock observed by Voyager 1 and compare with observations. We calculate the collisionality of the thermal particles and the dissipation terms such as heat conduction and viscosity that are associated with Coulomb collisions in the VLISM. The VLISM is collisional with respect to the thermal plasma and the VLISM shock structure is determined by thermal proton-proton collisions, which is the dominant thermal collisional term. The VLISM shock is controlled by particle collisions and not mediated by PUIs since they do not introduce significant dissipation through the shock transition. As a result, we find that the extremely broadness of the weak VLISM shock is due to the thermal collisionality.

1. Introduction

The HP is a tangential discontinuity that separates plasma in the VLISM from solar wind plasma. Voyager 1 passed the HP and entered the VLISM at a distance of 121 AU on August 25, 2012, becoming the first human-made spacecraft to enter the VLISM and make in situ measurements. The Low Energy Charge Particle (LECP) instrument on Voyager 1 measures differential intensities of low energy particles over a range of energies. The LECP instrument showed an abrupt decrease in the number density of anomalous cosmic rays (ACRs) and termination shock particles (TSP), which are dominant in the inner heliosheath (IHS) [3]. The galactic cosmic ray intensity increases at the position of HP. These are the characteristics of the VLISM region [4–6]. We are unable to directly measure the plasma quantities of the VLISM since Voyager 1's plasma instrument stopped working after the Jovian encounter. However, the plasma wave instrument can be used to indirectly measure the electron number density from the frequency of electron plasma oscillations [3].

Gurnett et al. (2015) [2] found that abrupt changes in the magnetic field and electron density detected by Voyager 1 in the VLISM corresponded to shock waves. The observed frequency of shock waves in the VLISM is about one per year. The observed VLISM shocks are due to interplanetary shocks propagating outward from energetic solar events such as coronal mass ejections (CMEs). The CME-driven shocks propagate through the supersonic solar wind, collide with the HTS, causing it to move outward, and then propagate through the inner heliosheath, until the HP, where the shock is partially reflected to propagate back into heliosheath and partially transmitted into the VLISM [7–10]. Burlaga et al. (2013) [1] studied the first shock observed in the VLISM, finding that interstellar shocks possess properties quite different than heliospheric shocks. The VLISM shock was observed to have a weak jump in magnetic field, B, and appeared to correspond to a weak, low beta, and subcritical shock [11]. Using the coplanarity theorem Burlaga et al. (2013) [1] showed that the shock is quasi-perpendicular (the angle between shock normal and **B** is about 85°). Being highly perpendicular the very weak change in magnetic field across the shock corresponds to a compression ratio of about $B_2/B_1 \approx 1.4$ (here 2/1 refers to the downstream/upstream state of the shock). The shock wave is remarkable because of its smooth transition and very large width, being about 10^4 times broader than a shock with similar properties at 1 AU. The third shock observed in the VLISM in 2014 had similar properties as the first observed shock, being a weak, laminar, smooth, quasi-perpendicular, and broad shock [2]. Thus far no theoretical explanation has been advanced to explain such a broad shock structure in the VLISM.

Collisional mean free paths in the heliosphere are very large. Collisionality can be neglected and therefore wave-particle interactions are important inside the heliosphere. Collisionless quasi-perpendicular shocks dissipate energy over a length scale (i.e., ion inertial length scale) that is much less than the collisional length scale [12]. Collisionless shocks have been observed throughout the heliosphere. Here we show that the electron and proton collisional time scale in the VLISM is much smaller than the characteristic dynamical time scale in the VLISM and the corresponding collisional mean free paths are very small. Therefore, the VLISM is a collisional environment and shock waves propagating in it are collisional with respect to the thermal plasma. Thermal Coulomb collisions introduce dissipation terms such as heat conduction and viscosity into the system and these processes are responsible for determining the structure of VLISM shocks. Hot and fast neutral atoms created in the supersonic solar wind and IHS are deposited in the VLISM [13–15], and can experience secondary charge exchange with interstellar charged particles, which leads to the creation of suprathermal PUIs in the VLISM. The very small number density of VLISM PUIs compared to that of the thermal gas introduces a very small pressure in the VLISM. We show that the physical process determining the structure of a VLISM shock and the HTS is different. The HTS is mediated by PUIs and controlled by wave-particle interactions whereas VLISM shocks are controlled by particle collisions. In the following section, we derive the basic equations describing the VLISM environment and the structure of shock waves therein. We model the VLISM shock and compare with Voyager 1 observations in Section 3. Finally, we conclude and summarize our results.

2. The VLISM

The VLISM consists of interstellar neutral atoms, thermal electrons and protons, and suprathermal PUIs. The charge exchange mean free path of protons and hydrogen atom in the VLISM is significantly larger than the scales of interest and we can neglect their coupling. Zank et al. (2014) [14] calculated collisional time scales between PUIs that are created by secondary charge exchange and thermal VLISM electrons and protons in the VLISM and showed that PUIs are not equilibrated with background thermal electrons and protons on a scale less than 75 AU. PUIs should therefore be treated as a separate component in the system. PUIs scatter off magnetic field fluctuations and drive streaming instabilities and experience pitch angle scattering that introduces a collisionless heat flux and a collisionless viscosity in the system. After using a collisionless Chapman-Enskog expansion, one can show the pressure tensor, P_I, and heat flux, q_I, terms assume the form,

$$\mathbf{P_I} = P_I\mathbf{I} + \mathbf{\Pi_I}, \tag{1}$$

$$\mathbf{\Pi_I} = \begin{pmatrix} 1 & 0 & 0 \\ 0 & 1 & 0 \\ 0 & 0 & -2 \end{pmatrix} \frac{\eta_{k\ell}}{2}\left(\frac{\partial U_k}{\partial x_\ell} + \frac{\partial U_\ell}{\partial x_k} - \frac{2}{3}\delta_{k\ell}\frac{\partial U_m}{\partial x_m}\right), \tag{2}$$

$$q_I(x,t) = -\frac{1}{2}\mathbf{K_I}.\nabla P_I, \tag{3}$$

where $\mathbf{\Pi_I}$ is the stress tensor (i.e., collisionless viscosity) and $\mathbf{K_I}$ the diffusion coefficient (i.e., collisionless heat flux). The inclusion of $\mathbf{\Pi_I}$ and $\mathbf{K_I}$ is due to the nearly non-isotropic PUI distribution. Here, P_I is the PUI isotropic pressure, η is the PUI viscosity and $\mathbf{U}$ is the bulk flow velocity.

Thermal VLISM protons and electrons are assumed to have a nearly Maxwellian distribution function. The streaming collisional time scale of thermal charged particle "a" colliding with a stationary background population of thermal charged particles "b" can be calculated as

$$(\tau_s^{ab})^{-1} = \frac{n_b q_a^2 q_b^2 ln\Lambda}{2\pi\epsilon_0^2 m_a^2 V_{Ta}^3}\frac{m_b}{m_a}(1+\frac{m_a}{m_b})\frac{T_a}{T_b}\frac{G(x_b)}{x_a}, \tag{4}$$

where $G(x)$ is the Chandrasekhar function,

$$G(x) \equiv \frac{f(x) - xf'(x)}{2x^2}, \quad f(x) = \frac{2}{\sqrt{\pi}}\int_0^x e^{-z^2}dz = erf(x), \tag{5}$$

$f(x)$ is the error function and $x_{a/b} \equiv v/V_{Ta/b}$. $V_{Ta/b} = \sqrt{2K_B T_{a/b} m_{a/b}^{-1}}$ is the thermal speed, K_B is Boltzmann's constant, and $\ln\Lambda$ is the Coulomb logarithm. The n, q, m, and T for particles a and b denote number density, charge, mass, and temperature, respectively.

The VLISM thermal protons and electrons are collisionally equilibrated and have the same temperature of about 7500 K [16]. Therefore, the VLISM proton and electron thermal speeds are 11 $km\ s^{-1}$ and 477 $km\ s^{-1}$, respectively. The plasma wave instrument on Voyager 1

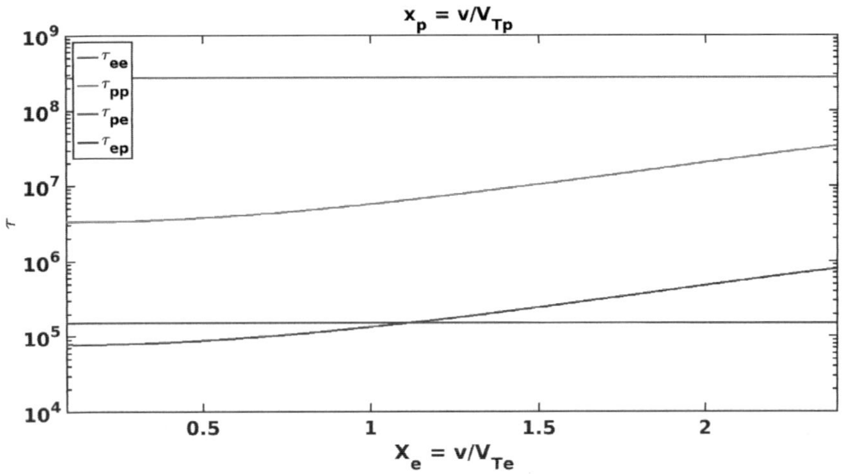

Figure 1: The *pp*, *ee*, *ep*, and *pe* collisional time-scales for the thermal VLISM plasma. Note that these collisional time scales do not apply to PUI collisional interactions with the thermal plasma.

measured the electron density upstream of the observed 2012 shock to be 0.06 cm^{-3} [2, 17, 18]. The magnetic field magnitude upstream of the VLISM shock measured by the magnetometer on Voyager 1 was about 0.38 nT [1]. We can evaluate the proton-proton (*pp*), electron-electron (*ee*), electron-proton (*ep*), and proton-electron (*pe*) collisional frictional time-scales and the mean free paths [10] for the region upstream of the observed VLISM shock using these parameters. Figure (1) shows the *pp*, *ee*, *ep*, and *pe* collisional time-scales for the thermal VLISM plasma as a function of velocity. Fig. (1) shows that the *ep* and *pe* collisional time-scales are independent of the velocity and constant,

$$\tau^{ep} = 1.5 \times 10^5 s, \quad \lambda^{ep} = 0.47AU; \quad \tau^{pe} = 2.7 \times 10^8 s, \quad \lambda^{pe} = 20.4AU, \tag{6}$$

where τ is the collisional time-scale and λ is the mean free path. The time-scales for *ee* and *pp* collisions are not constant and depend on velocity. The collisional heat conduction and viscous coefficients depend on the collisional scattering time scales, τ_s, (see for example section §4.8 of [19]). We should estimate τ_s for *pp* and *ee* collisions and therefore we approximate an appropriate collisional time, $< \tau_s >$ as follows. The coefficient of viscosity, η, can be estimated as [14]

$$\eta \equiv \frac{4\pi}{3} \int c^2 \tau_s f_0 m c^2 dc, \tag{7}$$

where $c = v - U$ is the random velocity. τ_s^{ee} and τ_s^{pp} can be approximated as 5th degree polynomial curves from Fig. (1) and f_0 is a Maxwellian distribution. On integrating from 0 to $2.5V_T$, we obtain the viscosity coefficient, η, for ee and pp collisions. Using the pressure upstream of the VLISM shock and the relation $\eta \simeq P < \tau_s >$, we derive the most probable estimates

$$\tau^{pp} = 4.2 \times 10^6 s, \quad \lambda^{pp} = 0.31AU; \quad \tau^{ee} = 7.5 \times 10^4 s, \quad \lambda^{ee} = 0.24AU. \tag{8}$$

Since the electron and proton collisional mean free paths are small compared to the almost featureless VLISM, we conclude that the VLISM is collisional with respect to the thermal plasma. The thermal gas Coulomb collisions introduce dissipation terms such as thermal heat conduction, K_g, and viscosity, η_g, (see [19] pages 159 – 161) in the VLISM region and therefore these collisional dissipation terms determine the structure of VLISM shocks.

Shock waves can be ideally modeled by considering a multi-fluid model for non-thermal PUIs, thermal protons, and thermal electrons. However, in most cases solving the multi-fluid model is very complicated and by making some assumptions it can be reduced to a single-fluid-like model (see [14, 15] for more details). The single-fluid-like system of equations comprises an equilibrated thermal plasma (electrons and protons) and a non-equilibrated PUI component. As was shown by Zank et al (2014) [14] PUIs should be treated as a separate pressure component, P_I, in the VLISM since they are not equilibrated with the background plasma. Taking into account that the bulk flow speed is dominated by protons and the electron mass is much smaller than the proton mass (details in [14, 15, 20, 21]), we can introduce the total density $\rho = m_e n_e + m_p n_p$, where the number density, n_p, includes both thermal protons and PUIs. The thermal gas pressure, P_g, contains both thermal proton and electron contributions, $P_g = P_e + P_p = n_e K_B T_e + n_p K_B T_p$. Here $n_{e/p}$, $P_{e/p}$, and $T_{e/p}$ are the thermal electron and proton number density, pressure, and temperature, respectively. The reduced single-fluid model comprises the continuity, momentum, and energy equations, together with Maxwell's equations,

$$\frac{\partial \rho}{\partial t} + \nabla \cdot (\rho \mathbf{U}) = 0; \tag{9}$$

$$\rho(\frac{\partial \mathbf{U}}{\partial t} + \mathbf{U} \cdot \nabla \mathbf{U}) = -\nabla(P_g + P_I) - \nabla \cdot (\mathbf{\Pi_I} + \mathbf{\Pi_g}) + \mathbf{J} \times \mathbf{B}; \tag{10}$$

$$\frac{\partial P_g}{\partial t} + \mathbf{U} \cdot \nabla P_g + \gamma_g P_g \nabla \cdot \mathbf{U} = \frac{1}{3}\nabla \cdot (\mathbf{K_g} \cdot \nabla P_g) - (\gamma_g - 1)\mathbf{\Pi_g} : (\nabla \mathbf{U}); \tag{11}$$

$$\frac{\partial P_I}{\partial t} + \mathbf{U} \cdot \nabla P_I + \gamma_I P_I \nabla \cdot \mathbf{U} = \frac{1}{3}\nabla \cdot (\mathbf{K_I} \cdot \nabla P_I) - (\gamma_I - 1)\mathbf{\Pi_I} : (\nabla \mathbf{U}); \tag{12}$$

$$\begin{aligned} &\frac{\partial}{\partial t}\left(\frac{1}{2}\rho \mathbf{U^2} + \frac{P_I}{\gamma_I - 1} + \frac{P_g}{\gamma_g - 1} + \frac{1}{2\mu_0}B^2\right) + \nabla \cdot (\frac{1}{2}\rho \mathbf{U} U^2 + \frac{\gamma_I}{\gamma_I - 1}P_I \mathbf{U} + \frac{\gamma_g}{\gamma_g - 1}P_g \mathbf{U} \\ &+ \frac{1}{\mu_0}B^2 \mathbf{U} - \frac{1}{\mu_0}\mathbf{U} \cdot \mathbf{BB} + \mathbf{\Pi_I} \cdot \mathbf{U} - \frac{1}{\gamma_I - 1}\frac{\mathbf{K_I}}{3} \cdot \nabla P_I + \mathbf{\Pi_g} \cdot \mathbf{U} - \frac{1}{\gamma_g - 1}\frac{\mathbf{K_g}}{3} \cdot \nabla P_g) = 0; \end{aligned} \tag{13}$$

$$\mathbf{E} = -\mathbf{U} \times \mathbf{B}; \ \frac{\partial \mathbf{B}}{\partial t} = -\nabla \times \mathbf{E}; \ \mu_0 \mathbf{J} = \nabla \times \mathbf{B}; \ \nabla \cdot \mathbf{B} = 0. \tag{14}$$

B and **E** denote the magnetic and electric fields. $\gamma_{I/g}$ are the adiabatic indices of the PUIs/thermal gas. All quantities pertaining to PUIs and thermal gas are denoted with the subscripts I and g, respectively. The one-dimensional form of the collisional thermal plasma stress tensor and collisionless PUI stress tensor may be expressed as $\Pi_{g/I} = -\frac{1}{3}\eta_{g/I}\frac{dU_x}{dx}$, where η_g is the thermal collisional viscosity coefficient and η_I is the collisionless PUI viscosity coefficient associated with magnetic fluctuations [14, 22] and is typically assumed to be constant.

To establish that we need to consider the collisional viscosity in the system, we assume initially that the only dissipation term in the VLISM is the thermal heat flux and we neglect thermal viscosity. The PUIs are assumed to behave adiabatically. The 1-D form of the perpendicular shock (i.e., $\mathbf{U} = (U_x, 0, 0)$, $\mathbf{B} = (0, 0, B_z)$) structure equation is a first-order ordinary differential equation

$$\frac{dy}{dx} = (\gamma_g - 1)\frac{U_{x1}}{3K_g}(y - 1)(y - y_A)\frac{M(y)}{N(y)}, \tag{15}$$

where

$$M(y) = \frac{1}{2}y_A y_B(1-y) - \frac{1}{2}(y+1)(y-y_A)^2 + \frac{1}{M_{I1}^2(\gamma_I - 1)}\frac{(y^{1-\gamma_I}-1)}{(1-y)}(y-y_A)^2$$
$$-\frac{1}{\gamma_I M_{I1}^2}\frac{\gamma_g}{\gamma_g - 1}y\frac{(y^{-\gamma_I}-1)}{(1-y)}(y-y_A)^2 - \frac{1}{(\gamma_g-1)M_{s1}^2}(y-y_A)^2 + \frac{\gamma_g}{\gamma_g-1}y(y-y_A)^2 \quad (16)$$
$$+y_B(y-y_A) - \frac{1}{2}\left(\frac{\gamma_g}{\gamma_g-1}\right)y_B y(y+1-2y_A);$$
$$N(y) = (y-y_A)^3 - y_B(1-y_A)^2 - \frac{1}{M_{I1}^2}(y-y_A)^3 y^{-\gamma_I - 1}. \quad (17)$$

Here y is the inverse compression ratio, y_A and y_B are the inverse Alfvénic Mach numbers squared along the x and z directions, respectively. Equation (15) has two solutions. The first solution yields smooth transitions from an upstream state to a downstream state. These solutions represent the class of smoothed shocks mediated completely by the thermal gas heat flux. The second set corresponds to double-valued solutions that are not physical, and a smooth transition from upstream to downstream is impossible. In this case, heat conduction associated with the thermal gas is not sufficient to smooth the shock structure, and an extra thermal gas dissipation term (thermal viscosity) must be added to affect the shock transition. Figure (2) shows a plot of M_{s1}^{-2} as a function of y_B for the case of quasi-perpendicular shocks ($y_A = B_x^2/(\rho_1 U_{x1}^2 \mu_0) \to 0$). The diagram identifies the two regimes i.e., smoothed and a regime that requires a sub-shock. The observed VLISM shock has the following parameters, $M_{s1}^2 = 0.06$, and $y_B = 0.3$, which shows that the VLISM shock is not in the smooth transition region. Therefore, thermal viscosity has to be added to ensure a smooth shock structure and hence transition (see also [20, 21]).

Here, a one-dimensional, steady-state model in which all physical quantities depend on the x Cartesian coordinate system is used. The 1-D background plasma velocity and magnetic field are $U = (U_x, 0, 0)$ and $B = (0, 0, B_z)$. The one dimensional form of equations (10)-(12) that includes both thermal gas viscosity and heat conduction in normalized form is

$$\gamma_g M_{s1}^2 \frac{\partial y}{\partial \bar{x}} = -\frac{\partial}{\partial \bar{x}}(P_g' + P_1 P_I') - y_B \gamma_g M_{s1}^2 B_z' \frac{\partial B_z'}{\partial \bar{x}} + \frac{1}{3}\gamma_g M_{s1}^2 (Sch_g + Sch_I \frac{K_I}{K_g}\frac{\rho_I}{\rho_g})\frac{\partial^2 y}{\partial \bar{x}^2}; \quad (18)$$
$$y P_1 \frac{\partial P_I'}{\partial \bar{x}} + \gamma_I P_I' P_1 \frac{\partial y}{\partial \bar{x}} = \frac{1}{3} P_1 \frac{K_I}{K_g}\frac{\partial^2 P_I'}{\partial \bar{x}^2} + \frac{1}{3}(\gamma_I - 1)\gamma_g M_{s1}^2 Sch_I \frac{K_I}{K_g}\frac{\rho_I}{\rho_g}(\frac{\partial y}{\partial \bar{x}})^2; \quad (19)$$
$$y\frac{\partial P_g'}{\partial \bar{x}} + \gamma_g P_g' \frac{\partial y}{\partial \bar{x}} = \frac{1}{3}\frac{\partial^2 P_g'}{\partial \bar{x}^2} + \frac{1}{3}(\gamma_g - 1)\gamma_g M_{s1}^2 Sch_g (\frac{\partial y}{\partial \bar{x}})^2, \quad (20)$$

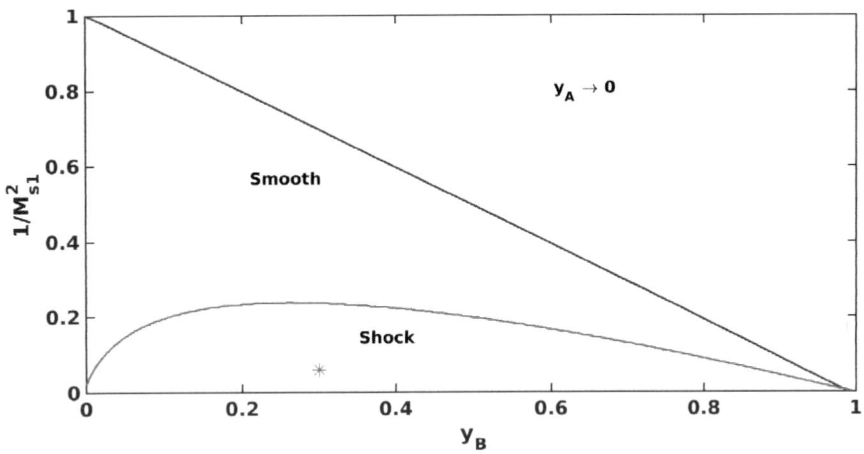

Figure 2: Initial upstream values M_{s1}^{-2} plotted as a function of y_B, for a quasi-perpendicular shock. The two regimes for which there is a smooth transition or a non-smooth (sub-shock) transition between the upstream and downstream states are identified. Here, $\gamma_I = \gamma_g = 5/3$ and $P_{I1}/P_{g1} = 0.05$. The green star corresponds to the parameters of the 2012 VLISM shock.

where $M_{s1}^2 = \rho_1 U_{x1}^2/\gamma_g P_{g1}$ is the thermal gas Mach number of the flow, $\bar{x} \equiv x/L$, $L \equiv K_g/U_1$, is distance normalized to the diffusion length scale, and $Sch_{g/I} \equiv \eta_{g/I}/(\rho_{1g/I} K_{g/I})$ is the Schmidt number of the thermal plasma/PUI. The inverse compression ratio is $y = U_x/U_{x1}$ and other normalized terms are defined as $P_g' = P_g/P_{g1}$, $P_I' = P_I/P_{I1}$, $B_z' = B_z/B_{z1}$, and $P_1 = P_{I1}/P_{g1}$. The inverse Alfvénic Mach number squared of the flow far upstream is expressed as $y_B = B_{z1}^2/(\rho_1 U_{x1}^2 \mu_0) = M_{Az0}^{-2}$.

The dominant collisional terms in a collisional magnetized plasma can be estimated by comparing the calculated thermal collisional mean free paths. We can neglect the electron-electron and electron-proton collisions. Since electrons have a very small mass compared to protons, they therefore do not contribute significantly to the dissipation process (i.e., thermal heat conduction or thermal viscosity) in the system. The large mean free path of proton-electron collisions compared to that of proton-proton collisions means that we can neglect proton-electron collisions. Thus the dominant collisional terms derive from proton-proton collisions and by using the expressions in Zank (2014) [19] (pages 159-161), the diffusion coefficient associated with

IOP Conf. Series: Journal of Physics: Conf. Series **1100** (2018) 012018 doi:10.1088/1742-6596/1100/1/012018

proton-proton collisions, K_{pp}, upstream of the observed VLISM shock can be calculated to be 10^{15} $m^2 s^{-1}$, the thermal viscosity $2.5 \times 10^{-8} kg m^{-1} s^{-1}$, and the thermal Schmidt number is therefore about 0.25. The number density of PUIs inside the VLISM is very small [23] and they contribute a very small pressure compared to the thermal plasma ($P_{I1}/P_{g1} \approx 0.05$). The pitch-angle scattering timescale in the VLISM is taken to be $3 \times 2\pi \times \Omega_g^{-1}$, giving 5.17×10^2 s and the collisionless PUI heat conduction is calculated to be about 10^{14} m^2 s^{-1}. The scattering time scale gives a value for the VLISM PUI viscosity, η_I, as $\sim 3.57 \times 10^{-13}$ kg $m^{-1} s^{-1}$, making the PUI Schmidt number about 10^{-3}. The thermal gas Schmidt number is much greater than the PUI Schmidt number ($Sch_g >> Sch_I$). Therefore as was shown by [20, 21], PUIs and the dissipation associated with them cannot mediate the VLISM shock and the PUI dissipation terms can be neglected compared to thermal gas dissipation terms. PUIs will therefore behave adiabatically and $P_I \rho^{-\gamma_I}$ is constant through a broad VLISM shock, and the thermal gas will be heated preferentially. However, this is completely different at the HTS. Zank et al. (1996) [13] predicted that the thermal gas remains relatively cold, with PUIs being heated through the HTS. The HTS observed by Voyager 2 confirmed this prediction and showed that the HTS is mediated by PUIs and reflected thermal protons do not provide a dissipation mechanism and are only heated adiabatically [24]. Thus the physical process determining the structure of the HTS and the VLISM shocks is very different.

The one-dimensional form of the model equations has been used to determine the structure of shocks in the VLISM, specifically the broad shock observed by Burlaga et al. (2013) [1]. One second-order ordinary differential equation in normalized form can be obtained from equations (18)-(20) for the structure of the VLISM shock,

$$\frac{d^2 y}{d\bar{x}^2} + \frac{3}{Sch_g}\left[\left(\frac{\gamma_I}{\gamma_g}\frac{1}{M_{s1}^2}\left(\frac{1}{y}\right)^{\gamma_I+1} - 1\right) - Sch_g y + \frac{y_B}{y^3}\right]\frac{dy}{d\bar{x}} = \frac{9}{Sch_g}\frac{1}{M_{s1}^2}(1-y)Z(y), \quad (21)$$

where

$$Z(y) \equiv M_{s1}^2 \frac{\gamma_g + 1}{2}\left(y - \frac{\gamma_g - 1}{\gamma_g + 1}\right) - \left(1 + P_1 - \frac{(\gamma_I - \gamma_g)}{\gamma_g(\gamma_I - 1)}\frac{(1 - y^{1-\gamma_I})}{(1-y)} - \gamma_g M_{s1}^2 \frac{y_B}{2y}\left(\frac{\gamma_g - 2}{\gamma_g} - y\right)\right). \quad (22)$$

Equation (21) is the shock structure equation in the presence of both thermal gas heat flux and viscosity. Further discussion related to the structure of VLISM shock waves was presented by

Mostafavi & Zank (2018) [25]. A related shock structure equation in the absence of the viscosity term was investigated by Webb (1983) [26].

3. Results

The angle between the shock normal and the magnetic field of the observed VLISM shock was 85 degrees which means the shock is highly perpendicular [1]. We assume the shock is perpendicular to model the VLISM shock. With the parameters observed upstream of the VLISM shock, the sound speed is $C_{s1} = 14\ km\ s^{-1}$ and the Alfvén speed is $V_{A1} = 34\ km\ s^{-1}$ ($C_s < V_A$). The magnetoacoustic speed of the ambient medium is 36.7 $km\ s^{-1}$ which is consistent with the estimated magnetosonic speed given by [17], and yields a subcritical shock. The plasma beta upstream of the VLISM shock, which is the ratio between thermal gas pressure and magnetic field pressure, is about 0.2. Mellott (1985) showed that a low beta, low magnetosonic Mach number perpendicular shock is a laminar shock (see Fig. 1 of [27]). Burlaga et al. (2013) [1] found a weak jump of about 1.4 in the magnetic field intensity at the shock of late 2012 and showed that Voyager 1 (which is moving relative to the Sun at 17 $km\ s^{-1}$) moved past the broad VLISM shock in about 8.7 days. There are no observations of the VLISM shock propagation speed and therefore to simulate the VLISM shock we need to estimate a shock propagation speed. Since a subcritical shock has a magnetosonic Mach number less than about two, the upstream flow speed in the shock frame should be considered as less than 72 $km\ s^{-1}$ and therefore, the shock speed in the stationary Sun frame should be less than 52 $km\ s^{-1}$ (assuming the VLISM flows toward the Sun with a speed of 20 $km\ s^{-1}$). Kim et al. (2017) [28] used near-Earth solar wind data to model interplanetary shock propagation beyond the HP with the Multi-Scale Fluid-Kinetic Simulation Suite (MS-FLUKSS) code. Their simulation shows that the shock observed at the end of 2012 had a compression ratio of about 1.65 and a shock speed of about 50 $km\ s^{-1}$ in a frame in which the Sun is stationary (private communication with Dr. Tae Kim). For our model, we adopt a shock speed of 40 $km\ s^{-1}$ with respect to the stationary Sun. The thickness of the VLISM shock is about 0.12 AU which corresponds to Voyager 1's traversal of the shock in 8.7 days. As noted by Burlaga et al (2013) [1] the thickness of the VLISM shock is much greater than that expected for a shock with similar properties in the solar wind at 1 AU. The VLISM proton-proton collisions upstream of the observed shock give values of the thermal heat conduction length scale, K/U_1, and the viscous length scale, $\eta/\rho_1 U_1$, of about 0.115 AU and

IOP Conf. Series: Journal of Physics: Conf. Series **1100** (2018) 012018 doi:10.1088/1742-6596/1100/1/012018

0.03 AU, respectively.

We solve equation (21) in the stationary frame of the shock to model the VLISM shock. Figure 3 shows a smooth transition connecting the upstream to the downstream state. Here the VLISM shock compression ratio is about 1.67, indicating a weak shock transition. The shock thickness is determined by proton-proton collisions and is about 0.12 AU. The change in magnetic field through the VLISM shock is shown in Fig. 4. Figure 5 shows the normalized pressure as a function of unnormalized distance (AU). The dominant components downstream of the VLISM shock are thermal gas and magnetic field pressure. The shock is not mediated by PUIs since they do not contribute a large pressure through the shock. The fast magnetosonic Mach number shows that the flow changes from supersonic to subsonic (Fig. 6). Here one can see, because PUIs do not contribute a large pressure through the shock, that the inclusion of PUIs in the sound speed does not change the magnetosonic Mach number much and they are are essentially the same. The last figure (Fig. 7) shows the change of thermal gas entropy and PUI entropy through the VLISM shock. Since PUIs are effectively heated adiabatically, the entropy is constant through the shock layer. However, the thermal gas is collisionally dissipated and its associated entropy increases at the shock position.

4. Conclusions

Since the thermal proton and electron collisional mean free paths in the VLISM are very small compared to the almost featureless VLISM, we conclude the VLISM is a collisional environment with respect to the thermal plasma. The thermal collisions, which are dominated by proton-proton collisions, yield thermal collisional dissipation terms (heat flux and viscosity). PUIs outside the heliosphere are generated by secondary charge exchange and experience pitch angle scattering by magnetic fluctuations. They contribute a collisionless heat flux and collisionless viscosity (i.e., introduced by wave-particle interactions). However, we can neglect the PUI dissipation terms because the PUI Schmidt number is small compared to the thermal gas Schmidt number. We have shown that the dominant collisional term in the VLISM is proton-proton collisions and thus the structure of the broad interstellar shocks observed by Voyager 1 is determined by interstellar proton-proton collisions. The VLISM shock is therefore the first in situ observed example of a classical collisional subcritical shock, which means the shock structure is not controlled by wave-particle interactions but by particle collisions. The weak shock is

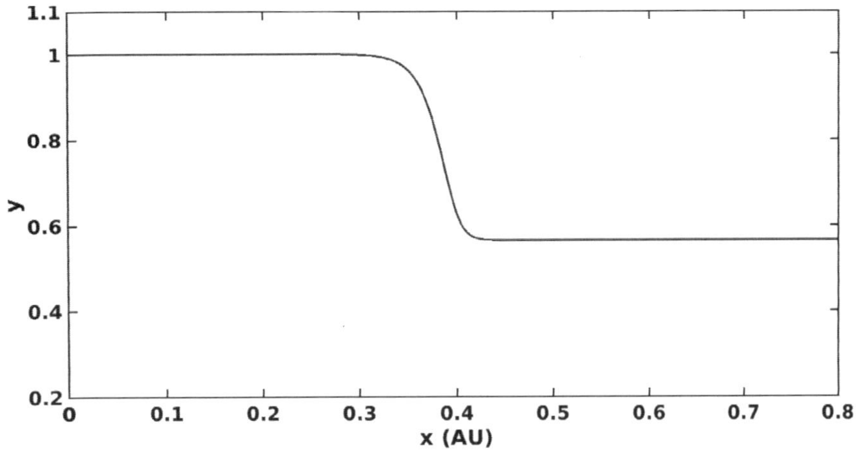

Figure 3: Inverse compression ratio as a function of unnormalized distance (AU) showing that the shock is smoothed. Here $\gamma_g = \gamma_I = 5/3$, $P_{I1}/P_{g1} = 0.05$, $y_B = 0.31$, and $M_{s1} = 4.17$.

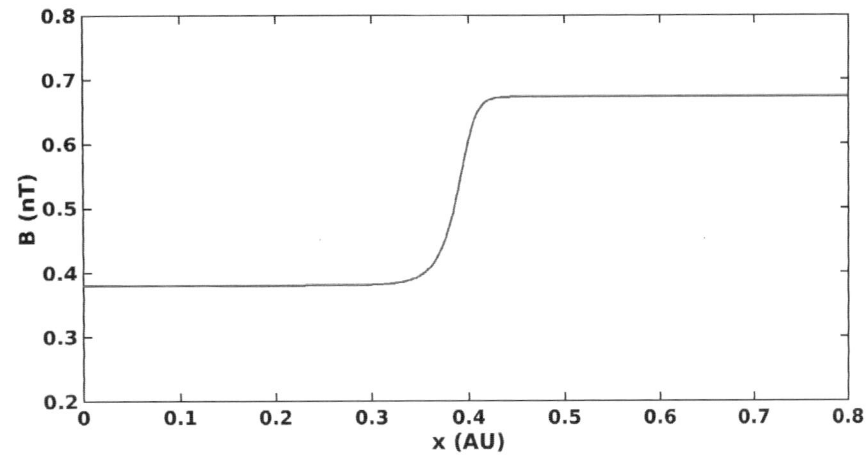

Figure 4: The unnormalized magnetic field (nT) through the shock as a function of unnormalized distance (AU) showing a weak compression in the magnetic field. Here $\gamma_g = \gamma_I = 5/3$, $P_{I1}/P_{g1} = 0.05$, $y_B = 0.31$, and $M_{s1} = 4.17$.

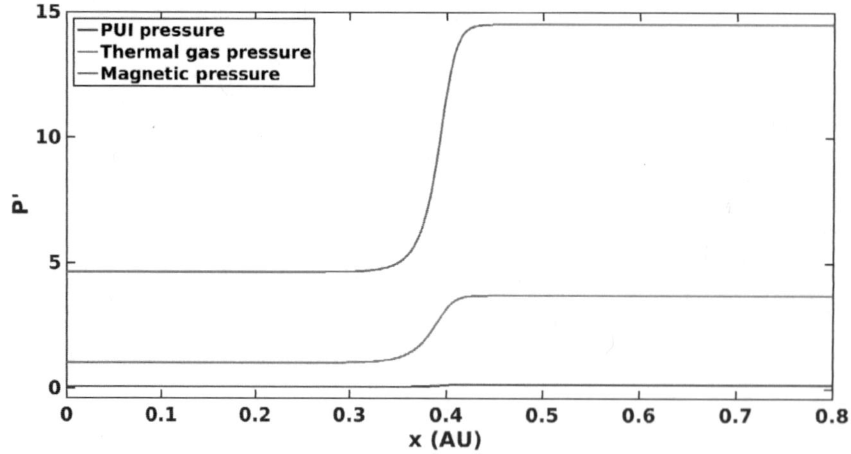

Figure 5: PUI, thermal gas, and magnetic pressure normalized to the thermal gas pressure far upstream as a function of unnormalized distance (AU).

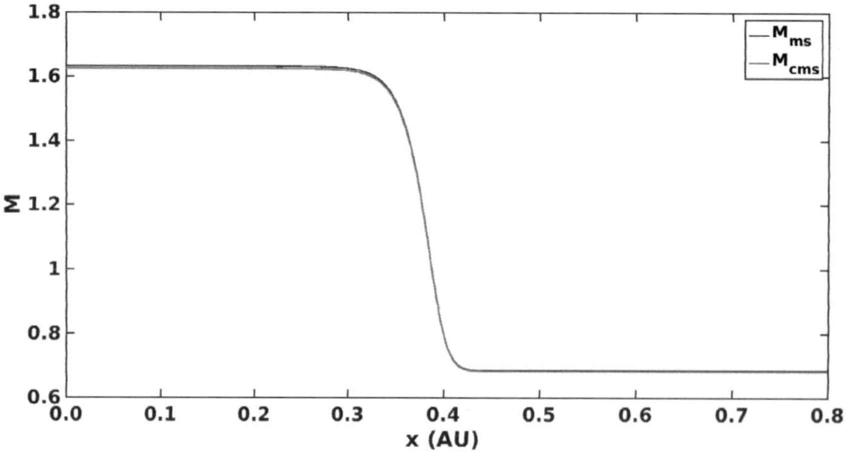

Figure 6: The magnetosonic Mach number as a function of unnormalized distance (AU). The thermal gas magnetosonic Mach number (blue line) and the combined effective thermal gas and PUI magnetosonic Mach number (red line) show that the flow changes from supersonic to subsonic.

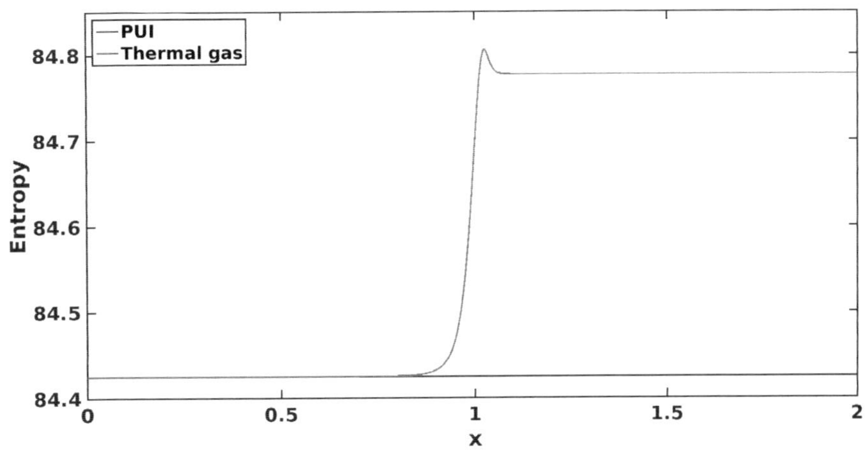

Figure 7: Entropy of thermal gas and PUIs through the VLISM shock as a function of normalized distance.

dominated by the magnetic field and thermal gas pressure whereas PUIs do not contribute a large pressure through the shock transition. Both the thermal and the combined thermal-PUI fast magnetosonic Mach number through the VLISM shock show that the flow changes from supersonic to subsonic. The overall thickness of the shock transition is about $0.12AU$, which corresponds to the VLISM heat conduction length scale. Entropy associated with the thermal gas is not constant and increases at the position of a VLISM shock wave.

Acknowledgments

P.M. acknowledges the support of a NASA Earth and Space Science Fellowship Program-Grant 16-HELIO16F-0022. G.P.Z. acknowledges the support of NASA grants NNX14AC08G, NNX15AI65G, NNX17AB04G, NNX14AJ53G, and the NSF EPSCoR RII-Track-1 Cooperative Agreement OIA-1655280.

References

[1] Burlaga L F, Ness N F, Gurnett D A and Kurth W S 2013 *The Astrophysical Journal Letters* **778** L3

[2] Gurnett D A, Kurth W S, Stone E C, Cummings A C, Krimigis S M, Decker R B, Ness N F and Burlaga L F 2015 *The Astrophysical Journal* **809** 121

[3] Gurnett D A, Kurth W S, Burlaga L F and Ness N F 2013 *Science* **341** 1489–1492

[4] Stone E C, Cummings A C, McDonald F B, Heikkila B C, Lal N and Webber W R 2013 *Science* **341** 150–153

[5] Krimigis S M, Decker R B, Roelof E C, Hill M E, Armstrong T P, Gloeckler G, Hamilton D C and Lanzerotti L J 2013 *Science* **341** 144–147

[6] Webber W R and McDonald F B 2013 *Geophysical Research Letters* **40** 1665–1668

[7] Whang Y C and Burlaga L F 1994 *JGR* **99** 21

[8] Zank G P and Müller H 2003 *JGR* **108** 1240–+

[9] Washimi H, Zank G P, Hu Q, Tanaka T, Munakata K and Shinagawa H 2011 *MNRAS* **416** 1475–1485

[10] Zank G P 2015 **53** 449–500

[11] Mellott M M and Greenstadt E W 1984 *JGR* **89** 2151–2161

[12] De Hoffmann F and Teller E 1950 *Phys. Rev.* **80**(4) 692–703 URL `https://link.aps.org/doi/10.1103/PhysRev.80.692`

[13] Zank G P, Pauls H L, Cairns I H and Webb G M 1996 *JGR* **101** 457–478

[14] Zank G P, Hunana P, Mostafavi P and Goldstein M L 2014 *The Astrophysical Journal* **797** 87

[15] Zank G P, Mostafavi P and Hunana P 2016 *Journal of Physics Conference Series* **719** 012014

[16] McComas D J, Bzowski M, Frisch P, Fuselier S A, Kubiak M A, Kucharek H, Leonard T, Mbius E, Schwadron N A, Sok J M, Swaczyna P and Witte M 2015 *The Astrophysical Journal* **801** 28 URL `http://stacks.iop.org/0004-637X/801/i=1/a=28`

[17] Gurnett D A, Kurth W S, Burlaga L F and Ness N F 2013 *Science* **341** 150–153

[18] Gurnett D A, Kurth W S, Stone E C, Cummings A C, Krimigis S M, Decker R B, Ness N F and Burlaga L F 2014 *AGU Fall Meeting Abstracts*

[19] Zank G P 2014 *Transport Processes in Space Physics and Astrophysics* (*Lecture Notes in Physics, Berlin Springer Verlag* vol 877)

[20] Mostafavi P, Zank G P and Webb G M 2017 *The Astrophysical Journal* **841** 4

[21] Mostafavi P, Zank G P and Webb G M 2017 *Journal of Physics: Conference Series* **900** 012016 URL `http://stacks.iop.org/1742-6596/900/i=1/a=012016`

[22] Jokipii J R and Williams L L 1992 *The Astrophysical Journal* **394** 184–187

[23] Zirnstein E J, Heerikhuisen J, Zank G P, Pogorelov N V, McComas D J and Desai M I 2014 *The Astrophysical Journal* **783** 129

[24] Richardson J D, Kasper J C, Wang C, Belcher J W and Lazarus A J 2008 *Nature* **454** 63–66

[25] Mostafavi P and Zank G P 2018 *The Astrophysical Journal Letters* **854** L15 URL `http://stacks.iop.org/2041-8205/854/i=1/a=L15`

[26] Webb G M 1983 *American Astronomical Society* **127** 97–112

[27] Mellott M M 1985 *Washington DC American Geophysical Union Geophysical Monograph Series* **35** 131–140

[28] Kim T K, Pogorelov N V and Burlaga L F 2017 *The Astrophysical Journal Letters* **843** L32 URL `http://stacks.iop.org/2041-8205/843/i=2/a=L32`

IOP Conf. Series: Journal of Physics: Conf. Series **1100** (2018) 012019 doi:10.1088/1742-6596/1100/1/012019

Voyager 2 Observations of Plasma and Pressure Pulses

John D. Richardson[1], John W. Belcher[1], Alan C. Cummings[2], Robert Decker[3], Edward C. Stone[2]

[1]Massachusetts Institute of Technology, Cambridge, MA 02139 USA
[2]California Institute of Technology, Pasadena, CA 91125 USA
[3]Applied Physics Laboratory, John Hopkins University, Laurel, MD 20723 USA

jdr@space.mit.edu

Abstract. This paper provides the latest data from Voyager 2 on plasma characteristics in the heliosheath including the observations of pressure waves in the plasma and particle data. Models and observations show that solar transients drive pressure waves through the heliosphere. Pressure pulses that could drive heliosheath waves are observed near the previous solar maximum upstream of the termination shock. We show that the most recent data is consistent with the presence of pressure waves and compare the heliosheath waves with the pressure increases in the heliosheath. The magnetic field is better correlated with density and galactic cosmic ray intensities in the supersonic solar wind than in the heliosheath. The galactic cosmic rays are correlated with the plasma and particles with a ~30-day lag in both the supersonic wind and heliosheath.

1. Introduction

Voyager 2 (V2) was in the heliosheath, 118 AU from the Sun, in April 2018 at a heliolatitude of 31° S. The Voyager 1 (V1) plasma instrument failed in 1980, so V2 is providing the first plasma data from this region. Observations show that the heliosheath flows are much faster at V2 than V1 and that the flows turn tailward more quickly [1]. Several studies predict that solar transients drive pressure waves in the heliosheath [2,3,4,5,6,7,8]. These pressure pulses are partially transmitted and partially reflected near the heliopause. The reflected waves again encounter the termination shock, moving it inward [4,5]. The transmitted waves drive weak shock waves into the local interstellar medium [3]. Since 2011, the heliosheath structure has been dominated by pressure pulses with correlated changes in the plasma density, temperature and speed and in the particle intensities at all energies but a with less strong correlation with the magnetic field B [9]. The source of these pressure pulses is likely MIRs (merged interaction regions) that dominate the outer heliosphere near solar maximum. This paper shows V2 data through April 2018 including the plasma parameters and the pressure pulses observed by both the plasma and particle instruments. The properties of the pressure pulses in the heliosheath are compared to those in MIRs in the supersonic solar wind in the outer heliosphere.

2. Recent Voyager 2 Data

V2 measures the plasma currents in the low-energy resolution L-mode that is best suited for heliosheath observations in four Faraday cups every 192 s. These spectra are fit with convected isotropic proton Maxwellian distributions to derive the plasma parameters. Figure 1 shows 25-day running averages of V2 plasma data in the heliosheath. The solar wind speed has remained on average constant at about 145 km/s. The radial speed has slowly decreased; a linear fit gives a decrease of 4.4 km/s/yr and the average V_R is now about 80 km/s. The flow angles initially rotated fairly rapidly tailward with the RT and RN angles (using the standard RTN coordinate system) reaching almost 60° and 30°, respectively, in 2014. Since 2014 the RT angle has not rotated further and the RN angle has gotten smaller. The density profile has shown numerous changes of 25-50% on time scales of several months, with larger increases in 2015 and 2018. The average density has increased since the beginning of 2016. The temperature of the thermal protons has remained steady and averages about 50,000 K.

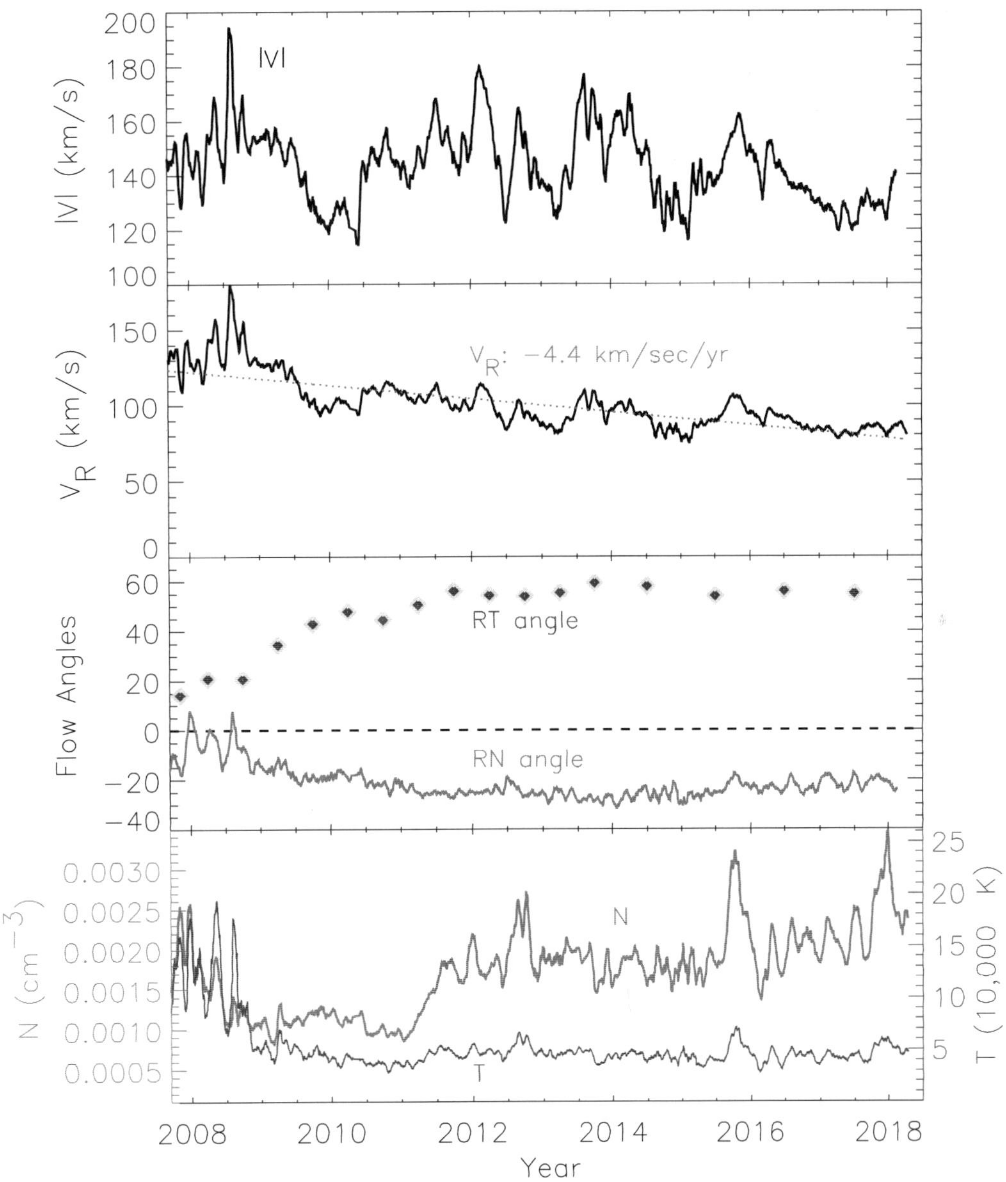

Figure 1. Plasma speed |V|, radial velocity V_R, flow angles, density N and temperature T.

3. Pressure Waves in the Heliosheath

The density (pressure) increases after 2011 are often observed in the particle data as well in the plasma. The right panel of Figure 2 shows heliosheath data from 2010 to 2018; magnetic field data are so far available only through 2014. The plasma dynamic pressure profile is dominated by a series of enhancements starting in 2011, with the largest increase in 2015. The pressure pulses at 2012.6 and 2013.6 are associated with MIRs reported by the V2 magnetometer instrument [10, 11]. The 28-43 keV ion intensities from the Low Energy Charged particle (LECP) instrument and the >0.5 MeV/nuc ion and >70 MeV/nuc GCR (galactic cosmic ray) Cosmic Ray Subsytem (CRS) counting rates track the plasma pressure profile very well, with almost every pressure peak from 2011 into 2018 observed in the particle data as well. As discussed later, the lower energy particle and pressure peaks occur nearly simultaneously but the GCR peaks occur before the pressure peaks arrive at V2.

The sources of these pressure waves in the heliosheath are probably the MIRs that dominate the outer heliosphere at and after solar maximum [12]. The left panel of Figure 2 shows the pressure pulses observed by V2 during the previous solar cycle in the supersonic solar wind. These pressure increases were observed roughly twice per year and are associated with simultaneous increases in the magnetic field indicating that they are MIRs. The lower energy particles are not present in this region, but the GCRs again have peaks that precede the pressure and magnetic field increases.

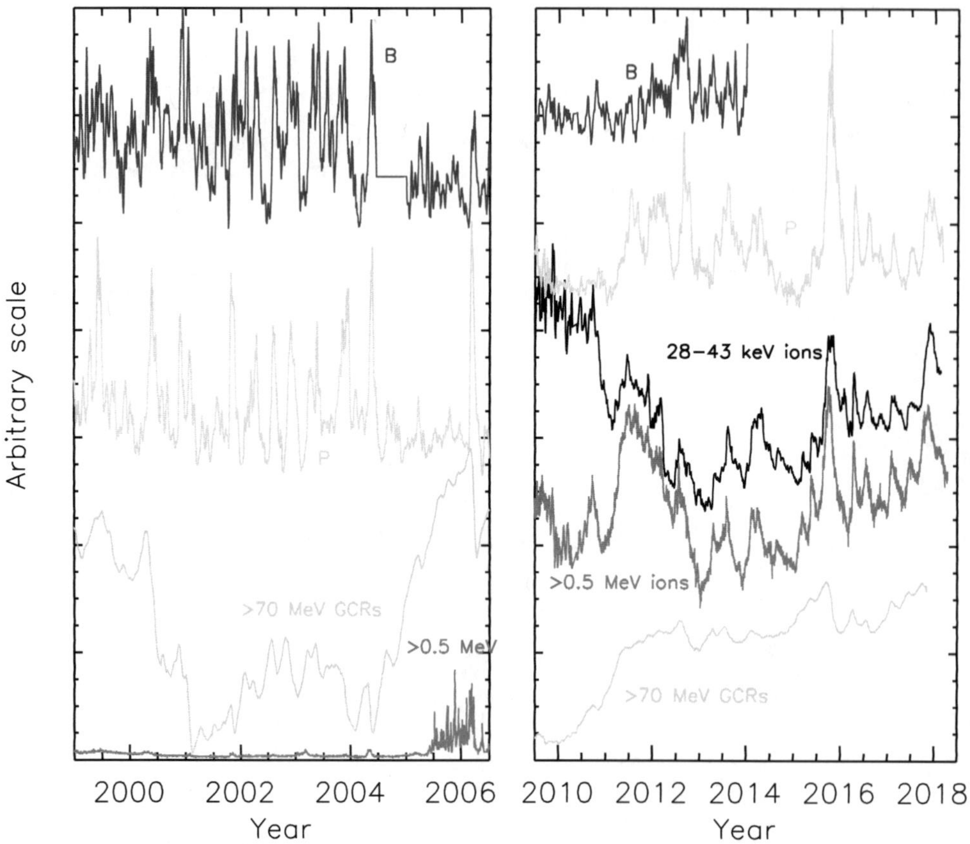

Figure 2. Fluctuations observed in MIRs in the supersonic solar wind (left) and in pressure waves in the heliosheath (right).

Figures 3 and 4 compare in more detail the pressure pulses in these two regions. Figure 3 shows the fluctuation amplitudes of the plasma density with other plasma, field, and particle measurements. Each panel has 25-day (1 solar rotation) running averages of (x - <x>)/<x> where x is the measured value and <x> is the 301-day average. The top panel compares the thermal proton density and temperature, which for a pressure wave would vary in phase since Tperp (to which V2 predominantly responds) changes due to conservation of the first adiabatic invariant as B increases. The correlation coefficient for N and T variations over the time period shown in the plot is 0.88 and peaks at a lag of zero. N and T are derived from the same fits to PLS spectra so are not independent quantities, but this correlation is not seen in the supersonic solar wind, only in the heliosheath.

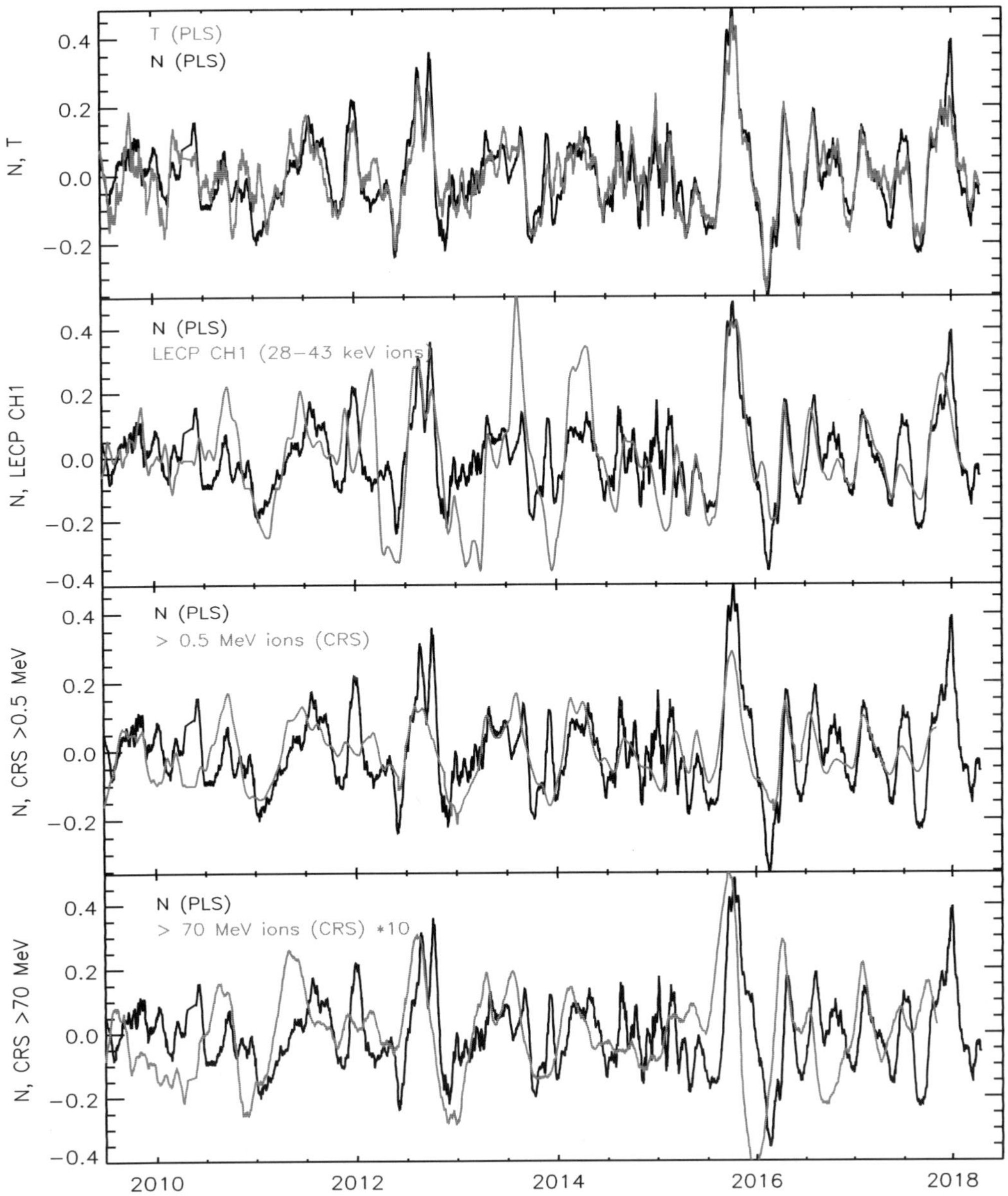

Figure 3. Changes in density (black curves) compared to changes in temperature and particle intensities (red curves) . The plots for each parameter x show (x-<x>)/<x> where <x> is averaged over 301 days.

IOP Conf. Series: Journal of Physics: Conf. Series **1100** (2018) 012019 doi:10.1088/1742-6596/1100/1/012019

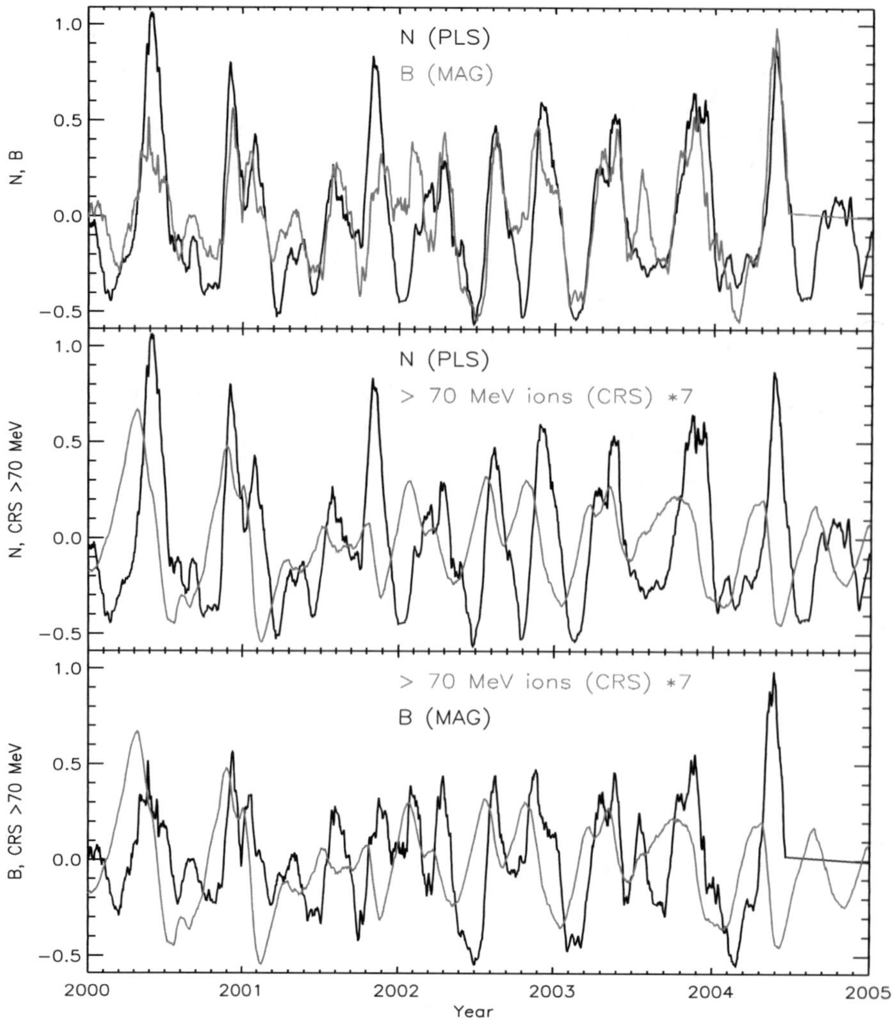

Figure 4. Comparison of fluctuations in N, B, and GCRs in MIRs in the outer heliosphere.

The second and third panels of Figure 3 compare fluctuations of density with the LECP 28-43 keV ion intensities and CRS >0.5 MeV ion count rates, respectively. If these changes are driven by pressure waves then the compressions (rarefactions) should increase (decrease) the energetic ion intensities roughly the same amount as the plasma density. Figure 3 shows that the amplitude of these changes are correlated and the magnitudes are similar; the >0.5 MeV/nuc rates vary less than the 28-43 keV intensities. The correlation between N and the 28-43 keV ion intensities is 0.60 with a lag near zero and between N and the >0.5 MeV/nuc ion count rates is 0.66 also with a lag near zero. The correlations are better in the recent data; for data after 2014.5 these two correlations are 0.77 and 0.80, respectively, indicating pressure waves are producing more of the observed variations.
The bottom panel compares N and the >70 MeV/nuc GCRs. The offset in the peaks is clear, with the GCR peaks occurring before the density peaks. The magnitude of the GCR variations is much less than those of the N variations (the GCR profile in Figure 3 is multiplied by 10). The correlation is 0.58 with a lag of 33 days. The GCR variations are not directly due to the pressure waves; they are likely due to the increased magnetic field associated with these waves inhibiting the transport of these waves

causing the GCRs to pile up in front of the pressure pulses, a mechanism which has been called the snow plow effect [13]. The correlation between N and B was found to be weak, 0.34, but magnetic after 2014 are not yet available.

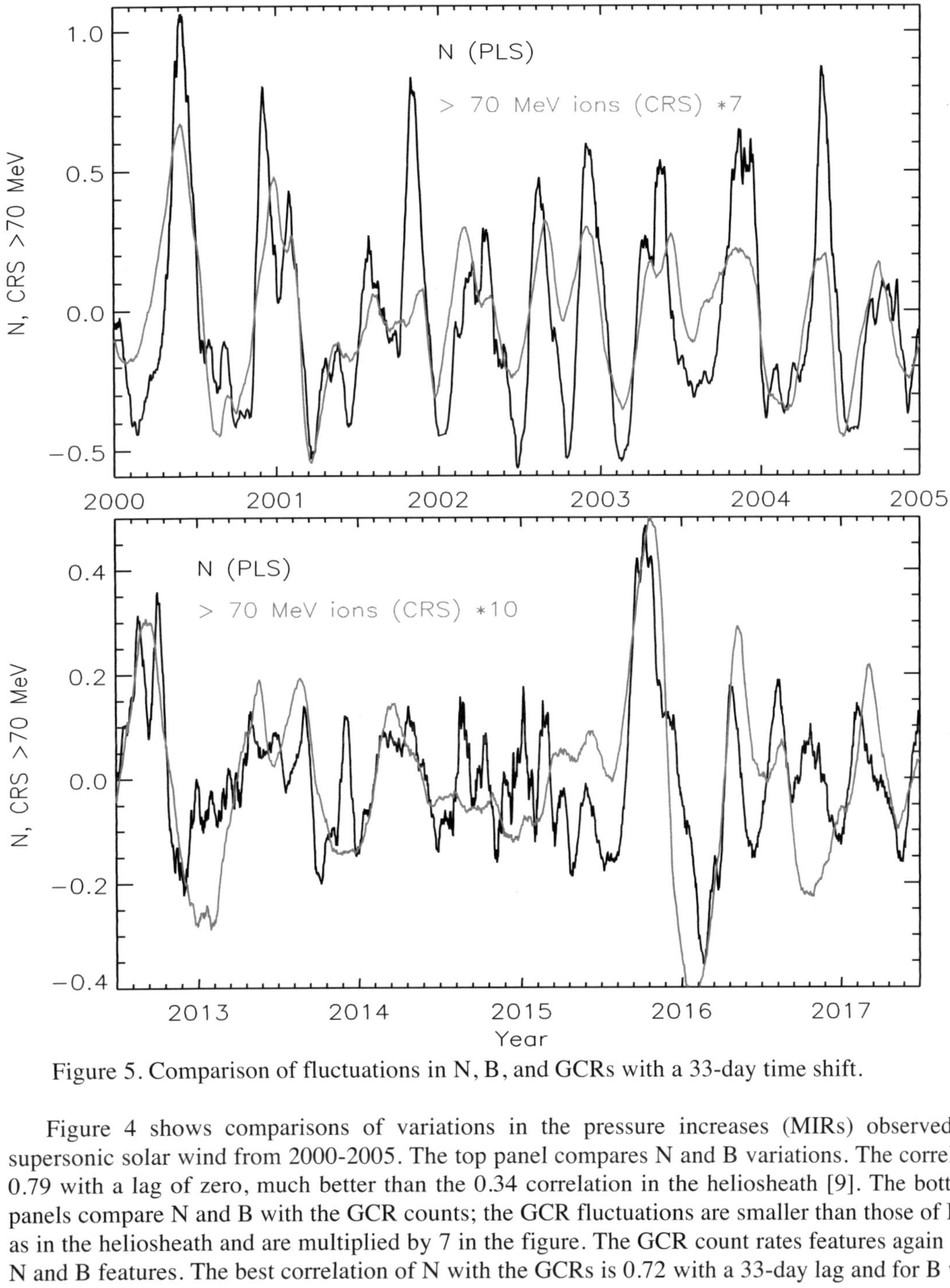

Figure 5. Comparison of fluctuations in N, B, and GCRs with a 33-day time shift.

Figure 4 shows comparisons of variations in the pressure increases (MIRs) observed in the supersonic solar wind from 2000-2005. The top panel compares N and B variations. The correlation is 0.79 with a lag of zero, much better than the 0.34 correlation in the heliosheath [9]. The bottom two panels compare N and B with the GCR counts; the GCR fluctuations are smaller than those of B and N as in the heliosheath and are multiplied by 7 in the figure. The GCR count rates features again lead the N and B features. The best correlation of N with the GCRs is 0.72 with a 33-day lag and for B with the GCRs it is 0.65 with a 30-day lag. Figure 5 shows the comparison of N with the GCRs in both the supersonic solar wind from 2000-2005 and in the heliosheath from 2012.5-2017.5 with the GCR data shifted by 33 days to show that this same shift best fits the data in both regions. The reason the same 33-day shift in the N and GCR profiles gives the best correlation is not known. The MIRs are propagating outward at 400-500 km/s in the supersonic solar wind. The heliosheath radial speed is about 100 km/s, although the pressure pulses could be propagating significantly faster [12]. The gyroradius of a GCR is about 0.1 AU; the solar wind moves this distance in roughly 10 hours so the 33-day shift seems not directly related to the gyroradius. The outward moving pressure and field increases are very large structures so their interaction with GCRs is complex.

4. Summary

This paper presents and discusses recent Voyager 2 observations. The plasma speed remains constant but V_R is decreasing slowly across the heliosheath. The flow directions, after initially turning tailward after the termination shock, have remained steady the past 4 years. Pressure waves continue to dominate the density profile; these waves are observed with similar magnitudes and phases in the keV and MeV particle data. The GCRs show changes correlated with the pressure waves but with smaller amplitudes and with a time shift of 33 days. Pressure increases in the last solar cycle in the supersonic solar wind show a better correlation between the magnetic field and other parameters (density and GCR intensity) than in the heliosheath. For reasons not understood, the GCR intensities lead the plasma density changes by 33 days in both the supersonic solar wind and heliosheath.

5. Acknowledgments

We thank SPDF for the Voyager 2 magnetometer data and CRS (P.I., E. C. Stone) and LECP (P.I., S. M. Krimigis) for providing data through their web sites. JDR was supported under NASA contract 959203 from the Jet Propulsion Laboratory to the Massachusetts Institute of Technology. RBD was supported by NASA contract NNN06AA01C. ECS and ACE were supported by NASA grant NNN12AA01C

References

[1] Richardson J D and Decker R B 2015 *ApJ* **792** 126
[2] Story T R and Zank G P 1997 *J. Geophys. Res.* **102** 17381
[3] Steinolfson R S and Gurnett D A 1995 *Geophys. Res. Lett.* **22** 651
[4] Zank G P and Muller H-R 2003 *J. Geophys. Res.* **108** 1240
[5] Washimi H, Zank G P, Hu Q, Tanaka T and Munakata, K. 2007 *ApJ Lett.* **670** L139A
[6] Washimi H, Zank G P and Hu, Q. 2011 *MNRAS* **416** 1475
[7] Washimi H, Webber W, Zank G P, Hu Q, Florinski V, Adams J and Kubo Y 2012 *ApJ Lett.* **757** L2
[8] Zank G. P. 2015 *Ann. Rev. Astron. Astrophys.* **53** 449
[9] Richardson J D 2017 J. Phys.: Conf. Series **900** (1), 012017
[10] Burlaga L F, Ness N F, Richardson J D, Decker R B and Krimigis S M 2016 *ApJ* **818** 147
[11] Burlaga L F, Ness N F and Richardson J D 2017 *ApJ* **841** 47.
[12] Richardson J D, Wang C, Liu Y D, Šafránková J, Němeček Z and Kurth W S 2017 *ApJ* **834** 190
[13] McDonald F B, Burlaga L F, McGuire R E and Ness N F 2000 *J. Geophys. Res.* **105** 20997

Heliospheric magnetic field structures: Predictions & Implications

Ilan Roth

7 Gauss Way, Space Sciences, UC Berkeley, CA 94720, USA
ilanr@berkeley.edu

Abstract. The observational implication of braided solar magnetic fields opens a new venue for an interpretation of various solar and interplanetary phenomena. Direct imaging of the coronal fields at frequencies corresponding to atomic transition of hot (~1MK) plasma pinpoints to their braiding structure, while solar wind measurements of magnetized plasma parcels with distinct orientation, large field deviation and intermittent fading of energetic flare ions suggest that coronal braided field may have been carried by the solar wind to 1AU. The interconnection between the mathematical braids and knots is applied to the topologically non-trivial magnetized structures and their dynamics, from solar corona to the interplanetary medium. The analysis of braided magnetic configurations results in conjectures regarding (i) their stability under large oscillations of magnetic loops, (ii) the structure and stability of the coronal field through successive emergence/decay of heated magnetic braids, and (ii) their evolution into the solar wind. It is anticipated that the resulting conjectures will add to the understanding of the physical processes in the magnetosphere (Van Allen, Cluster satellites) and particularly will attain confirmation with the launch of the Parker Solar Probe.

1. Introduction

Most of the heliospheric magnetized plasmas evolve slowly over typical plasma time-scales, coupling currents due to the motion of charged particles with magnetic fields, in a variety of geometrical configurations. The heliospheric field is formed mainly due to convection in solar and planetary interiors via the dynamo process through conversion of kinetic into magnetic energy. Some of the newly emerging solar fields with a large curl appear in geometrical forms compatible with current sheets. Small-scale magnetic features hold the key to various facets of solar magnetism on larger scales: (a) formation of small-scale helicity in dynamo process and its subsequent transformation into large-scale fields (1-2), (b) local Alfven waves (3-4) or nanoflares (5-6) heating coronal and chromospheric plasma over extended spatial domain, (3) emergence of magnetic flux from the solar interior with the appearance of small-scale events (compact flare, plasmoid, X-ray brightening) leading to a large-scale CME eruption. Generally, stable, topologically non-standard small-scale magnetic coronal configuration may be carried into the heliosphere expanding into large-scale, entangled magnetic structure.

Recent improvements in the spatial resolution of solar magnetic fields allows one to advance the high resolution analysis of small-scale structures in the coronal field, inferring their evolution as they are fed by the solar dynamo and eventually carried into the interplanetary medium by the solar wind. This ability resulted in remote observations of inhomogeneous bundles of arched solar fields with braiding features, i.e., axial bending of strands of enhanced field with twisting and over/under crossing of adjacent field line strands, forming shapes resembling braids. The bundling and braiding of the fields indicate that the solar dynamo may result in configurations with topological intricacies which, when extended into the interplanetary medium could modify, among others, (a) the cascade rates and

IOP Conf. Series: Journal of Physics: Conf. Series **1100** (2018) 012020 doi:10.1088/1742-6596/1100/1/012020

the electromagnetic turbulent wave spectra and (b) the connectivity to flaring sites during active solar times, both leading to measurable results. Hence, nontrivial magnetic topologies with large field deviation within the solar corona or solar wind (7) and repeated drop-outs and reappearance of energetic ion fluxes from flaring sites (8) are consistent with an emergence of magnetized structures in the form of braids or knots. The observation of braided magnetic fields allows us also to cast them into a framework of mathematical braids and their equivalent knots, facilitating a new description of magnetic fields as they emerge from the photosphere and are transported into the interplanetary medium. Combining theoretical considerations, remote images and soon to be realized in situ satellite observations at solar vicinity, one may construct new characteristics of the braided/knotted magnetic structures to constrain their detectable oscillations and specify the conditions of solar magnetic fields stability, presenting an additional perspective into their topological evolution.

2. Braids in coronal fields

Solar braids were coined as such due to the inferred geometry of the observed solar emissions, which were deduced as concentrated along segments of the arched magnetic field lines anchored at the photospheric foot-points, indicating adjacent strands passing over or under, consistent with a 2-D projection of a braid (9). The braiding of the field lines is partly due to the ubiquitous small-scale motion at the photospheric leg(s) of the magnetic field. Part of the motional energy is transmitted to coronal loops through up-flows and propagating Alfven waves whose damping along the field lines heats up the coronal plasma (10-11); additionally, small-scale reconfigurations may form new active magnetic field lines, such that braid formation may be inherently connected to heating processes. The outflows or jets from an active region observed via motion of braided field may be used to assess the necessary additional coronal heating through the nano-flare reconnection processes (12-14). Since transition line intensity of the coronal ions at given equilibrium temperature are inherently related to their charging states, filters at specific wavelength band indicate the approximate temperature to which the particular charge state was heated. Hence, the "flaring" of loops and the intensity of the emission, captured at a specific frequency band are indicators of the heating process.

While the solar atmosphere has been observed from ground in the visible/infrared frequencies down to coherent structures of a fraction of arcsec, the ultraviolet/X ray spectra require observations above the terrestrial atmosphere. The NASA Atmospheric Imaging Assembly (AIA) instrument on the Solar Dynamics Observatory (SDO), with a continuous imaging of the full Sun in extreme ultraviolet (EUV), ultraviolet and visible channels, incurred a resolution in spatial scales of <1.0 arcsec (11,15), while the High resolution Coronal Imager (Hi-C) reached a resolution of 0.2 arcsec (~150 km) at wavelength centred at 193A (16-17) indicating coronal plasma at a temperature of 1.5MK with transition lines of Fe XII. Both measurements revealed braided magnetic structures, with the appearance and decay of field lines without affecting the stability of the whole arched structure. High resolution observations at various frequencies are instrumental to an understanding of the intricacies of coronal magnetic field evolution.

Following the observations (16) and considering a small section of the coronal field line (i.e., ignoring the large scale field curvature), magnetic braid may be depicted as a disjoint collection of crossings among a number of strands attached to foot-points at two photospheric (parallel) planes ("top" and "bottom"), with one-to-one correspondence between the foot-points. Large number of these magnetic field braids can be envisioned as a small scale braid-like topology (18). The observed loops are undergoing constant modifications due to the dynamic motion of the underlying photosphere, as new loops light up, fade out or connect into more extended compact structures (11,15,19).

3. Magnetized plasma as a knot

Both mathematical knots and field lines of magnetized plasma in the magnetohydrodynamics (MHD) approximation are described as closed loops or bundles in three-dimensional (3D) space, transformed dynamically via continuous deformation of 3D upon itself, pushed smoothly in the surrounding

IOP Conf. Series: Journal of Physics: Conf. Series **1100** (2018) 012020 doi:10.1088/1742-6596/1100/1/012020

viscous (plasma) fluid, respectively, without self-intersection. Since these magnetic bundles of enhanced magnetic field usually form the most active sites for plasma processes, the dynamics of the magnetic configuration can be approximated as the evolution of finite width knots. The physical "frozen in" condition for magnetic knots is controlled by the three Reidemeister diagrammatic moves (e.g. 20), as shown on Figure 1: R_1 - twist (green), R_2 - poke (orange) and R_3 - slide (yellow or blue). Trivial deformations like (wave) field line wiggling, which clearly preserve the topology, are denoted by R_o. R_j moves reduce then the complicated topological problem to a simpler diagrammatic one, relating the changes in the observed projection to the relations between the crossings in the magnetic configuration. The invariance under R_j assures that any quantity which characterizes the magnetic field knot must preserve its value while undergoing the R_j transformations, becoming a topological invariant which assigns uniqueness to each knotted magnetic configuration. Helicity is one of these invariants. Non-equivalent magnetized knots are characterized by many additional invariants in the form of various polynomials (e.g., Alexander, Conway, Jones).

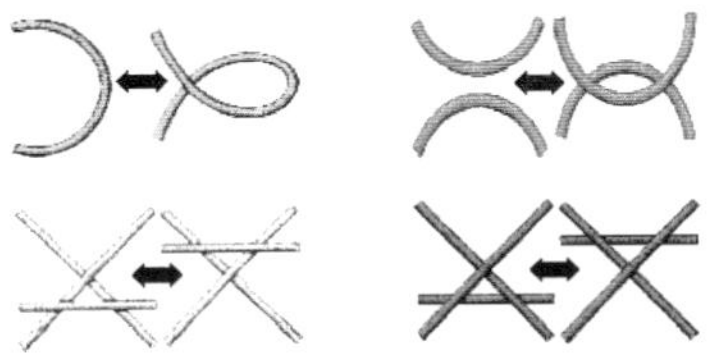

Figure 1. Reidemeister moves R_j, j=1,3

4. Braid Structure – Stable Coronal Field Oscillations

Mathematical (Artin) braid is a set of n disjoint strings in 3-space attached to two (horizontal) bars at the top and at the bottom (21). In many aspects they are similar to the solar magnetic braids observed in high resolution. Conventionally, the braids (strings) head downwards as one moves from the top to the bottom bar (footpoints), indicating the direction of the magnetic vector field, with a condition that each string never "turns back"; no string intersects any horizontal plane more than once. Pictorially, in this vertical configuration, the path of each string in a braid could be traced out by a falling object if acted upon only by gravity and horizontal forces (like ExB). Projecting the configuration on 2D and aligning the foot-points into a line results in a form shown in Figure 2 for two examples of simple braids; these forms have remarkable similarity to the remotely observed magnetic solar braids. Perturbations of the photospheric legs may wiggle the braids as Alfvenic-like oscillations without changing their topological characteristics, similarly to the knot R_o, R_2, R_3 moves. In more general moves, magnetic or mathematical braid with the same number of strings may change its configuration to an equivalent braid if the strings of one can be continuously deformed (without intersections) into the strings of the other braid. Additionally, large successive braid reconfiguration due to interaction with new strings occurs when plasma processes heat adjacent magnetic fields lines.

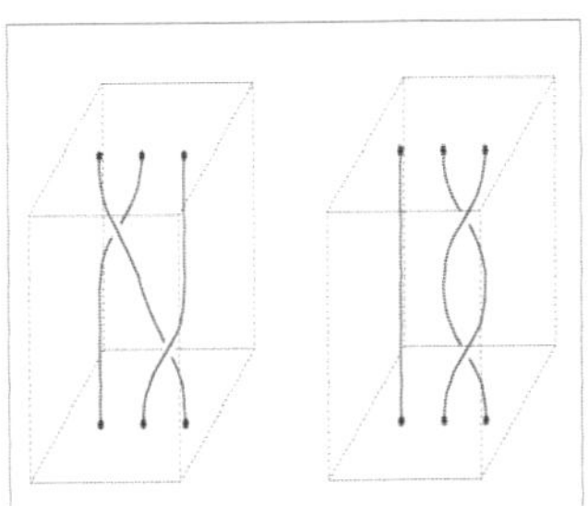

Figure 2. Examples of mathematical braids.

The classification of braids is done in a projection diagram given by elementary crossings of strands, similarly to coronal observations. The i-th strand passing from above over the (i-1)-th strand defines the crossing generator s_i, while passing below it forms s_i^{-1}, as shown for a braid with n=4 strings at Figure. 3. The examples include also two unperturbed, straight strings I_2. These mathematical generators constitute the basis of the physical magnetic braid structure and its oscillations.

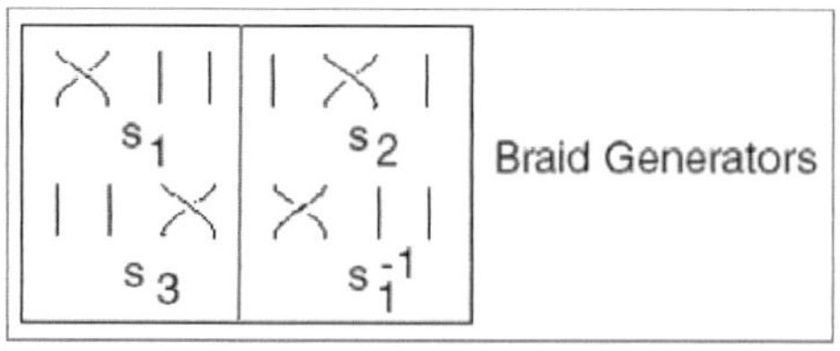

Figure 3. Elementary generators of braid oscillations for 4 strings.

Following this description, the braid can be characterized by its successive crossing generators; moving from the top down the braids of Figure 2 can be described as $s_1s_2^{-1}$ and $s_2^{-1}s_2^{-1}$, respectively. More generally, any (magnetic or mathematical) braid can be formulated as a combination of a set of elementary generators s_i, s_i^{-1}, forming a braid *word.* These words translate the braiding images into an algebraic entity, and each mathematical braid is characterized by its own word. Similarly, each magnetic braid in high resolution can be classified by its crossings, forming magnetic braid word which undergoes temporal modifications due to the motion of its strands. The present resolution of coronal braids is able to observe these structures and their oscillations imprecisely; an improved resolution will be able to assign to each braid fully its magnetic braid word, allowing to validate the braid operations in physical environment.

The connection between the mathematical and the magnetic braids facilitates to conjecture the conditions for coronal field moves, which form equivalent braids. The equivalency of two braids indicate stability of the magnetic configuration which could be seen by an observer. These moves can be then translated into constrains on its generators to satisfy the following relations:

a) $s_1 s_1^{-1}=1$.

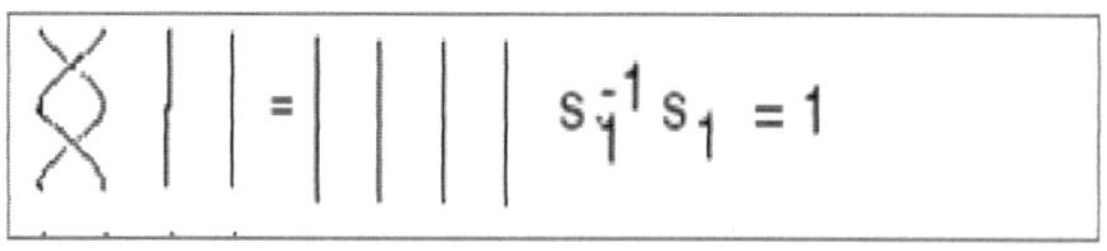

Figure 4a. Two adjacent magnetic strands affected by a local ExB or large Alfvenic force.

This relation describes a local braid motion resembling the R_2 knot move, when two strands interchange positions (Figure 4a) at some segment of the braid. In the observed solar braids this may happen when part of one (or both) of adjacent strand(s) is (are) subjected to (different) local cross ExB motion. The resulting braid word acquires $s_1s_1^{-1}$ at the observed level, while the braid preserves its characteristics; Figure 4a shows its equivalency for n=4.

b) $s_is_{i+1}s_i = s_{i+1}s_i s_{i+1}$.

This relation resembles the R_3 knot move when one strand moves with respect to a pair of crossed strands, e.g., the middle one in Figure 4b. In the corona it may happen when an active magnetic strand is anchored at different foot-point of its neighbours, and its perpendicular motion is independent of the motions of other strands.

IOP Conf. Series: Journal of Physics: Conf. Series **1100** (2018) 012020 doi:10.1088/1742-6596/1100/1/012020

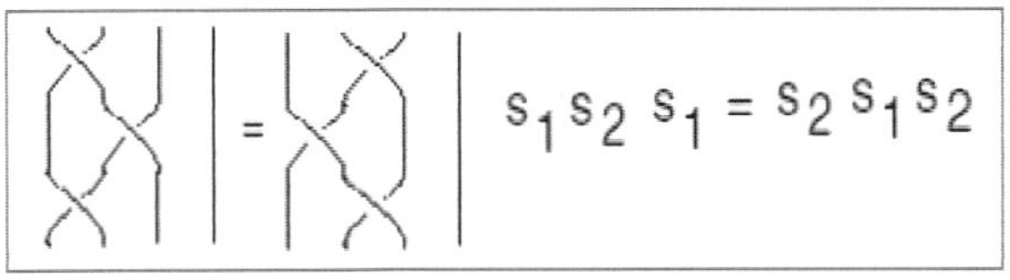

Figure 4b. Braid equivalency under separate motion of a strand vs two intertwined strands (i=1, j=2).

c) $s_i s_j = s_j s_i$ $(|i-j|>1)$

This braid relation occurs when motion along one part of the braid does not affect the other on a non-adjacent field line. This can happen at solar corona when a small scale (e.g., Alfvenic) pulse is applied only to a segment of the magnetic field braid (Fig. 4c), forming local commutative operation.

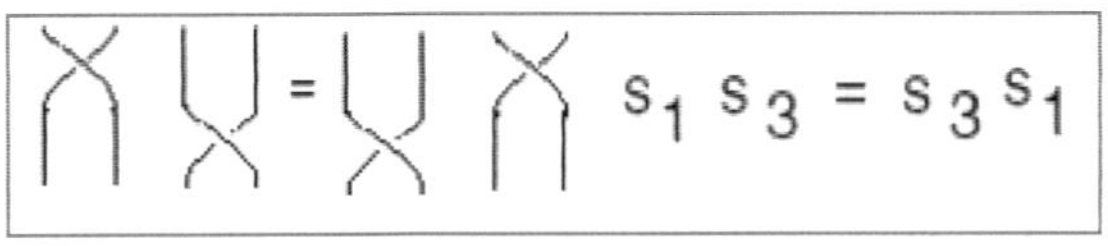

Figure 4c. Pulse along a segment of the braid, (i=1, j=3).

Clearly, the above moves which could become observable in very fine resolution, can be repeated numerous times. They preserve the topological structure of the braids, hence their successive observation at the corona is consistent with the stable magnetic braid activity. Therefore, time dependent braid oscillations due to photospheric footpoint moves, Alfvenic pulses or local ExB forces become natural eigenmodes of the magnetic braid. In mathematical terms these conditions, with the existence of the generators s_i and s_i^{-1} and unit strings I_k (k straight lines) create the (Artin) braid group B_n of n strings: the composite field lines of magnetic braids satisfy associativity – do not change when split into segments in different orders.

The changes in brightness with an emergence or decay of stable solar braids indicate that the system is dynamically active with significant structural braid modifications. The local repeated braid moves of the Artin group form the stable modes which should be observed as basic solar oscillations at fine resolution. At times the emergence of new braids and major modifications in the braid configuration well beyond the Artin group still may preserve its basic topological structure and stability, while other modifications may indicate global, unstable magnetic reconfigurations. The mathematical braids can shed light on these aspects of magnetic braid evolution.

5. Braid Closure - Large Stable Coronal Field Modifications

Both braids and knots and their physical analogues at the solar corona and in the interplanetary medium form entities which may be related topologically. An important relation between braids and knots/links was given through the Alexander (22) theorem, stating that any knot may be obtained from a closure of some braid, joining orderly through simple arcs the endpoints at the upper loose ends of a braid with the lower ones. Conversely, there exist many braids which through closure become the same knot. There exists simple physical interpretation of this theorem: since the magnetic braid consists of a bundle of coronal magnetic field lines anchored at the photospheric foot points, these field lines must be connected through the (turbulent) convection region such that the closure has physical validity, irrespectively of the details of the current system in the photosphere or convective region. Therefore, each magnetic braid is related to some knot or link – two or more intertwined knots. Figure 5 shows the closure of the Borromean ring braid into Borromean ring knot (5a) and through regular planar R_o isotopy movements, into its standard form of three mutually intertwined circles (5b). Hence, if a closure of two different braids results in the same knot, magnetic deformation leading from one braid to another describes magnetic reconfigurations preserving the knot invariant. Since knot invariant is a robust characteristic quantity these deformations may be used as a stability criterion for significant modifications of magnetic braids, forming a broad range of stable coronal deformations.

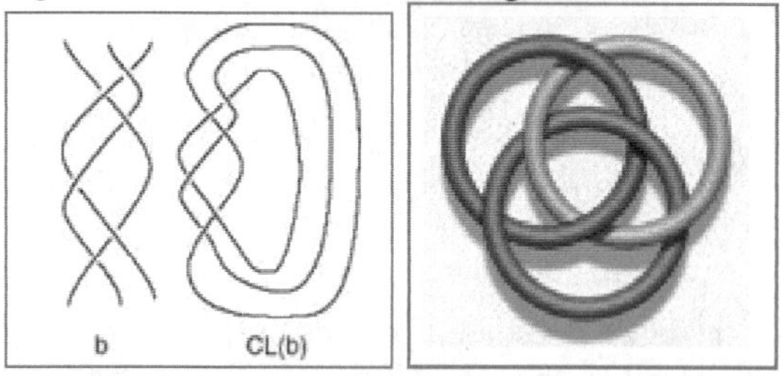

Figure 5. (a) Closure of a braid into a Borromean ring. (b) Final form of Borromean ring – three mutually intertwined circles.

The closure of a braid into a knot is not a one-on-one transformation. The convergence of multiple (infinite) braids into the same knot is based on Markov theorem (23), stating that two braids under closure become the same knot if they are related by a sequence of Markov moves (below). Physically it indicates that a large number of magnetic braids may undergo significant modifications, changing their braid word well beyond the above group B_n moves and still converge by closure to the same knot. This redundancy offers an important observational criterion, indicating that large changes in the structure of the magnetic braid, with an excitation (or fading) of magnetic strands, including nanoflare/microflare disruptions, are constrained by the preservation of knot invariant, keeping the whole active field structurally stable. Therefore, long term coronal heating may modify significantly the active coronal fields while keeping them stable over very long times, as often observed. Conversely, modification of a braid into another one with a different knot closure violates the knot invariants, indicating configuration instability, which may occur over a very short time scale.

The two Markov moves include:

1) Conjugation: combination of a braid with two complementary mirror parts of another braid. Dynamic fluctuations of a braid with straight magnetic lines (i.e., unit n-braid I) may undergo eigenmode oscillations with a series of generators s_i, s_i^{-1}. When reconfiguration process splits this braid into its mirror components B and B^{-1} (with $I=BB^{-1}$) which later join another braid A on both sides, one obtains a new equivalent braid $C=BAB^{-1}$; this braid (Figure 6a) is related to the same knot as A at braid closure. Hence, sudden observation of an elongated solar magnetic braid at both ends by symmetric mirror pieces of another braid indicates a major stable braid reconfiguration which preserves the whole magnetic structure. This operation may be repeated numerous times, indicating that significant observable modifications in magnetic braids may preserve the stable configuration.

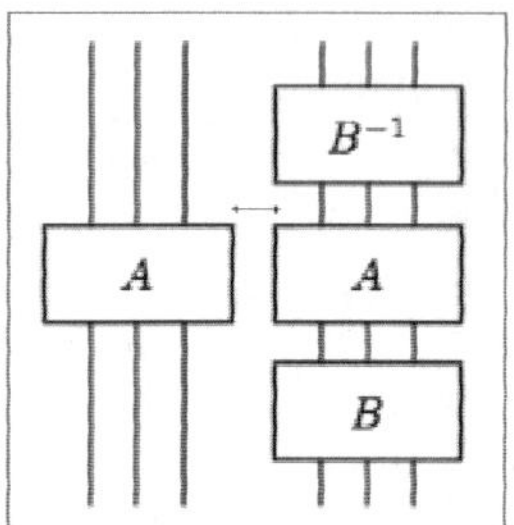

Figure 6a. (First) Markov conjugation move.

2) Stabilization: addition of a new strand with crossing. When a new, single active magnetic field line reconnects with n-th string of a braid A adding a new s_{n+1} crossing string, the resulting knot of n+1

strings formed under closure is isotopic to the closure of A – the new braid converges to the same knot under a closure (Figure 6b). The heating of magnetic field lines and addition of string generators s to A may be repeated many times, forming new braids which do not change the equivalent knot structure. Similar result applies to reconfiguration adding multiple times the generator $(s_{n+1})^{-1}$

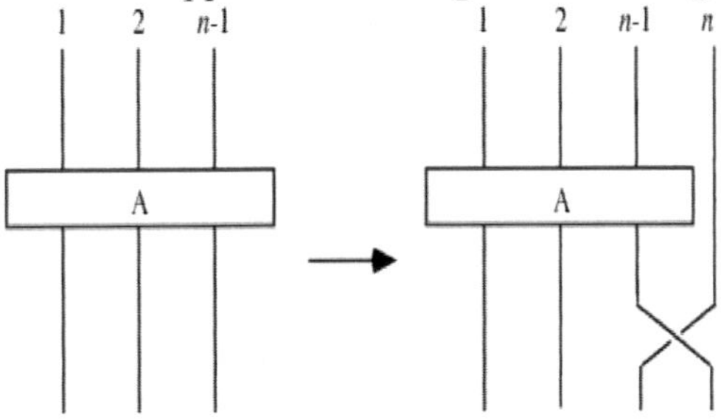

Figure 6b. (Second) Markov stabilization move.

As a corollary of the Markov moves, a sequence of the above operations which modify significantly the structure of magnetic braid due to coronal heating and emergence of new active field lines, converge after braid closure into an identical knot preserving the stability of the magnetic structure. In contrast, those large braid moves which do not satisfy these conditions are candidates for the dissolution of the magnetic configuration.

6. Discussion

Fluid motion in the solar convection region results through dynamo processes in an emergence of magnetic fields in various geometrical forms. The measured fields rely on remote observations of photons which are emitted by the heated plasma in the presence of magnetic fields. These observations depict the magnetic fields as simply connected arched lines and intermittently as magnetic braids. The braids preserve their topology while being carried later by the Solar Wind into the heliosphere in the form of magnetic knots. Temporal changes in the observed braid structure are due to the motion of field line foot-points, local perturbations or formation of newly heated magnetic strands. The present work promotes the idea of correlating high resolution solar observations of small scale coronal magnetic field geometry with the topological properties of mathematical braids and knots. This analogy allows us to extend the realm of magnetized plasma at the Sun and in the interplanetary medium into new, not often explored magnetic structures. Following the mathematical axioms one may impose stability constrains on the dynamics of heliospheric fields and offer predictions of topological configurations in soon to be realized measurements at the inner heliosphere when the Parker Solar Probe satellite, equipped with modern instrumentation, will reach the unexplored regime of the solar system well beyond the region which was intermittently traversed 40 years ago by the Helios satellites. The interconnection between the magnetic and mathematical braids/knots is performed partly as a *gedanken* experiment in which mathematical properties are transcribed into the anticipated physical measurements.

The formulation of the Artin (21) braid group as a set of disjoint strings with crossings attached to two bars can be uplifted directly to magnetic braids, when the basic coronal braid oscillations transform one braid into an equivalent one. The notion of a (mathematical) group allows operations among its members which keep the braid properties intact and by analogy to magnetic braids, indicate various observable coronal oscillations. Local **ExB** drift of magnetic strand looping around another and returning to its previous site forms two crossings resulting in observable small scale oscillations at solar corona. Similarly, local wandering of a strand with respect to two other strands as well as independent Alfvenic pulses of two non-adjacent crossings along their field lines are part of the group definition, and as such become additional observable features. These moves form the basic eigenmodes which could be measured in high resolution as braid oscillations. Hence, the existence of a mathematical group which transforms one braid to another under its operations can be translated into detectable physical features in coronal excitations.

Magnetic braid stability under large perturbations can be partly addressed through its equivalent conversion into a knot, following its closure according to Alexander theorem (22). Each (magnetic) knot possesses its unique topological invariants which are dynamically preserved. All braids undergoing successive Markov moves are related to the same knot, indicating stability of the braid under these large distortions. These modifications may include major reconfigurations of the magnetic field with cutting and pasting of field lines due to resistivity or plasma kinetic interactions.

Therefore, magnetic reconfiguration through attaching a braid and its mirror to another braid at both ends, respectively, (Markov move I), or successive crossing of newly heated strands with an existing braid (Markov move II) are sources of significant perturbations which preserve the relevant closure knot. Hence, these operations, which can be repeated numerous times, qualify as a wide range of stable, large braid perturbations, observable at high resolution. Heating and nano-flare events fall under this category of magnetic reconfigurations. Alternatively, large magnetic braid modifications which alter (after closure) the knot invariants, form observable conditions of coronal destabilization.

The evolution of the coronal field is envisioned as follows: magnetic field emerges from the photosphere either as an arched non-intersecting strand (trivial braid) or as a braid with multiple crossings. The small-scale braid crossings are supported by current sheets while their oscillations are due to photospheric input or interaction with the hot coronal plasma currents. Any braid is equivalent to some knot, hence magnetic field observations include bundles of unknots (distorted loops) or prime knots – knots which as a result of any cut in a strand disintegrate into an unknot. The stable modifications which are included in the Artin equivalency group (21) or in Markov moves (23) may modify significantly the braids but preserve their knot invariants. If braid modification violates the knot invariants, the magnetic structure may become unstable and cause eruption observed as energy release through X-ray brightening on a small scale or CME on a large scale. Non-erupting knotted configurations may be dragged by the solar wind resulting in very elongated geometrical forms in the interplanetary medium, which at some distance from the Sun may become structurally unstable. Therefore, predictions for the observation of knotted structures is balanced by their formation probability (lower crossing number) vs probability of survival (larger crossing number). A satellite encountering such a structure will measure intermittently different plasmas, especially when the knotted field lines are connected to active coronal region. Due to the geometrical stretching the probability to encounter a topologically non-trivial magnetic structures decreases with the increasing solar distance, hence it is expected that the incursion of Parker Solar Probe into a region inside Mercury orbit and in the vicinity of solar corona will increase the probability of encountering a variety of knotted features in magnetic field. It is believed that the proposed scenario can be extended to magnetospheric plasmas with observations on Van Allen, Cluster and Themis satellites.

7. References.

[1] Frisch U, Poquet A, Leorat J and Mazure A 1975 *Jour. Fluid Mech,* **68**, 769-78
[2] Brandenburg A & Subramanian, K, 2005 *Phys. Rep.* **417**, 1-209
[3] De Pontieu B, McIntosh S, Carlsson M, Hansteen V, Tarbell T, Boerner P, Martinez-Sykora J, Schrijver C and Title A 2011 *Science* **331** 55-58
[4] Parnell E C and De Moortel I 2012 *Phil. Trans. R. Soc.* A **370** 3217-3240
[5] Parker E N 1988 *Astroph. Jour.* **330** 474-479
[6] Klimchuk J 2006 *Sol. Phys.* **234** 41-77
[7] Borovsky J 2008 *Jour. Geoph. Res.* **113** A08110
[8] Mazur J, Mason G, Dwyer J, Giacolone J, Jokipii J and Stone E 2000 *Astroph. Jour.* **532**, L79-82
[9] Berger M A and Asgari-Targhi M 2009 *Astroph. Jour.* **705** 347–355

[10] De Pontieu B et al, 2007 *Science* **318** 1574-1577
[11] McIntosh S, De Pontieu B, Carlsson M, Hansteed V, Boerner P and Goossens M 2011 *Nature* **475** 477-480
[12] Parker E N 1983 *Astroph. Jour.* **264** 635-647.
[13] Schrijver K 2007 *Astroph. Jour.* **662** L119-122
[14] Klimchuk J 2006 *Sol. Phys.* **234** 41–77
[15] Tomczyk S, McIntosh S, Keil, S, Judge P, Schad D, Seeley D and Edmondson J 2007 *Science*, **317** 1192-1196
[16] Cirtain J W et al. 2013 *Nature* **493** 501–503
[17] Pant V, Datta A and Banerjee D 2015 *Astroph. Jour. Lett.* **801** L2 1-7
[18] De Wijn, AG, Stenflo J, Kolanki S and Tsuneta S 2009 *Space Sci Rev* **144** 275-315
[19] Thalmann J K, Tiwari S K & Wiegelmann T 2014 *Astroph. Jour.* **780**, 102
[20] Adams C 2004 *The Knot Book*, American Mathematical Society, W H Freeman and Company
[21] Artin E 1947 *Theory of braids*, Ann. of Math. **4** 101-126
[22] Alexander J 1928 *Trans Amer. Math Soc.* **30** 275-306
[23] Markov A A 1935 *Recueil Math. Moscou*, **1** 73–78

The Local Interstellar Magnetic Field Observed by Voyager 1 and IBEX

N. A. Schwadron
University of New Hampshire, Durham, NH 03824
E-mail: nschwadron@unh.edu

M. Bzowski
Space Research Centre PAS (CBK PAN)
E-mail: bzowski@cbk.waw.pl

D. J. McComas
Princeton University, Princeton, NJ, 08544
E-mail: dmccomas@princeton.edu

E. Moebius
University of New Hampshire, Durham, NH 03824
E-mail: eberhard.moebius@unh.edu

Abstract. Observations by the Interstellar Boundary Explorer (IBEX) have revealed the presence of the IBEX Ribbon, an almost circular band of increased ENA emissions that is believed to be centered on the direction of the interstellar magnetic field. IBEX measures a range of interstellar species including H, He, and O. IBEX has been able to distinguish primary He coming directly from the interstellar medium from secondary He that is created through neutralization of He^+ atoms in the heliosheath. Together with the observations of Lyman-α radiation from SOHO/SWAN, which is resonantly absorbed and re-emitted from interstellar H atoms that move into the heliosphere, the directions of interstellar flow for each of the observed interstellar species (H, primary He, secondary He, and primary O) line-up along a symmetry plane — the interstellar B–V plane — that should contain the interstellar magnetic field direction (B) and the interstellar flow direction (V). The B–V plane also contains the center of the IBEX Ribbon, which supports the concept that the Ribbon center is close to the direction of the interstellar magnetic field. Further, the deflection of the various interstellar species along the B–V plane are consistent with the compression of the heliosheath by the interstellar magnetic field pressure. Lastly, the interstellar magnetic field in the outer heliosheath is directly observed by Voyager 1 (V1). The direction of the interstellar magnetic field observed by V1 is offset from the Ribbon center, consistent with draping around the heliopause, and time variations in the interstellar magnetic field are consistent with the effects of large-scale compression and rarefaction in the outer heliosheath. Thus, we summarize the array of observations that suggest a consistent direction of the interstellar magnetic field, and the effects of the interstellar magnetic field on the structure and time variations in the heliosheath.

1. Introduction

The heliosphere is defined by the Sun's solar wind expansion and its interaction with the local interstellar plasma. The solar wind transmits the plasma from the Sun to beyond the termination shock (TS) prior to being diverted back toward the heliotail. The heliopause (HP) separates the local the solar wind from the local interstellar medium (LISM).

Neutral interstellar atoms move on direct trajectories 100's of astronomical units (AU) through the heliosphere, affected only by solar gravitation and radiation pressure (in the case of H) prior to ionization. Interstellar neural atoms are critical for supplying information about the interstellar flow. Interstellar He atoms supply the best available information on the local interstellar flow direction, speed and temperature [1–6].

The direction and strength of the local interstellar magnetic field (LISMF) has remained an enigma for many years due to a lack of direct observations in the outer heliosheath and LISM. The Interstellar Boundary Explorer (IBEX) mission provided new information about the LISMF. With the discovery of the existence of the Ribbon, a narrow semi-circular band of high flux energetic neutral atom (ENA) emissions [7–10], it was believed that the IBEX Ribbon center is near the direction of the LISMF [7, 9]. The strength of the LISMF was also estimated to be $\sim 3\mu$G from IBEX observations based on pressure balance [11] and modeling of the IBEX Ribbon [12].

This paper summarizes recent observations of interstellar neutral atom species, the IBEX observations of the Ribbon, and Voyager 1 (V1) observations of time variations in the outer heliosheath. Together these observations indicate a consistent direction and strength of the interstellar magnetic field and effects of the LISMF on the global heliosheath.

The paper is structured as follows. Observations that are used to define the B–V plane containing the interstellar magnetic field direction and the interstellar flow direction are discussed in §2. §3 describes observations made by V1 indicating compression and rarefaction regions in the outer heliosheath, which affect the observed draping of the interstellar magnetic field around the heliopause. Concluding remarks are made in §4.

2. The Interstellar B–V plane

The B–V symmetry plane is defined to contain both the direction of the LISMF and the local interstellar flow vector [13]. Observational studies defined the B–V plane by comparing the velocities of interstellar He [1–6, 13–15] and interstellar H [16, 17] measured in the heliosphere. The H flow direction is determined from observations of the Lyman-α radiation scattered from interstellar hydrogen that moves into the heliosphere. Therefore, the H-direction contains influences from both primary H atoms from the LISM and secondary H atoms generated through charge-exchange interactions in the heliosheath, which on average deflects the LISM H flow vector along the B–V plane, slows the H LISM flow, and increases its temperature. In contrast, LISM He undergoes very little charge-exchange in the heliosheath and has a flow vector that more accurately reflects the LISM conditions far from the heliosphere. The He and H velocity vectors uniquely define the B–V plane (Figure 1).

In addition to measurements of LISM H and He that define the B–V plane, more recent LISM measurements by IBEX include primary LISM O [18] and secondary He [19], which have improved the definition of the B–V plane. The separation of secondary He and primary He provide one of the best measures of the B-V plane (Figure 1, 2, grey curve).

In Figure 1, we also show recent magnetic field measurements of V1 [21]. Over the period from May 10, 2013 to Aug. 20, 2014, the LISMF shows steady undraping in the outer heliosheath. However, later, from May 26, 2015 to Aug 25, 2016 the LISMF is affected by a rarefaction region in the outer heliosheath, as detailed by [21] and summarized in the next section.

Figure 2 shows expanded ranges of the B–V plane (left) to better resolve the velocity directions of LISM species. Primary and secondary species show systematic deflections along the B–V

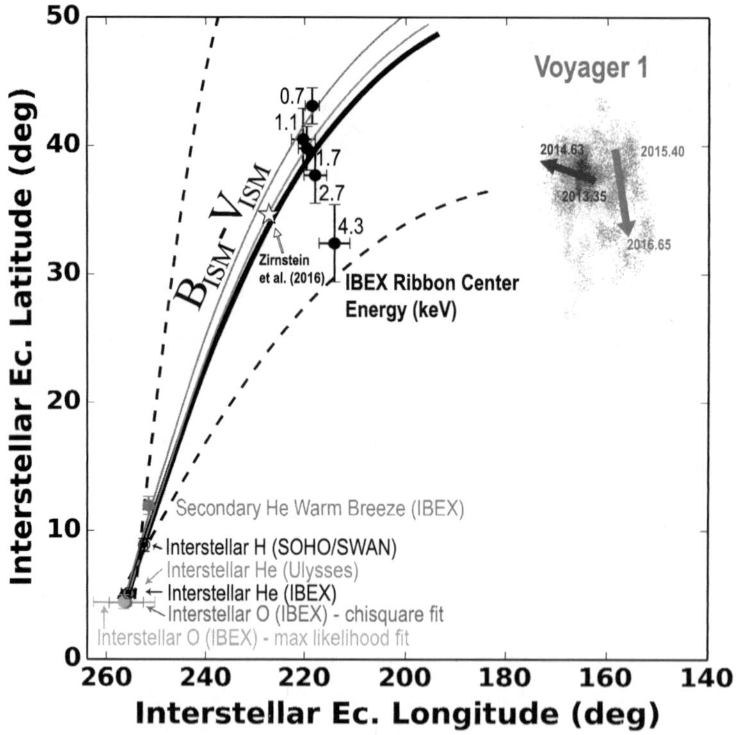

Figure 1. The LISM B–V symmetry plane (black, magenta and grey curves) is defined by the directions of LISM neutral O, He, and H from IBEX, Ulysses and SOHO/SWAN. The deflection of LISM H compared to LISM He provides a way to measure the B–V plane (black curve) [13, 17–19]. Dashed black curves show uncertainty limits of the LISM B–V plane. [19] provided an analysis of secondary He, which was compared to primary He both measured by IBEX to provide another important determination of the LISM B–V plane (grey curve). A third estimate of the LISM B–V plane (magenta curve) from measurements from SOHO/SWAN for H and Ulysses for He. Also shown are measurements of the LISMF direction from Voyager 1 (grey, blue and red data points) that shows departures from the IBEX ribbon center (black points) and from the LISM B–V plane. The black points from the Ribbon center are defined by fitting the ribbon [20]. A model-based fit (yellow star) was provided by [12]. The direction of the LISMF measured by Voyager 1 changes with time: from May 10, 2013 to Aug. 20, 2014 (blue arrow fit to blue points) the LISMF tends toward the B-V plane and the IBEX ribbon center, consistent with undraping of the LISMF; LISMF directions from May 26, 2015 to Aug. 25, 2016 (red points) show the opposite trend (red arrow fit to red data points) away from B-V plane and away from the IBEX Ribbon center. Figure taken from [21].

plane. LISM primary He represents the most definitive direction for the LISM flow far from the heliosphere. Secondary He, and H, which contains a secondary component, are both shifted northward with respect to primary He. Primary O is shifted southward with respect to primary He due to stronger filtration north of the nose. These directional deviations of primary and secondary components are caused by the heliospheric asymmetry due to stronger LISMF pressure south of the nose (Figure 2, right).

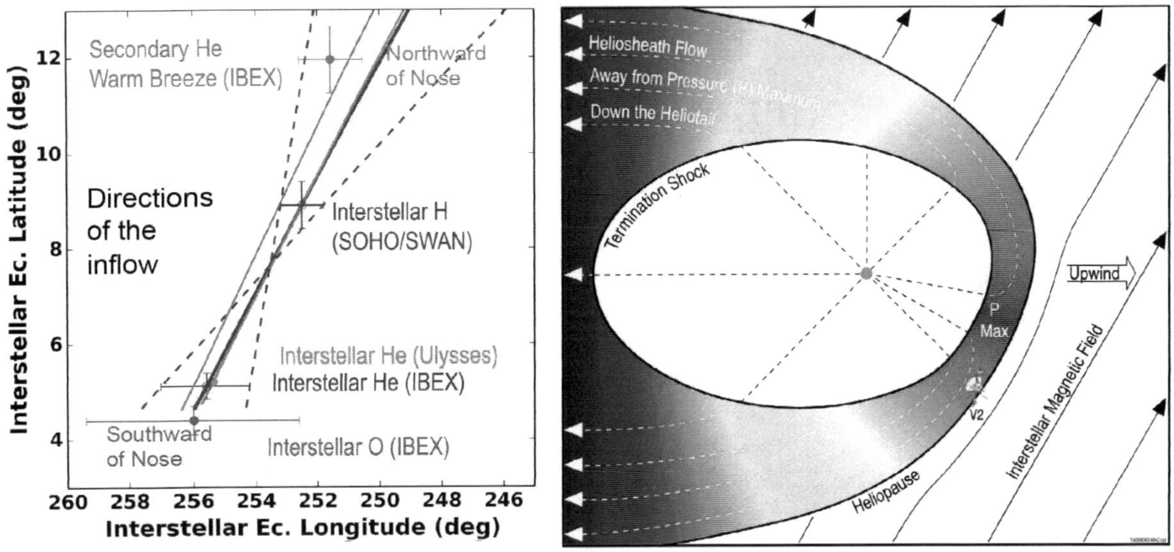

Figure 2. Expanded ranges to show the directions of the LISM flow for LISM species defining the B–V plane (left). The asymmetry of the heliosphere (right) driven by interstellar pressure, which thins the heliosheath south of the nose. Left panel from [18] and right panel from [22].

3. Voyager 1 observations in the outer heliosheath

The direction of the LISMF measured by V1 in the outer heliosheath shows deviation from the LISM B–V plane. This deviation is consistent with draping of the LISMF around the heliopause [23]. Draping of the LISMF around the heliopause creates a significant difference between the field direction measured by V1 and the direction of the ribbon center, as shown in Figure 1. [21] compared V1's measurements of the LISMF with measurements of the solar wind made at 1 AU. Solar events in 2012 and 2013 create merged interaction regions that drive pressure changes. These variations are transmitted though the heliopause to drive compression and rarefaction within the outer heliosheath. The effect of the rarefaction region at V1 causes the expansion of the magnetic field (Figure 3), which leads to stronger draping of the LISMF near V1.

The magnetic observations of V1 provide direct evidence of the draping of the LISMF. The deflection of the magnetic field around the heliopause increases the magnetic tension, and is balanced by enhanced magnetic pressure. The deflection of the magnetic field measured by V1 has a direct connection to the magnetic pressure that drives asymmetries in the heliosphere (Figure 2, right) and, in turn, the deflections of primary and secondary neutral atom components (He, H, and O) that are used to determine the LISM B–V plane.

4. Conclusions

Neutral atoms and Lyman-α observations from Ulysses, SOHO/SWAN, and IBEX were used to determine the LISM B–V plane. This symmetry plane intersects the center of the IBEX Ribbon, showing the consistency of the concept that the center of the IBEX Ribbon is near the direction of the LISMF.

Voyager 1 provides direct observations of the LISMF in the outer heliosheath. The deflection of the magnetic field direction measured by V1 is consistent with the draping of the LISMF around the heliopause. Time variations measured by Voyager 1 indicate the presence of compressions and rarefactions driven by pressure variations transmitted through the heliopause

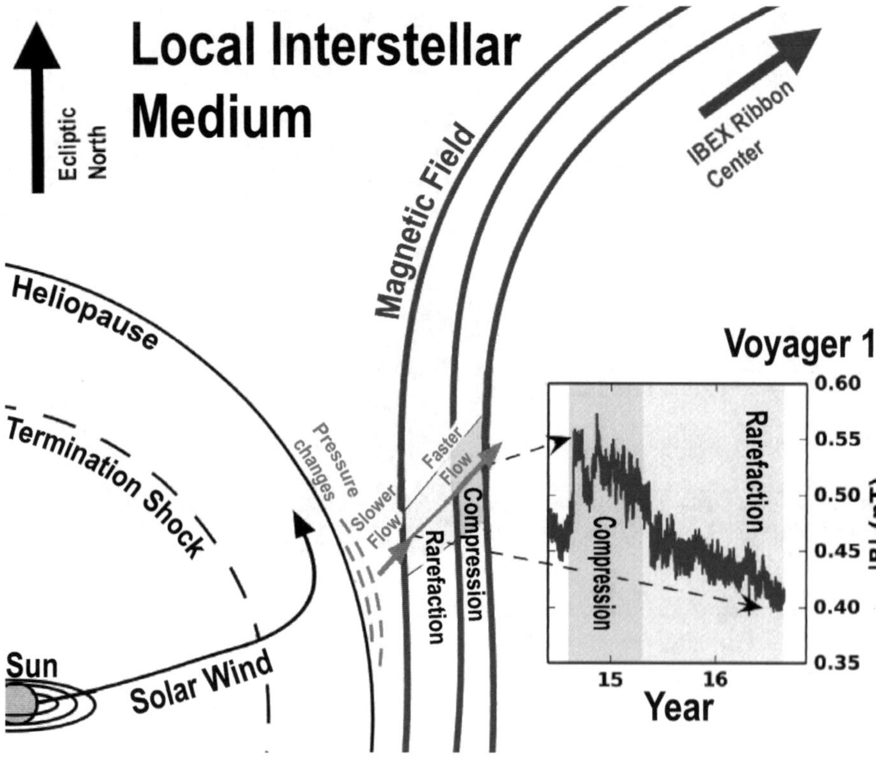

Figure 3. Global Merged Interaction Regions from solar events in the 2012-2013 peak activity period of solar cycle 24 create pressure changes that are transmitted into the outer heliosheath. The compression and rarefaction regions cause further changes in the draped LISMF. Figure from [21].

from merged interaction regions in the solar wind. These time variations are consistent with the excursions in the magnetic field due to draping in the presence of expansion and compression of the outer heliosheath plasma.

Datasets from Ulysses, IBEX, SOHO/SWAN, and Voyager are now allowing determination of the direction and strength of the local interstellar magnetic field. The draping of the local interstellar magnetic field around the heliopause is associated with increased magnetic pressure in the outer heliosheath that drives asymmetries in heliosheath structure.

4.1. Acknowledgments
We are very grateful to the many individuals who have made the Voyager, SOHO, and IBEX projects possible. This work is supported by the Interstellar Boundary Explorer mission as a part of NASAs Explorer Program and partially by NASA SR&T Grant NNG06GD55G. NAS was also supported by the Sun-2-Ice (NSF grant number AGS1135432) project. M.B. acknowledges support from the Polish National Science Center grants 2015-18-M-ST9-00036 and 2015-19-B-ST9-01328.

References
[1] Witte M, Banaszkiewicz M, Rosenbauer H and McMullin D 2004 *Advances in Space Research* **34** 61–65
[2] Möbius E, Bzowski M, Frisch P C, Fuselier S A, Heirtzler D, Kubiak M A, Kucharek H, Lee M A, Leonard T, McComas D J, Schwadron N A, Sokół J M, Swaczyna P and Wurz P 2015 *Astrophys. J. Suppl.* **220** 24

[3] Leonard T W, Möbius E, Bzowski M, Fuselier S A, Heirtzler D, Kubiak M A, Kucharek H, Lee M A, McComas D J, Schwadron N A and Wurz P 2015 *Astrophys. J.* **804** 42

[4] McComas D J, Bzowski M, Frisch P, Fuselier S A, Kubiak M A, Kucharek H, Leonard T, Möbius E, Schwadron N A, Sokół J M, Swaczyna P and Witte M 2015 *Astrophys. J.* **801** 28

[5] Bzowski M, Swaczyna P, Kubiak M A, Sokół J M, Fuselier S A, Galli A, Heirtzler D, Kucharek H, Leonard T W, McComas D J, Möbius E, Schwadron N A and Wurz P 2015 *Astrophys. J. Suppl.* **220** 28 (*Preprint* 1510.04835)

[6] Schwadron N A, Möbius E, Leonard T, Fuselier S A, McComas D J, Heirtzler D, Kucharek H, Rahmanifard F, Bzowski M, Kubiak M A, Sokół J M, Swaczyna P and Frisch P 2015 *Astrophys. J. Suppl.* **220** 25

[7] McComas D J, Allegrini F, Bochsler P, Bzowski M, Christian E R, Crew G B, DeMajistre R, Fahr H, Fichtner H, Frisch P C, Funsten H O, Fuselier S A, Gloeckler G, Gruntman M, Heerikhuisen J, Izmodenov V, Janzen P, Knappenberger P, Krimigis S, Kucharek H, Lee M, Livadiotis G, Livi S, MacDowall R J, Mitchell D, Möbius E, Moore T, Pogorelov N V, Reisenfeld D, Roelof E, Saul L, Schwadron N A, Valek P W, Vanderspek R, Wurz P and Zank G P 2009 *Science* **326** 959–

[8] Fuselier S A, Allegrini F, Funsten H O, Ghielmetti A G, Heirtzler D, Kucharek H, Lennartsson O W, McComas D J, Möbius E, Moore T E, Petrinec S M, Saul L A, Scheer J A, Schwadron N and Wurz P 2009 *Science* **326** 962

[9] Schwadron N A, Bzowski M, Crew G B, Gruntman M, Fahr H, Fichtner H, Frisch P C, Funsten H O, Fuselier S, Heerikhuisen J, Izmodenov V, Kucharek H, Lee M, Livadiotis G, McComas D J, Moebius E, Moore T, Mukherjee J, Pogorelov N V, Prested C, Reisenfeld D, Roelof E and Zank G P 2009 *Science* **326** 966–

[10] Funsten H O, Allegrini F, Crew G B, DeMajistre R, Frisch P C, Fuselier S A, Gruntman M, Janzen P, McComas D J, Möbius E, Randol B, Reisenfeld D B, Roelof E C and Schwadron N A 2009 *Science* **326** 964–

[11] Schwadron N A, Allegrini F, Bzowski M, Christian E R, Crew G B, Dayeh M, DeMajistre R, Frisch P, Funsten H O, Fuselier S A, Goodrich K, Gruntman M, Janzen P, Kucharek H, Livadiotis G, McComas D J, Moebius E, Prested C, Reisenfeld D, Reno M, Roelof E, Siegel J and Vanderspek R 2011 *Astrophys. J.* **731** 56

[12] Zirnstein E J, Heerikhuisen J, Funsten H O, Livadiotis G, McComas D J and Pogorelov N V 2016 *Astrophys. J. Lett.* **818** L18

[13] Lallement R, Quémerais E, Bertaux J L, Ferron S, Koutroumpa D and Pellinen R 2005 *Science* **307** 1447–1449

[14] Bzowski M, Kubiak M A, Hłond M, Sokół J M, Banaszkiewicz M and Witte M 2014 *Astron. Astrophys.* **569** A8

[15] Wood B E, Müller H R and Witte M 2015 *Astrophys. J.* **801** 62 (*Preprint* 1501.02725)

[16] Wood B E, Izmodenov V V, Linsky J L and Alexashov D 2007 *Astrophys. J.* **659** 1784–1791 (*Preprint* arXiv:astro-ph/0701274)

[17] Lallement R, Quémerais E, Koutroumpa D, Bertaux J L, Ferron S, Schmidt W and Lamy P 2010 *Twelfth International Solar Wind Conference* **1216** 555–558

[18] Schwadron N A, Möbius E, McComas D J, Bochsler P, Bzowski M, Fuselier S A, Livadiotis G, Frisch P, Müller H R, Heirtzler D, Kucharek H and Lee M A 2016 *Astrophys. J.* **828** 81

[19] Kubiak M A, Swaczyna P, Bzowski M, Sokół J M, Fuselier S A, Galli A, Heirtzler D, Kucharek H, Leonard T W, McComas D J, Möbius E, Park J, Schwadron N A and Wurz P 2016 *Astrophys. J. Suppl.* **223**

[20] Funsten H O, DeMajistre R, Frisch P C, Heerikhuisen J, Higdon D M, Janzen P, Larsen B A, Livadiotis G, McComas D J, Möbius E, Reese C S, Reisenfeld D B, Schwadron N A and Zirnstein E J 2013 *Astrophys. J.* **776** 30

[21] Schwadron N A and McComas D J 2017 *Astrophys. J.* **849** 135

[22] McComas D J and Schwadron N A 2014 *Astrophys. J. Lett.* **795** L17

[23] Czechowski A and Grygorczuk J 2017 *Journal of Physics Conference Series* (*Journal of Physics Conference Series* vol 900) p 012004

IOP Conf. Series: Journal of Physics: Conf. Series **1100** (2018) 012022 doi:10.1088/1742-6596/1100/1/012022

Heating the Local Interstellar Cloud

Jonathan D. Slavin

Harvard-Smithsonian Center for Astrophysics, 60 Garden Street, MS 83, Cambridge, MA USA 02138

E-mail: jslavin@cfa.harvard.edu

Abstract. The Local Interstellar Cloud (LIC) is the region of the interstellar medium (ISM) that surrounds and helps to shape the heliosphere. The LIC is part of a collection of nearby low density warm clouds known as the Complex of Local Interstellar Clouds (CLIC), all of which exist inside the hot Local Bubble. Observations of interstellar neutral He atoms flowing into the heliosphere by the IBEX mission and Voyager have constrained the temperature of the LIC to be roughly 7500 K. This temperature is consistent with that derived from absorption line measurements toward nearby stars. Such observations also indicate that the LIC is partially ionized with elemental abundances consistent with a moderate level of depletion onto dust grains as might occur for low density ISM that has been subject to a shock that partially destroyed the dust. The temperature of the cloud is not unusual for the warm ionized medium in the ISM, but it is less ionized than typical. We discuss the various processes that may be important for heating the LIC. We show that the only viable heat source for the ongoing heating of the CLIC is photoionization. Equilibrium models of the ionization and heating of the LIC allow for solutions that match the observations, but the likely origins of the local interstellar medium suggest that the situation is more complex. We propose an evolution scenario in which the clouds were formerly cold and were heated by shocks to reach their current warm state. We present new magneto-hydrodynamical calculations of the evolution of the local ISM. Multiple supernova models can match the parameters of the Local Bubble and heat the cold clouds as desired, though the many constraints on the clouds and bubble have yet to be fully satisfied by the models.

1. Introduction

The Local Interstellar Medium (LISM) is dominated by the large low density bubble or cavity that surrounds the solar location known as the Local Bubble (LB). The dimensions of the bubble have been mapped out both by absorption lines, e.g. using the lack of Na I absorption toward nearby stars [1], and by the lack of dust extinction [2]. The existence of the Local Bubble has also been inferred by observations of soft X-ray (~ 0.25 keV) emission [e.g. 3, 4] which has been observed in every direction over the sky and from which it is inferred that the low density cavity is filled with hot, $T \approx 10^6$ K gas. From the intensity of the emission and its inferred temperature as well as the dimensions of the cavity and models for the emissivity of the plasma, the pressure and density of the hot gas has been determined. The relatively high value of $P/k_B \approx 1.5 \times 10^4$ cm^{-3} K has been found, though details about possible contributions from distant backgrounds and foregrounds (discussed below) create uncertainties in this value. Such a pressure value leads to an estimated density of ~ 0.005 cm^{-3}.

Despite the very low average density of the LB, there are some denser, cooler regions detected in the bubble. In particular the Complex of Local Interstellar Clouds (CLIC) is a collection of

IOP Conf. Series: Journal of Physics: Conf. Series **1100** (2018) 012022 doi:10.1088/1742-6596/1100/1/012022

15 identified clouds that are within 15 pc of the Sun [5]. The cloud properties do vary from cloud to cloud, but all are warm, $T \sim 4000 - 9000$ K and partially ionized, $X(H^+) \sim 0.25$. The clouds have been detected using primarily UV absorption lines toward nearby stars and most lines of sight show multiple absorption components (i.e. velocities). Combinations of line column densities and models for the ionization levels of different elements yield estimates of the gas phase elemental abundances. The results indicate significant depletion onto dust for silicate dust elements including Fe, Si and Mg but less depletion than is typical for so-called diffuse H I clouds. Another relevant property of the CLIC clouds is that they apparently have a low level of turbulence. We discuss this further below.

The closest cloud in the CLIC, the Local Interstellar Cloud (LIC), is believed to surround the Solar System and, with the solar wind, creates the heliosphere. In fact we might want to define the LIC as that region immediately outside the heliosphere, though the name LIC has been associated with a particular velocity component of gas seen in absorption toward nearby stars. The fact that the velocity, density, temperature and ionization of the LIC and the ISM inflowing into the heliosphere appear to be consistent makes us confident that the gas seen in absorption and that detected *in situ* are coming from the same "cloud". As we discuss below, there are good reasons to believe that small scale variations are common in the CLIC and that regions that appear as a single velocity vector may be complex in their spatial morphology [see also 6]. Nevertheless we will assume from here on that the LIC is the source of the inflowing ISM. The combination of *in situ* measurements and line-of-sight observations in many directions makes the LIC the most tightly constrained cloud there is and gives us the ability to probe the physical processes at work in such low density ISM with a level of completeness that is not available anywhere else.

2. Heating Mechanism for Diffuse Warm Gas

The local ISM is like the ISM as a whole in that it exists at a wide range of temperatures and densities. In the ISM it is seen that certain temperatures, densities and ionization states seem to be more common and these have come to be referred to as "phases" [7]. One of the primary phases of the ISM is known as the warm ionized medium (WIM) and was first found to be widespread by its faint diffuse Hα emission [8, 9]. This emission is observed over the entire sky implying that the WIM is present throughout the ISM as a low density intercloud medium rather than being confined to the environs of hot stars in H II regions. The characteristics of the WIM have been determined by those Hα observations as well as associated optical [N II] and [S II] emission [see 10, for a review]. The source of the ionization and heating of the WIM is not certain though there have been several proposals put forward. The power needed to balance the cooling associated with the Hα emission and to provide its ionization is substantial, roughly equivalent to the power associated with supernova explosions in the Galaxy.

The WIM, however, differs significantly from the warm partially ionized medium of the CLIC. First, because of the weakness of [O I] λ6300Å emission relative Hα associated with the WIM, it is clear that in the WIM H is highly ionized, $X(H^+) > 67\%$ [11]. The weakness of He I λ5876Å line shows that the He ionization fraction is small, $X(He^+) < 0.27X(H^+)$ [12]. This contrasts to the conditions in the LISM wherein observations of H I and He I continuum absorption toward nearby stars observed by *EUVE* indicate that He is more ionized than H [13]. Since the ionization of the WIM differs from that in the CLIC, it is likely that its heating differs as well. Nevertheless the temperatures and density of the WIM and the CLIC are similar and thus it is worth considering the heating mechanisms put forward for heating the WIM as possible heat sources for the CLIC.

IOP Conf. Series: Journal of Physics: Conf. Series **1100** (2018) 012022 doi:10.1088/1742-6596/1100/1/012022

3. Heating Sources Proposed for the Diffuse ISM

Most of the heating sources we discuss here are expected to mostly heat the warm ionized ISM, though some could also be sources for diffuse neutral gas in the ISM, the so-called warm neutral medium (WNM) and cold neutral medium (CNM). The mechanisms that have been gotten the most attention are the following:

- photoionization heating from the FUV/EUV/Soft X-ray radiation field
- photoelectric emission from dust grains
- dissipation of MHD turbulence
- low energy cosmic rays
- shocks (mostly from SNRs, though stellar winds and outflows could also contribute).

The most obvious source of heating for the WIM is photoionization heating since the photoionization that produces the ionization of the WIM will necessarily be accompanied by heating. This heating comes from the excess energy of the electrons that have been freed from H atoms (primarily) which are then thermalized. O and B stars generate plenty of photons and energy to account for the ionization and heating of the WIM. However such early type stars are comparatively rare and it is unclear if the photons can escape the H II regions surrounding the stars in adequate numbers to account for the relatively smooth distribution of WIM seen via Hα emission. For the CLIC, the problem is that there is a dearth of early type stars that are near enough to provide the ionization, only two B stars known to be inside the Local Bubble. Hot white dwarfs also contribute to the ionizing radiation field, though they have much lower flux per star.

Photoelectric emission from grains can be effective as a heat source for the gas if the dust abundance is large enough [14]. It is in fact believed to be the primary source of heat for the WNM [15]. Radiative transfer is a less important effect in this case because the radiation field that is effective is that longward of the Lyman limit, i.e. the far UV, which is only effectively absorbed by dust and thus has a long mean free path even in the Galactic plane. Grain heating becomes less effective, however, as the temperature rises above ~ 6000 K since cooling of the gas via collisions with the grains becomes important. In the case of the CLIC, this is compounded by the fact that there is evidence for significant grain destruction with a gas-to-dust mass ratio is higher than the typical value of 100 by 50–100%. PAHs are believed to be important sources of heat in some regions but can be destroyed in warm ionized regions [16]. Detailed photoionization models (discussed below) find that dust heating is not significant for the LIC.

Dissipation of turbulence was proposed as a heating source for the WIM by [17] and [18]. This is a means of transferring the energy in the magnetic field and the kinetic energy in non-thermal motions to the thermal energy of the gas. [18] found that only certain types of dissipation lead to a heating rate that would balance the radiative cooling rate in the WIM and that Landau damping would create too much heating. Details of that model were later disputed by [19] who claimed that the heating and cooling rates could be reconciled with the correct cut-offs in wavenumber space. This debate has continued, but for our purposes of looking at the heating of the CLIC, it appears that it is not relevant and that turbulent heating cannot be a significant contributor to the heating. The reason is that the absorption line observations have shown that there is very little energy in turbulence in the CLIC. By using lines from elements with a large difference in atomic mass, e.g. D and Fe, the non-thermal (presumably turbulent) width and the thermal width of the lines can be determined. Using this method it has been found [5] that the turbulent widths in the CLIC are $\sim 1-3$ km s^{-1} implying a low turbulent dissipation rate.

The amount of heating that might be available from low energy (~ 1 MeV) cosmic rays has been very uncertain due to modulation of those cosmic rays by the heliospheric magnetic field. Recently [20] proposed that low energy cosmic rays could be an important source for heating the WIM based on new values for the inferred rate of ionization in diffuse molecular gas in the solar neighborhood. The mechanisms for this heating are collisional ionization of neutral gas and direct heating of electrons via Coulomb collisions. Recent observations by Voyager I in the outer heliosheath have finally constrained the lower end of the cosmic ray energy spectrum, however, and calculations by [21] indicate that the cosmic ray ionization rate is a factor of ~ 10 below the "local" rate cited by [20]. Since [20] was already taking the high end of estimates for the cosmic ray ionization rate, it would appear that cosmic rays are not a significant contributor to ionization or heating of the CLIC.

The final suggested heating source above was shocks. For the CLIC, it is likely that in the past the clouds were shocked by the same shock that heated the surrounding gas to its current high temperature ($T \sim 10^6$ K). However, there is no clear signature for a strong shock passing through the CLIC at present, though weaker shocks may be present [6]. Since the post-shock cooling time for such shocks is short, they do not present a viable mechanism for maintaining the clouds at their current temperature, though their ionization could be significantly affected by past and current (weak) shocks.

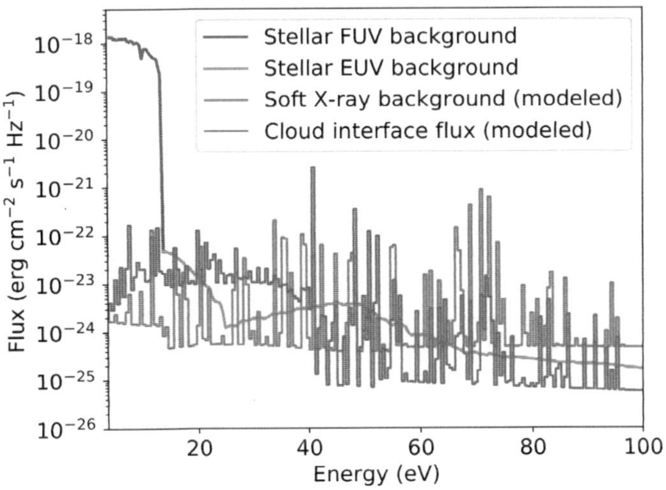

Figure 1. The components of the model interstellar radiation field used for calculating the ionization in the LIC. The stellar parts are based on observations, though some modeling is needed as well. The soft X-ray background part, though modeled, is constrained by the broad band observations. There is no direct evidence for the cloud interface component, though, if present it contributes to the lowest energy bands of the soft X-ray background that have been observed.

4. The Interstellar Radiation Field

Given the discussion above, it seems clear that the most viable source of heating for the CLIC is photoionization by the interstellar radiation field. However, to determine if that works in detail requires a closer look at the radiation field. In [22] we presented photoionization models of the LIC. The radiation field that was constructed included radiation from known nearby hot stars observed by *EUVE* and the FUV background that is due to many A and B stars within a few kpc of the Solar System. In addition we included modeled soft X-ray background emission from the hot gas in the Local Bubble and from the edge of the cloud that was assumed to be evaporating via thermal conduction. In Figure 1 we show an example radiation field. Using radiation fields like the one in the figure we calculated the ionization created assuming thermal and ionization equilibrium, making use of the code Cloudy. As detailed in [22] we found satisfactory fits to the data on the ionization and temperature of the LIC. Since the publication of [22] there were questions raised about the soft X-ray background and the nature of the Local Bubble [see e.g. 23]. Understanding the origins of the Local Bubble and the CLIC may be essential to understanding the current observed state of the bubble and clouds.

4.1. Properties of the Local Bubble

The model of low density hot gas within the local cavity as the source of the observed soft-X-ray diffuse background, i.e. the Local Bubble model, was first put forward in 1983 [3], though it has been refined over the years [e.g. 4]. The model assumes that hot gas of uniform temperature and density fills the neutral gas cavity and thus the emission brightness gives the path length in any given direction. The size of the cavity thus derived matches fairly well with that derived from absorption line measurements, though the required thermal pressure, $P/k_B \sim 1.5 \times 10^4$ cm^{-3} K, is substantially higher than that derived for the warm clouds of the CLIC [24]. More recently though the model has faced other, potentially more serious, challenges.

It was recognized soon after the discovery of X-rays from a Comet Hyakutake [25] that charge exchange of highly ionized solar wind ions with the neutrals flowing out from the comet is the best explanation for the emission. In this process the highly ionized atom, e.g. $O^{+}7$, gains an electron in an excited state and the cascade that follows results in emission of soft X-ray photons. Cox [26] first recognized that interstellar neutrals flowing into the heliosphere, specifically neutral H and He, should also undergo similar solar wind charge exchange (SWCX) reactions and emit soft X-rays, possibly contributing significantly to the observed soft X-ray background and thus altering our view of the nature of the Local Bubble. Lallement [27] showed that indeed the amount of soft X-ray emission attributable to hot gas could be significantly reduced if one took into account SWCX emission. More modeling of the emission followed that work [e.g. 28], though substantial uncertainties remained because of uncertainties in the atomic data regarding the charge exchange process and the solar wind parameters.

Recently the *Diffuse X-rays in the Local Galaxy* (DXL) mission carried out new observations with the observing geometry designed with the specific goal of separating the SWCX emission from the hot gas emission [29]. Results from DXL were that in the Galactic plane, roughly 60% of the emission comes from the hot gas. At high Galactic latitudes, where the emission is brightest, that fraction is higher. Based on the DXL results, [30] used SWCX models along with the data from *ROSAT* to create "cleaned" all-sky maps with only emission from the hot gas included. They found a fairly uniform temperature distribution with $T = (1.12 \pm 0.17) \times 10^6$ K (where the statistical and systematic errors have been combined). Assuming a uniform electron density of 4.68×10^{-3} cm^{-3} leads to a cavity size and shape that is a fairly good match to that derived by extinction measurements with typical radius in the Galactic Plane of $\sim$ 100 pc and more like 200 pc out of the plane. This combination of temperature and electron density implies a thermal pressure in the bubble of $P/k_{\rm B} \sim 10^4$ cm^{-3} K. Note that the thermal pressure in the LIC is estimated to be closer to 3000 cm^{-3} K. It has been suggested that the magnetic pressure in the cloud can balance the excess pressure in the hot gas [31, 32].

5. Simulating the LISM

The picture of the CLIC located within the old, hot Local Bubble appears to be consistent with the data, generally speaking, but we may still ask if it is physically realistic. The most curious aspect of this scenario is the existence of these low density clouds within the bubble. If we presume that the bubble was created by fast shocks that heated the gas to temperatures above 10^6 K, then the question presents itself, why weren't the clouds swept up or heated as was the rest of the medium? Warm gas with a density of ~ 0.3 cm^{-3}, such as is inferred for the clouds in the CLIC, is typical of the lowest density ISM in the Galactic plane other than hot gas. Therefore we conclude that the gas that currently makes up the CLIC was substantially denser at the time the shock (or shocks) that created the Local Bubble swept past. In particular the gas needed to be denser than its surroundings then, as it is now. Thus we posit that the CLIC originated as a cloud or clouds of relatively high density gas, probably CNM type gas, embedded in a lower density medium. With this as an initial state, it is to be expected that the clouds would be left behind as fast shocks swept past, creating the hot gas that is seen now via its soft X-ray emission. Such an origin for the CLIC could have significant impacts on its current thermal and dynamical state.

While this scenario appears to be plausible, it is unclear whether it can work in detail and if it does, what constraints it imposes on the nature and history of the LISM. We report here on our current progress in simulating the LISM using numerical magnetohydrodynamical calculations. Among our goals in this work are to match:

- temperatures in the clouds–which requires conversion of initially cold gas into warm gas,
- densities of the warm cloud gas, $n \sim 0.1 - 0.3$ cm^{-3},
- velocity field of the clouds that allows them to be seen as separate velocity components as in the work Redfield & Linsky,
- not so much small scale turbulence as to violate observed constraints of $v_{\rm turb} \sim 1 - 3$ km s^{-1}.

For the Local Bubble the observational constraints are:

- a fairly uniform apparent distribution in temperature with $T \sim 10^6$K,
- the absolute level on average of soft X-ray intensity, while allowing for variability,
- the size of cavity with radius of $\sim$ 100 pc.

The combination of all of these constraints on the bubble and clouds is difficult to match and no model put forward to date has attempted to match them all. Smith & Cox [33] created 1D numerical hydrodynamical models (though with an approximate magnetic pressure term) with multiple SNRs with the primary goals being replicating the magnitude, spectrum (as seen with broad band proportional counters) and variation of the soft X-ray background. Schulreich et al. [34] used fully 3D simulations employing a hydrodynamical code with adaptive mesh refinement to simulate the bubble including examining how the observed ^{60}Fe enhancements found on Earth could be attributed to the same supernovae that created the Local Bubble. Breitschwerdt & de Avillez [35] used a similar code in a large scale simulation that included the gravitational field and ongoing supernova activity to establish a global dynamical equilibrium. Their focus was to match both the Local Bubble attributes and those of the nearby Loop I bubble. Our simulations are the first to attempt to match both the Local Bubble and CLIC characteristics at the same time.

5.1. Methods

While a full description of our results and methods is beyond the scope of this paper, we give here a basic overview. We have used the multidimensional numerical magnetohydrodynamical (MHD) code FLASH [36] with adaptive mesh refinement. We have enhanced the code by including optically thin radiative cooling and heating due to photoionization, both via lookup tables that contain the cooling and heating coefficients vs. temperature. Cooling in the diffuse ISM is due primarily to the collisional excitation of lines, radiative recombination and bremsstrahlung and is thus proportional to $n_e n_{\rm H}$, where n_e is the electron density and $n_{\rm H}$ is the H density. Heating by photoionization, on the other hand, is proportional to the photon flux (assumed uniform) and the $n_{\rm H}$. This leads to the well-known result that, for a range of thermal pressures, there are two stable thermal phases (see equilibrium curve in Figure 4). In our calculations the initial setup has clouds in the stable cold phase with $T = 100$ K and $n = 25$ cm^{-3} and the surrounding medium is in the warm stable phase, $T = 7500$ K, $n = 0.358$ cm^{-3}. The magnetic field is assumed to be uniform with value $B_z = 5\,\mu$G. We use adaptive mesh refinement with 4 levels of refinement and the unsplit staggered mesh method for the MHD. Our calculations have been done in 2D with cylindrical symmetry $(r - z)$ because of the need for high spatial resolution. A collection of clouds was placed near the origin $(10 - 20$ pc) and supernova explosion(s) were

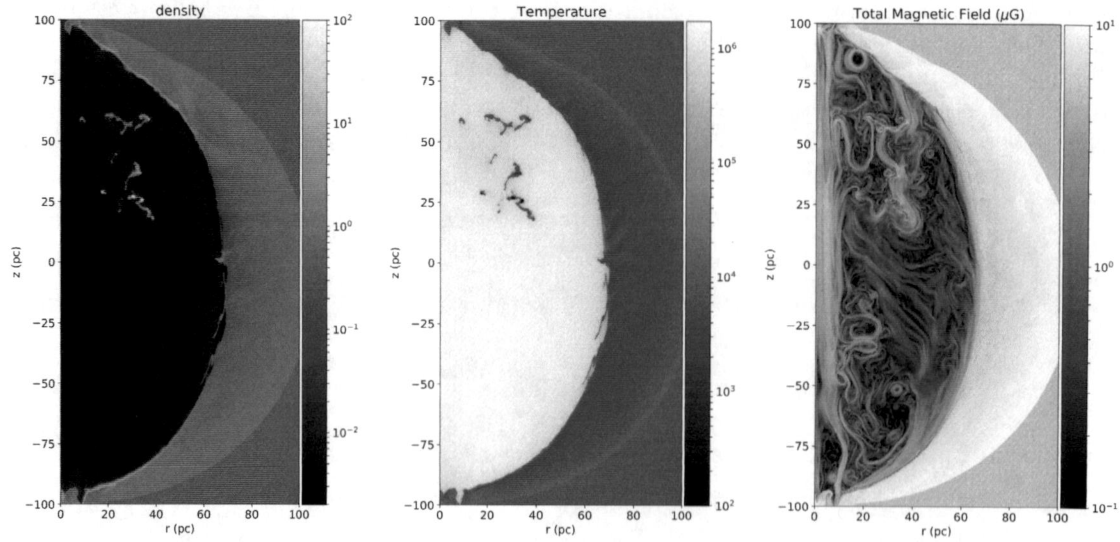

Figure 2. Density, temperature and magnetic field for run with two explosions, 1.1×10^6 yr after the second explosion. It can be seen that the initially compact cloud complex has split into two groups and that most of the cloud material is relatively warm. The temperature appears relatively uniform, consistent with the observations of the soft X-ray diffuse background. The magnetic field appears to be, on average, stronger in the clouds though it is also strong in certain regions of the hot gas. The multiple shocks including reverse shocks and reflections have created a complex magnetic field morphology.

set off with an explosion energy of 10^{51} erg at the origin. We have included thermal conduction with saturation but have not included its anisotropy due to the magnetic field – though we have tested uniform reduction of the conductivity.

5.2. Results

It seems clear from our calculations that reaching a state similar to that of the hot gas in the Local Bubble is attainable, though we have found that allowing for thermal conduction is necessary. We find that reducing the conductivity relative to the Spitzer value is helpful in achieving the desired size of the bubble and the thermal pressure, temperature and rough uniformity of temperature in the bubble. We are able to achieve the correct size and temperature of the bubble with just two explosions and have not tested the effects of more. It appears that the temperature and size of the bubble depends on a combination of the number of explosions, the conductivity, and the radiative cooling rate (set by a combination of the cooling curve and pressure in the ambient medium). In Figure 2 we show the density, temperature and pressure at a time of 1.1×10^6 yr after the most recent explosion for a run with two explosions, the second one occurring 5×10^5 yr after the first one.

A general result for the clouds is that they are heated by the shocks and contain mostly warm gas at late times, i.e. several 10^5 yr to more than 10^6 yr after the most recent SN explosion. The velocity field seen in the clouds is somewhat complex, though motion of complexes of clouds that remain at late times are not highly turbulent. We are still evaluating to what extent the velocity fields in the clouds in the simulations are consistent with observations of the CLIC, though it is clear that, despite substantial motions of the clouds relative to the grid, their relative motions

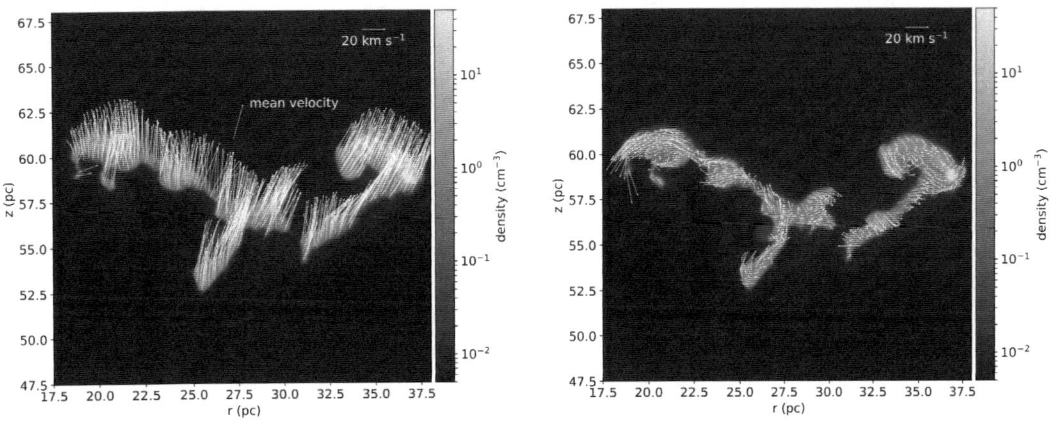

Figure 3. Velocity field for a cloud complex for the same run and time as in Figure 2. Velocities relative to the ambient medium (*left*) show that the cloud group has a substantial velocity, though when the mean velocity is subtracted (*right*) it is clear that the relative velocities of different parts of the clouds and the different clouds is fairly small.

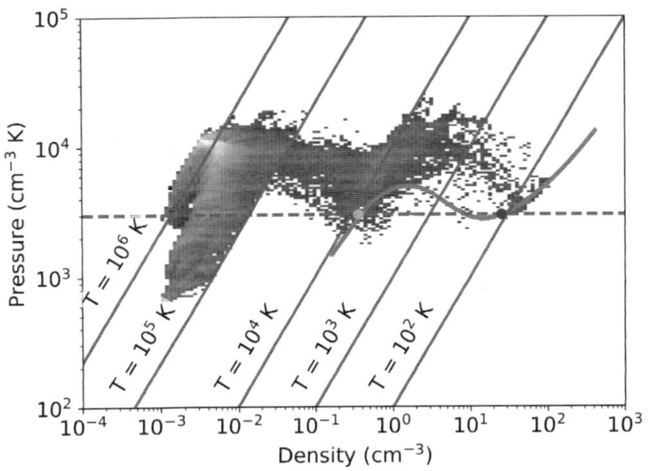

Figure 4. Phase diagram for the same run and at the same time as Figure 2. This is a 2D histogram (volume weighted) of density and pressure of only the gas inside the cavity. The red curve is the locus of points at which there exists an equilibrium between heating and cooling. The dashed magenta line shows the ambient thermal pressure before the SN explosions occur. The hot gas has the desired temperature and pressure, but the the cloud gas ($T \sim 10^4$ K) is denser than typical for the CLIC.

are relatively slow. This can be seen in Figure 3 where the absolute motion and relative cloud motions are shown. It can be seen that there is at least one high density knot in the cloud at right while most of the cloud is at a relatively low density. While the figure with the mean motion removed shows that there are gradients within clouds and velocity differences between clouds, the velocity field appears to be coherent on scales of ~ 0.5 pc.

In Figure 4 we show a 2D histogram of the density and thermal pressure for the same time and run as in Figure 2. Here we have only included points inside the hot bubble. Also plotted is the equilibrium curve on which heating and cooling balance. The blue and orange points indicate the two phases, cloud and intercloud respectively, in the initial ambient medium. The dashed magenta line is the thermal pressure in the ambient medium. It can be seen that the hot gas does concentrate around 10^6 K and a pressure of $\sim 10^4$ cm^{-3} K, as inferred for the Local Bubble. The cloud gas, which concentrates near 10^4 K has roughly the temperature of the CLIC, though its thermal pressure is apparently too high. It remains to be seen if other parameter choices will lead to a better match with the thermal pressure for the clouds. Nevertheless, it is clear that the shock heating has lead to the creation of warm clouds that persist inside the hot bubble.

6. Summary

Having reviewed the possibilities it seems clear that the only heat source capable of maintaining the nearby clouds of the CLIC at their warm temperatures is photoionization by the interstellar radiation field. However, shocks must have played a role in the past in heating the clouds up to their current temperature. We have shown via our numerical simulations that the shocks that created the hot gas of the Local Bubble are capable of heating the initially cold gas clouds up to the warm temperatures they have now. The ongoing heating via the interstellar radiation field helps to maintain the clouds at a warm temperature.

The low current level of turbulence in the CLIC, which is substantially subsonic, does not contain enough energy to allow for turbulent dissipation to be a significant source of heating at this time. At the same time, we see substantial flows within clouds, which are filamentary, and suggests that the distinct velocity components of the CLIC may not correspond to physically distinct clouds.

The long thermal equilibration time of the cloud gas, illustrated by the large spread in density and temperature in the phase diagram (Figure 4), and the ongoing motions in the hot gas make it clear that non-equilibrium effects are important. Therefore understanding the history of the LISM is necessary for assessing the roles of the various physical processes at work, such as radiative cooling, photoionization heating and thermal conduction.

Acknowledgments

I wish to thank the organizers for the invitation to present this work at the AIAC 2018 and for the stimulating conference. I also thank my collaborators Priscilla Frisch and Seth Redfield for interesting conversations. This work was supported by NASA Astrophysics Theory Program grant no. NNX17AH80G.

References

[1] Welsh B Y, Lallement R, Vergely J L and Raimond S 2010 *A&A* **510** A54

[2] Lallement R, Vergely J L, Valette B, Puspitarini L, Eyer L and Casagrande L 2014 *A&A* **561** A91

[3] McCammon D, Burrows D N, Sanders W T and Kraushaar W L 1983 *ApJ* **269** 107

[4] Snowden S L, Egger R, Finkbeiner D P, Freyberg M J and Plucinsky P P 1998 *ApJ* **493** 715–729

[5] Redfield S and Linsky J L 2008 *ApJ* **673** 283

[6] Gry C and Jenkins E B 2014 *A&A* **567** A58

[7] McKee C F and Ostriker J P 1977 *ApJ* **218** 148–169

[8] Reynolds R J 1990 Faint optical line emission from the diffuse interstellar medium - Observations and implications *The Galactic and Extragalactic Background Radiation* (*IAU Symposium* vol 139) ed Bowyer S and Leinert C pp 157–169

[9] Haffner L M, Reynolds R J, Tufte S L, Madsen G J, Jaehnig K P and Percival J W 2003 *ApJS* **149** 405–422

[10] Haffner L M, Dettmar R J, Beckman J E, Wood K, Slavin J D, Giammanco C, Madsen G J, Zurita A and Reynolds R J 2009 *Reviews of Modern Physics* **81** 969–997

[11] Reynolds R J, Hausen N R, Tufte S L and Haffner L M 1998 *ApJL* **494** L99–L102

[12] Reynolds R J and Tufte S L 1995 *ApJL* **439** L17–L20

[13] Dupuis J, Vennes S, Bowyer S, Pradhan A K and Thejll P 1995 *ApJ* **455** 574

[14] Reynolds R J and Cox D P 1992 *ApJL* **400** L33

[15] Wolfire M G, McKee C F, Hollenbach D and Tielens A G G M 2003 *ApJ* **587** 278–311

[16] Madden S C, Galliano F, Jones A P and Sauvage M 2006 *A&A* **446** 877–896

[17] Minter A H and Balser D S 1997 *ApJL* **484** L133–L136

[18] Minter A H and Spangler S R 1997 *ApJ* **485** 182–194

[19] Lerche I, Spanier F, Schlickeiser R and Hoffmann P 2007 *Journal of Physics G Nuclear Physics* **34** 2691–2706

[20] Walker M A 2016 *ApJ* **818** 23

[21] Cummings A C, Stone E C, Heikkila B C, Lal N, Webber W R, Jóhannesson G, Moskalenko I V, Orlando E and Porter T A 2016 *ApJ* **831** 18

[22] Slavin J D and Frisch P C 2008 *A&A* **491** 53 ISSN 0004-6361

[23] Welsh B Y and Shelton R L 2009 *Ap&SS* **323** 1

[24] Jenkins E B 2009 *Space Science Reviews* **143** 205–216

[25] Lisse C M, Dennerl K, Englhauser J, Harden M, Marshall F E, Mumma M J, Petre R, Pye J P, Ricketts M J, Schmitt J, Trumper J and West R G 1996 *Science* **274** 205–209

[26] Cox D P 1998 Modeling the Local Bubble *IAU Colloq. 166: The Local Bubble and Beyond* (*Lecture Notes in Physics, Berlin Springer Verlag* vol 506) ed Breitschwerdt D, Freyberg M J and Truemper J pp 121–131

[27] Lallement R 2004 *A&A* **418** 143–150

[28] Koutroumpa D, Lallement R, Kharchenko V, Dalgarno A, Pepino R, Izmodenov V and Quémerais E 2006 *A&A* **460** 289–300

[29] Galeazzi M, Chiao M, Collier M R, Cravens T, Koutroumpa D, Kuntz K D, Lallement R, Lepri S T, McCammon D, Morgan K, Porter F S, Robertson I P, Snowden S L, Thomas N E, Uprety Y, Ursino E and Walsh B M 2014 *Nature* **512** 171

[30] Liu W, Chiao M, Collier M R, Cravens T, Galeazzi M, Koutroumpa D, Kuntz K D, Lallement R, Lepri S T, McCammon D, Morgan K, Porter F S, Snowden S L, Thomas N E, Uprety Y, Ursino E and Walsh B M 2017 *ApJ* **834** 33

[31] Slavin J D 2014 IBEX, SWCX and a Consistent Model for the Local ISM *Outstanding Problems in Heliophysics: From Coronal Heating to the Edge of the Heliosphere* (*Astronomical Society of the Pacific Conference Series* vol 484) ed Hu Q and Zank G P p 204

[32] Snowden S L, Chiao M, Collier M R, Porter F S, Thomas N E, Cravens T, Robertson I P, Galeazzi M, Uprety Y, Ursino E, Koutroumpa D, Kuntz K D, Lallement R, Puspitarini L, Lepri S T, McCammon D, Morgan K and Walsh B M 2014 *ApJL* **791** L14

[33] Smith R K and Cox D P 2001 *ApJS* **134** 283–309

[34] Schulreich M M, Breitschwerdt D, Feige J and Dettbarn C 2017 *A&A* **604** A81

[35] Breitschwerdt D and de Avillez M A 2006 *A&A* **452** L1–L5

[36] Fryxell B, Olson K, Ricker P, Timmes F X, Zingale M, Lamb D Q, MacNeice P, Rosner R, Truran J W and Tufo H 2010 FLASH: Adaptive Mesh Hydrodynamics Code for Modeling Astrophysical Thermonuclear Flashes Astrophysics Source Code Library (*Preprint* 1010.082)

IOP Conf. Series: Journal of Physics: Conf. Series **1100** (2018) 012023 doi:10.1088/1742-6596/1100/1/012023

Correlation Scales of the Turbulent Cascade at 1 AU

Charles W. Smith[1], Jesse T. Coburn[2], Bernard J. Vasquez[1], Miriam A. Forman[3], and Julia E. Stawarz[4]

[1] Space Science Center, Institute for the Study of Earth, Oceans and Space, University of New Hampshire, Durham, NH 03824

[2] Dipartimento di Fisica, Università della Calabria, 87036 Rende (CS), Italia

[3] Department of Physics and Astronomy, State University of New York at Stony Brook, Stony Brook, New York, USA

[4] Department of Physics, Imperial College London, London, United Kingdom

E-mail: Charles.Smith@unh.edu

Abstract. We collect and review the conclusions of several recent papers that address the transport of energy within the inertial range of interplanetary turbulence at 1 AU. Techniques have been developed that provide a means of measuring energy transport within the spectrum that do not rely on an assumption of specific dynamics other than the adoption of the complete single-fluid, incompressible magnetohydrodynamic equations. These techniques lead to expressions that do not depend on the form of the power spectrum other than requiring scale separation such that there exists an inertial range where the dynamics are energy conserving. We compute the correlation function for the derived expressions and find that the dynamics of turbulent transport decorrelate on the scale size of interest. We also show that the rate of energy transport at any given scale and any given time or location can be either positive or negative and that the typical rate of energy transport is $\sim 10\times$ the average rate. It is the average rate that agrees well with the computed rate of thermal proton heating in the solar wind. This leads to a model of the turbulence where local transport is stronger than normally considered.

1. Introduction

There are numerous complimentary and conflicting theories that attempt to describe inertial-range dynamics governing the spectral transport of energy in the solar wind. They include magnetohydrodynamic (MHD) extensions of traditional Navier-Stokes (NS) theory [14, 38], MHD weak turbulence theory [13, 16, 12], hybrid models [10, 6], and other theories [3]. See reviews [25, 19, 4, 5]. All these theories possess the common traits that they predict, or someday will predict, (1) an asymptotic spectral anisotropy for both fluctuations and the underlying wave vector orientations, (2) a power spectral form that can be compared against observations, and (3) an average rate of energy transport that can be compared against the heating of the plasma. These average energy transport rates are most often cited as the nonlinear transport of turbulence.

Kolmogorov's prediction for the omnidirectional power spectrum of the total energy fluctuations for isotropic hydrodynamics (HD) is [14]:

$$P(k) = A_K \epsilon^{2/3} k^{-5/3} \tag{1}$$

where $A_K = 1.6$ [2, 30], ϵ is the average energy transport in the inertial range, and k is the wave number. Comparison of the above extended to an MHD formalism and early weak turbulence theory [13, 16] against the average thermal proton heating rates at 1 AU found that the Kolmogorov prediction provided the best fit to the observed heating rate once the constant A_K was modified [35].

It is possible to perform direct measurement of the energy transport at each scale in the inertial range. Third-moment theory provides this formalism [15, 23, 24] without the need to apply any particular theory for the specific and relevant nonlinear dynamics. The derived third-moment expressions require only the MHD or NS equations and a short list of statistical properties: homogeneity, incompressibility, stationarity, isotropy or another geometry, and scale separation. Each of these assumptions have been extended in more advanced derivations [15, 18, 36, 37, 31, 33, 11, 1]. The measured average energy transport rate derived from third-moment theory has been shown to provide good agreement with the average thermal plasma heating rate [18, 31, 7, 20, 11].

We are now learning that the average energy transport rate is a poor measure of the overall nonlinearity of the inertial range [8, 9, 27]. While the average rate of energy transport agrees well with the average rate of plasma heating, the inertial range is not well described as the transport of energy with a uniform and fixed rate at all scales. The inertial range is not a pipeline for energy moving steadily from large to small scales. It is, instead, a highly variable nonlinear system where each scale appears to act independently in moving energy either to large or small scales at any given time and location with the net average transport of energy in agreement with theories of the type described above.

We will review below the results of several papers [8, 9, 27]. The correlation functions and correlation scales for the nonlinear dynamics are computed. Third-moment theory will be used to show that the correlation scale for inertial range dynamics is scale dependent with the implication that dynamics at different scales cannot remain correlated. A view of MHD turbulence dynamics will develop that suggests the true measure of nonlinearity within the inertial range is an order of magnitude greater than the above average transport rates would suggest.

2. Mean Energy Transport Rates

While various theories based on specific transport dynamics yield predictions for the turbulent spectrum and the average rate of energy transport through the inertial range, a direct measure of the transport rate can be obtained using third-moment theory [15, 23, 24]. For incompressible, homogeneous, stationary, isotropic MHD turbulence we can write:

$$D_3^{\pm}(n,\lambda) = \left[\Delta Z_L^{\mp}(n,\lambda)\right] \sum_{i=1}^{3} \left[\Delta Z_i^{\pm}(n,\lambda)\right]^2 . \tag{2}$$

where $\Delta \mathbf{Z}^{\pm}(n,\lambda) \equiv \mathbf{Z}^{\pm}(n+\lambda) - \mathbf{Z}^{\pm}(n)$ and $\mathbf{Z}^{\pm} \equiv \mathbf{V} \pm \mathbf{B}/\sqrt{(4\pi\rho)}$. This yields the local energy transport rate when averaged over a suitable ensemble:

$$\begin{aligned} \left\langle D_3^{\pm}(\lambda)\right\rangle &= \frac{1}{(N-\lambda)} \sum_{n=1}^{N-\lambda} \left\{ \Delta Z_L^{\mp}(n,\lambda) \sum_{i=1}^{3} \left[\Delta Z_i^{\pm}(n,\lambda)\right]^2 \right\} \\ &= -(4/3)\epsilon^{\pm}|\mathbf{L}| \end{aligned} \tag{3}$$

where $L = V_{SW}\lambda\tau$ and τ is the time resolution of the data. Note that $\left\langle D_3^{\pm}(\lambda)\right\rangle$ scales linearly with lag L indicating uniform transport at all inertial range scales. The total energy transport rate is given by $\epsilon^T = (\epsilon^+ + \epsilon^-)/2$ from which it follows that $\left\langle D_3^T\right\rangle = \left\langle (D_3^+ + D_3^-)/2\right\rangle = -(4/3)\epsilon^T L$.

Tests of third-moment theory against the measured average rate of thermal proton heating at 1 AU using data from the ACE spacecraft have provided good agreement [31, 7]. Figure 1

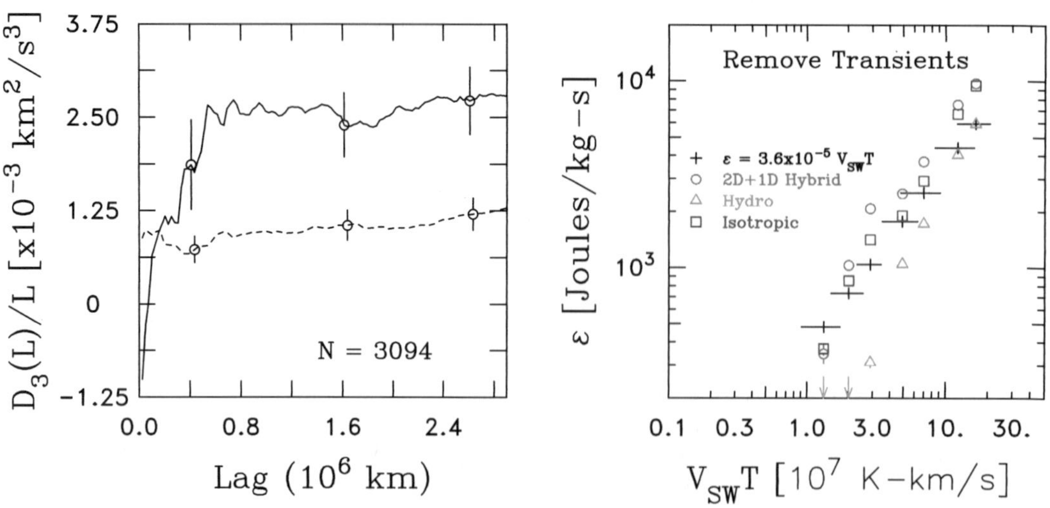

Figure 1. Taken from Figures 1 and 2 of [31]. (left) $D_3(L)/L$ for MHD and HD analysis is proportional to the energy transport rate which is seen as constant in this analysis. (right) Comparison of energy transport rate to average heating rate is favorable.

shows results from that analysis using an ensemble of 3094 12 hr samples. The sample duration was chosen because it is long compared to the published values of the correlation scale for the primitive variables, thereby allowing Gaussian statistics to be used. Note that $D_3(L)/L$ yields a constant function of lag in Figure 1 (left). Figure 1 (right) compares the average heating rate [35] to the computed energy transport rate derived from hydrodynamic theory, isotropic MHD and a hyrbid MHD model that includes both field-aligned and perpendicular wave vectors. The computed rates of energy cascade is in good agreement with the local proton heating rates as shown in Figure 1 (right) when the ensemble is divided according to the parameter $V_{SW}T$ where V_{SW} is the solar wind speed and T is the proton thermal temperature.

3. Local Energy Transport Rates

What Figure 1 omits is the underlying distribution of values that are averaged in order to obtain the means shown. We can reduce the sample size to 1 hr, which remains a reasonable estimate for the correlation scale of the primitive variables and subset the data to a limited range of fluctuation energy and cross helicity. We can then evaluate eq. 3 for each hourly sample. Both fluctuation energy and cross helicity values are limited so as to create a distribution of values from within a common ensemble of expected mean heating rates. The average proton heating rate depends on both [35, 32, 17]. Data intervals from 12 hrs upstream to 36 hrs downstream of interplanetary shocks are omitted to avoid including transient dynamics.

If we subset the ACE data to include only those samples with mean fluctuation energy in the range $300 < E < 700\,\mathrm{km^2\,s^{-2}}$ and mean normalized cross helicity $|\sigma^C| \leq 0.75$, we obtain 7075 1 hr samples. We compute D_3^T across a range of lags $2 - 20$ points, corresponding to $128 - 1280$ s, and fit the function D_3^T with a straight line for each hourly sample. Since 1 hr is comparable to the nominal correlation length of primitive variables, we adopted this sample size as a proxy for the unknown correlation scale of the underlying dynamics. This made it possible to employ Gaussian statistics in those analyses. We obtain the distribution of R^2 values for the straight line fit shown in Figure 2. A total of 4787 samples (approx. 68%) had $R^2 \geq 0.7$. The implication is that a good linear fit can be obtained when the sample size is comparable to the correlation

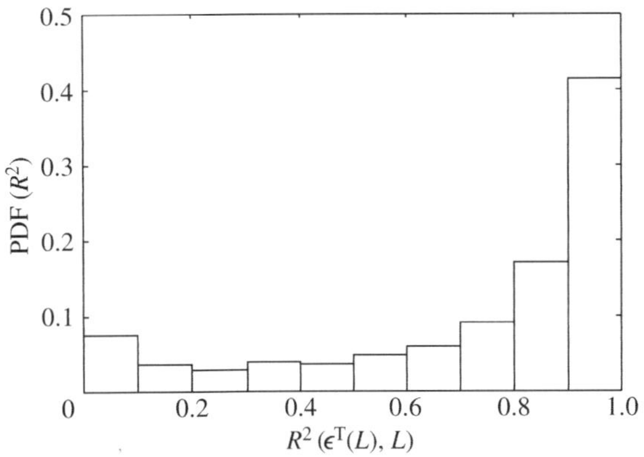

Figure 2. Taken from Figure 1 of [9]. Analysis of 7075 1 hr samples using 12 yr of ACE data showing the distribution of R^2, the square of the correlation coefficient for $D_3^T(L)$ vs. L, for lags 2 – 20 corresponding to 128 – 1280 s. The data has τ = 64 s cadence and transients are removed. The analysis is limited to samples with mean fluctuation energy $300 < E < 700\,\mathrm{km}^2\,\mathrm{s}^{-2}$ and mean normalized cross helicity $|\sigma^C| \leq 0.75$. The selection reduces possible distribution spread due to a too broadly defined sample population.

scale of the primitive variables, and in so doing we obtain the necessary averaging to produce a viable estimate for the local rate of cascade.

We examine the underling distribution of ϵ values by defining ϵ^{in} and ϵ^{out} to be either $\epsilon^{\pm}$ according to the mean magnetic field direction such that ϵ^{in} (ϵ^{out}) is the transport of energy associate with inward (outward) propagating fluctuations. We make this division because at 1 AU it is typical to find the energy associated with outward-propagating fluctuations exceeds the energy associated with the inward propagating fluctuations. We also define $\epsilon^C = (\epsilon^+ - \epsilon^-)/2$ to be the difference between the two cascade rates. The parameter ϵ^C provides a measure of the relative cascade rates of the two components. Figure 3 shows the distribution functions of ϵ^{in}, ϵ^{out}, ϵ^T and ϵ^C values derived from the same subset of ACE data under the condition that $R^2 \geq 0.7$. In so doing, we use only those intervals for which D_3^T is well-fit by a linear function of lag. Note that the standard deviation for each panel is 10× the mean value and that the mean value is positive. Not shown here, there is a general anti-correlation between the transport (cascade) of ϵ^{in} and ϵ^{out} [9].

If our interpretation of this result is correct, then several questions must be addressed:

(i) What is the energy reservoir that supports energy transport from small to large scales (negative values of ϵ^T)?

(ii) Does the inertial range act as a coordinated dynamical process that moves energy in the same sense at all scales at a given time?

(iii) What is the minimum sample size that will provide statistically independent samples of ϵ at a prescribed scale?

The mean rate of energy transport through the inertial range must be positive, indicating transport from large to small scales, in order to accomplish a net, average heating of the background plasma. However, average negative cascade rates have been reported for high cross-helicity (imbalance), shear, and expanding flows [26, 32, 33, 28]. We believe we now understand this as the neglect of an error term that arises when the gradient in the mean flow that provides a significant contribution to the third-moment analysis in some circumstances. Efforts to demonstrate this are ongoing at this time. It appears at this time that this explains away the need for an energy reservoir at small scales so long as negative energy cascade rates are short-lived. Locally negative energy transfer rates have been seen in hybrid Vlasov-Maxwell simulations and shell-shell energy transfer Hall-MHD simulations [21, 29]. The rest of this paper and [27] address questions 2 and 3 above.

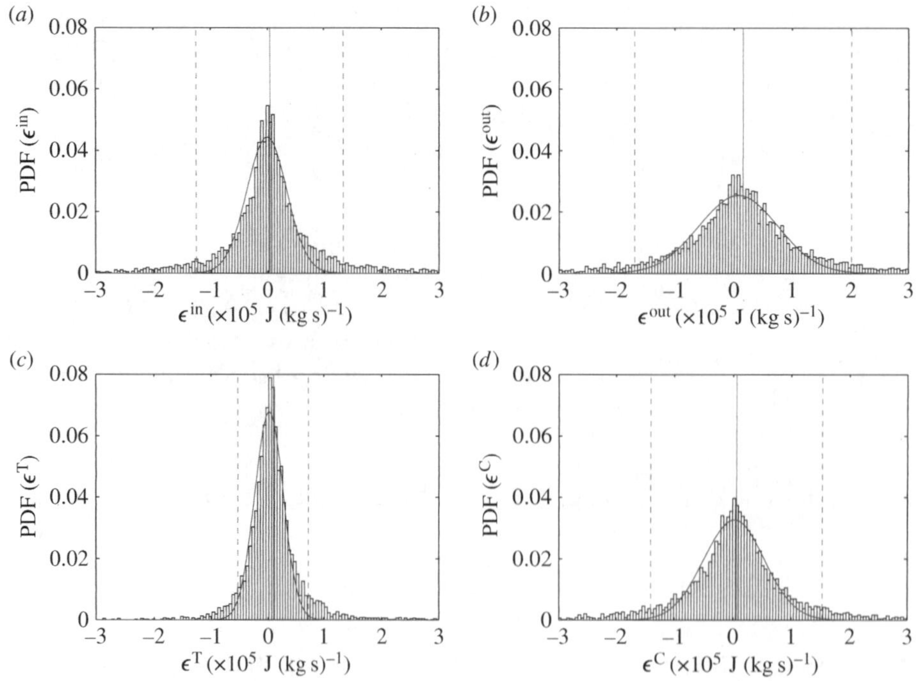

Figure 3. Taken from Figure 2 of [9]. PDF for ϵ^{in}, ϵ^{out}, ϵ^{T} and ϵ^{C} computed from 4787 1 hr samples that had $R^2 \geq 0.7$. Solid red line marks the mean. Dashed lines are the standard deviation. In each case, the standard deviation is $\sim 10\times$ the mean value with almost half of the results being negative.

4. Correlation Function of D_3

Single-spacecraft measurements in the solar wind are unable to provide detailed measurements of the time scale for solar wind dynamics such as the duration and variability of the local energy cascade rate in the plasma frame. What they can provide is an assessment of spatial scales. Using single-spacecraft measurements and the Taylor frozen-in-flow assumption [34], we can show that specific values of the energy transport are confined to limited regions of space and that the energy transport at different scales is unlikely to be correlated. This creates a model for the turbulence where each scale in the inertial range acts independently with large variations in energy transport. We obtain this result by measuring the correlation function for the energy transport terms D_3.

The autocorrelation function for the fluctuation variable δF as a function of lag δn can be written

$$A(\delta F, \delta n) = \frac{1}{(N - \delta n)} \sum_{n=1}^{N-\delta n} \delta F(n) \delta F(n + \delta n) \tag{4}$$

where the sum is over points in the data set of length N. This creates a correlation function A for a variable δF that is a function of the separation variable δn.

We can apply this formalism to the expressions for $D_3^{\pm}$ and D_3^{T} to compute the correlation function at lag λ as a function of separation δn:

$$A\left(D_3^{\pm,T}(\lambda), \delta n\right) \equiv \langle D_3^{\pm,T}(n,\lambda) D_3^{\pm,T}(n+\delta n, \lambda)\rangle_t$$

IOP Conf. Series: Journal of Physics: Conf. Series **1100** (2018) 012023 doi:10.1088/1742-6596/1100/1/012023

$$= \frac{1}{(N - \lambda - \delta n)} \sum_{n=1}^{N-\lambda-\delta n} D_3^{\pm,T}(n, \lambda) D_3^{\pm,T}(n + \delta n, \lambda). \tag{5}$$

The anticipated ensemble average becomes $\langle ... \rangle_t$, which denotes an average over time, which becomes $\sum_{n=1}^{N-\lambda-\delta n}$ when applied to the data.

There are two techniques we have used to define the correlation scale for $A(\delta n)$. The first is based on the normalized integral of the correlation function:

$$\lambda_c^i \equiv \left\{ \int_0^\infty A(\delta n) d(\delta n) \right\} / A(0) \tag{6}$$

and the second is λ_c^e defined to be that scale at which the correlation function falls to 1/e of $A(0)$. Both give comparable values in this analysis [27].

Figure 4 shows our analysis of ACE magnetic field data for Carrington solar rotation 2303 (DOYs 101-127 of 2002). The correlation function of $D_3^T(\delta n)$ is shown for four values of lag $\lambda = 5$, 10, 25, and 100 pts ($\lambda\tau = 320$, 640, 1600, and 6400 s). The correlation scale as determined from the exponential definition is consistently $\sim$ 20% of the lag value. The definition of the correlation scale that is derived from integration of the correlation function yields comparable results. If both large and small scales within the inertial range each decorrelate differently, it is difficult to see how the associated dynamics could remain correlated across those scales.

5. Discussion and Summary

We have examined the functions that yield the rate of energy transport through the inertial range for incompressible MHD when averaged over a suitably defined ensemble. That average is critical to the derivation and we must stress that in the formal sense we have not computed the correlation scale of the nonlinear dynamics. We have only computed the correlation scale

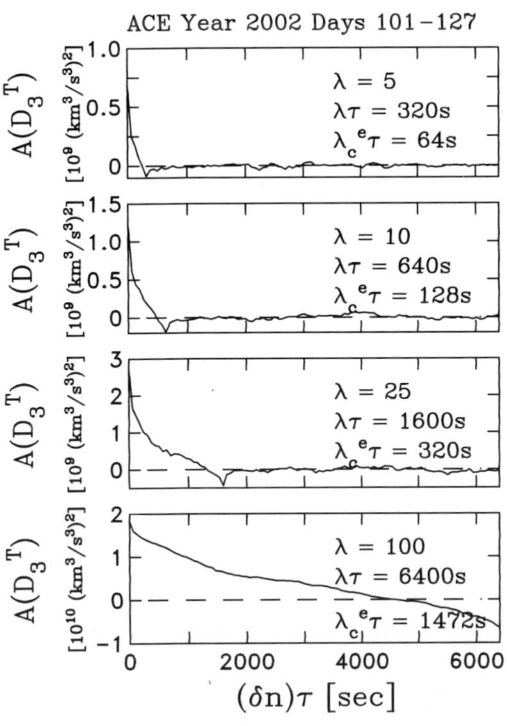

Figure 4. Taken from Figure 7 of [27]. Correlation function for D_3^T computed for four different values of lag $\lambda = 5$, 10, 25, and 100 data points ($\lambda\tau = 320$, 640, 1600, and 6400 s). The correlation scales as determined by the exponential formalism are 64, 128, 320, and 1472 s, respectively. The correlation scale is consistently $\sim$ 20% of the lag. This implies that the transport dynamics decorrelate over relatively short distances and that the dynamics of two different scales within the inertial range are generally uncorrelated. However, this is only an implication as computation of the nonlinear dynamics as determined by third-moment theory requires an ensemble average of D_3^T to represent energy transport. This is not possible in the present analysis. $D_3^{\pm}$ possess comparable correlation functions and scales.

of the function that, when suitably averaged, yields the energy transport rate of the nonlinear dynamics. Still, it is an indication of the potential correlation scale for the nonlinear dynamics and it indicates that the correlation scale is $\sim 20\%$ of the lag value. This means that the inertial range dynamics decorrelate over relatively short distances. When the nonlinear dynamics at two different scales decorrelate differently it is difficult to impossible to argue they may remain correlated across those two scales. The view of turbulence that derives from this result is that the inertial range nonlinear dynamics are composed of seemingly random variations in sign and magnitude that transport energy at $\sim 10\times$ the average rate of cascade. That average rate, when a suitably large ensemble of samples is employed to produce convergence, is in good agreement with the averge rate of thermal proton heating in the solar wind. However, heating is not the result of a systematic dynamic that moves energy from large to small scales throughout the inertial range. Rather, it is the result of many random dynamics moving energy at a much faster rate that, only on average, moves energy at a rate that is consistent with plasma heating. The degree of uncorrelated transport within the inertial range is much greater than traditional theory would suggest.

Last, we should note that the application of eq. 3 to the data interval used here yields a negative value of ϵ^+ and ϵ^{out}. As this stands in contradiction to the published results using larger ensembles [31, 7], we conclude that the sample is insufficient to obtain convergence [22]. This is not a concern to this analysis as our only interest here is to obtain insights into the underlying distribution of values and the scale on which those values decorrelate. It appears from our analyzes [8, 9, 27] that the third-moment expression converges to a linear scaling with lag with much less data than is required for convergence to the mean of the ensemble.

Acknowledgments

The authors thank the ACE/SWEPAM team for providing the thermal proton data used in this study. C.W.S. is supported by Caltech subcontract 44A-1062037 to the University of New Hampshire in support of the ACE/MAG instrument. C.W.S., B.J.V., and M.A.F. are supported by NSF/SHINE grant 1622413 to the University of New Hampshire. J.T.C. is a graduate student at Università della Calabria. Data used here can be obtained at the ACE Science Center and National Space Science Data Center. The shocks identified in this study are described at http://www.ssg.sr.unh.edu/mag/ace/ACElists/obs_list.html and are available online through the ACE Science Center.

References

[1] Andrés N, Galtier S, & Sahraoui F 2018 Phys. Rev. E97(1) 013204
[2] Batchelor, G K 1953 The Theory of Homogeneous Turbulence, Cambridge Univ. Press, New York
[3] Boldyrev S Astrophys. J.626 L37 2005
[4] Bruno R, & Carbone V 2013 Living Rev. Sol. Phys. 10 2
[5] Chen C H K 2016 J. Plasma Phys.82 535820602
[6] Cho J, & Vishniac E T 2000 Astrophys. J.539 273
[7] Coburn J T, Smith C W, Vasquez B J, Stawarz J E, & Forman M A 2012 Astrophys. J.754 93
[8] Coburn J T, Smith C W, Vasquez B J, Forman M A, & Stawarz J E 2014 Astrophys. J.786 52
[9] Coburn J T, Forman M A, Smith C W, Vasquez B J, & Stawarz J E 2015 Phil. Trans. Royal Soc. A 373 20140150
[10] Goldreich P, & Sridhar S 1995 Astrophys. J.438 763
[11] Hadid L Z, Sahraoui F, & Galtier S 2017 Astrophys. J.838 9
[12] Howes G G, Cowley S C, Dorland W, Hammett G W, Quataert E, & Schekochihin A A 2008 J. Geophys. Res.113 A05103
[13] Iroshnikov P S 1964 Sov. Astron. 7 566
[14] Kolmogorov A N 1941a Dokl. Akad. Nauk SSSR 30 301-305 (Reprinted in Proc. R. Soc. London A 434 9-13 1991)
[15] Kolmogorov A N 1941b Dokl. Akad. Nauk SSSR 32 16 (Reprinted in Proc. R. Soc. London A 434 15 1991)
[16] Kraichnan R H 1965 Phys. Fluids8 1385

[17] Lamarche L J, Vasquez B J, & Smith C W 2014 J. Geophys. Res.119 3267
[18] MacBride B T, Smith C W, & Forman M A 2008 Astrophys. J.679 1644
[19] Matthaeus W H, & Velli M 2011 Space Sci. Rev.160 145
[20] Marino R, Sorisso-Valvo L, D'Amicis R, Carbone V, Bruno R, & Veltri P 2012 Astrophys. J.750 41
[21] Mininni P D, Alexakis A, & Pouquet A 2007 J. Plasma Phys.73 377
[22] Podesta J J, Forman M A, Smith C W, Elton D C, & Malécot Y 2009 Nonlin. Proc. Geophys. 16 99
[23] Politano H, & Pouquet A 1998a Phys. Rev. E 57(1) R21
[24] Politano H, & Pouquet A 1998b Geophys. Res. Lett. 25 273
[25] Smith C W 2009 Chapter 7 in Heliophysics I. Plasma Physics of the Local Cosmos, edited by Schrijver C J and Siscoe G, Cambridge University Press
[26] Smith C W, Stawarz J E, Vasquez B J, Forman M A, & MacBride B T 2009 Phys. Rev. Lett.103 201101
[27] Smith C W, Vasquez B J, Coburn J T, Forman M A, & Stawarz J E 2018 Astrophys. J.in press
[28] Sorriso-Valvo L, Carbone F, Perri S, Greco A, Marino R, & Bruno R 2018a Solar Phys.293 10
[29] Sorriso-Valvo L, Perrone D, Pezzi O, Valentini F, Servidio S, Zouganelis I, & Veltri P 2018 arXiv:1803.01614
[30] Sreenivasan K R 1995 Phys. Fluids7, 2778
[31] Stawarz J E, Smith C W, Vasquez B J, Forman M A, & MacBride B T 2009, Astrophys. J.713 920
[32] Stawarz J E, Smith C W, Vasquez B J, Forman M A, & MacBride B T 2010 Astrophys. J.697 1119
[33] Stawarz J E, Vasquez B J, Smith C W, Forman M A, & Klewicki J 2011 Astrophys. J.736 44
[34] Taylor G I 1938 Proc. R. Soc. London Ser. A 164 476
[35] Vasquez B J, Smith C W, Hamilton K, MacBride B T, & Leamon R J 2007 J. Geophys. Res.112 A07101
[36] Wan M, Sevidio S, Oughton S, & Matthaeus W H 2009 Phys. Plasmas 16 090703
[37] Wan M, Sevidio S, Oughton S, & Matthaeus W H 2010 Phys. Plasmas 17 052307
[38] Zhou Y, & Matthaeus W H 1990 J. Geophys. Res.95, 14,881

IOP Conf. Series: Journal of Physics: Conf. Series **1100** (2018) 012024 doi:10.1088/1742-6596/1100/1/012024

Coronal Jets, and the Jet-CME Connection

Alphonse C. Sterling
NASA/Marshall Space Flight Center, Huntsville, AL, 35806, USA
E-mail: alphonse.sterling@nasa.gov

Abstract.
Solar coronal jets have been observed in detail since the early 1990s. While it is clear that these jets are magnetically driven, the details of the driving process has recently been updated. Previously it was suspected that the jets were a consequence of magnetic flux emergence interacting with ambient coronal field. New evidence however indicates that often the direct driver of the jets is erupting field, often carrying cool material (a "minifilament"), that undergoes interchange magnetic reconnection with preexisting field ([1]). More recent work indicates that the trigger for eruption of the minifilament is frequently cancelation of photospheric magnetic fields at the base of the minifilament. These erupting minifilaments are analogous to the better-known larger-scale filament eruptions that produce solar flares and, frequently, coronal mass ejections (CMEs). A subset of coronal jets drive narrow "white-light jets," which are very narrow CME-like features, and apparently a few jets can drive wider, although relatively weak, "streamer-puff" CMEs. Here we summarize these recent findings.

1. Introduction

Coronal jets are transient, long and relatively narrow features seen in EUV and X-ray images that shoot out from the solar surface and into the corona. While there were a smattering of earlier observations [2], jets were first studied in great detail with the Soft X-ray Telescope (SXT) on the *Yohkoh* spacecraft, launched in 1991. Because it was relatively sensitive to comparatively hot coronal emissions, the jets that appeared in the vicinity of active regions tended to be easiest to detect [3, 4]. SXT observations found jets to reach lengths of 1.5×10^4 km [4], and temperatures to range over 3 MK—8 MK with an average of 5.6 MK ([5])

Launched in 2006, the *Hinode* satellite carries the X-Ray Telescope (XRT), which is sensitive to comparatively cooler coronal plasmas than was SXT; in particular, it is sensitive to emissions of 1 MK—2 MK, a characteristic temperature range for many coronal hole jets ([6, 7]). XRT found jets to be plentiful in polar coronal holes [8]; according to [9], jets occur at a rate of about 60 day^{-1} in the two polar coronal holes, have lifetimes of $\sim$10 min, lengths of 50,000 km, and widths of 8,000 km.

A jet consists of a spire, which extends into the corona, and a base region that is generally much brighter than the spire. Moreover, the brightest part of the base is often positioned asymmetrically with the spire, being located off to one side of the base region. These jet-base brightenings (hereafter, JBP, following [1]) have long been suspected as holding a clue to the cause of jets ([3]; also see [10]).

IOP Conf. Series: Journal of Physics: Conf. Series **1100** (2018) 012024 doi:10.1088/1742-6596/1100/1/012024

2. The Cause of Coronal Jets

It was quickly realized that the energy source for jets was likely magnetic. Magnetic field that permeates the solar atmosphere is fed by new magnetic flux that almost continuously "bubbles up" (i.e. emerges) through the photospheric surface. Upon reaching the surface, this emerging field spreads horizontally on the surface and expands outward into the higher atmosphere. An early suggestion was that jets resulted when flux emerging in this manner into the chromosphere and low corona runs into nearby pre-existing magnetic field in those regions. Interactions between the two flux systems would result in magnetic reconnection, at the interface between the emerging bipole and the pre-existing (e.g. open) field. According to this emerging-flux model, this reconnection would result in a small closed magnetic loop field, and also new open magnetic field, where the new closed loop would form the JBP and the jet spire plasma would flow out along newly open field.

New studies became possible with the launch of the Solar Dynamics Observatory (*SDO*) satellite in 2010. Its Atmospheric Imaging Assembly (AIA) instrument images the Sun in multiple full-disk EUV and UV wavelengths ([11]) at 12 s cadence (in the EUV bands) and with $0.''6$ pixels, allowing for a much broader range of coverage of jet emissions than was possible with the X-ray instruments alone (both SXT and XRT had somewhat lower resolution than AIA, and variable cadences and fields of view). Also critical for jet studies is *SDO*'s Helioseismic and Magnetic Imager (HMI), which takes full-disk line-of-sight magnetograms at a cadence of 45 s, among other capabilities ([12]). Using these instruments and other resources, several jets were observed on the solar disk. They frequently showed the jets to emanate from locations where flux *cancelation* was occurring, without emergence (e.g., [13, 14, 15, 16, 17]). In some cases emergence was seen at the base locations of jets, but usually in conjunction with flux cancelation (e.g., [18, 19]). Some pre-*SDO* observations also showed emergence occurring together with cancelation at jet bases ([20]). Therefore it seemed that flux emergence was not essential for jet formation.

Using AIA EUV images, [17] found that a jet they observed appeared to form as the result of an eruption of a small-scale filament, which they called a minifilament, and that the JBP was a miniature flare that occurred in conjunction with eruption of the minifilament. In an extensive study of 20 randomly selected near-limb jets, [1] verified that the minifilament picture holds generally. They concluded that at least many (if not all) jets are a miniature version of the familiar full-scale eruptions that include an erupting filament, a solar flare, and often a coronal mass ejection (CME). Although their data were too near the limb for reliable comparisons with magnetograms, they argued that, on balance, the existing studies suggested that flux cancelation was often crucial to jet occurrence.

To investigate this further, [21] and [22] studied substantial numbers ($\sim$10) of on-disk jets, respectively in quiet Sun and coronal holes. For all of those examined jets, they found that cancelation occurred at the base of the jets. In some of those cases, flux emergence was also occurring, but in those cases the jet occurred at the neutral line where one pole of the emerging flux was canceling with pre-existing opposite-polarity field; this was also found by, e.g., [23] and [14]. Thus this strongly supports that magnetic flux cancelation is the trigger for onset of at least many jets. Moreover, for the jets of the quiet region study ([21]), the study of [24] found that the minifilaments that erupted to cause the jets also formed via flux cancelation, between about two hours and two days prior to the jet occurrence ([24]).

In the case of active regions ([25, 26]), the situation is not as clear cut. One reason is that minifilaments are sometimes but not always visible in the lead up to the jets. And secondly, the evolutionary magnetic changes are very rapid in the jetting locations, often with both flux cancelations and flux emergence occurring either concurrently or within a short time of each other. Nonetheless, all but one of several active region jets studied in [25] and [26] clearly occurred at the site of flux cancelation; one event of [25] was unclear in this regard. [26]

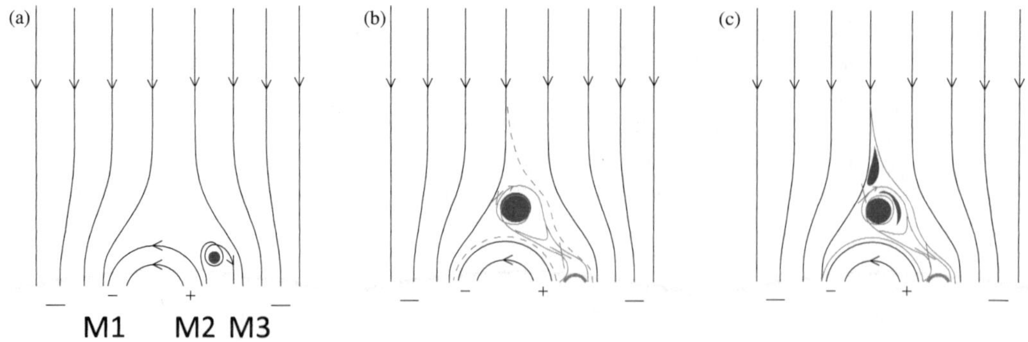

Figure 1. Schematic showing the *minifilament eruption model for coronal jets*; originally presented in [1], this version is from [30]. (a) A minifilament (blue disk) is in a magnetic bipole M2-M3, adjacent to a larger bipole M1-M2 (this is a 2D cut of a 3D negative-polarity background region with an intruding positive-polarity source at M2). The looped field line encircling the minifilament indicates non-potential twist in the field holding the minifilament. (b) The minifilament erupts, running into and undergoing interchange reconnection with opposite-polarity open (or far-reaching) field, leading to new field lines (dotted) including new open field that guides a hot jet (shaded region), and with field below the minifilament reconnecting "internal reconnection" [1]) and forming the JBP (thick red semicircle). (c) As the minifilament eruption progresses, eventually the interchange reconnection erodes the minifilament field, releasing the cool minifilament material to travel outward along the jet spire. See [1, 25, 26, 27] for more details. This picture has been modeled numerically by [31].

considers possibilities for why minifilaments are not always seen in active region events; these reasons include that pre-eruption minifilaments might be obscured by surrounding material, or in some cases a flux rope might erupt without a cool-material minifilament, as is the case with some large-scale eruptions.

Figure 1 summarizes several basic aspects of the view put forth for coronal jets in recent studies by the author and his colleagues ([17, 1, 25, 26, 21, 24, 27]).

(This team's earlier work pointed out two "flavors" of jets: "standard" and "blowout" ([28, 29]). While the observational aspects of those jets are described appropriately in those two papers, that was prior to our findings on the connection between minifilament eruptions and jets. Therefore we have revised our view on the cause of standard and blowout jets; the updated description is presented in [1] and [27].)

3. Jets and CMEs

Given the evidence that coronal jets are miniature versions of CME-producing large-scale eruptions, one might expect that jets could also result in some version of CMEs. Indeed ejections visible in visible-light coronagraph images that originate from coronal jets have been detected. These coronagraph-imaged features seem to be of two basic types: narrow events and (somewhat) broader events, in reference to the heliocentric angle subtended by the ejected feature.

Narrow coronagraph-imaged ejections have been called "narrow CMEs" by some workers (e.g. [32, 33, 25]), while other workers refer to them as "white-light jets" ([34, 35]). Depending on the study, the observed events were of width $\lesssim 10^\circ$ or $\lesssim 15^\circ$, and so in either case of much smaller angular size than that of the average (non-halo) CME of $\sim 44^\circ$ ([36]). Other observations showing narrow CMEs and/or a jet-CME connection include [37, 38, 39, 13] and [18].

In addition to white-light-jet CMEs, somewhat broader CMEs can also result from jets

([38, 40, 41]). In these cases, the jets occur in the base region of the location that eventually erupts outward as the CME. In the cited studies, the base regions of the CMEs are bipolar active regions, and the jets occur along one edge of the active regions. These active regions form the bases of streamer structures that extend far into the corona, and the ejected CMEs travel outward along the streamers. These CMEs tend to be weak in intensity compared to the coronagraph-bright streamer, and the CMEs do not totally disrupt the streamer; for these reasons we refer to these events as "streamer-puff" CMEs ([38]). For the six streamer-puff events analyzed in [40], widths ranged over 25°—50°. Thus they are somewhat broad; broader than the above-mentioned narrow CMEs ($\lesssim$15°), but only reaching about the above-mentioned average width for general CMEs ($\sim$44°).

There is evidence that both of these types of CMEs are driven by the jets, but the driving mechanism differs between the two. Moreover, a key component to the driving appears to be non-potential magnetic twist contained in minifilaments, where the twist either is added to the minifilament as it is in the process of erupting to form the jet, or the twist exists on the minifilament prior to its eruption (in which case more twist might be added during eruption); Figure 1 shows schematically this twist on the minifilament.

Regarding the mechanism in narrow CMEs, [42] found that the jets that formed such narrow CMEs all had a propensity to contain a relatively large amount of twist, compared to the average jet that did not make a (narrow) CME. The picture envisioned by [42] for these narrow CMEs is basically that of Figure 1, but where the reconnection in Figure 1(b) transfers a relatively large amount of twist onto the open field ([43]). That twist that has been added to the open field then helps propel the jet material far enough away from the Sun so that it can show enhanced density in *SDO*/LASCO C2 coronagraph images.

Regarding the mechanism for the streamer-puff CMEs, [38] provided an initial suggestion. That study however was prior to our new understanding about jets as expressed in the minifilament eruption model (Fig. 1). [40] now provides a revised view that supersedes the explanation given in [38]. According to [40], streamer-puffs result as shown in Figure 2; basically, the setup is as in Figure 1, but where the apparently vertical field is actually a part of a larger-scale loop, and that larger-scale loop forms the base region of a coronal streamer. Upon eruption, twist contained in the erupting minifilament is transferred to the larger-scale loop. That injected twist drives the larger-scale loop outward along the streamer, forming the streamer-puff CME.

4. In Closing

Much work has been done to advance our understanding of coronal jets over the past two decades since their (re)discovery with the *Yohkoh* satellite. Paper [2] reviews much of this progress. But that review was largely written during 2015, just as an understanding of a connection between minifilament eruptions and jets ([1]) was coming to the fore. Thus, in the author's view, [2] under-emphasizes the importance of minifilament eruptions in jet production. Of course more studies are required to establish with certainty (or to challenge) this idea. Similarly, it now appears that the importance of flux cancelation in triggering jets [15, 16, 21, 22, 25, 26]) has until recently been under appreciated.

There are still however some suggestions that jets can occur when there is flux emergence in the absence of cancelation, as argued by [44] for 6 of 20 events they studied (the remaining 14 events all clearly included cancelation). While we recognize that the Sun may choose to make jets in this fashion sometimes, we also caution that for jets near active regions, it can be difficult to isolate emergence from cancelation ([25, 26]). Because a large majority of jets are clearly triggered by cancelation in quiet Sun and coronal holes, where cancelation and emergence can be more easily differentiated ([21, 22]), one might expect cancelation to trigger almost all jets in active regions too. This, however, must be resolved by careful study of more jet events.

Finally, it is now clear that some jets can reach the outer corona and make ejections in the

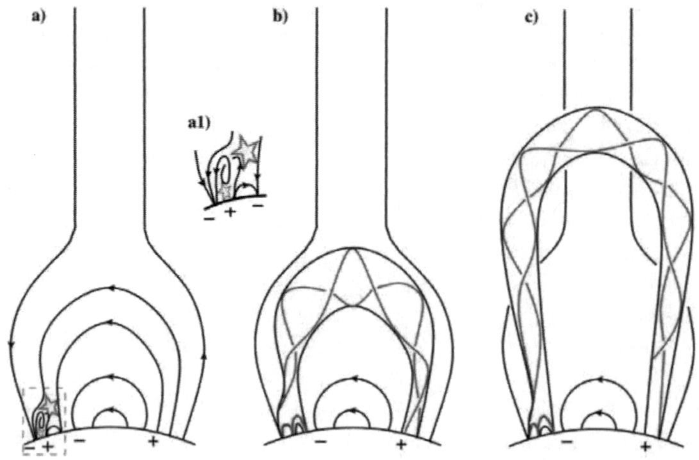

Figure 2. Jet eruption driving a *streamer-puff* CME, from [40]. (a) A minifilament-field (with or without a cool minifilament) erupts as in Figs. 1(a) and 1(b) (see details in insert a1). In this case the minifilament (with a twisted-field magnetic envelope) erupts inside a larger-loop field (in Fig. 1(a), the corresponding field was either open or far-reaching; here is is "far reaching," extending to the "+" on to the right of the erupting minifilament system). This larger loop is part of a loop system that forms the base of a surrounding and overlying streamer field. (b) Twist from the erupting-minifilament field is transferred to the larger loop. (c) This renders the larger loop unstable, causing it to erupt along the streamer to form the streamer-puff CME.

form of narrow CMEs or steamer-puff CMEs. There is even the thrilling possibility that some of this material might be observed *in situ* by the Parker Solar Probe, which in turn could tell us more about the nature of jets.

Acknowledgments
The author thanks R. L. Moore, N. K. Panesar, and D. A. Falconer for their contributions to the discussed content. This work was supported by funding from the NASA Heliophysics Guest Investigators program, and the MSFC Hinode project.

References

[1] Sterling A C, Moore R L, Falconer D A and Adams M 2015 *Nature* **523** 437–440
[2] Raouafi N E, Patsourakos S, Pariat E, Young P R, Sterliing A C and et al 2016 *Space Sci. Rev.* **201** 1–53
[3] Shibata K, Ishido Y, Acton L W and et al 1992 *PASJ* **44** L173–L179
[4] Shimojo M, Hashimoto S, Shibata K, T H, Hudson H S and Acton L W 1996 *PASJ* **48** 123–136
[5] Shimojo M and Shibata K 2000 *ApJ* **542** 1100–1108
[6] Pucci S, Poletto G, Sterling A C and Romoli M 2013 *ApJ* **776** 16
[7] Parachiv A R, Bemporad A and Sterling A C 2015 *A&A* **579** 96
[8] Cirtain J W, Golub L, Lundquist L and et al 2007 *Science* **318** 1580
[9] Savcheva A, Cirtain J, Deluca E E and et al 2007 *PASJ* **59** S771–S778
[10] Heyvaerts J, Priest E R and Rust D M 1977 *ApJ* **216** 123
[11] Lemen J R, Title A M, Akin D and et al 2012 *Solar Phys.* **275** 17
[12] Scherrer P H and et al 2012 *Solar Phys.* **275** 207
[13] Hong J, Jiang Y, Zheng R, Yang J, Bi Y and Yang B 2011 *ApJL* **738** 20
[14] Huang Z, Madjarska M S, Doyle J G and Lamb D A 2012 *A&A* **548** A62
[15] Young P R and Muglach K 2014a *Solar Phys.* **289** 3313–3329
[16] Young P R and Muglach K 2014b *PASJ* **66** S129
[17] Adams M, Sterling A C, Moore R L and Gary G A 2014 *ApJ* **783** 11
[18] Shen Y, Liu Y, Su J and Deng Y 2012 *ApJ* **745** 164

[19] Li X, Yang S, Chen H and Zhng J 2015 *ApJ* **814** 13L
[20] Shimojo M, Shibata K, T H and Harvey K L 1998 *Solar Phys.* **178** 379–392
[21] Panesar N K, Sterling A C, Moore R L and Chakrapani P 2016b *ApJL* **832** L7
[22] Panesar N K, Sterling A C and Moore R L 2018 *ApJ* **853** 189
[23] Liu C, Deng N, Liu R, Ugarte-Urra I, Wang S and Wang H 2011b *ApJL* **735** L18
[24] Panesar N K, Sterling A C and Moore R L 2017 *ApJ* **844** 131
[25] Sterling A C, Moore R L, Falconer D A, Panesar N K and et al 2016 *ApJ* **821** 100
[26] Sterling A C, Moore R L, Falconer D A, Panesar N K and Martinez F 2017 *ApJ* **844** 28
[27] Moore R L, Sterling A C and Panesar N K 2018 *ApJ* **in press**
[28] Moore R L, Cirtain J W, Sterling A C and Falconer D A 2010 *ApJ* **720** 757–770
[29] Moore R L, Sterling A C, Falconer D A and Robe D 2013 *ApJ* **769** 134
[30] Sterling A C, Moore R L and Panesar H K 2018 *ApJ* **submitted**
[31] Wyper P F, Antocohos S K and DeVore C R 2017 *Nature* **544** 452
[32] Gilbert H R, Serex E C, Holzer T E, MacQueen R M and McIntosh P S 2001 *ApJ* **550** 1093
[33] Dobrzycka D, Raymond J C, Biesecker D A and et al 2003 *ApJ* **588** 586
[34] Wang Y M, Sheeley N R J, Socker D G and et al 1998 *ApJ* **508** 899–907
[35] Wang Y M and Sheeley N R J 2002 *ApJ* **575** 542
[36] Gopalswamy N and et al 2009 *Earth, Moon, & Planets* **104 (1-4)** 295
[37] S Yashiro N, Gopalswamy, Michalek G and Howard R A 2003 *AdSpR* **32**
[38] Bemporad A, Sterling A C, Moore R L and Poletto G 2005 *ApJL* **635** 189
[39] Nisticò G, Bothmer V, Patsourakos S and Zimbardo G 2009 *Solar Phys.* **259** 87–108
[40] Panesar N K, Sterling A C and Moore R L 2016a *ApJL* **822** L23
[41] Alzate N and Morgan H 2016 *ApJ* **823** 129
[42] Moore R L, Sterling A C and Falconer D A 2015 *ApJ* **806** 11
[43] Shibata K and Uchida Y 1986 *Solar Phys.* **103** 299–310
[44] Mulay S M, Tripathi D, Zanna G D and Mason H 2016 *A&A* **589** A79

IOP Conf. Series: Journal of Physics: Conf. Series **1100** (2018) 012025 doi:10.1088/1742-6596/1100/1/012025

Numerical Modeling of Electron Transport in Solar Wind: Effects of Whistler Turbulence and Coulomb Collisions

Bofeng Tang[1], **Gary P. Zank**[1,2], **Vladimir Kolobov**[3]

[1] Department of Space Science, University of Alabama in Huntsville, Huntsville, AL 35899, USA

[2] Center for Space Plasma and Aeronomic Research (CSPAR), University of Alabama in Huntsville, Huntsville, AL 35805, USA

[3] CFD Research Corporation, Huntsville, AL 35805, USA

E-mail: bt0008@uah.edu, garyp.zank@gmail.com, vik0001@uah.edu

Abstract. The electron distribution function (eVDF) in the solar wind deviates significantly from an equilibrium Maxwellian distribution, and is comprised of a Maxwellian core, a suprathermal halo, a field-aligned component strahl, and a higher energy superhalo. Charged particle Coulomb collisions are ineffective in relaxing such a velocity distribution beyond a few solar radii. Therefore wave-particle interactions need to be considered. A wave-particle interaction term was introduced into the kinetic equation that describes the interaction of electrons with whistler waves, as well as particle collision terms. The kinetic equation has the form of an advection-diffusion-like equation in which the advection and diffusion coefficients describe the scattering and drag of electrons in whistler turbulence. A reliable numerical method has been developed to solve a full form of the advection-diffusion-like kinetic equation. Preliminary applications of the numerical method to the solar wind electron problem are presented. Comparison and analysis of the electron VDFs in the presence of Coulomb collisons and resonant wave-partcicle interactions are made.

1. Introduction

The sun constantly emits a flux of electrically charged particles into space, mostly protons and electrons, known as the solar wind. This plasma affects the entire solar system, including the Earth's magnetic field and is therefore crucial to our understanding of space weather. In situ observations show that the solar wind electron distribution function (eVDF) deviates significantly from a Maxwellian distribution, which is the thermal equilibrium form. The solar wind electron VDF is comprised of a Maxwellian core, halo, and a field-aligned component called the strahl which is typically associated with the high-speed solar wind. The Maxwellian core is characterized by energies of $<\sim$ 10 eV, the halo and strahl falls in the energy range of about 10^2 - 10^3 eV, and the super-halo in the range 10^3 - 10^5 eV. At 1 AU the Maxwellian component comprises about 95% of the total electron number density, the halo approximately 4%, and the strahl the remaining 1%.

The strahl is an anti-sunward magnetic field-aligned beam, although "counterstreaming" strahls, which are directed towards the sun, are also observed [1]. Observations show the strahl has a finite width. The strahl vanishes at electron energies above approximately 1 keV. At higher energies, above 2 keV, the superhalo appears to be isotropic. The relative number density of strahl electrons decreases with increasing heliocentric distance, whereas the relative number density of halo electrons increases [2,3]. Although strahl electrons comprise a part of the electron number density, they are responsible for most of the heat flux transported away from the Sun due to their high energy [2]. Štverák et al. [2] combined measurements obtained from three spacecraft (*Helios*, *Cluster II* and *Ulysses*) for low ecliptic latitudes covering a heliocentric distance from 0.3 to 4 AU. They confirm that the halo and the strahl relative densities vary in an opposite sense (Figure 1).

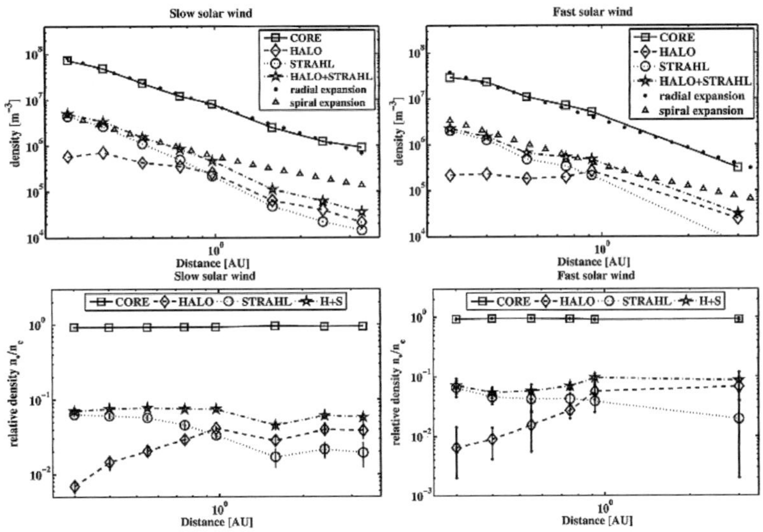

Figure 1. Radial evolution of the number densities profiles in the slow wind (top left) and fast wind (top right). Radial evolution of the relative number densities (i.e. the ratio of the desity of individual eVDF components to the total electron density) of the eVDF components for the slow (bottom left) and fast (bottom right) solar wind. [2]

However the connection between the strahl and the halo electron population remains poorly understood. Both strahl and halo electrons fall in a similar energy regime and it is frequently speculated that the halo corresponds simply to strahl electrons that have been scattered gradually in pitch angle towards an isotropic distribution. There are two primary scattering mechanisms for electrons in the solar wind. Charged particle collisions are effective in the low corona, leading to the formation of an equilibrium Maxwellian distribution thus. Salem et al. [4] show that the electron temperature anisotropy actually depends mainly on Coulomb collisions. They also show that the role of Coulomb collisions in regulating the electron heat flux is not as negligible as has been suggested by other authors. For a large number of collisions, the more the distribution function is isotropized, smaller the heat flux. Horaties et al. [5, 6] proposed a self-similar kinetic theory of thermal conductivity in a magnetized collisional plasma. Their kinetic equation included only Coulomb collisions, and the plasma parameters of the solar wind (electron number density, temperature and magnetic field) were approximated as power laws in heliocentric distance. They developed a model for the strahl and compared their results to *Wind* observations at 1 AU.

Beyond a few solar radii, intra-particle collisions are rare, and only wave-particle interactions can scatter electrons. The basic effect of wave-particle interactions on the electron VDF is pitch-angle diffusion in the reference frame of the waves. Vocks et al. [7, 8] studied the formation of the halo and strahl in the electron VDF in the solar corona and solar wind due to whistler turulence wave-particle interactions at electron energies below 1 keV. Vocks et al. [9] subsequently extended their work to electron energies of more than 100 keV, thus including superhalo electrons. Their studies suggest that the quiet solar corona is capable of producing suprathermal electron VDFs by whistler turulence wave-particle interactions. They asserted that such an electron population should appear in the solar wind. Kim et al. [10, 11], and Yoon et al. [12] considered the interaction between electrons and spontaneously emitted electronmagnetic fluctuations. They assumed that halo electrons interact with not only whistler turbulence but also Langmuir turbulence, and superhalo electrons ineract with Langmuir turbulence. They show that the Langmuir fluctuation terms are orders of magnitude higher than the whistler related terms, and thus, whistler fluctuations and cyclotron resonance terms contribute virtually nothing to the overall halo electron VDF [12]. They hence proposed a new model that only Langmuir turbulence interacts with suprathermal electrons [13]. They treated the suprathermal tail found in the electron VDF as a whole, rather than dividing the suprathermal tail into three different parts. However they presumed that the wave-particle interaction is a very rapid temporal processes, and ignored all interactions associated with electric and magnetic fields. This assumption renders their equation incomplete. In examining their model [10–13], we found that the diffusion coefficients were too small to affect the electron VDF. Pierrard et al. [14] devloped a kinetic model that includes both Coulomb collisions and wave-particle resonant interactions with whistler turbulence.

In this study, we first introduce a wave-particle interaction term into the kinetic equation that describes the interaction of electrons with whistler waves, as well particle collision terms. We developed a numerical method using adaptive mesh refinement to solve an advection-diffusion-like Fokker-Planck kinetic equation in 3D phase space. We apply the numerical method to analyze the effects of Coulomb collisions and resonant wave-partcle interactions on the strahl found in the solar wind. By studying the temporal and radial evolution of the electron velocity distribution function, we confirm that for large electron energies and away from the corona Coulomb collisons are weak. We also show that resonent wave-particle interactions associated with whislter turbulence can diffuse the electron VDF significantly. We hence propose that whistler turbulence in the solar wind diffuses the electron VDF from an anisotropic strahl into an isotropic halo.

2. Description of the Model

In the case of non-relativistic charged particles, the general kinetic equation for the evolution of the velocity distribution function $f(\mathbf{r}, \mathbf{v}, t)$ is:

$$\frac{\partial f}{\partial t} + \mathbf{v} \cdot \nabla f + \frac{q}{m}\left(\mathbf{E} + \frac{\mathbf{v}}{c} \times \mathbf{B}\right) \cdot \frac{\partial f}{\partial \mathbf{v}} = \left(\frac{\delta f}{\delta t}\right)_{\text{coll}} + \left(\frac{\delta f}{\delta t}\right)_{\text{wp}}, \tag{1}$$

where q is charge of particle, $\mathbf{v}$ the particle velocity, $\mathbf{r}$ the position and m the mass of particle, $\mathbf{E}$ the electric field and $\mathbf{B}$ the magnetic field. The right side $(\delta f/\delta t)_{\text{coll}}$ denotes the collisional operator, and $(\delta f/\delta t)_{\text{wp}}$ refers to the wave-particle interaction term associated with various plasma waves.

Since the electron gyroperiod is much smaller than any other characteristic timescale, it is reasonable to assume gyrotropy of the electron VDF. The gyrophase averaged kinetic equation (1) for solar wind electron distribution function $f(r, v, \mu)$ has the form [15]:

$$\frac{\partial f}{\partial t} + \mu v \frac{\partial f}{\partial r} + \left[\frac{v(1-\mu^2)}{r} - \frac{eE_{\parallel}}{m}\frac{1-\mu^2}{v}\right]\frac{\partial f}{\partial \mu} - \frac{eE_{\parallel}}{m}\mu\frac{\partial f}{\partial v} = \left(\frac{\delta f}{\delta t}\right)_{\text{coll}} + \left(\frac{\delta f}{\delta t}\right)_{\text{wp}}, \tag{2}$$

where (v, μ) is the magnitude of velocity and pitch-angle measured in the solar wind frame, r the heliocentric distance, e the electron charge and $E_{\parallel}$ the parallel electric field with respect to the magnetic field. Note that we assume $\mu v >> U$, where U is the bulk velocity of the background plasma flow

We assume the collisional operator used by Horaites et al. [5,6]:

$$\left(\frac{\delta f}{\delta t}\right)_{\text{coll}} = \frac{4\pi n e^4 \Lambda}{m^2}\left\{\frac{1}{v^3}\frac{\partial}{\partial \mu}\left[(1-\mu^2)\frac{\partial f}{\partial \mu}\right] + \frac{1}{v^2}\frac{\partial f}{\partial v} + \frac{v^2_{\text{th-core}}}{2v^2}\left(\frac{1}{v}\frac{\partial^2 f}{\partial v^2} - \frac{1}{v^2}\frac{\partial f}{\partial v}\right)\right\}, \quad (3)$$

where Λ is the Coulomb logarithm and $v^2_{\text{th-core}}$ the thermal velocity of the Maxwellian core electrons.

The wave-particle interaction term associated with whistler turbulence is given by Pierrard et al. [14]:

$$\left(\frac{\delta f}{\delta t}\right)_{\text{wp}} = \frac{\partial}{\partial \mu}\left(D_{\mu\mu}\frac{\partial f}{\partial \mu} + D_{\mu p}\frac{\partial f}{\partial p}\right) + \frac{1}{p^2}\frac{\partial}{\partial p}\left[p^2\left(D_{p\mu}\frac{\partial f}{\partial \mu} + D_{pp}\frac{\partial f}{\partial p}\right)\right] \quad (4)$$

where p is the particle's momentum, $D_{\mu p} = D_{p\mu}$, and for nonrelativistic electrons:

$$\frac{D_{pp}}{\Omega_e (mc)^2} = \frac{\pi}{3}\frac{A}{a}\left(\frac{\beta|\mu|}{a}\right)^{\frac{s-1}{3}}(1-\mu^2), \quad (4a)$$

$$\frac{D_{\mu p}}{\Omega_e (mc)} = -\frac{\pi}{3}\frac{A}{a}\left[\frac{\mu}{|\mu|}\left(\frac{\beta|\mu|}{a}\right)^{\frac{s-2}{3}} + \frac{\mu}{\beta}\left(\frac{\beta|\mu|}{a}\right)^{\frac{s-1}{3}}\right](1-\mu^2), \quad (4b)$$

$$\frac{D_{\mu\mu}}{\Omega_e} = \frac{\pi}{3}\frac{A}{a}\left[\left(\frac{\beta|\mu|}{a}\right)^{\frac{s-3}{3}} + \frac{2\mu}{|\mu|}\frac{\mu}{\beta}\left(\frac{\beta|\mu|}{a}\right)^{\frac{s-2}{3}} + \left(\frac{\mu}{\beta}\right)^2\left(\frac{\beta|\mu|}{a}\right)^{\frac{s-1}{3}}\right](1-\mu^2). \quad (4c)$$

Here $\beta = v/c$, $a = \omega^2_{\text{pe}}/\Omega^2_e$, $A = 0.1$, $s = 3/2 - 5/3$ and the electron plasma and gyrofrequencies are given by

$$\omega^2_{\text{pe}} = \frac{4\pi n e^2}{m}, \quad \Omega_e = \frac{eB}{mc} \quad .$$

The background plasma parameters are assumed have the form of power laws, as used by Horaites et al. [5,6]:

$$B = B_0\left(\frac{r}{R}\right)^{-2}; \quad n = n_0\left(\frac{r}{R}\right)^{-1.8};$$
$$T = T_0\left(\frac{r}{R}\right)^{-0.4}; \quad E_{\parallel} = 2.2\frac{T_0}{eR}\left(\frac{r}{R}\right)^{-1.4}.$$

where B_0, n_0 and T_0 are the magnetic field, electron number density and electron temperature at 1 AU, r is the helocentric distance and R is 1 AU.

3. The Fokker-Planck equation for solar wind electrons

We developed a numerical method using adaptive mesh refinement [16–20] in a 3D Cartesian coordinates system to solve an equation that has the form of:

$$\frac{\partial f}{\partial t} + \nabla \cdot (\boldsymbol{V} f) = \nabla \cdot (\mathbb{D} \cdot \nabla f) + \beta f + r. \quad (5)$$

where $f(x_1, x_2, x_3)$ is a general function, $\boldsymbol{V} = (V_1, V_2, V_3)$ is the phase space advection velocity vector and the phase space diffusion tensor $\mathbb{D}$ here has only the simplest diagonal form in which there are no cross terms. A reduced form of diffusion coefficient is sufficient for our purposes.

IOP Conf. Series: Journal of Physics: Conf. Series **1100** (2018) 012025 doi:10.1088/1742-6596/1100/1/012025

$$\begin{pmatrix} \mathbb{D}_{11} & 0 & 0 \\ 0 & \mathbb{D}_{22} & 0 \\ 0 & 0 & \mathbb{D}_{33} \end{pmatrix}.$$

On combining equation (2) and equation (4), changing the variable from p to v and ignoring the two cross terms ($D_{p\mu}$, $D_{\mu p}$), we then obtain the kinetic equation with wave-particle interaction terms associated with whistler turbulence for solar wind electrons:

$$\frac{\partial f}{\partial t} + \mu v \frac{\partial f}{\partial r} + \left[\frac{v(1-\mu^2)}{r} - \frac{eE_\parallel}{m} \frac{1-\mu^2}{v} \right] \frac{\partial f}{\partial \mu} - \frac{eE_\parallel}{m} \mu \frac{\partial f}{\partial v}$$
$$= \frac{\partial}{\partial \mu} \left(D_{\mu\mu} \frac{\partial f}{\partial \mu} \right) + \frac{1}{v^2} \frac{\partial}{\partial v} \left(\frac{v^2 D_{pp}}{m^2} \frac{\partial f}{\partial v} \right). \tag{6}$$

On setting $f = Y/(v^2 r^2)$ allows the kinetic equation above to be rewritten as:

$$\frac{\partial Y}{\partial t} + \frac{\partial}{\partial r}(\mu v Y) + \frac{\partial}{\partial \mu} \left\{ \left[\frac{v(1-\mu^2)}{r} - \frac{eE_\parallel}{m} \frac{1-\mu^2}{v} \right] Y \right\}$$
$$+ \frac{\partial}{\partial v} \left\{ \left[-\frac{eE_\parallel}{m} \mu + \frac{\pi c^2 \Omega_\mathrm{e}^3}{15 v \omega_\mathrm{pe}^2} \left(\frac{v}{c} \frac{\Omega_\mathrm{e}^2}{\omega_\mathrm{pe}^2} \right)^{1/6} |\mu|^{1/6} (1-\mu^2) \right] Y \right\}$$
$$= \frac{\partial}{\partial v} \left[\frac{\pi c^2 \Omega_\mathrm{e}^3}{30 \omega_\mathrm{pe}^2} \left(\frac{v}{c} \frac{\Omega_\mathrm{e}^2}{\omega_\mathrm{pe}^2} \right)^{1/6} |\mu|^{1/6} (1-\mu^2) \frac{\partial Y}{\partial v} \right]$$
$$+ \frac{\partial}{\partial \mu} \left\{ \frac{\pi \Omega_\mathrm{e}^3}{30 \omega_\mathrm{pe}^2} \left[\left(\frac{v}{c} \frac{\Omega_\mathrm{e}^2}{\omega_\mathrm{pe}^2} \right)^{\frac{-1}{2}} |\mu|^{\frac{-1}{2}} + \frac{c}{v} \left(\frac{v}{c} \frac{\Omega_\mathrm{e}^2}{\omega_\mathrm{pe}^2} \right)^{\frac{-1}{6}} \frac{2\mu^2}{|\mu|^{7/6}} \right. \right.$$
$$\left. \left. + \frac{c^2}{v^2} \left(\frac{v}{c} \frac{\Omega_\mathrm{e}^2}{\omega_\mathrm{pe}^2} \right)^{\frac{1}{6}} |\mu|^{\frac{1}{6}} \mu^2 \right] (1-\mu^2) \frac{\partial Y}{\partial \mu} \right\}. \tag{7}$$

On combining equation (2) and equation (3), we also obtain the kinetic equation for solar wind electrons with the diffusion term associated with Coulomb collisions:

$$\frac{\partial f}{\partial t} + \mu v \frac{\partial f}{\partial r} + \left[\frac{v(1-\mu^2)}{r} - \frac{eE_\parallel}{m} \frac{1-\mu^2}{v} \right] \frac{\partial f}{\partial \mu} - \frac{eE_\parallel}{m} \mu \frac{\partial f}{\partial v}$$
$$= \frac{4\pi n e^4 \Lambda}{m^2} \left\{ \frac{1}{v^3} \frac{\partial}{\partial \mu} \left[(1-\mu^2) \frac{\partial f}{\partial \mu} \right] + \frac{1}{v^2} \frac{\partial f}{\partial v} + \frac{v_\mathrm{th}^2}{2v^2} \left(\frac{1}{v} \frac{\partial^2 f}{\partial v^2} - \frac{1}{v^2} \frac{\partial f}{\partial v} \right) \right\}. \tag{8}$$

Similarly, let $f = Y/(v^2r^2)$, allowing the above equation to be rewritten as:

$$\frac{\partial Y}{\partial t} + \frac{\partial}{\partial r}(\mu v Y) + \frac{\partial}{\partial \mu}\left\{\left[\frac{v(1-\mu^2)}{r} - \frac{eE_\parallel}{m}\frac{1-\mu^2}{v}\right]Y\right\}$$
$$+ \frac{\partial}{\partial v}\left[\left(-\frac{eE_\parallel}{m} - \frac{4\pi ne^4\Lambda}{m^2}\frac{1}{v^2} + \frac{4\pi ne^4\Lambda}{m^2}\frac{v_{\text{th-core}}^2}{v^4}\right)Y\right]$$
$$= \frac{\partial}{\partial v}\left(\frac{1}{v^3}\frac{4\pi ne^4\Lambda v_{\text{th-core}}^2}{2m^2}\frac{\partial Y}{\partial v}\right) + \frac{\partial}{\partial \mu}\left(\frac{4\pi ne^4\Lambda}{m^2}\frac{1-\mu^2}{v^3}\frac{\partial Y}{\partial \mu}\right). \tag{9}$$

Here we have rewritten equaton (6) and equation (8) for an electron VDF $f(r, v, \mu)$ in the form of an advection-diffusion-like equation for $Y(r, v, \mu)$ in the 3D phase space (r, v, μ), i.e. eqautions (7) and (9) which have the same form as equation (5). The phase space advection velocities and phase space diffusion coefficients associated with whistler turbulence and Coulomb collision separately are as follows.

For the advection-diffusion-like kinetic equation with whistler turbulence diffusion terms we have:

$$V_r = \mu v;$$
$$V_v = -\frac{eE_\parallel}{m}\mu + \frac{\pi c^2\Omega_{\text{e}}^3}{15v\omega_{\text{pe}}^2}\left(\frac{v}{c}\frac{\Omega_{\text{e}}^2}{\omega_{\text{pe}}^2}\right)^{1/6}|\mu|^{1/6}(1-\mu^2);$$
$$V_\mu = \frac{v(1-\mu^2)}{r} - \frac{eE_\parallel}{m}\frac{1-\mu^2}{v};$$
$$\mathbb{D}_{rr} = 0;$$
$$\mathbb{D}_{vv} = \frac{\pi c^2\Omega_{\text{e}}^3}{30\omega_{\text{pe}}^2}\left(\frac{v}{c}\frac{\Omega_{\text{e}}^2}{\omega_{\text{pe}}^2}\right)^{1/6}|\mu|^{1/6}(1-\mu^2);$$
$$\mathbb{D}_{\mu\mu} = \frac{\pi\Omega_{\text{e}}^3}{30\omega_{\text{pe}}^2}\left[\left(\frac{v}{c}\frac{\Omega_{\text{e}}^2}{\omega_{\text{pe}}^2}\right)^{\frac{-1}{2}}|\mu|^{\frac{-1}{2}} + \frac{c}{v}\left(\frac{v}{c}\frac{\Omega_{\text{e}}^2}{\omega_{\text{pe}}^2}\right)^{\frac{-1}{6}}\frac{2\mu^2}{|\mu|^{7/6}} + \frac{c^2}{v^2}\left(\frac{v}{c}\frac{\Omega_{\text{e}}^2}{\omega_{\text{pe}}^2}\right)^{\frac{1}{6}}|\mu|^{\frac{1}{6}}\mu^2\right](1-\mu^2).$$

For the advection-diffusion-like kinetic equation with Coulomb collision diffusion terms we have:

$$V_r = \mu v;$$
$$V_v = -\frac{eE_\parallel}{m} - \frac{4\pi ne^4\Lambda}{m^2}\frac{1}{v^2} + \frac{4\pi ne^4\Lambda}{m^2}\frac{v_{\text{th-core}}^2}{v^4};$$
$$V_\mu = \frac{v(1-\mu^2)}{r} - \frac{eE_\parallel}{m}\frac{1-\mu^2}{v}d;$$
$$\mathbb{D}_{rr} = 0;$$
$$\mathbb{D}_{vv} = \frac{4\pi ne^4\Lambda}{m^2}\frac{v_{\text{th-core}}^2}{2v^3};$$
$$\mathbb{D}_{\mu\mu} = \frac{4\pi ne^4\Lambda}{m^2}\frac{1-\mu^2}{v^3}.$$

4. Numerical solutions

Equations (7) and (9) can be solved numerically by the method we developed for appropriate boundary and initial conditions. The solutions yield the temporal and radial evolution of the function Y (hence the electron VDF $f = Y/(v^2 r^2)$). At the left boundary $r = 0$, a narrow beam of particles is injected into interplanetary space. The beam is comprised of a δ-function in v and a δ-function in μ. For simplicity we use normal distributions to substitute for the two δ-functions. In this case, the boundary condition at $r = 0$ has the form:

$$f(r=0,v,\mu) = \frac{n_0}{2\pi v^2}\frac{1}{\sqrt{4\pi^2\sigma_\mu^2\sigma_v^2}}\exp\left[-\frac{(\mu-\mu_0)^2}{2\sigma_\mu^2}\right]\exp\left[-\frac{(v-v_0)^2}{2\sigma_v^2}\right] \tag{10}$$

where n_0 is electron number density at 1 AU, and the following parameters are used:

$$\sigma_\mu^2 = 0.02; \quad \sigma_v^2 = 0.2;$$
$$\mu_0 = 0.95; \quad v_0 = 3.$$

The initial distribution is plotted in Figure (2), shown in normalized phase space coordinates $(v_\parallel, v_\perp)$, $v_\parallel$ is the velocity component parallel to the mean magnetic field and $v_\perp$ the perpendicular component.

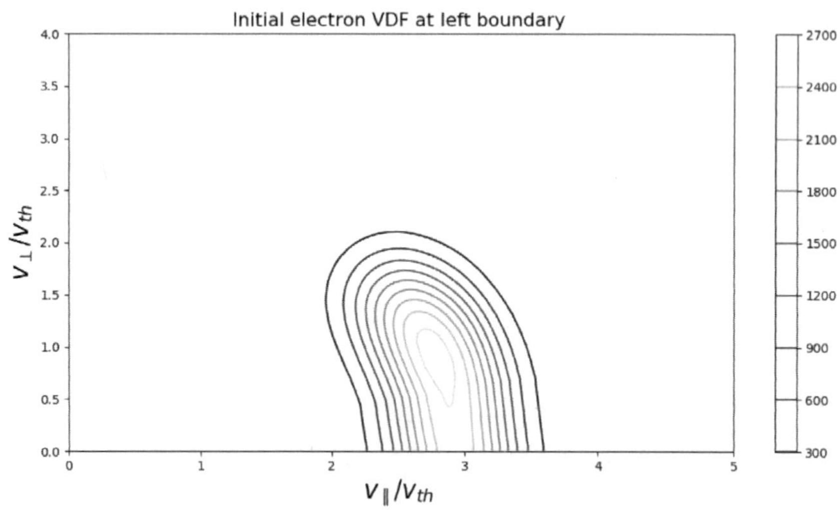

Figure 2. The left boundary condition for the injected electron VDF, plotted in $(v_\parallel, v_\perp)$, $v_\parallel$ phase space.

We obtained the numerical solution of electron VDF after a steady state was reached. Figure (3) shows the VDF at 1 AU after being subjected to whistler turbulence and Coulomb collisions. Figure (3) illustrates that whistler turbulence has a much stronger diffusive effect on the narrow beam of particles than Coulomb collisions. Figure 3a and 3b are the electron VDFs at 0.3 and 1 AU assuming only Coulomb collisions from Horaites' formulation [5, 6]. Figure 3c and 3d are the electron VDFs at 0.3 and 1 AU assuming only whistler turbulence from Pierrard's formulation [14].

IOP Conf. Series: Journal of Physics: Conf. Series **1100** (2018) 012025 doi:10.1088/1742-6596/1100/1/012025

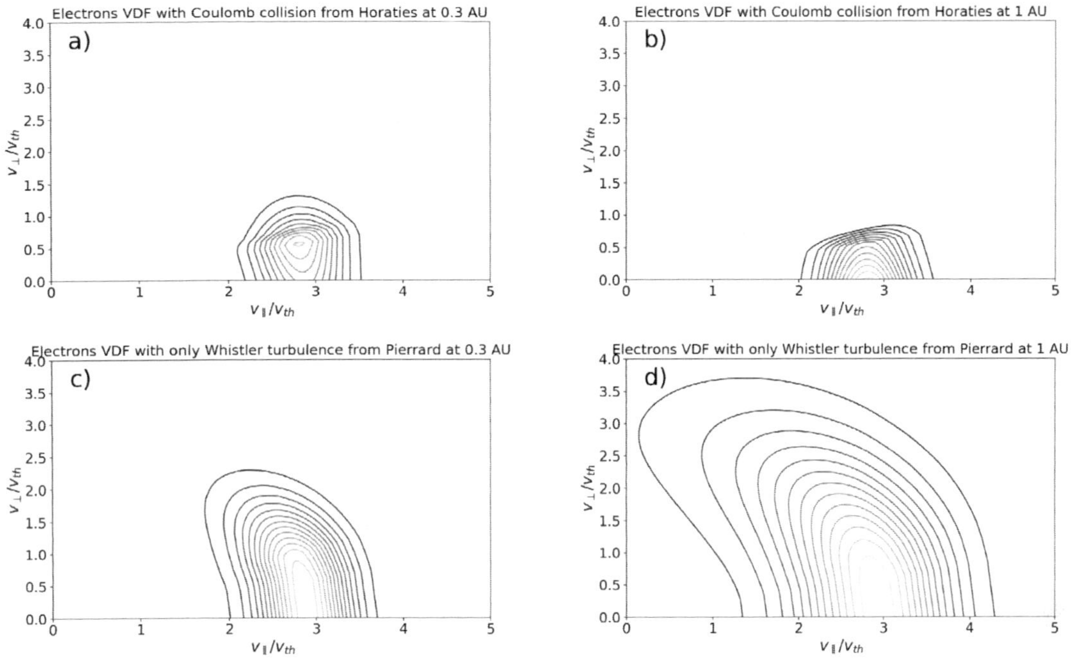

Figure 3. Comparison of the electron VDFs. The left column shows results at 0.3 AU and the right column at 1 AU. The top row shows the evolution of the electron VDF when subjected to Coulomb collisions only. The bottom row correspond to whistler turbulence only.

5. Summary and Conclusion

We have investigated the influence of whistler turbulence and Coulomb collisions separately on the temporal and radial evolution of the electron distribution function by numerically solving the Fokker-Planck kinetic equations. From the viewpoint of the method of characteristics, it is not difficult to identify that these phase space advection velocities are actually the equations for the characteristics. Therefore, in the 3D phase space (r, v, μ), the function Y (hence the electron VDF or electrons) moves along these characteristics. Electrons will move to regions in phase space where $\mu \approx 1$ and move in the r directon. In the solar wind, electrons therefore focused and aligned with the magnetic field and the beam is narrowed. This effect is illustrated in Figures (2) and (3a,b). The left boundary condition distribution is narrowed and electrons are therefore focused and aligned with the magnetic field. However, there are diffusion terms on the right side of both equations (7) and (9). Besides moving along the characteristics, electrons will also experience diffusion. From the equations, diffusion in phase space is with respect to v and pitch-angle μ. Because of diffusion associated with whistler turbulence and Coulomb collisions, the injected narrow beam of electrons broadens in velocity space (v, μ). Pitch-angle diffusion causes the beam to expand (or diffuse) in the μ direction, which will increase the width of beam. This effect can explain the observation that the strahl has a finite width.

From Figures (2), (3c) and (3d), we see that the diffusion in $v_\parallel$ is a little larger than $v_\perp$. For the whistler turbulence, the isotropization is due to $\mathbb{D}_{\mu\mu}$ term, but there also appears to be quite significant energization due to $\mathbb{D}_{vv}$. It is clear that whistler turbulence is far more important than Coulomb collisions. Comparing with the initial beam, the beam is broadened along $v_\parallel$. While the beam is broadened along $v_\perp$ too, such broadening is more slowly. Therefore, the electron VDF retains some anisotropic with a beam along the prallel direction. This tendency is a consequence of the focusing term.

Although both Coulomb collisions and whistler turbulence can cause diffusion of electrons, their efficiency or ability to diffuse is different. When the heliocentric distance is large or the energies of the electrons are high enough, Coulomb collisons are a much weaker effect and wave-particle interactions becomes important. Figure (3b) is the electron VDF at 1 AU after electrons have been subjected to Coulomb collisions only. Figure (3d) is the electron VDF at 1 AU after electrons have been subjected to whistler turbulence only. Diffusion associated with whistler turbulence is indeed much stronger than that associated with Coulomb collisions.

Our present results confirm that whistler turbulence is much much efficient than Coulomb collisions in scattering particles in phase space in the supersonic solar wind.

References

[1] Anderson B, Skoug R, Steinberg J and McComas D 2012 *Journal of Geophysical Research: Space Physics* **117**

[2] Štverák Š, Maksimovic M, Trávníček P M, Marsch E, Fazakerley A N and Scime E E 2009 *Journal of Geophysical Research: Space Physics* **114**

[3] Tao J, Wang L, Zong Q, Li G, Salem C S, Wimmer-Schweingruber R F, He J, Tu C, Bale S D, Wang L *et al.* 2016 Solar wind 0.1-1.5 kev electrons at quiet times *AIP Conference Proceedings* vol 1720 (AIP Publishing) p 070006

[4] Salem C, Hubert D, Lacombe C, Bale S, Mangeney A, Larson D and Lin R 2003 *The Astrophysical Journal* **585** 1147

[5] Horaites K, Boldyrev S, Krasheninnikov S, Salem C, Bale S and Pulupa M 2015 *Physical review letters* **114** 245003

[6] Horaites K, Boldyrev S, Wilson III L B, Viñas A F and Merka J 2017 *Monthly Notices of the Royal Astronomical Society* **474** 115–127

[7] Vocks C and Mann G 2003 *The Astrophysical Journal* **593** 1134

[8] Vocks C, Salem C, Lin R and Mann G 2005 *The Astrophysical Journal* **627** 540

[9] Vocks C, Mann G and Rausche G 2008 *Astronomy & Astrophysics* **480** 527–536

[10] Kim S, Yoon P H, Choe G and Wang L 2015 *The Astrophysical Journal* **806** 32

[11] Kim S, Yoon P H and Choe G 2016 *Physics of Plasmas* **23** 022111

[12] Yoon P H, Kim S, Choe G *et al.* 2016 *The Astrophysical Journal* **826** 204

[13] Kim S, Yoon P H, Choe G *et al.* 2016 *The Astrophysical Journal* **828** 60

[14] Pierrard V, Lazar M and Schlickeiser R 2011 *Solar Physics* **269** 421–438

[15] Zank G P 2014 *Transport Processes in Space Physics and Astrophysics* (*Lecture Notes in Physics, Berlin Springer Verlag* vol 877) (Springer-Verlag)

[16] Kolobov V I 2003 *Computational materials science* **28** 302–320

[17] Kolobov V and Arslanbekov R 2012 *Journal of Computational Physics* **231** 839–869

[18] Arslanbekov R R, Kolobov V I and Frolova A A 2013 *Physical Review E* **88** 063301

[19] Kolobov V I 2013 *Physics of Plasmas* **20** 101610

[20] Zabelok S, Arslanbekov R and Kolobov V 2015 *Journal of Computational Physics* **303** 455–469

IOP Conf. Series: Journal of Physics: Conf. Series **1100** (2018) 012026 doi:10.1088/1742-6596/1100/1/012026

STEREO Observations of Non-linear Plasma Processes in Solar Type III Radio Bursts

G. Thejappa[1], **R. J. MacDowall**[2]

[1] Department of Astronomy, University of Maryland, College Park, MD 20742, USA
[2] NASA/Goddard Space Flight Center, Greenbelt MD 20771, USA

Abstract. Solar type III radio bursts are the fast drifting radio emissions excited by the flare accelerated electron beams in the corona and interplanetary medium. In this paper, we report the observations of several nonlinear processes, obtained by the time domain sampler (TDS) of the WAVES experiment of the STEREO spacecraft in the source region of one of these bursts. These include: (1) a four wave-wave interaction process called the oscillating two stream instability (OTSI) $L + L \xrightarrow{S} L_U + L_D$, where L is the beam excited Langmuir wave, S is the ion sound wave, and L_U and L_D are the up- and down-shifted Langmuir waves, respectively, and (2) two wave-wave merging processes $L_U + L_D \rightarrow T_{2f_{pe}}$ and $L + T_{2f_{pe}} \rightarrow T_{3f_{pe}}$, where $T_{2f_{pe}}$ and $T_{3f_{pe}}$ are the electromagnetic waves at the second and third harmonic of the electron plasma frequency, f_{pe} and L can be either a beam-excited Langmuir wave, or an up- or down-shifted sideband. The newly developed higher order spectral (HOS) techniques have been used to identify the signatures of these wave-wave interactions in the waveform data. The implication of these findings is that strong Langmuir turbulence processes play key roles in excitation of solar type III radio bursts, especially, in the beam stabilization and conversion of Langmuir waves into electromagnetic waves at the fundamental and higher harmonics of the electron plasma frequency, f_{pe}.

1. Introduction

The most intense radio emissions from the Sun, which are referred to as the type III radio bursts occur as the fast drifting emission bands in the dynamic spectrum. Usually, these bursts are characterized by the fundamental and harmonic bands. Ginzburg and Zhelenyakov [9] were the first to suggest that these bursts are probably excited by a mechanism called the plasma mechanism, which involves: (1) formation of bump-on-tail distributions by the flare accelerated electrons, (2) excitation of Langmuir waves by the bump-on-tail distributions, and (3) conversion of these Langmuir waves into electromagnetic waves at the fundamental and second harmonic of the electron plasma frequency, $f_{\rm pe}$. This suggestion has been well confirmed by the in situ detection of bump-on-tail distributions of electron beams [19] and Langmuir waves [11, 20, 12, 27, 29, 30] in type III radio bursts. However, answers to some science questions remain unsettled, which include: (1) How does the electron beam preserve its bump-on-tail over distances of the order of 1 AU and more against the quasi-linear relaxation, and (2) What mechanisms are responsible for conversion of Langmuir waves into radio emissions at the fundamental and higher harmonics of the electron plasma frequency, f_{pe}.

The electron beam can maintain its bump-on-tail against the quasi-linear relaxation if the resonance between Langmuir waves and the beam is disrupted well before they back react and relax the beam. The induced scattering off ion clouds, which can pump the Langmuir waves toward lower wave numbers is suggested to accomplish such a task [14]. If the electron temperature T_e is higher than the ion temperature T_i, the induced scattering is known to act as the electrostatic decay (ESD) $L \rightarrow L^{'} + S$, where L and $L^{'}$ are the oppositely propagating beam-excited and back-scattered Langmuir waves, respectively, and S is the ion sound wave. The ESD also can facilitate the excitation of second harmonic electromagnetic waves $T_{2f_{pe}}$ through the three wave interaction: $L+L^{'} \rightarrow T_{2f_{pe}}$. In some type III events, evidence for ESD has been reported [20, 12, 28, 29, 31, 13].

According to some authors [25, 26, 10], the intensities of type III burst associated Langmuir waves are expected to well exceed the strong turbulence thresholds, and, therefore, the oscillating two stream instability (OTSI), and, related soliton formation and Langmuir collapse [44, 24] probably are the relevant processes. As far as the stabilization of the electron beam is concerned, the strong turbulence processes can disrupt the resonance between the Langmuir waves and the beam by pumping the Langmuir waves towards higher wave numbers. The OTSI generates an ion sound wave S of frequency and wave number (f_S, k_S), which in turn beats with two beam-driven Langmuir waves of frequency and wave number (f_{pe}, k_L) and excites two oppositely propagating up- and down- shifted side bands with ($f_{pe} \pm f_S$, $k_L \pm k_S$), respectively. Thus, the OTSI can be represented as the four wave interaction

$$L + L \xrightarrow{S} L_U + L_D, \tag{1}$$

where L is the beam-excited Langmuir wave, and L_U and L_D are the oppositely propagating up- and down-shifted sidebands, respectively. The OTSI also facilitates the excitation of $T_{2f_{pe}}$ through wave-wave merging [25]

$$L_U + L_D \rightarrow T_{2f_{pe}}. \tag{2}$$

The observations of Langmuir waves by the Ulysses spacecraft have provided some indication that the strong turbulence processes probably occur in type III burst sources [27, 29, 30, 32]. The much superior STEREO high time resolution in situ wave data have provided evidence for various nonlinear processes in type III bursts [6, 21, 15, 16], which also include strong Langmuir turbulence processes [33, 34, 35, 36, 37, 38, 39, 40, 41, 42].

In this paper, we present the observations of a one-dimensional field aligned localized wave packet, obtained by the time domain sampler (TDS) of the WAVES experiment of the STEREO A spacecraft [4] in the source region of a solar type III radio burst. We show that (1) the peak intensity and short scale length of this wave packet satisfy the threshold conditions of OTSI and soliton formation, and (2) its spectrum contains: (a) sidebands, which satisfy the frequency and wave number resonance conditions of OTSI, and (b) two harmonic peaks, one at $2f_{\rm pe}$ and another one at $3f_{\rm pe}$ excited probably by three wave interactions. We show that the resonance conditions including the phase coherence required by the four- and three wave interactions are easily satisfied. These findings strongly suggest that the strong Langmuir turbulence processes probably play key roles in the beam stabilization as well as conversion of Langmuir waves into the solar type III radio emissions. In section 2, we describe the observations, in section 3, we describe higher order spectral analysis techniques, and in sections 4 and 5, we present the discussion, and the summary and conclusions, respectively.

2. Observations

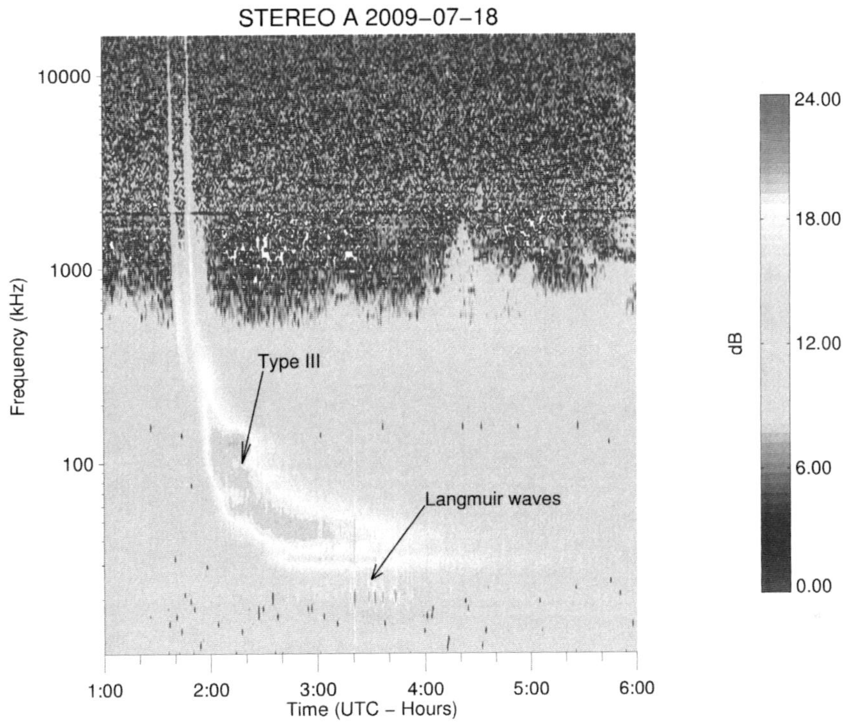

Figure 1. Dynamic spectrum of a local type III radio burst (fast drifting emission from ≈ 5 MHz down to ≈ 26 kHz) and associated Langmuir waves (non-drifting emissions in the frequency interval 19-22 kHz).

In Figure 1, we present the dynamic spectrum of a local type III radio burst and its associated in situ waves, observed by the STEREO WAVES experiment [4], which uses three orthogonal monopole antennas with effective lengths of ∼ 1 m. The emission drifting fast from $\approx 5MHz$, all the way to ≈ 26 kHz is the type III radio burst. Using its fast negative frequency drift, we estimate the velocity of the electron beam responsible for this event as $v_b \approx 0.37c$ (c is the velocity of light). This estimation is based on the assumptions that the mode of emission is the second harmonic and the ambient electron

IOP Conf. Series: Journal of Physics: Conf. Series **1100** (2018) 012026 doi:10.1088/1742-6596/1100/1/012026

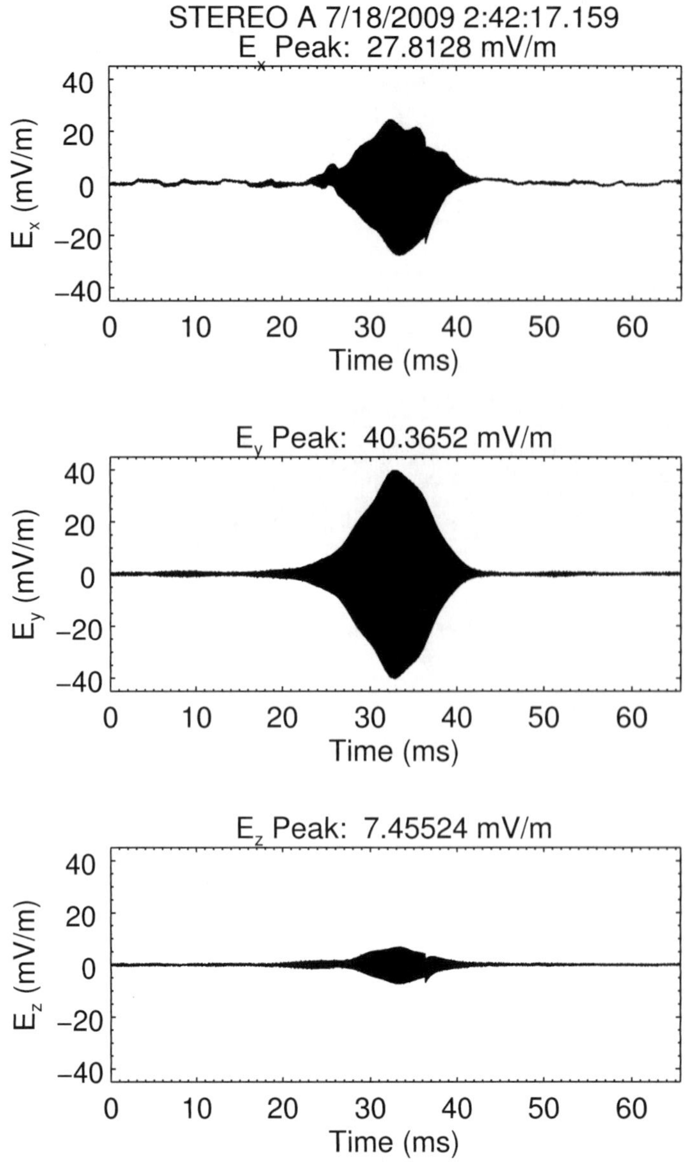

Figure 2. The X, Y and Z-electric field components of one of the intense Langmuir wave packets associated with the type III burst in the spacecraft frame of reference.

density n_e is described by the Radio Astronomy Explorer (RAE) model [7]. We identify the non-drifting emissions seen in the 19-22 kHz interval as the Langmuir waves. The Langmuir wave burst, shown by the arrow in this figure is of interest in the present study.

The TDS of the WAVES experiment provides the high time resolution electric field

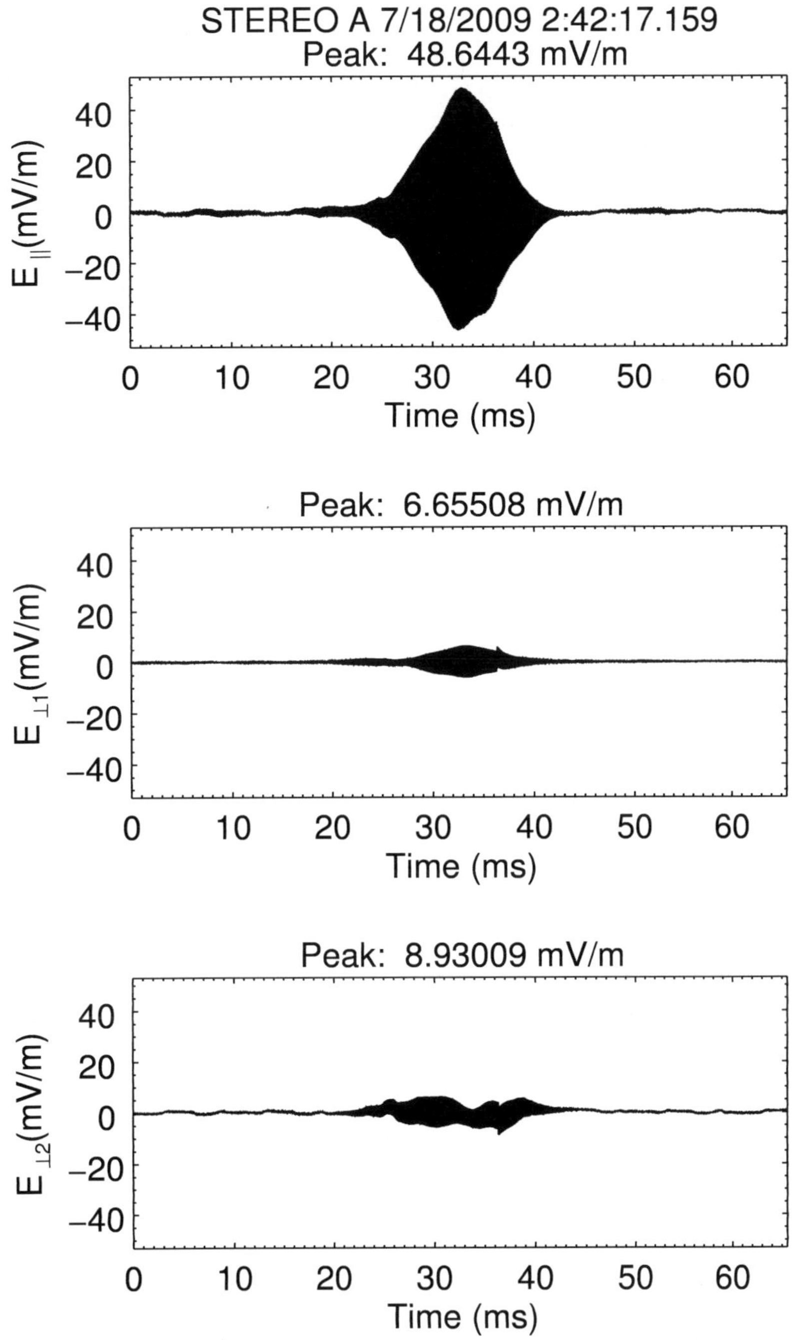

Figure 3. The parallel and perpendicular electric field components of the Langmuir wave packet with respect to the magnetic field. The unit vectors of magnetic field and solar wind velocity used in the coordinate transformation are $\vec{b}$ = [-0.60747, -0.79512, 0.005553] and $\vec{v}$ = [0.99167, -0.12865, 0.0060756], respectively.

potential data of all three components, which can be converted into the X- Y- and Z-components of the wave electric field in the spacecraft frame using the transformation matrix given by [2]. TDS has resolved the Langmuir wave bursts seen in Figure 1 into several intense waveforms. In Figure 2, we present the time profiles of E_X, E_Y and E_Z components of one of the waveforms, whose peak amplitudes are 27.8 mVm^{-1}, 40.4 mVm^{-1} and 7.5 mVm^{-1}, respectively. This waveform contains 16384 samples, acquired at the rate of 250,000 samples per second (a time step of 4 μs for a total duration of 65 ms). We transform the E_X, E_Y and E_Z components from the spacecraft into a more useful magnetic field ($\vec{B}$)-aligned coordinate system with X-, Y- and Z-axes aligned along $\vec{b}$, $\vec{b} \times \vec{v}$ and $\vec{b} \times (\vec{v} \times \vec{b})$, respectively ($\vec{b}$ and $\vec{v}$ are the unit vectors directed along the magnetic field, and the solar wind velocity, respectively). The relevant unit vectors in this case are $\vec{b} = (-0.60747, -0.79512, 0.005553)$ and $\vec{v} = (0.9917, -0.12865, 0.0060756)$, and the angle between them is $\approx$ 120°. In Figure 3, we present the components of this waveform in the B-aligned coordinate system, where, the peak amplitude of the parallel component is $\approx$ 48.6 mVm^{-1}, and those of the perpendicular components are $\approx$ 6.7 mVm^{-1} and $\approx$ 8.9 mVm^{-1}, respectively. Since, these peak amplitudes indicate that this is a one-dimensional field aligned wave packet, we mainly focus on the parallel component.

In Figure 4, we present the spectrum of the parallel component. The total spectrum from 0 to 65 kHz (Figure 4a) clearly shows peaks at (approximate) 20 kHz, 40 kHz and 60 kHz, where, the base frequency, 20 kHz, corresponds to the local electron plasma frequency, $f_{\rm pe}$, and the 40 kHz corresponds to the second harmonic and 60 kHz corresponds to the third harmonic. Figure 4a also shows that the second harmonic is weaker than the fundamental, and the third harmonic is weaker than the second harmonic, which is typical of the natural signals [43, 16, 38]. The logarithmic spectrum in the frequency interval from 19 to 21 kHz (Figure 4b) also shows an intense peak (L) of the beam excited Langmuir waves at $f_{\rm pe} \approx$ 20 kHz, and two side bands, namely a Stokes peak (D) at $\approx$ 19.8 kHz and an anti-Stokes peak (U) at $\approx$ 20.2 kHz. As seen from this spectral plot, the anti-Stokes peak is weaker than that of the Stokes peak. The linear spectrum from 0 to 1.5 kHz, presented in Figure 4c shows a low frequency enhancement below $\approx$ 400 Hz with a peak $\sim$ 100 Hz corresponding probably to the ion sound waves.

In Figure 5, we plot the time profile of total electric field of the wave packet $E_t = \sqrt{E_X^2 + E_Y^2 + E_Z^2}$, which clearly shows that this is a localized wave packet with peak $E_t \sim$ 49 mVm^{-1} and $\frac{1}{2}$- power duration $\tau_{0.5}$ of 8.46 ms. By assuming that this wave packet is stationary in the solar wind, we can convert this time scale $\tau_{0.5}$ into spatial scale $L_{0.5}$ by interpreting it in terms of Doppler shift in the solar wind

$$L_{0.5} = \tau_{0.5} v_{sw} \cos\theta. \quad (3)$$

Here v_{sw} is the solar wind speed and θ is the angle between the solar wind velocity and the electric field vector, which is nothing but the angle between the ambient magnetic field and the electric field of the wave packet since it is parallel to the magnetic field.

For electron temperature ($T_{\rm e}$), we assign a typical value of 10^5 K; the measurements of $T_{\rm e}$ are not available. Assuming that the intense peak (L) in the spectrum of the parallel component corresponds to Langmuir waves excited at $f_{\rm pe} \approx$ 20 kHz (Figure 4b),

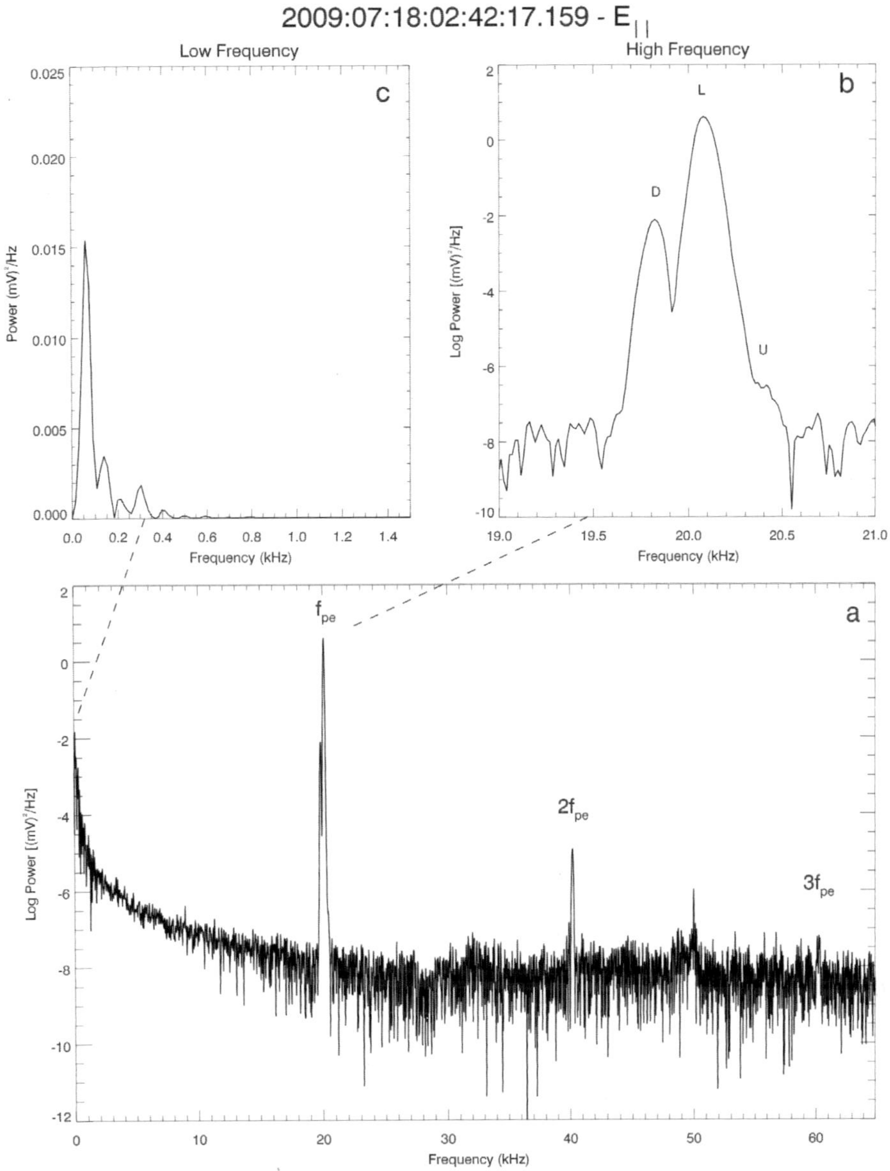

Figure 4. (a): The complete spectrum of the parallel component of the wave packet, which shows harmonic peaks at $f_{\rm pe}$, $2f_{\rm pe}$ and $3f_{\rm pe}$ ($f_{\rm pe}$ is the electron plasma frequency), (b) the narrow spectrum around $f \approx f_{\rm pe} \approx 20.0$ kHz, where L, D, and U correspond to the beam excited Langmuir wave, down-shifted sideband at ≈ 19.8 kHz, and up-shifted sideband at ≈ 20.2 kHz, respectively, and (c) the low frequency spectrum, which shows the enhancement below 400 Hz corresponding to ion-sound waves.

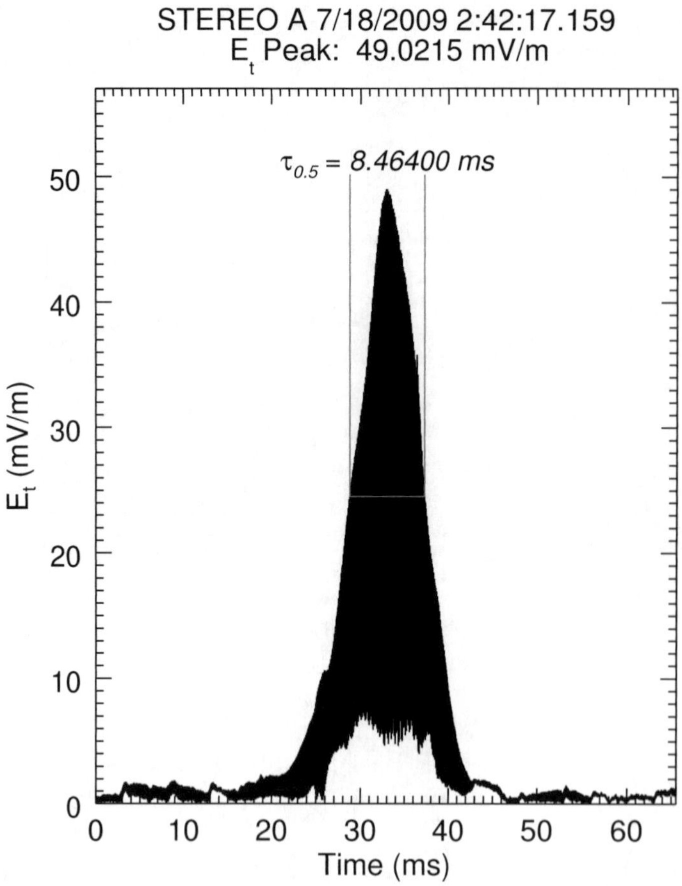

Figure 5. The time profile of the total electric field $E_t = \sqrt{E_X^2 + E_Y^2 + E_Z^2}$. The $\frac{1}{2}$-power duration of 8.464 ms is equivalent to the spatial scale of $\sim 227\lambda_{De}$.

we estimate the electron density as $n_e \approx 5 \times 10^6$ m^{-3} using the relation $f_{\rm pe}[Hz] = 9n_{\rm e}^{1/2}$. These values yield Debye length, $\lambda_{\rm De} = 69T_{\rm e}^{1/2}n_{\rm e}^{-1/2} \approx 9.8$ m. We estimate the wave number of Langmuir waves as $k_{\rm L} = \frac{2\pi f_{\rm pe}}{v_{\rm b}} \approx 1.1 \times 10^{-3}$ m^{-1}, and $k_{\rm L}\lambda_{\rm De} \approx 1.1 \times 10^{-2}$ for $v_{\rm b} \approx 0.37c$. The STEREO/PLASTIC experiment [8] has measured $v_{\rm sw} \approx 525.7$ kms^{-1}. Thus, we can estimate the normalized peak energy density as $\frac{W_{\rm L}}{n_{\rm e}T_{\rm e}} = \frac{\epsilon_0 E_t^2}{4n_e T_e} \approx 7.7 \times 10^{-4}$ for $E_t = \sqrt{E_X^2 + E_Y^2 + E_Z^2} = 49$ mVm^{-1}. The measured $\tau_{0.5} \approx 8.46$ ms yields $L_{0.5} \approx \tau_{\rm e} v_{\rm sw} \cos\theta \approx 227\lambda_{\rm De}$.

3. Higher Order Spectral Techniques

In a turbulence process, energy transfer occurs between interacting modes, which is rather difficult to identify in the time series data. However, interacting waves obey energy and momentum conservation laws, which exhibit certain statistical properties of observations. The HOS techniques are used to extract the phase information of the interacting waves. The bi-spectral analysis is used to extract the nonlinear characteristics between three spectral components with frequency sum and difference [17]. The bi-spectrum, which is related to the skewness of the signal can detect the asymmetric non-linearity in three wave interactions, such as the electrostatic decay [13] and harmonic generation [37, 38]. The FFT based bi-spectrum is defined as

$$B(f_1, f_2) = E[X(f_1)X(f_2)X(f_1 + f_2)^*], \tag{4}$$

where E denotes the expected value, $X(f)$ is the Fourier transform of $x(t)$ and $*$ denotes complex conjugation. This equation indicates that the bi-spectrum is zero unless the phase coherence between three waves at frequencies f_1, f_2 and $f_1 + f_2$ present in the time series data is non-zero. For example, for spontaneously generated non-interacting modes, the phases would be random and therefore the statistical averaging will be zero. The bi-coherence, which is the normalized bi-spectrum

$$b^2(f_1, f_2) = \frac{|B(f_1, f_2)|^2}{(E[|X(f_1)X(f_2)X^*(f_1 + f_2)|])^2} \tag{5}$$

provides the quantitative measure of the phase coherence. A unit value for the bi-coherence indicates perfect coupling, a zero value indicates no coupling, and any value between zero and one indicates partial coupling. Generally, the bi-spectrum is estimated using the periodogram (interval segmentation) method. The approximate variance of the bi-coherence estimator is defined as [17]:

$$var(b^2) \simeq \frac{M}{2N}[1 - b^2(f_1, f_2)], \tag{6}$$

where N is the number of total data points and M is the number of data records into which the time series is divided. To assess the low bi-coherence levels, one needs long and stationary time series with $N >> 1$.

The wavelet based bi-spectral analysis technique, which can handle both short and non-stationary time series is used to study non-linear processes in space plasmas [5, 13]. The wavelet based bi-spectrum is defined as

$$B(a_1, a_2) = \int W(a_1, \tau)W(a_2, \tau)W^*(a_3, \tau)d\tau, \tag{7}$$

where $W(a, \tau)$ is the CWT at scale a and time τ, and $a_3^{-1} = a_1^{-1} + a_2^{-2}$ (frequency sum rule). Here the integral is taken over a finite time interval T: $\tau_0 \leq \tau \leq \tau_1$. The wavelet bi-spectrum measures the amount of phase coupling between the wavelet components of the time series $x(t)$ with scale lengths a_1, a_2 and a_3, such that the sum rule is satisfied. It can be interpreted as the coupling between wavelets of frequencies, such that $f_3 = f_1 + f_2$, within the frequency resolution, where $f = f_0/a$. The squared wavelet bi-coherence, which is the normalized squared wavelet bi-spectrum is defined as

$$b^2(a_1, a_2) = \frac{|B(a_1, a_2)|^2}{[\int |W(a_1, \tau)W(a_2, \tau)|^2 d\tau][\int |W(a_3, \tau)|^2 d\tau]}. \tag{8}$$

For digital signals, these integrals are replaced by summations over N points. The bi-spectrum and bi-coherence are usually computed over a triangle of frequencies defined by the relations:

$$0 < f_1 < \frac{f_{\mathrm{N}}}{N}, f_1 < f_2 < \frac{f_{\mathrm{N}}}{2} - f_1 \tag{9}$$

where f_{N} is the Nyquist frequency.

The tri-spectral analysis technique can yield the higher order coherence among four spectral components and modulation sideband. A cumulant based tri-spectral method has been applied to synthetic [18], and in situ wave data [33, 34, 36, 37, 38]. The tri-spectrum is defined in terms of the complex Fourier components of the signal X_1, X_2, X_3 and X_4 at frequencies f_1, f_2, f_3 and f_4 as [18]

$$T(1,2,3) = E[X_1X_2X_3^*X_4^*] - G(1,2,3,4), \tag{10}$$

where $G(1,2,3,4) = E[X_1X_2]E[X_3^*X_4^*] + E[X_1X_3^*]E[X_2X_4^*] + E[X_1X_4^*]E[X_2X_3^*]$ and $f_4 = f_1 + f_2 - f_3$, and $E[]$ is the expectation operator. The tri-coherence, which is the normalized tri-spectrum eliminates the dependence of tri-spectrum on the amplitude of the signals. Most importantly, it quantifies the phase coherence amongst the spectral components of the wave packet. The squared tri-coherence is expressed as [18]:

$$t^2(1,2,3) = \frac{|T(1,2,3)|^2}{(E[|X_1X_2X_3^*X_4^*|] + |G(1,2,3,4)|)^2}. \tag{11}$$

A unit value of the tri-coherence indicates perfect coupling, a zero value indicates no coupling, and a value between zero and one indicates partial coupling. The tri-coherence quantifies the fraction of the total product of powers at the frequency quartet, (f_1, f_2, f_3, $f_1 + f_2 - f_3$), that is owing to cubically phase-coupled modes.

The tri-spectrum and tri-coherence are estimated using the method of periodograms, which involves (1) the division of the data record into M segments, (2) application of an appropriate window to each segment to reduce leakage, (3) computation of the tri-spectrum and tri-coherence for each segment by using the DFT, and (4) taking the average of the tri-coherence across segments in order to reduce the variance of the estimator. The tri-spectrum estimator is symmetric with respect to permutations of its arguments f_1, f_2 and f_3. The principal domain for the interaction of the type $f_1 + f_2 = f_3 + f_4$ is determined as [18] $0 \le f_1 \le f_{\mathrm{N}}$, $0 \le f_2 \le f_1$, $0 \le f_3 \le f_2$, and $f_3 \le f_1 + f_2 - f_3 \le f_{\mathrm{N}}$, where f_{N} is the Nyquist frequency.

4. Discussion

The dispersion relations of the relevant Langmuir and ion sound waves are

$$\omega_{\mathrm{L}} = \omega_{\mathrm{pe}}(1 + \frac{3}{2}k_L^2\lambda_{\mathrm{De}}^2), \tag{12}$$

and

$$\omega_S = k_{\mathrm{S}}c_{\mathrm{S}}, \tag{13}$$

respectively, where $\omega_{\rm pe} = 2\pi f_{\rm pe}$ and $c_{\rm S}$ is the ion sound speed. If

$$\frac{W_{\rm L}}{n_{\rm e}T_{\rm e}} > (k_L\lambda_{\rm De})^2, \tag{14}$$

the weak turbulence approximation is not valid, and the strong turbulence processes, such as the OTSI, soliton formation and spatial collapse become important [44]. In such a case, there won't be one to one correspondence between $\omega_{\rm L}$ and k, since $\frac{W_{\rm L}}{n_{\rm e}T_{\rm e}}$ determines the bandwidth. Since the Langmuir oscillations get trapped in the self-generated density cavities, the turbulence becomes highly inhomogeneous consisting of localized collapsing soliton-caviton pairs. The energy transfer in this case occurs in the direction of higher values of k.

The threshold for the supersonic collapse is [44]

$$\frac{W_{\rm L}}{n_{\rm e}T_{\rm e}} \geq \frac{m_{\rm e}}{m_{\rm i}}, \tag{15}$$

where $m_{\rm e}$ and $m_{\rm i}$ are the electron and ion masses, respectively. This condition signifies that the compression due to ponder-motive force and self focusing overwhelms the wave packet spreading due to dispersion. The criterion for the formation of the collapsing wave packet is [23]

$$\frac{W_{\rm L}}{n_{\rm e}T_{\rm e}} \geq (\Delta k\lambda_{\rm De})^2, \tag{16}$$

where $\Delta k = \frac{2\pi}{L}$ is the wave number characteristic of the envelope. In the present case, the conditions (14-16) are easily satisfied, since $\frac{W_{\rm L}}{n_{\rm e}T_{\rm e}} \approx 7.7 \times 10^{-4}$ and, $(k_L\lambda_{\rm De})^2$ is $\approx 1.2 \times 10^{-4}$ for $k = k_L \sim 1.1 \times 10^{-3}$ m^{-1}, $\frac{m_{\rm e}}{m_{\rm i}} \approx 5.5 \times 10^{-4}$ and $(\Delta k\lambda_{\rm De})^2 \approx 7.7 \times 10^{-4}$ estimated for the spatial scale L of $\approx 227\lambda_{\rm De}$.

4.1. Four Wave Interactions

The OTSI is a four wave interaction process, in which two Langmuir waves with frequencies and wave numbers ($f_{\rm L}$, k_L) serve as pump waves, and the up- and down-shifted Langmuir sidebands with ($f_{\rm U}$, $k_{\rm U}$) and ($f_{\rm D}$, $k_{\rm D}$) serve as the daughter waves. The non-linear coupling between these four modes occurs through a purely growing ion sound mode with ($f_{\rm S}$, $k_{\rm S}$). The relevant frequency, wave number and phase matching conditions are

$$2f_{\rm L} = f_{\rm D} + f_{\rm U} \tag{17}$$
$$2k_L = k_{\rm D} + k_{\rm U} \tag{18}$$
$$2\phi_{\rm L} = \phi_{\rm D} + \phi_{\rm U}, \tag{19}$$

where, the subscripts L, D and U correspond to the beam-excited Langmuir wave, down- and up-shifted sidebands, respectively.

The spectrum of the parallel component (Figure 4b) shows a strong Langmuir wave peak with up- and down-shifted sidebands, together with low frequency enhancement (Figure 4c). We can verify, whether these observed spectral peaks correspond to the beam-excited Langmuir waves, and to the OTSI excited sidebands and low frequency waves. Here, we assume that $k \approx k_L$. As far as the frequency matching condition $2f_{\rm L} = f_{\rm D} + f_{\rm U}$ is concerned, it is easily satisfied, since the frequency shifts Δf of the down- and up-shifted side bands are symmetric with respect to the Langmuir wave pump, being ≈ 300 Hz and ≈ 300 Hz, respectively. Moreover, the frequency differences $\Delta f = |f_{\rm L} - f_{\rm U,D}|$ are also in good agreement with the frequency of the ion sound waves of < 400 Hz. The matching condition $k_{\rm U,D} = k_{\rm L} \pm k_{\rm S}$ is also reasonably well satisfied, for the wave numbers of the ion sound waves $k_{\rm S}\lambda_{\rm De} \approx 0.04$, estimated using the expression for the Doppler shift, $k_{\rm S} = \frac{2\pi f_{\rm S}}{v_{\rm sw}}$ and for $f_{\rm S} = 300$ Hz and $v_{\rm sw} = 527$ kms^{-1}, which yields $|k_{\rm U,D}| \approx |k_{\rm S}|$, since $k_{\rm L}\lambda_{\rm De} \approx 1.1 \times 10^{-2}$ is three to four times less than $k_{\rm S}\lambda_{\rm De} \approx 0.04$. Thus, the frequency and wave number matching conditions of OTSI are easily satisfied.

As far as the phase coherence between the Langmuir pump waves and side bands is concerned, we have to calculate the tri-coherence spectrum as a function of three frequencies. Using equation (11), we have computed the tri-coherence spectrum of the parallel component in the frequency ranges [20 kHz $\leq F_{\rm k} \leq$ 20.75 kHz, 20 kHz $\leq F_{\rm l} \leq$ 20.25 kHz, 19.75 kHz $\leq F_{\rm p} \leq$ 20.25 kHz], for $N = 1000$ (0.004s) and $M = 16$ with a Hamming window, where N is the segment length, and M is the number of segments. We display the computed three dimensional tri-coherence spectrum in Figure 6. As seen from this Figure, the 3-D tri-coherence spectrum shows a peak at ($\approx$ 20.25, $\approx$ 20, $\approx$ 19.75) kHz with a statistically significant peak value of $t^2 \approx 0.33$. This tri-coherence spectral peak is indicative of the phase relation $2\phi_{\rm L} = \phi_{\rm D} + \phi_{\rm U}$.

The computed tri-coherence spectrum can also be displayed as 2-D cross-sections at different frequencies. We present such cross sections in Figure 7. The cross-section at $F_{\rm p} = f_{\rm D} = 19.75$ kHz (top panel) shows the tri-coherence spectral peak at ($\approx$ 20, $\approx$ 20, $\approx$ 19.75) kHz with $t^2 \approx 0.33$; this value agrees with that of 3-D spectrum. This peak tri-coherence again quantifies the phase relation $2\phi_{\rm L} = \phi_{\rm D} + \phi_{\rm U}$, where $\phi_{\rm L}$, $\phi_{\rm D}$ and $\phi_{\rm U}$ are the phases of the beam-excited Langmuir wave ($\approx$ 20 kHz), Stokes ($\approx$ 19.75 kHz) and anti-Stokes ($\approx$ 20.25 kHz) modes, respectively, i.e., the frequency quartet in this case is ($\approx$ 20, $\approx$ 20, $\approx$ 19.75, $\approx$ 20.25) kHz. The cross-section at $F_{\rm p} = f_{\rm L} = 20.25$ kHz (second panel) shows a relatively weaker spectral feature, with peak value of ≈ 0.30 at ($\approx$ 20.25 kHz, $\approx$ 20.5 kHz, $\approx$ 20.5 kHz). In the bottom panel, we present the cross-section at $F_{\rm p} = f_{\rm U} = 20.75$ kHz, which also shows a much weaker peak of $t^2 \approx 0.1$ at ($\approx$ 20.75, $\approx$ 21, $\approx$ 21) kHz. These tri-coherence spectral peaks reflect the four wave interactions involving the sidebands themselves. However, the tri-coherence spectral peak $t^2 \approx 0.33$ (top panel) is more significant in the statistical sense, which indicates that the four wave interaction $2f_{\rm L} \to f_{\rm D} + f_{\rm U}$ is the dominant one. Thus, the significant tri-coherences corresponding to $2\phi_{\rm L} = \phi_{\rm D} + \phi_{\rm U}$ in the tri-coherence spectrum at Stokes, anti-Stokes and beam-excited Langmuir wave frequencies provide evidence for the OTSI type of four-wave interaction.

4.2. Wave-Wave Merging

As seen in Figure 4a, the spectrum of the parallel component contains harmonics. One can interpret these spectral peaks at $2f_{\rm pe}$ and $3f_{\rm pe}$ in terms of second and third harmonic electromagnetic waves excited by wave-wave interactions, as discussed by several authors

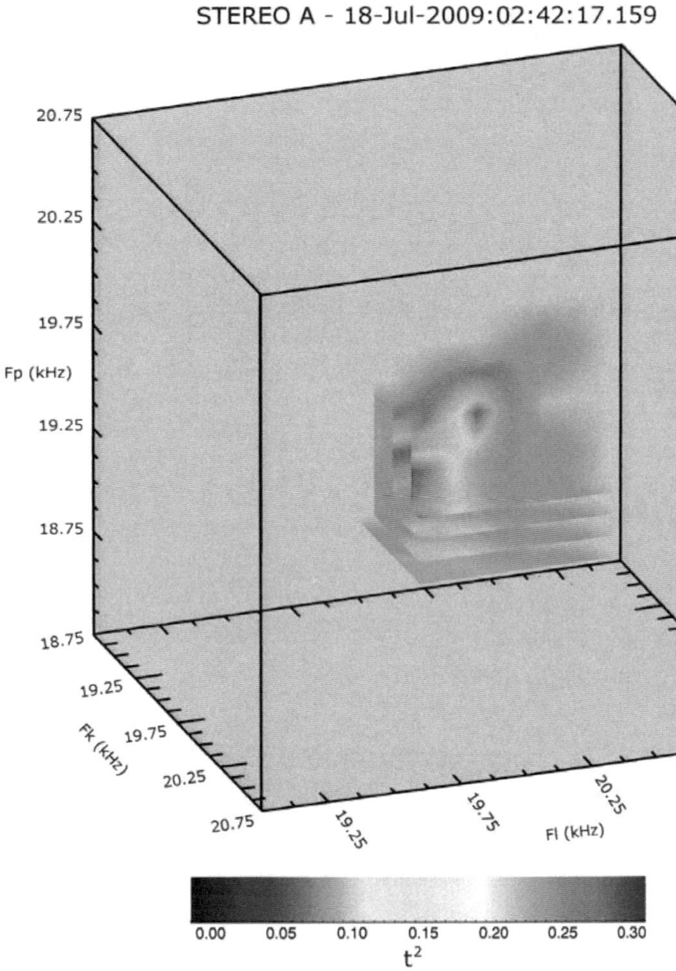

Figure 6. The three dimensional tri-coherence spectrum $t^2(F_\mathrm{k}, F_\mathrm{l}, F_\mathrm{p})$ of the $E_\|$ component of Figure 3a. The peak tri-coherence of ≈ 0.33 at (≈ 20.25, ≈ 20, ≈ 19.75, ≈ 20) kHz quantifies the phase matching condition $2\phi_\mathrm{L} = \phi_\mathrm{D} + \phi_\mathrm{U}$, where L, D and U correspond to beam-excited, Stokes and anti-Stokes modes, respectively

(see for example, [45]). This can be possible only if the waves corresponding to these spectral peaks satisfy the following matching rules

$$L_1 + L_2 \rightarrow T_{2\mathrm{f}_\mathrm{pe}}, \tag{20}$$

and

$$L + T_{2f_\mathrm{pe}} \rightarrow T_{3\mathrm{f}_\mathrm{pe}}. \tag{21}$$

As far as the frequency and wave number resonance conditions are concerned, they are easily satisfied for these two type of three wave interactions (see, for example, [38]).

IOP Conf. Series: Journal of Physics: Conf. Series **1100** (2018) 012026 doi:10.1088/1742-6596/1100/1/012026

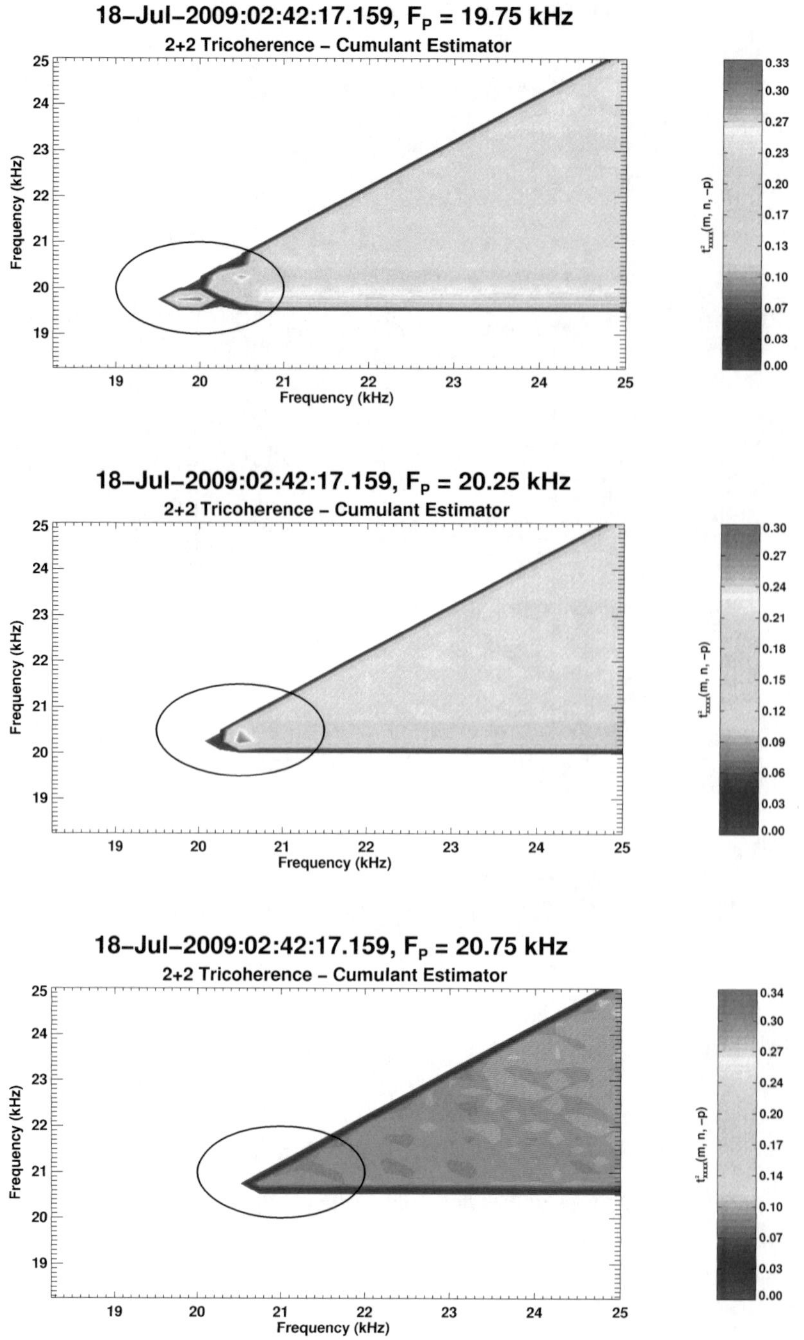

Figure 7. The cross-sections of the tri-coherence spectrum of the $E_{||}$ component (Figure 3a). The top, middle and bottom panels show the cross-sections at $F_{\rm p} \approx F_{\rm D} \approx 19.75$ kHz, $F_{\rm P} \approx f_{\rm L} \sim 20.25$ kHz, and $F_{\rm P} \approx f_{\rm U} \approx 20.75$ kHz, respectively (the subscripts L, D and U correspond to beam-excited, Stokes and anti-Stokes modes, respectively.)

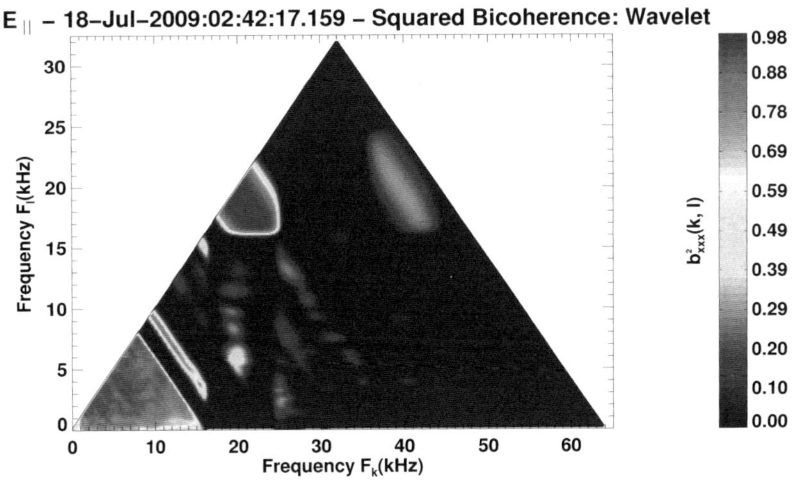

Figure 8. The wavelet based bicoherence spectrum of the $E_{\|}$-component. The bicoherence spectral peaks at (20, 20) kHz and (40, 20) kHz correspond to the three wave interactions $L_U + L_D \rightarrow T_{2f_{pe}}$ and $L + T_{2f_{pe}} \rightarrow T_{3f_{pe}}$, respectively. Here, L_U and L_D correspond to up- and down-shifted side bands excited by the OTSI, and $T_{2f_{pe}}$ and $T_{3f_{pe}}$ correspond to the second and third harmonic electromagnetic waves, respectively.

The bi-spectral analysis of the wave packet can provide definite clues whether these processes are actually occurring in the present case or not. If these non-linear processes are responsible for the observed spectral peaks, the corresponding waves should be non-linearly coupled to each other, i.e., one would expect peaks at (f_{pe}, f_{pe}) and at ($2f_{pe}$, f_{pe}) in the bi-coherence spectrum; these bi-coherence spectral peaks are indicative of the three wave interactions (20) and (21), respectively.

We have computed the wavelet- based bi-coherence spectrum of the parallel component of the waveform. As seen from Figure 8, this clearly shows two intense peaks, one at (20, 20) kHz and another one at (40, 20) kHz, with peak bi-coherences of b(20 kHz, 20 kHz)$\simeq$ 0.98 and b(40 kHz, 20 kHz)$\simeq$ 0.30, respectively. These bi-coherence spectral peaks provide unambiguous evidence for the non-linear interactions $(20 + 20 \rightarrow 40)$ kHz and $(20 + 40 \rightarrow 60)$kHz. Thus, the observed second and third harmonic spectral peaks probably correspond to electromagnetic waves, generated as a result of coalescence of two oppositely propagating Langmuir waves, and coalescence of a Langmuir wave with an electromagnetic wave at $2f_{pe}$, respectively.

As far as the three wave interaction $L_1 + L_2 \rightarrow T_{2f_{pe}}$ is concerned, the sidebands L_U and L_D excited by the OTSI are well suited to play the role of the oppositely propagating Langmuir waves L_1 and L_2. In the case of the third harmonic $T_{3f_{pe}}$ is concerned, it could be excited either by a four wave interaction $L_1 + L_2 + L_3 \rightarrow T_{3f_{pe}}$ or by a three wave interaction $L + T_{2f_{pe}} \rightarrow T_{3f_{pe}}$, where L_1, L_2 and L_3 are the Langmuir waves with different wave vectors. However, the intense bi-spectral peak at ($2f_{pe}$, f_{pe}) supports only the later process, where the Langmuir wave could be either the beam-excited Langmuir wave, or any one of the side bands.

5. Summary and Conclusions

We have presented the observations of an intense localized one dimensional magnetic field aligned Langmuir waveform obtained by the STEREO/TDS in the source region of a local solar type III radio burst. The theories predict that the bump-on-tail distributions of electron beams propagating along the magnetic field excite one-dimensional field aligned Langmuir wave packets. The observations presented in this study represent one of such wave packets. Our analysis has revealed that the peak intensity and short spatial scale of this wave packet easily satisfy the threshold condition for excitation of OTSI and related strong turbulence processes. Our spectral analysis has shown that its spectrum exhibits the characteristic signatures of OTSI, namely, a resonant peak at f_{pe}, Stokes and anti-Stokes peaks at $f_{\mathrm{pe}} \pm f_{\mathrm{S}}$, and a low frequency enhancement below f_{S}. We have clearly demonstrated that these spectral components easily satisfy the frequency and wave number resonance conditions of the OTSI type of four wave interaction. In order to verify whether the phases of these spectral components satisfy the resonance condition required by the OTSI, we have computed FFT based tri-coherence spectrum of the wave packet. We have shown that the computed tri-coherence spectrum contains the needed peak, which is indicative of phase coherence amongst the spectral components.

The spectrum of the wave packet is also found to contain two harmonic peaks, one at $2f_{\mathrm{pe}}$ and another one at $3f_{\mathrm{pe}}$ in addition to an intense peak at f_{pe}. We have calculated its wavelet based bi-coherence spectrum and found that it contains two bi-coherence spectral peaks, one at $(f_{\mathrm{pe}}, f_{\mathrm{pe}})$ and another one at $(2f_{\mathrm{pe}}, f_{\mathrm{pe}})$. These bi-coherence spectral peaks provide clear evidence for two kinds of three wave interactions, namely, $L_{\mathrm{U}} + L_{\mathrm{D}} \rightarrow T_{2\mathrm{f}_{\mathrm{pe}}}$ and $L + T_{2\mathrm{f}_{\mathrm{pe}}} \rightarrow T_{3\mathrm{f}_{\mathrm{pe}}}$, respectively. Those observations are consistent with simulation results (see, for example [1]). The one dimensional wave packet presented in this study with its characteristic sideband spectral structure is in the linear regime of OTSI, and the second harmonic electromagnetic wave is most probably excited as a result of merging of oppositely propagating sidebands excited by the OTSI.

Thus, the observations of the one dimensional wave packets presented in this study confirm that the OTSI and related strong turbulence processes play a significant role in the stabilization of the electron beam as well as excitation of second and third harmonic electromagnetic waves.

Acknowledgment

The SWAVES instruments include contributions from the Observatoire of Paris, University of Minnesota, University of California, Berkeley, and NASA/GSFC. G. T. acknowledges the NASA STEREO and Wind projects for the financial support

6. References

[1] Akimoto, K., Rowland, H. L., and Papadopoulos, K 1988, *Phys. Fluids*, 31, 2185, 1988.
[2] Bale, S. D. et al., 2008, *Space Sci. Rev.,* 136, 529, 2008.
[3] Bardwell, S. and Goldman, M. V 1974, *Astrophys. J.,* 209, 912
[4] Bougeret, J. L. et al. 2008, *Space Sci. Rev.*, **136**, 487
[5] Dudok de Wit, T., and Krasnosel'skikh, V. V 1995, *Phys. Plasmas*, 2, 4307
[6] R. E. Ergun, R. E. et al. 2008, *Phys. Rev. Letts.*, 101, 051101
[7] Fainberg,J., and Stone, R. G. 1971, *Solar Phys.*, 17, 392
[8] Galvin, A. B. *et al.* 2008, *Space Sci. Rev.*, 136, 437

[9] Ginzburg, V. L., and Zheleznyakov, V. V. 1958, *Soviet Astron.*, 2, 623
[10] Goldstein, M. L., Smith, R. A., and Papadopoulos, K. 1979, *Astrophys. J.*, 237, 683
[11] Gurnett, D. A., and Anderson, R. R 1977, *J. Geophys. Res.*, 82, 632
[12] Gurnett, D. A., Hospodarsky, G. B., Kurth, W. S., D. J. Williams, D. J., and S. J. Bolton, S. J 1993, *J. Geophys. Res.*, 98, 5631
[13] Henri, P., Briand, C., Mangeney, A., Bale, S. D., Califano, F., Goetz, K., and Kaiser, M. L 2009, *J. Geophys. Res.*, 114, A03103, doi:10.1029/2008JA013738, 2009.
[14] Kaplan, S. A., and V. N. Tsytovich, V. N 1968, *Soviet Astron.*, 11, 956
[15] Kellogg, P. J., Goetz, K., Monson, S. J., Bale, S. D., Reiner, M. J., and M. Maksimovic, M 2009, *J. Geophys. Res.*, 114, A02107, doi:10.1029/2008JA013566
[16] Kellogg, P. J., Goetz, K., and Monson, S. J 2010, *J. Geophys. Res.*, 115, A06107, doi:10.1029/2009JA014635
[17] Kim, Y. C., and Powers, E. J 1979, *IEEE Transactions on Plasma science*, PS-7, 120-131
[18] Kravtchenko-Berejnoi, Lefeuvre, V. F., Krasnoselskikh, V. V., and D. Lagoutte, D 1995, *Signal Processing* , 42, 291
[19] Lin, R. P., Potter, D. W., Gurnett, D. A., and Scarf, F. L 1981, *Astrophys. J.*, 251, 364
[20] Lin, R. P., Levedahl, W. K., Lotko, W., Gurnett, D. A., and Scarf, F. L 1986, *Astrophys. J.* , 308, 954
[21] Malaspina, D. M., and Ergun, R. E 2008, *J. Geophys. Res.* , 113, A12108
[22] Malaspina, D. M., Cairns, I. H., and Ergun, E. R. 2011, *Geophys. Res. Lett.*, 38, L13101
[23] Nezlin, M. V. 1993, (in) Physics of intense beams in plasmas, p. 186-249, Institute of Physics Publications, Philadelphia.
[24] Nicholson, D. R., Goldman, M. V., Hoyang, P., and Weatherall, J. C 1978, *Astrophys. J.*, 223, 605, 1978.
[25] Papadopoulos, K., Goldstein, M. L., and R. A. Smith, R. A 1974, *Astrophys. J.*, 190, 175
[26] Smith, R. A., Goldstein, M. L., and Papadopoulos, K 1979, *Astrophys. J.* , 234, 348
[27] Thejappa, G., Lengyel-Frey, D., Stone, R. G., and M. L. Goldstein, M. L 1993, *Astrophys. J.*, 416, 831
[28] Thejappa, G., Wentzel, D., and Stone, R. G. 1995, *J. Geophys. Res.*, 100, 3417
[29] Thejappa, G., and MacDowall, R. J 1998, *Astrophys. J.*, 498, 465
[30] Thejappa, G., Goldstein, M. L., MacDowall, R. J., Papadopoulos, K., and Stone R. G 1999, *J. Geophys. Res.*, 104, 28279
[31] Thejappa, G., MacDowall, R. J., Scime, E. E., and Littleton, J. E 2003, *J. Geophys. Res.* , 108, 1139
[32] Thejappa, G., and MacDowall, R. J. 2004, *Nonlinear Processes in Geophysics*, 11, 411
[33] Thejappa, G., MacDowall, R. J., Bergamo, M., and K. Papadopoulos, K 2012a, *Astrophys. J.* , 747, L1
[34] Thejappa, G., MacDowall, R. J., and Bergamo, M 2012b, *Geophys. Res. Lett.*, 39, LXXXX, doi:10.1029/2012GL051017
[35] Thejappa, G., MacDowall, R. J., and Bergamo, M 2012c, *J. Geophys. Res.* , 117, A08111, doi:10.1029/2012JA017695
[36] Thejappa, G., MacDowall, R. J., and Bergamo, M 2013a, *Ann. Geophys.*, 31, 1417-1428
[37] Thejappa, G., MacDowall, R. J., and Bergamo, M 2013b, AIP Conference Proceedings, 1539, 70
[38] Thejappa, G., MacDowall, R. J., and Bergamo, M 2013c, *J. Geophys. Res.* , 118, 114, doi:10.1002/jgra.50441
[39] Thejappa, G., and MacDowall, R. J 2014a, AIP Conference Proceedings 1582, 201
[40] Thejappa, G., and MacDowall, R. J 2014b, ASP Conference Series, 484, 210
[41] Thejappa, G., and MacDowall, R. J. 2018a, *Astrophys. J.*, Submitted
[42] Thejappa, G., and MacDowall, R. J. 2018b, *Astrophys. J.*, Submitted
[43] Walker, S. N., Balikhin, M. A., Bates, I., and Huff, R 2002, *Adv. Space Res.*, 30, 2815
[44] Zakharov, V. E. 1972, *Sov. Phys.-JETP*, 35, 908
[45] Zheleznyakov, V. V. 1976, "Electromagnetic Waves in a Cosmic Plasma (in Russian)", Nauka, Moscow

The Heating of Coronal Loops in Solar Active Regions

A A van Ballegooijen and M Asgari-Targhi

Harvard-Smithsonian Center for Astrophysics, 60 Garden Street, Cambridge, MA 02138, USA

E-mail: avanballegooijen@cfa.harvard.edu

Abstract. The active region corona is believed to be heated by magnetic disturbances that propagate into the corona from the convection zone below. A reduced magnetohydrodynamic (MHD) model of Alfvén waves in a coronal loop is presented. The waves are launched in the photosphere from a collection of kilogauss flux tubes, and they reflect at various positions along the loop, leading to counter-propagating waves and turbulence. It is found that turbulent Alfvén waves can produce only enough heat to maintain a peak temperature of about 2.5 MK, less than the temperatures typically observed in active regions ($\sim$ 4 MK). We consider an alternative model in which the flux tubes are subject to slow random footpoint motions, but we find that such braiding motions produce less heating than the waves inside the flux tubes. Therefore, models of coronal heating based on small-scale random footpoint motions cannot readily explain the observed high temperatures loops; more energetic "nanoflare" heating events are required. We suggest that such strong heating events may be produced by disturbances associated with flux emergence and large-scale non-potential fields.

1. Introduction

The solar corona is heated to temperatures of several million degrees. The strongest heating occurs in active regions (ARs), which contain closed magnetic loops that confine the coronal plasma. In recent years, the EUV Imaging Spectrometer (EIS) on *Hinode* has made it possible to measure the intensities of spectral lines from many different ions and determine the differential emission measure (DEM) distributions [1–7]. It was found that the DEM has its maximum value at a temperature of 3 – 4 MK. AR loops undergo continuous heating and cooling, which cause temporal variations in their emission [8–10]. The heating may occur in the form of small, impulsive heating events called "nanoflares" [11–14]. Originally, the term nanoflare was used to describe the sudden release of stored magnetic energy by magnetic reconnection, but more recently the term is used more generally and does not imply any particular physical mechanism. According to nanoflare models, an observed coronal loop consists of multiple strands, each in a different state of its temperature evolution. An important parameter in such models is the time delay between successive heating events on a given strand. If the delay is much shorter than the cooling time, then the temperature and density are approximately constant between heating events, and the heating is effectively steady ("high-frequency heating"). In contrast, if the delay is much longer than the cooling time, the plasma is heated to high temperatures ($\sim$ 10 MK) and then cools and drains without being reheated ("low-frequency heating"). Whether ARs have high-frequency or low-frequency heating is a matter of intense debate [15].

The mechanisms of coronal heating have been reviewed by many authors [16, 17]. The brightest loops in ARs are generally rooted in plage areas and sunspot penumbrae, but the loops rooted in sunspot umbrae are usually not very bright. This suggests that interactions of the magnetic field with subsurface convective flows play an important role in coronal heating process (convection is suppressed in sunspot umbrae). Plage regions contain small, kilogauss flux concentrations ("flux tubes") surrounded by regions of weaker field [18, 19]. The interactions of these flux tubes with granule-scale convective flows are believed to create disturbances that propagate upwards along the magnetic field lines and inject energy into the corona. The details of these disturbances are still unclear. One possibility is that subsurface flows cause twisting and braiding of the coronal field lines, which leads to the build-up of magnetic free energy in the corona [20, 21]. Three-dimensional (3D) MHD simulations of braided fields have shown that thin current sheets are naturally formed in such fields [22–25]. Magnetic reconnection in current sheets may occur in a burst-like manner, producing "nanoflare" heating events [11, 26, 27]. Several authors have developed "realistic" models of magnetic braiding that are either based on observed magnetograms [28–31] or on results from flux emergence simulations [32]. Advanced radiative MHD models have been developed that extend from the upper convection zone into the corona [33–35]. Such models can reproduce many of the observed features of the corona, indicating that magnetic braiding plays an important role in coronal heating.

Theories based on wave heating of the corona also have a long history [36, 37]. Recently, observations have provided strong evidence for the existence of transverse and longitudinal MHD waves in the solar atmosphere [38–43]. The interactions of the photospheric flux tubes with granule-scale flows may create transverse MHD waves that propagate upward along the field lines and dissipate their energy in the corona [44–48]. Alfvén waves are of particular interest because they can propagate over large distances before giving up their energy. Transverse MHD waves can be dissipated in a variety of ways: (1) in the presence of density variations across the magnetic field the waves may be damped by resonant absorption [49–51]; (2) torsional Alfvén waves can interact nonlinearly with the background plasma to produce parallel flows and shocks [52–55]; (3) counter-propagating Alfvén waves interact nonlinearly to produce a turbulent cascade of wave energy [56, 57]. Alfvén wave turbulence (AWT) is believed to play an important role in the heating and acceleration of the solar wind [58–60], and reduced MHD models for the fast solar wind have been developed [61, 62]. In this approximation only the incompressible Alfvén waves are simulated, and other MHD waves are neglected.

Coronal loops may also be heated by AWT. Reduced MHD models have been developed in which the coronal loop is treated as a single magnetic flux tube that extends from the photosphere at one end of the loop to the photosphere at the other end [63–66]. The Alfvén waves are generated by small-scale footpoint motions inside the flux tube. Recently, we presented a more realistic version in which the loop is anchored in multiple flux tubes [67]. Full 3D MHD models of AWT have also been developed [68, 69], and such models show that the turbulent waves can drive strong plasma flows along the loop. In the present paper we review the results from recent loop modeling and discuss the implications for coronal heating. The main difference between the magnetic braiding model and the AWT model is the time scale of the footpoint motions: short-period motions with correlation times less than about 1 minute tend to produce transverse waves, whereas longer-period motions tend to produce magnetic braiding.

2. Coronal Loop Models

The dynamics of transverse waves in AR loops have been simulated using 3D reduced MHD models [63–67]. Horizontal "footpoint" motions are assumed to be present inside the kilogauss flux tubes in the photosphere on length scales of about 100 km, below the resolution of present-day telescopes. The random footpoint motions are caused by granule-scale convective flows, and are assumed to have velocities of about 1.5 km s^{-1} and dynamical time scales of about 1 minute.

The footpoint motions create transverse MHD waves that travel upward along the flux tubes. For simplicity the waves are approximated as incompressible Alfvén waves that are confined to the interiors of the flux tubes, so that we can use the reduced MHD approximation. As the waves propagate upward in height, they are significantly amplified due to the density stratification in the photosphere and chromosphere. The upward propagating waves reflect due to gradients in Alfvén speed and create downward propagating waves. Nonlinear interactions between the counter-propagating waves cause a turbulent cascade of wave energy to small perpendicular length scales, where the energy is dissipated. Such dissipation occurs everywhere along the loop, and is assumed to be the main mechanism for coronal heating. At the chromosphere-corona transition region (TR) about 10% of the wave energy is transmitted into the corona and the remainder is reflected back down. Despite this strong reflection, the transmitted energy is sufficient to heat the coronal part of the loop to temperatures of about 2.5 MK.

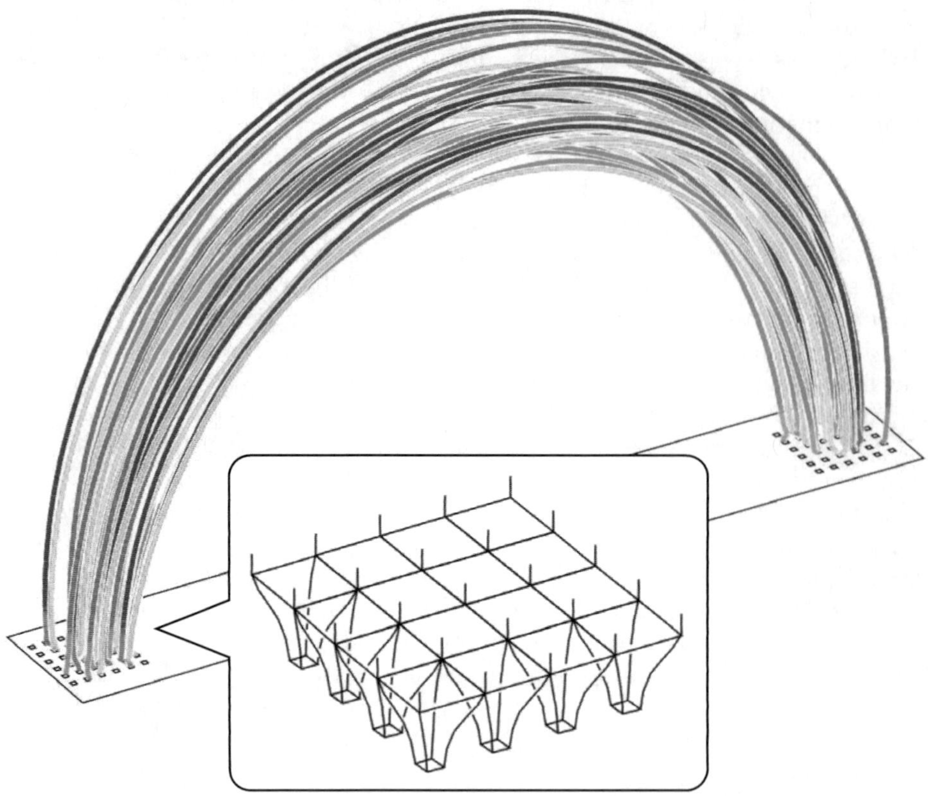

Figure 1. Magnetic structure of a coronal loop heated by AWT. Note that field lines (colored curves) deviate only slightly from the background dipolar field. The inset shows the magnetic flux tubes in the photosphere.

Recently, a version of the AWT model was developed in which the coronal loop is anchored in a collection of photospheric flux tubes [67]. Figure 1 shows the magnetic geometry of the model: the colored curves show the magnetic field lines in the corona, and the inset shows the array of 4×4 flux tubes in the photosphere at each end of the loop. The flux tubes have square cross-sections and expand with height in the photosphere. Neighboring flux tubes merge at a height of about 520 km in the low chromosphere. The merged field extends from the chromosphere at one end of the loop to the chromosphere at the other end, so the TRs are located within the merged field (the coronal loop length $L_c = 98.4$ Mm). Alfvén waves with complex transverse wave patterns are generated by imposing random footpoint motions inside each of the 32 flux tubes. The waves are described using the reduced MHD approximation: the magnetic- and velocity perturbations are given by $\delta\mathbf{B}_\perp = \nabla h \times \mathbf{B}_0$ and $\delta\mathbf{v}_\perp = \nabla f \times \hat{\mathbf{s}}$, where $\mathbf{B}_0(\mathbf{r})$ is the undisturbed background field, $\hat{\mathbf{s}}$ is the unit vector along $\mathbf{B}_0$, $h(\mathbf{r},t)$ is the magnetic flux function, and $f(\mathbf{r},t)$ is the velocity stream function. The side boundaries of the flux tubes are fixed, and the normal components of $\delta\mathbf{B}_\perp$ and $\delta\mathbf{v}_\perp$ vanish at these boundaries; in the merged field we use periodic side boundary conditions. At the merging height the waves can travel from the flux tubes into the merged field or vice versa.

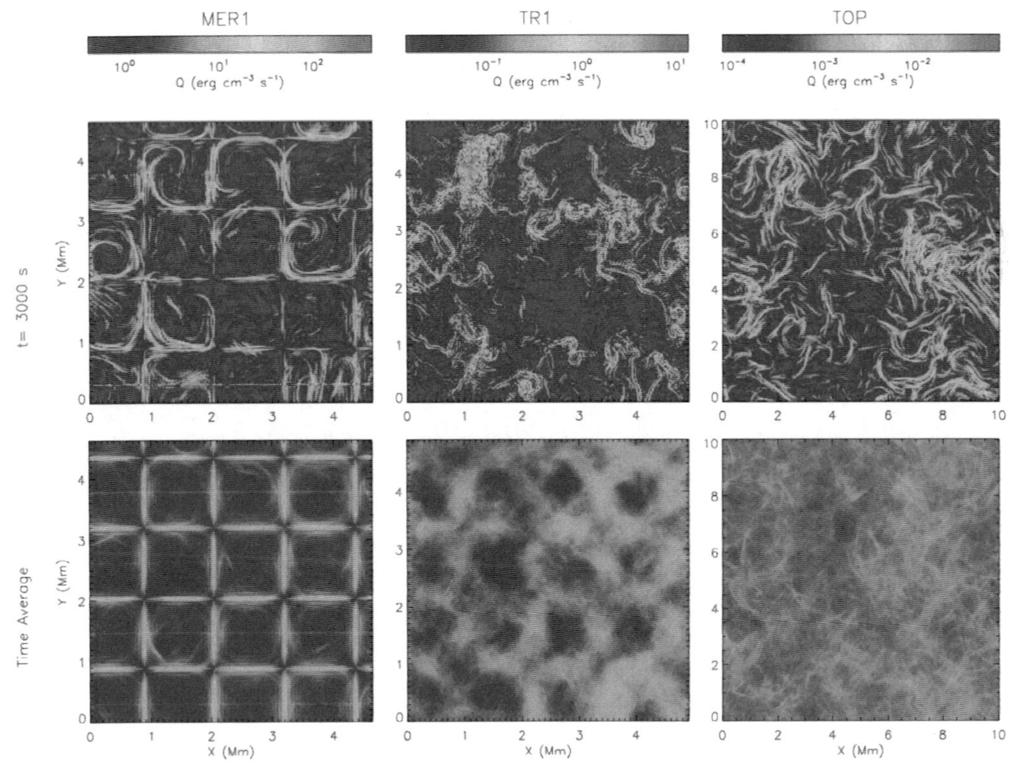

Figure 2. Heating rate in the AWT model, plotted as function of position (x, y) in three cross-sections of the loop: at the merging height (MER1), the chromosphere-corona transition region (TR1) and the loop top (TOP). The upper row shows the heating rate at one instant of time ($t = 3000$ s), and the bottom row shows the time average over 2800 s. The color bars at the top show the heating rates on logarithmic scales.

The nonlinear dynamics of the waves are simulated for a period of 3000 s. As in models with a single flux tube, AWT develops everywhere along the loop and produces significant heating in the corona. The magnetic perturbations at the loop top are only about 1 G, while the background field at the top is about 38 G. Therefore, the magnetic perturbations are relatively small and the field lines have small tilt angles relative to the background potential field (see Figure 1). Magnetic braiding does occur in the AWT model, but the braids are highly dynamic and not close to a force-free state. This is a consequence of the fact that the lower atmosphere is included in the model. It takes about 41 s for the waves to travel from the base of the photosphere to the TR, and a comparable time (47 s) to travel through the corona, so the waves spend most of the time in the lower atmosphere. Also, the high density of the photosphere compared to the corona implies that all perturbations tend to be wave-like [66]. Another feature of the AWT model is that the energy injected into the corona is dissipated very efficiently (on a time scale comparable to the coronal Alfvén travel time), hence there is no long-term energy storage, as is assumed to occur in the magnetic braiding model.

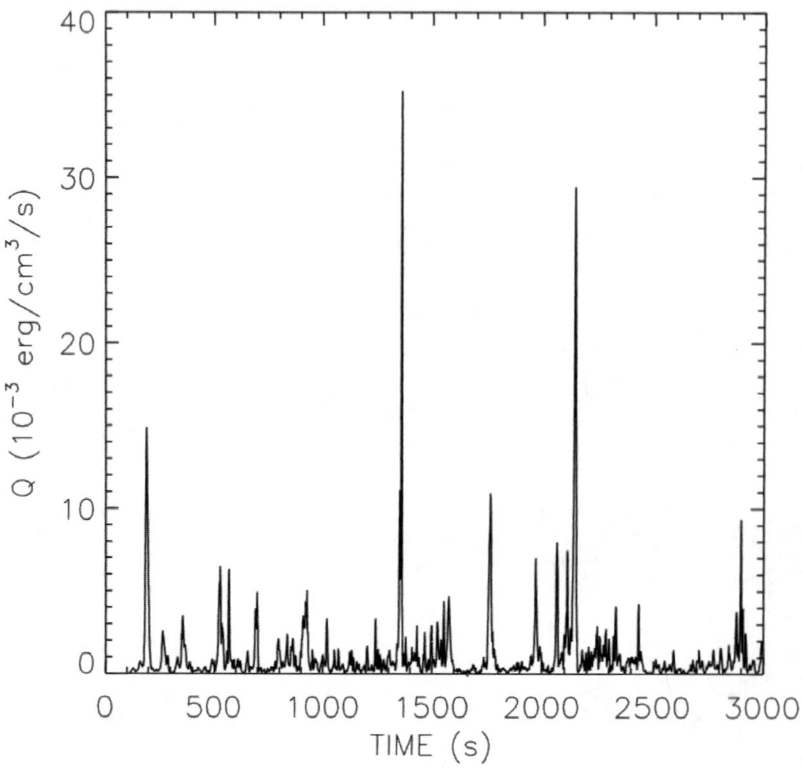

Figure 3. Heating rate as function of time for a co-moving point at the loop top.

Figure 2 shows the spatial distribution of the heating rate $Q(x, y)$ in cross-sections of the loop at the merging height, the TR and the loop top. Note that the instantaneous heating rates (top row) vary by more than 2 orders of magnitude, and most of the heating occurs in thin current sheets and shear layers. At the merging height the strongest heating occurs near the fixed boundaries between the flux tubes (square grid), but in the corona the shear layers are produced by turbulence and are not strongly correlated with the tube boundaries at the merging height. Therefore, the effect of these boundaries does not extend to large height. At the loop top the time-averaged heating rate is nearly uniform across the loop (see lower right panel in Figure 2). Figure 3 shows the heating rate as function of time for a point moving with the turbulent flow at the loop top. Note that the heating is highly variable with strong heating events that occur whenever the tracked point passes through a current sheet or shear layer. The energy dissipated in such "nanoflares" is a combination of magnetic- and kinetic energy. There are about 13 events with peak heating rates in excess of 0.004 erg cm^{-3} s^{-1} and durations of about 20 s, so each event produces about 0.08 erg cm^{-3}. This is small compared to the thermal energy density at the loop top ($E_{th} \approx 2.7$ erg cm^{-3} s^{-1}), so the simulated heating events are not expected to produce large temperature fluctuations in the corona.

The AWT model produces only enough heat to maintain the coronal loop at a pressure of 1.8 dyne cm^{-2} and peak temperature of 2.5 MK. As shown in Figures 2 and 3, the heating rate Q varies strongly in space and time, but the time-averaged rate is nearly constant across the loop and the temperature fluctuations are predicted to be small. In contrast, the observations show that ARs have broad DEM distributions and the peak DEM occurs at temperatures of 3 – 4 MK [2–4]. Therefore, the AWT model cannot explain the observed DEM distributions.

In the AWT model we assume that the side boundaries of the flux tubes are fixed in space, so the model does not take into account that the granule-scale flows also cause lateral displacements of the flux tubes. It is difficult to incorporate such displacements in a reduced MHD model. Therefore, we also constructed a "magnetic braiding" model in which the photospheric flux tubes are removed and the "footpoint" motions are applied directly at the merging height [67]. The imposed footpoint motions have an rms velocities of 1 km s^{-1}, a correlation time of about 400 s, and a perpendicular length scale of 1100 km, which corresponds to a photospheric diffusion constant $D \approx 200$ km^2 s^{-1}, consistent with observations [70]. We find that the imposed footpoint motions cause quasi-static braiding of the coronal field lines. The magnetic perturbations are again about 1 G at the loop top, similar to the AWT model. However, the magnetic energy is dissipated on longer time scales, and the heating rates involved are much smaller than those in the AWT model. We conclude that neither the AWT model nor the magnetic braiding model can reproduce the observed high-temperatures emissions.

3. Discussion

The broad DEM distributions observed in ARs are believed to be produced by "nanoflare" heating of plasma to temperatures of about 10 MK and subsequent cooling by radiative and conductive losses [15]. To heat a plasma with initial temperature $T \approx 2$ MK and electron density $n_e \approx 2 \times 10^9$ cm^{-3} to an electron temperature of 10 MK requires an energy input of at least 3.3 erg cm^{-3} per heating event. Equating this energy to the magnetic energy density $\delta B_\perp^2/8\pi$ we find that magnetic perturbations with amplitude $\delta B_\perp \sim 9$ G are required. Therefore, the magnetic perturbations needed to obtain high temperatures ($T > 4$ MK) are significantly larger than the values $\delta B_\perp \sim 1$ G predicted by the AWT and magnetic braiding models. Both models assume the energy input into the corona is provided by granule-scale convective flows acting on kilogauss flux tubes in the photosphere, and the small length scale of such flows (~ 1 Mm) compared to the loop length (~ 100 Mm) limits the amount of energy that can be injected. Therefore, it appears that granule-scale flows alone cannot produce the energy inputs for high-temperatures emissions, and other forms of energy input should be considered.

Bipolar ARs are formed by the emergence of Ω-shaped loops in the solar atmosphere. During the growth phase of an AR a large amount of magnetic energy is injected into the corona, and new ARs are generally very bright in X-rays. In some ARs smaller magnetic bipoles continue to emerge for several days after the initial growth phase of the region [71]. The opposite polarities of these bipoles move apart over a period of several hours and interact with the other flux concentrations in the region. In the corona the newly emerged field is generally not aligned with the preexisting field, creating electric currents at the boundaries between the old and new flux systems, and the magnetic free energy associated with these currents may be released by magnetic reconnection. We suggest that heating associated with emerging flux is important for some ARs. As an example we consider AR 11726, which was observed with the Extreme Ultraviolet Normal Incidence Spectrograph (EUNIS) rocket instrument [72]. The observers found pervasive faint emission in the Fe XIX 592.2 Å spectral line, which is formed

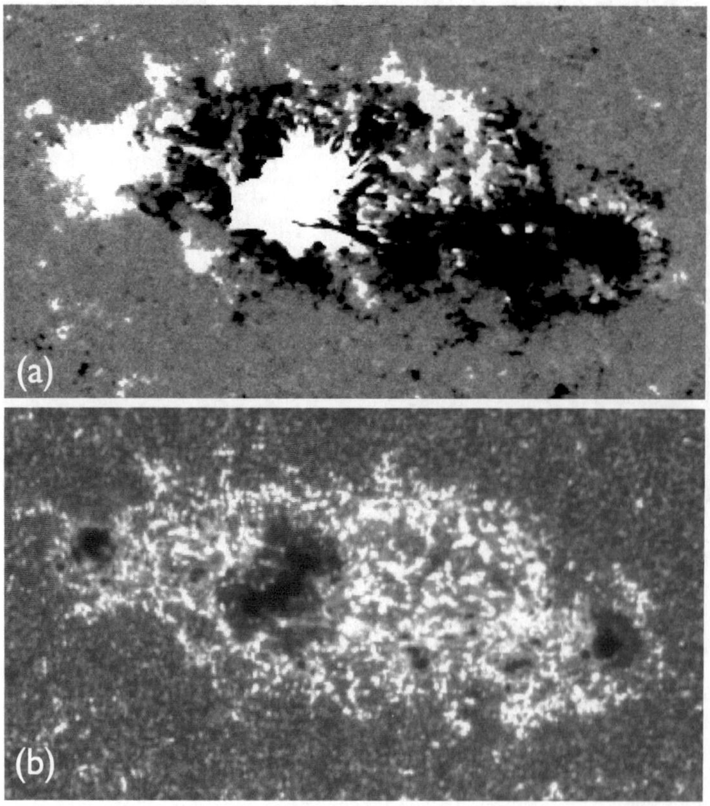

Figure 4. Active region 11726 on 2013 April 20, three days before the EUNIS flight: (a) magnetogram obtained with SDO/HMI; (b) image in the 1700 Å channel of SDO/AIA.

at a temperature of about 9 MK. Figure 4 shows this AR three days before the rocket flight when the region was near the central meridian. Figure 4(a) shows a magnetogram obtained with the Helioseismic and Magnetic Imager (HMI) on the Solar Dynamics Observatory (SDO), and Figure 4(b) shows a 1700 Å image obtained with the Atmospheric Imaging Assembly (AIA) on SDO[1]. Note that away from the large sunspots the magnetic field is complex and contains many small opposite polarity features, which are part of recently emerged bipoles. Therefore, the mixed-polarity field in Figure 4(a) indicates that new flux is emerging in this region. We suggest that magnetic reconnection associated with emerging flux may be responsible for the high-temperature emission observed by EUNIS.

When observed in X-rays, many ARs show the presence of bright S-shaped or inverse S-shaped structures called "sigmoids" [73–77]. The magnetic field of a sigmoid can be described as a sheared arcade and/or a coronal flux rope lying horizontally over the polarity inversion line (PIL). The sigmoid field deviates strongly from the potential field, and contains a large amount of magnetic free energy (10^{30} – 10^{32} erg). Sigmoids are often associated with Hα filaments, flares and coronal mass. The non-potential field can also be detected in photospheric vector magnetograms [78–80]. The energy build-up in the non-potential field may be due to photospheric flux cancellation [81] and/or magnetic reconnection in the corona [82]. It has been

[1] http://www.suntoday.lmsal.com/suntoday

suggested that some of the heating in ARs may be due to the release of magnetic free energy associated with coronal flux ropes [83]. In this case the energy is already present in the corona, and does not need to be injected into the corona through photospheric footpoint motions. The free energy in a coronal flux rope may be sufficient to explain both the heating of the coronal plasma and the occurrence of flares and coronal mass ejections. A possible example is AR 11520, which was observed with the High-resolution Coronal Imager (Hi-C), the X-ray Telescope (XRT) on Hinode, and instruments on SDO [84]. The observers found rapid variability in "moss" regions at the footpoints of hot loops seen in the 94 Å channel of SDO/AIA, and in XRT images. They interpret this variability as a signature of heating events associated with reconnection occurring in the overlying hot coronal loops, i.e., coronal nanoflares. Some of the hot loops in this region overlie the PIL and deviate strongly from the potential field (see Figure 1 in [84]). This suggests that the nanoflare heating in these loops may derive its energy from the non-potential field, not from perturbations that propagate upwards through the footpoints.

In summary, reduced MHD modeling suggests that granule-scale convective flows cannot provide enough energy to heat AR loops to high temperatures ($T > 4$ MK). To explain the observed broad DEM distributions, more energetic "nanoflare" heating events are required. Two possible mechanisms for producing strong heating events are discussed: (1) injection of energy by emergence of magnetic flux, and (2) the dissipation of energy associated with a large-scale non-potential field (e.g., in sigmoids). Further observational studies are needed to determine whether such processes play a significant role in producing the observed high-temperature emissions.

Acknowledgments
This project was supported under contract NNM07AB07C from NASA to the Smithsonian Astrophysical Observatory (SAO) and contract SP02H1701R from Lockheed Martin Space and Astrophysics Laboratory (LMSAL) to SAO.

References

[1] Tripathi D, Klimchuk J A and Mason H E 2011 *Astrophys. J.* **740** 111
[2] Winebarger A R, Schmelz J T, Warren H P, Saar S H and Kashyap V L 2011 *Astrophys. J.* **740** 2
[3] Warren H P, Brooks D H and Winebarger A R 2011 *Astrophys. J.* **734** 90
[4] Warren H P, Winebarger A R and Brooks D H 2012 *Astrophys. J.* **759** 141
[5] DelZanna G, Tripathi D, Mason H, Subramanian S and O'Dwyer B 2015 *Astron. & Astrophys.* **573** A104
[6] Schmelz J T, Asgari-Targhi M, Christian G M, Dhaliwal R S and Pathak S 2015 *Astrophys. J.* **806** 232
[7] Schmelz J T, Christian G M and Matheny P O 2016 *Astrophys. J.* **833** 182
[8] Ugarte-Urra I, Warren H P and Brooks D H 2009 *Astrophys. J.* **696** 642
[9] Ugarte-Urra I and Warren H P 2014 *Astrophys. J.* **783** 12
[10] Viall N M and Klimchuk J A 2011 *Astrophys. J.* **738** 24
[11] Parker E N 1988 *Astrophys. J.* **330** 474
[12] Cargill P 1994 *Astrophys. J.* **422** 381
[13] Klimchuk J A 2009 *The Second Hinode Science Meeting: Beyond Discovery-Toward Understanding* (*ASP Conference Series* vol 415) ed Lites B *et al.* (San Francisco: Astronomical Society of the Pacific) p 221
[14] Bradshaw S J and Klimchuk J A 2011 *Astrophys. J. Suppl.* **194** 26
[15] Cargill P J, Warren H P and Bradshaw S J 2015 *Phil. Trans. R. Soc.* A **373** 20140260

[16] Klimchuk J A 2006 *Solar Phys.* **234** 41
[17] DeMoortel I and Browning P 2015 *Phil. Trans. R. Soc.* A **373** 20140269
[18] Dudík J, Dzifčákov'a E and Cirtain J W 2014 *Astrophys. J.* **796** 20
[19] Buehler D, Lagg A, Solanki S K and van Noort M 2015 *Astron. & Astrophys.* **576** A27
[20] Parker E N 1972 *Astrophys. J.* **174** 499
[21] Parker E N 1983 *Astrophys. J.* **264** 642
[22] Mikić Z, Schnack D D and van Hoven G 1989 *Astrophys. J.* **338** 1148
[23] Galsgaard K and Nordlund Å 1996 *J. Geophys. Res.* **101** 13445
[24] Rappazzo A F, Velli M, Einaudi G and Dahlburg R B 2008 *Astrophys. J.* **677** 1348
[25] Dahlburg R B, Einaudi G, Taylor B D, Ugarte-Urra I, Warren H P, Rappazzo A F and Velli M 2016 *Astrophys. J.* **817** 47
[26] Pontin D I and Hornig G 2015 *Astrophys. J.* **805** 47
[27] Pontin D I, Janvier M, Tiwari S K, Galsgaard K, Winebarger A R and Cirtain J W 2017 *Astrophys. J.* **837** 108
[28] Gudiksen B V and Nordlund Å 2005 *Astrophys. J.* **618** 1020
[29] Bingert S and Peter H 2011 *Astron. & Astrophys.* **530** A112
[30] Bourdin P A, Bingert S and Peter H 2013 *Astron. & Astrophys.* **555** A123
[31] Bourdin P A, Bingert S and Peter H 2015 *Astron. & Astrophys.* **580** A72
[32] Chen F, Peter H, Bingert S and Cheung M C M 2014 *Astron. & Astrophys.* **564** A12
[33] Hansteen V, Hara H, Pontieu B D and Carlsson M 2010 *Astrophys. J.* **718** 1070
[34] Hansteen V, Guerreiro N, Pontieu B D and Carlsson M 2015 *Astrophys. J.* **811** 105
[35] Rempel M 2017 *Astrophys. J.* **834** 10
[36] Alfvén H 1947 *MNRAS* **107** 211
[37] Hollweg J V 1978 *Sol. Phys.* **56** 305
[38] Pontieu B D *et al.* 2007 *Science* **318** 1574
[39] Tomczyk S, McIntosh S W, Keil S L, Judge P G, Schad T, Seeley D H and Edmondson J 2007 *Science* **317** 1192
[40] Tomczyk S and McIntosh S W 2009 *Astrophys. J.* **697** 1384
[41] Tian H *et al.* 2014 *Science* **346** 1255711
[42] Morton R J, Tomczyk S and Pinto R 2015 *Nat. Comm.* **6** 7813
[43] Antolin P, Moortel I D, van Doorsselaere T and Yokoyama T 2017 *Astrophys. J.* **836** 219
[44] Matsumoto T and Shibata K 2010 *Astrophys. J.* **710** 1857
[45] Parnell C E and Moortel I D 2012 *Phil. Trans. R. Soc.* A **370** 3217
[46] Mumford S J, Fedun V and Erdélyi R 2015 *Astrophys. J.* **799** 6
[47] Arregui I 2015 *Phil. Trans. R. Soc.* A **373** 20140261
[48] Cranmer S R, Asgari-Targhi M, Miralles M P, Raymond J C, Strachan L, Tian H and Woolsey L N 2015 *Phil. Trans. R. Soc.* A **373** 20140148
[49] Heyvaerts J and Priest E R 1983 *Astron. & Astrophys.* A **117** 220
[50] Pascoe D J, Hood A W, de Moortel I and Wright A N 2012 *Astron. & Astrophys.* A **539** 37
[51] Goossens M, Andries J, Soler R, van Doorsselaere T, Arregui I and Terradas J 2012 *Astrophys. J.* **753** 111

[52] Kudoh T and Shibata K 1999 *Astrophys. J.* **514** 493
[53] Moriyasu S, Kudoh T, Yokoyama T and Shibata K 2004 *Astrophys. J. Letters* **601** L107
[54] Suzuki T K and Inutsuka S I 2005 *Astrophys. J. Letters* **632** L49
[55] Antolin P and Shibata K 2010 *Astrophys. J.* **712** 494
[56] Matthaeus W H, Zank G P, Oughton S, Mullan D J and Dmitruk P 1999 *Astrophys. J. Letters* **523** L93
[57] Dmitruk P and Matthaeus W H 2003 *Astrophys. J.* **597** 1097
[58] Cranmer S R, van Ballegooijen A A and Edgar R J 2007 *Astrophys. J. Suppl.* **171** 520
[59] Verdini A and Velli M 2007 *Astrophys. J.* **662** 669
[60] Chandran B D G, Dennis T J, Quataert E and Bale S D 2011 *Astrophys. J.* **743** 197
[61] Perez J C and Chandran B D G 2013 *Astrophys. J.* **776** 124
[62] van Ballegooijen A A and Asgari-Targhi M 2017 *Astrophys. J.* **835** 10
[63] van Ballegooijen A A, Asgari-Targhi M, Cranmer S R and DeLuca E E 2011 *Astrophys. J.* **736** 3
[64] Asgari-Targhi M and van Ballegooijen A A 2012 *Astrophys. J.* **746** 81
[65] Asgari-Targhi M, van Ballegooijen A A, Cranmer S R and DeLuca E E 2013 *Astrophys. J.* **773** 111
[66] van Ballegooijen A A, Asgari-Targhi M and Berger M A 2014 *Astrophys. J.* **787** 87
[67] van Ballegooijen A A, Asgari-Targhi M and Voss A 2017 *Astrophys. J.* **849** 46
[68] Matsumoto T 2016 *MNRAS* **463** 502
[69] Matsumoto T 2018 *MNRAS* **476** 3328
[70] Berger T E, Lofdahl M G, Shine R A and Title A M 1998 *Astrophys. J.* **506** 439
[71] Ortiz A, Bellot Rubio L R, Hansteen V H, de la Cruz Rodríguez J and Rouppe van der Voort L 2014 *Astrophys. J.* **781** 126
[72] Brosius J W, Daw A N and Rabin D M 2014 *Astrophys. J.* **790** 112
[73] Acton L *et al.* 1992 *Science* **258**(5082) 618
[74] Rust D M and Kumar A 1996 *Astrophys. J. Letters* **464** L199
[75] Canfield R C, Hudson H S and McKenzie D E 1999 *Geophys. Res. Lett.* **26** 627
[76] Sterling A C, Hudson H S, Thompson B J and Zarro D M 2000 *Astrophys. J.* **532** 628
[77] Gibson S E *et al.* 2002 *Astrophys. J.* **574** 1021
[78] Gary G A, Moore R L, Hagyard M J and Haisch B M 1987 *Astrophys. J.* **314** 782
[79] Pevtsov A A, Canfield R C and Metcalf T R 1995 *Astrophys. J. Letters* **440** L109
[80] Leka K D, Canfield R C, McClymont A N and van Driel-Gesztelyi L 1996 *Astrophys. J.* **462** 547
[81] Savcheva A S, Green L M, van Ballegooijen A A and DeLuca E E 2012 *Astrophys. J.* **759** 105
[82] James A W, Green L M, Palmerio E, Valori G, Reid H A S, Baker D, Brooks D H, van Driel-Gesztelyi L and Kilpua E K J 2017 *Solar Phys.* **292** 71
[83] van Ballegooijen A A and Cranmer S R 2008 *Astrophys. J.* **682** 644
[84] Testa P *et al.* 2013 *Astrophys. J. Letters* **770** L1

Implications of Recent Stellar Wind Measurements

Brian E. Wood[1]

[1] Naval Research Laboratory, Space Science Division, Washington, DC 20375, USA

E-mail: brian.wood@nrl.navy.mil

Abstract.

Very recent measurements of stellar winds are used to update relations between winds and coronal activity. New wind constraints include an upper limit of $\dot{M} < 0.1\ \dot{M}_\odot$ for τ Ceti (G8 V), derived from a nondetection of astrospheric H I Lyman-α absorption. This upper limit is reported here for the first time, and represents the weakest wind constrained using the astrospheric absorption technique. A high mass loss rate measurement of $\dot{M} = 10\ \dot{M}_\odot$ for δ Pav (G8 IV) from astrospheric Lyman-α absorption suggests stronger winds for subgiants than for main sequence stars of equivalent activity. A very low mass-loss rate of $\dot{M} \approx 0.06\ \dot{M}_\odot$ recently estimated for GJ 436 (M3 V) from Lyman-α absorption from an evaporating exoplanetary atmosphere implies inactive M dwarfs may have weak winds compared with GK dwarfs of similar activity.

1. Introduction

The first stellar wind ever detected was naturally that of our own Sun, by one of the earliest spacecraft [1]. Most stars are now known to possess stellar winds of some sort, but hot coronal winds like that of the Sun are among the hardest to detect. The solar wind is quite weak compared to the stronger and more easily detected winds of hot stars and cool giant/supergiant stars. There has still in fact been no unambiguous direct detection of a coronal wind streaming from another star.

The primary coronal wind detection method that has arisen relies on detecting not the wind itself, but its interaction with the surrounding interstellar medium (ISM), i.e., the stellar astrosphere. This technique relies on detecting H I Lyman-α absorption from the astrosphere, using high resolution UV spectra from the *Hubble Space Telescope* (HST), specifically from the Goddard High Resolution Spectrograph (GHRS) or Space Telescope Imaging Spectrograph (STIS) instruments. The first such detection was for the nearest star system, α Cen AB [2]. Numerous detections have since followed [3, 4], but not every observed line-of-sight (LOS) has led to a detection. In addition to the astrospheric absorption, absorption from our own heliosphere is also sometimes observed in upwind directions of the ISM flow seen by the Sun, depending on the amount of obscuring absorption from the ISM itself. It is worth noting that in addition to the HST Lyman-α studies of the astrospheres of coronal stars, the astrospheres of other types of stars have been studied using both imaging and spectrocopy [5, 6, 7].

Mass loss rates estimated from astrospheric Lyman-α absorption have been used to assess how coronal winds relate to coronal activity, as quantified by X-ray surface flux, F_X [3, 4]. From these results it is possible to infer what the solar wind was like when the Sun was younger and more active. Constraints on wind evolution are also important for understanding how coronal stars shed angular momentum [8]. Wind effects on exoplanets are another motivating factor for improving our understanding of coronal winds. Most of the exoplanets that have been discovered orbit very close to their stars, where they will likely see very high wind fluxes due to their close proximity [9, 10, 11].

2. New Wind Measurements

It has been over 20 years since the first astrospheric detection, and HST continues to be capable of providing relevant new data. Nevertheless, the number of detections remains relatively small. Table 1 lists the detections and resulting mass-loss rate measurements, $\dot{M}$, in solar units, with $\dot{M}_\odot = 2 \times 10^{-14}$ $M_\odot$ yr^{-1} [3, 4]. The $\dot{M}$ measurements require knowledge of the ISM wind vector seen in the rest frame of the star, so Table 1 also lists the wind speed seen by the star, V_{ISM}, and the angle between the upwind direction of the ISM flow vector and our LOS to the star, θ. The final two columns list X-ray luminosities (in erg s^{-1} units) and stellar radii.

Nearly all of the detections listed as "Old Observations" are from before 2005 [3], with the exception of π^1 UMa, which is from 2014 [4]. There are a number of reasons for the sparsity of detections post-2005, and the relatively paltry number of detections overall. The distance column in Table 1 provides an indication of one fundamental difficulty. There are only three detections outside 10 pc, and two of those are actually considered marginal [12]. Two major reasons for nondetections are high ISM column densities, meaning that ISM Lyman-α absorption obscures the astrospheric absorption signature; and a surrounding ISM that is fully ionized, meaning that there is no astrospheric Lyman-α absorption at all because no neutral H is injected into the astrosphere. Both of these are far more likely to occur for more distant targets, for reasons now described.

The Sun lies in a region of space called the Local Bubble (LB), a region of generally low ISM density extending roughly 100 pc from the Sun in most directions [13, 14]. Most of the LB is hot and fully ionized. However, there are small, warm, partially neutral clouds embedded within it. It so happens that the Sun resides within one of these, the Local Interstellar Cloud (LIC), which is one of a number of similar clouds in the solar vicinity [15]. Very nearby stars are likely to reside within either the LIC or one of the other nearby clouds. For such stars, there will probably be an astrospheric absorption signature, since the surrouding ISM is likely partly neutral, and there is also a good chance the ISM H I column will not be too high to obscure that signature due to the short LOS.

However, an undetectable astrospheric absorption signature can also arise from a low V_{ISM} value, a disadvantageous astrosphere orientation, or a weak stellar wind. Low V_{ISM} leads to cooler and less decelerated astrospheric H I, which makes the astrospheric

Table 1. Mass Loss Measurements for Coronal Winds

Star	Spectral Type	d (pc)	V_{ISM} (km/s)	θ (deg)	$\dot{M}^a$ ($\dot{M}_\odot$)	Log L_x	Radius ($R_\odot$)
OLD MEASUREMENTS							
Prox Cen	M5.5 Ve	1.30	25	79	< 0.2	27.22	0.14
α Cen AB	G2 V+K0 V	1.35	25	79	0.46+1.54	26.99+27.32	1.22+0.86
ϵ Eri	K2 V	3.22	27	76	30	28.31	0.74
61 Cyg A	K5 V	3.48	86	46	0.5	27.03	0.67
ϵ Ind	K5 V	3.63	68	64	0.5	27.39	0.73
EV Lac	M3.5 Ve	5.05	45	84	1	28.99	0.30
70 Oph AB	K0 V+K5 V	5.09	37	120	55.7+44.3	28.09+27.97	0.83+0.67
36 Oph AB	K1 V+K1 V	5.99	40	134	8.5+6.5	28.02+27.89	0.69+0.59
ξ Boo AB	G8 V+K4 V	6.70	32	131	0.5+4.5	28.91+28.08	0.86+0.61
61 Vir	G5 V	8.53	51	98	0.3	26.87	0.99
δ Eri	K0 IV	9.04	37	41	4	27.05	2.58
π^1 UMa	G1.5 V	14.4	43	34	0.5	28.99	0.97
λ And	G8 IV-III	25.8	53	89	5	30.82	7.40
DK UMa	G4 III-IV	32.4	43	32	0.15	30.36	4.40
NEW MEASUREMENTS							
τ Ceti	G8 V	3.65	56	59	< 0.1	26.69	0.77
δ Pav	G8 IV	6.11	29	72	10	27.29	1.22
GJ 892	K3 V	6.55	49	60	0.5	26.85	0.78
GJ 436	M3 V	9.75	79	97	0.059^b	26.76	0.44

[a]Mass loss rates for individual members of binaries estimated from mass-loss/activity relations (see text).
[b]Measured from absorption from an evaporating exoplanetary atmosphere instead of from astrospheric absorption.

absorption narrower and less separable from the ISM absorption. It is not a coincidence that all of the detections in Table 1 have V_{ISM} higher or comparable to the solar value of $V_{ISM} = 26.08 \pm 0.21$ km s^{-1} [16]. As for astrosphere orientation, it is easier to detect astrospheres in upwind directions (i.e., $\theta < 90°$). All but three of the detections in Table 1 have $\theta < 100°$. As for the stellar wind, a weak wind naturally means a smaller astrosphere with lower H I column densities, and therefore weaker and narrower astrospheric absorption. Measurements of mass-loss rates from astrospheric absorption rely on this effect.

The low detection fraction of astrospheres, particularly at distances over 10 pc, makes it very difficult to obtain HST observing time explicitly to detect more astrospheres. Most of the detections in Table 1 are actually based on archival HST spectra taken for reasons having nothing to do with astrospheres, or even with the H I Lyman-α line.

IOP Conf. Series: Journal of Physics: Conf. Series **1100** (2018) 012028 doi:10.1088/1742-6596/1100/1/012028

And this leads to the explanation for why there have been so few detections since 2005. The golden age for this kind of research was 1997–2004, after the STIS instrument was installed on HST (replacing GHRS), but before STIS suffered a failure in 2004. During this period, STIS was the primary UV spectrometer on HST, and many observers used its E140M grating to observe the entire 1150–1700 Å spectral region for various reasons, most having nothing to do with the H I Lyman-α line at 1216 Å. This filled the HST archives with spectra that could be used to search for astrospheric absorption. Except for α Cen and π^1 UMa, all of the "Old Measurements" of astrospheres listed in Table 1 are from data taken during this time period [12].

The STIS instrument was repaired during the last HST servicing mission in 2009, but this mission also installed another UV spectrometer onto HST, the Cosmic Origins Spectrograph (COS). The newer COS has much higher sensitivity than STIS and is therefore now the dominant FUV spectrometer. However, COS observations cannot be used for astrospheric studies, because COS has insufficient spectral resolution and its large apertures let in too much geocoronal Lyman-α emission. Thus, since 2004 the HST archive has not been accumulating high-resolution Lyman-α spectra usable for astrospheric research at the rate it was in 1997–2004.

Nevertheless, there are two very recent astrosphere detections that have been reported just in the past year, for δ Pav (G8 IV) and GJ 892 (K3 V), as listed in Table 1. For δ Pav, a mass-loss rate of $\dot{M} = 10\ \dot{M}_\odot$ is inferred [17], while for GJ 892 the measurement is $\dot{M} = 0.5\ \dot{M}_\odot$ (Fossati et al. 2018, in preparation). It is noteworthy that δ Pav and GJ 892 are both within 7 pc. For targets within 7 pc, there are now 13 independent lines of sight with HST Lyman-α spectra possessing sufficient spectral resolution for a definitive Lyman-α search for astrospheric absorption. (Proxima Cen is a distant companion of α Cen, and so is not considered an independent LOS from α Cen.) Ten of the 13 lines of sight within 7 pc lead to successful detections of astrospheric absorption.

This 76% detection fraction means that the ISM within 7 pc must be nearly or entirely filled with partially neutral ISM material. This has the important ramification that if all stars within 7 pc are surrounded by partially neutral ISM, then meaningful upper limits for $\dot{M}$ can in principle be measured for the nondetections of astrospheric absorption. The primary reason that upper limits are not generally estimated from nondetections is that a likely reason for most of the nondetections (especially beyond 10 pc) is that the surrounding ISM is fully ionized, meaning that there would be no astrospheric Lyman-α absorption regardless of wind strength. A meaningful upper limit can only be made if this explanation can be excluded, as it could for Proxima Cen given that astrospheric absorption is detected for Proxima Cen's companion, α Cen AB. The only previous $\dot{M}$ upper limit inferred from a nondetection is for Proxima Cen (see Table 1).

The three other nondetections within 7 pc are τ Ceti (G8 V), 40 Eri A (K1 V), and AD Leo (M3.5 Ve) [12]. Based on the preceding paragraph, meaningful upper limits for $\dot{M}$ can in principle be made from the nondetections of astrospheric absorption in the Lyman-α spectra. However, in practice I exclude 40 Eri A and AD Leo, for very different reasons. For AD Leo, the ISM flow speed is an unfavorably low $V_{ISM} = 13$ km s^{-1} value. When combined with the surprisingly high ISM column (in cm^{-2}) of $\log N_H = 18.47$ for this LOS, which is the highest yet observed within 15 pc, I conclude that any $\dot{M}$ upper limit that might be derived for AD Leo would be too high to be useful.

For 40 Eri A the problem is an extremely high ISM wind velocity of $V_{ISM} = 127$ km s^{-1}. It was emphasized above how a low V_{ISM} speed is bad for astrosphere detectability, but it turns out that an extremely high V_{ISM} value can be a problem as well, as it can make astrospheres so small and the astrospheric H I so hot that H I Lyman-α becomes optically thin [18]. This greatly complicates the task of inferring constraints on the wind from the absorption, so I simply ignore 40 Eri A for now. (See [18] for more details.)

This leaves only τ Ceti, which has very advantageous values of $V_{ISM} = 56$ km s^{-1} and $\theta = 59°$, and a low ISM column density of $\log N_H = 18.01$. Furthermore, τ Ceti is particularly close, with $d = 3.65$ pc, making it even more likely to be surrounded by partially neutral ISM material like that around the Sun. Thus, an $\dot{M}$ upper limit for τ Ceti is inferred here for the first time. This requires the assistance of hydrodynamical models of the astrosphere. However, rather than compute new models, which can be time-consuming, existing models for the 61 Vir astrosphere are used instead. These are relevant thanks to the similar $V_{ISM} = 51$ km s^{-1} value of 61 Vir compared to τ Ceti's $V_{ISM} = 56$ km s^{-1}. Note that the recent δ Pav and GJ 892 $\dot{M}$ measurements were similarly made using existing astrospheric models of other stars as proxies [17].

The full Lyman-α spectrum and ISM absorption fit to the τ Ceti data can be found in [12]. Figure 1 zooms into the blue side of the H I Lyman-α absorption profile where the astrospheric absorption would be observed. This is an astrospheric nondetection, meaning that the absorption could be reasonably well fit with ISM absorption alone. Predicted astrospheric absorption is shown for five values of $\dot{M}$, from $\dot{M} = 0.05-1.0\ \dot{M}_\odot$. Models actually only exist for $\dot{M} \geq 0.2\ \dot{M}_\odot$, so the $\dot{M} = 0.05\ \dot{M}_\odot$ and $\dot{M} = 0.1\ \dot{M}_\odot$ models are actually made by extrapolation from the higher $\dot{M}$ models. It is obvious that a $\dot{M} = 0.5 - 1.0\ \dot{M}_\odot$ wind should have yielded easily detectable absorption towards τ Ceti, but detectability naturally decreases for lower $\dot{M}$ values. It is dubious whether the $\dot{M} = 0.05\ \dot{M}_\odot$ model predicts a sufficient amount of absorption to be considered detectable. Relying mostly on subjective judgment, I conclude that it does not and settle on a more conservative $\dot{M} < 0.1\ \dot{M}_\odot$ upper limit for τ Ceti, which is what is reported in Table 1. This is the weakest wind ever measured using the astrospheric absorption technique, slightly lower than the $\dot{M} = 0.15\ \dot{M}_\odot$ value for the marginal DK UMa detection [3].

There is one final new $\dot{M}$ measurement listed in Table 1 that has not yet been mentioned, for the transiting exoplanet host star GJ 436 (M3 V). For this star, Lyman-α absorption has been observed during exoplanet transits, implying an evaporating planetary atmosphere, and the amount of absorption has been used to infer the strength of the stellar wind [19]. Thus, this $\dot{M} = 0.059^{+0.074}_{-0.040}\ \dot{M}_\odot$ constraint is listed in Table 1 as well, though the measurement is from exoplanetary instead of astrospheric Lyman-α absorption.

3. Revised Wind-Activity Relation

It is natural to look for correlations between wind properties and the properties of the stellar coronae that produce the winds. The most readily available coronal property for comparison is the coronal X-ray luminosity, so values of $\log L_X$ are provided in Table 1. Many of these are from ROSAT PSPC measurements [20], but newer *Chandra*

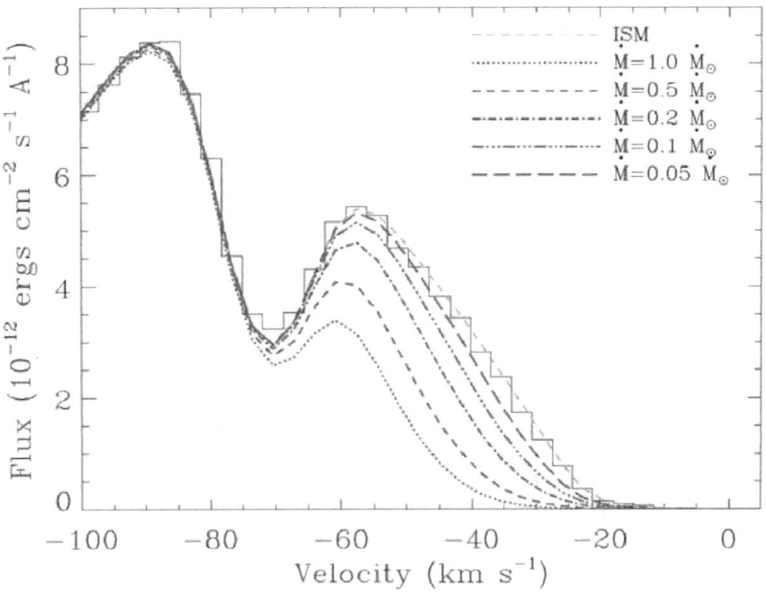

Figure 1. A close-up of the blue side of the H I Lyman-α absorption observed towards τ Ceti, plotted on a heliocentric velocity scale. The narrow absorption at -70 km s^{-1} is from interstellar deuterium. The green dashed line shows the ISM absorption, which provides a reasonably good fit to the data. Red lines show additional absorption predicted by astrospheric models assuming different mass-loss rates of $\dot{M} = 0.05 - 1.0\ \dot{M}_\odot$. The $\dot{M} = 0.1\ \dot{M}_\odot$ model is assumed to represent an $\dot{M}$ upper limit for τ Ceti.

measurements are used when available [21].

Figure 2 plots mass loss rate per unit surface area versus X-ray surface flux for the stars in Table 1. A correlation is seen for GK dwarfs with $\log F_X < 6.0$, and a power law relation is fitted to the data. This relation does not extend beyond the $\log F_X = 6.0$ "Wind Dividing Line," above which surprisingly weak winds are observed. This figure is very much analogous to similar figures that have been published in the past, including the power law fit and wind dividing line [3, 4]. However, the figure is updated in a number of ways. The most obvious is the inclusion of the four new measurements listed in Table 1. There have also been relatively minor revisions to the X-ray luminosities and stellar radii. The resulting change to the power law relation is insignificant: $\dot{M} \propto F_x^{1.29\pm0.16}$ compared to the previous $\dot{M} \propto F_x^{1.34\pm0.18}$ [3]. The new GJ 892 measurements is completely consistent with the relation. However, the new upper limit for τ Ceti looks anomalously low. The significance of this inconsistency is marginal, as the ϵ Ind data point is a similar distance below the power law relation.

Another difference from previous studies lies in our treatment of binary systems.

IOP Conf. Series: Journal of Physics: Conf. Series **1100** (2018) 012028 doi:10.1088/1742-6596/1100/1/012028

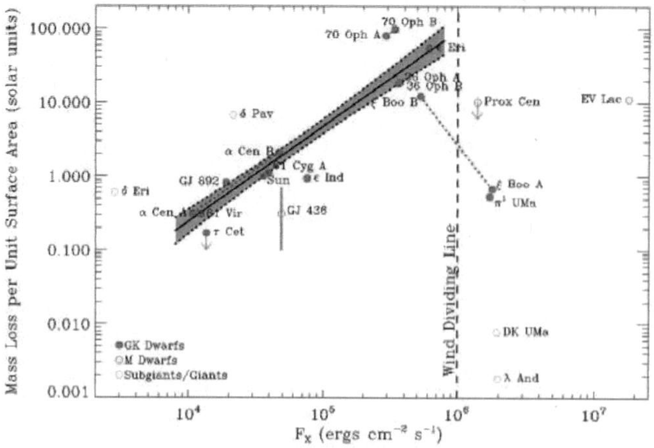

Figure 2. A plot of mass loss rate (per unit surface area) versus X-ray surface flux for all main sequence stars with measured winds, analogous to previously published figures [3, 4]. Different symbols are used to represent GK dwarfs, M dwarfs, and subgiant/giant stars. A power law, $\dot{M} \propto F_X^{1.29\pm0.16}$, is fitted to the GK dwarfs where a wind/X-ray relation seems to exist, but this relation seems to fail for stars to the right of the "Wind Dividing Line." Separate points are plotted for the two members of four binary systems, where for the three less active cases the power law relation is used to estimate the wind contributions of individual stars. In order to yield consistency with stars of similar F_X, for the ξ Boo binary ξ Boo B is assumed to account for 90% of the wind and ξ Boo A only 10%.

There are four systems listed in Table 1 where both members of the binary lie within the same astrosphere (α Cen AB, 70 Oph AB, 36 Oph AB, and ξ Boo AB), meaning both stellar winds will be contributing to the measured mass-loss rate. For the three binaries that lie entirely to the left of the wind dividing line, the power law relation can be used to estimate the contributions of each individual star to the binary's collective mass-loss rate, and the results are indicated in both Table 1 and Figure 2. Separate X-ray luminosities are listed for the two stars as well. The most interesting of these three cases is α Cen AB, where the system's wind is predicted to be dominated by the secondary star, α Cen B, with $\dot{M} = 1.54\ \dot{M}_\odot$ compared to $\dot{M} = 0.46\ \dot{M}_\odot$ for α Cen A. The ξ Boo AB system is more complicated, with the primary to the left of the dividing line and the secondary to the right. In order for the two stars to be consistent with other stars of similar F_X, ξ Boo's wind must be dominated by the secondary. Consistent with past work, a division with $\dot{M} = 4.5\ \dot{M}_\odot$ for ξ Boo B and $\dot{M} = 0.5\ \dot{M}_\odot$ for ξ Boo A is assumed [22].

The new δ Pav measurement has interesting implications for the winds of coronal subgiants/giants. Both δ Pav and δ Eri imply that inactive subgiants have significantly stronger winds per unit surface area compared to main sequence stars of similar F_X. This is not a surprising result given that subgiants have lower surface gravities and surface escape speeds than main sequence stars, which should make it easier for coronal material to escape [17]. In contrast, the very active subgiant/giant stars DK UMa and λ And seem to have remarkably weak winds, perhaps implying that these stars have strong, global fields that inhibit mass loss in some way.

The GJ 436 measurement in Figure 2 provides a first indication that the winds of inactive M dwarfs may be weaker per unit surface area than GK dwarfs with similar F_X. Obviously, additional measurements are required to support this conclusion. Fortunately, these measurements should be forthcoming, as a proposed HST Lyman-α survey of 10 M dwarfs within 7 pc has recently been accepted, and these observations should be carried out in the coming year. In the near future, these data will hopefully allow the M dwarf $\dot{M} - F_X$ relation to be characterized as well as the GK dwarf relation in Figure 2.

Acknowledgments
Support for this project was provided by NASA award NNH16AC40I to the Naval Research Laboratory.

References

[1] Neugebauer M, and Snyder C W 1962 *Science* **138** 1095
[2] Linsky J L, and Wood B E 1996 *ApJ* **463** 254
[3] Wood B E, Müller H -R, Zank G P, Linsky J L, and Redfield S 2005 *ApJ* **628** L143
[4] Wood B E, Müller H -R, Redfield S, and Edelman E 2014 *ApJ* **781** L33
[5] van Marle A J, Decin L, and Meliani Z 2014 *A&A* **561** A152
[6] Kobulnicky H A *et al* 2016 *ApJS* **227** 18
[7] Wood B E, Müller H -R, and Harper G M 2016 *ApJ* **829** 74
[8] Johnstone C P, Zhilkin A, Pilat-Lohinger E, Bisikalo D, Güdel M, and Lüftinger T 2015 *A&A* **577** A28
[9] Lammer H *et al* 2010 *Astrobiology* **10** 45
[10] Alvarado-Gómez J D*et al* 2016 *A&A* **594** A95
[11] Shaikhislamov I F *et al* 2016 *ApJ* **832** 173
[12] Wood B E, Redfield S, Linsky J L, Müller H -R, and Zank G P 2005 *ApJS* **159** 118
[13] Sfeir D M, Lallement R, Crifo F, and Welsh B Y 1999 *A&A* **346** 785
[14] Lallement R, Welsh B Y, Vergely J L, Crifo F, and Sfeir D M 2003 *A&A* **411** 447
[15] Redfield S, and Linsky J L 2008 *ApJ* **673** 283
[16] Wood B E, Müller, H -R, and Witte M 2015 *ApJ* **801** 62
[17] Zachary J, Redfield S, Linsky J L, and Wood B E 2018 *ApJ*, in press
[18] Wood B E, Linsky J L, Müller H -R, and Zank G P 2003 *ApJ* **591** 1210
[19] Vidotto A A, and Bourrier V 2017 *MNRAS* **470** 4026
[20] Schmitt J H M H, and Liefke C 2004 *A&A* **417** 651
[21] Wood B E, Laming J M, Warren H P, and Poppenhaeger K 2018, *ApJ* submitted
[22] Wood B E, and Linsky J L 2010 *ApJ* **717** 1279

IOP Conf. Series: Journal of Physics: Conf. Series **1100** (2018) 012029 doi:10.1088/1742-6596/1100/1/012029

Analytical investigation of turbulence quantities and cosmic ray mean free paths from 1995-2017

L L Zhao[1], L Adhikari[1], G P Zank[1,2], Q Hu[1,2] and X S Feng[3]

[1] Center for Space Plasma and Aeronomic Research (CSPAR), University of Alabama in Huntsville, Huntsville, AL 35899, USA

[2] Department of Space Science, University of Alabama in Huntsville, Huntsville, AL 35899, USA

[3] State Key Laboratory of Space Weather, National Space Science Center, Chinese Academy of Sciences, Beijing, 100190, People's Republic of China

E-mail: `lz0009@uah.edu`

Abstract. We use OMNI 1-minute resolution data sets from 1995 through 2017, covering about two consecutive solar cycles, to investigate the solar cycle dependence of various turbulence quantities and cosmic ray (CR) mean free paths. We employ quasi-linear theory (QLT) and nonlinear guiding center theory (NLGC) to evaluate the CR parallel and perpendicular diffusion. We find that in the ecliptic plane at 1 au (1) the fluctuating magnetic energy density $\langle z^{\pm 2} \rangle$, residual energy E_D, and corresponding correlation functions all have an obvious solar cycle dependence. The residual energy E_D is always negative, which indicates that the energy in magnetic fluctuations is larger than the energy in kinetic fluctuations, especially at solar maximum; (2) the correlation length λ for magnetic fluctuations does not show significant solar cycle variation; (3) the temporally varying shear source of turbulence, which is most important in the inner heliosphere, depends on the solar cycle; and (4) high level turbulence will increase CR perpendicular diffusion and decrease CR parallel diffusion, but this trend can be masked if the background interplanetary magnetic field (IMF) changes in concert with turbulence in response to solar activity. These results provide quantitative inputs for both turbulence transport models and CR diffusion coefficient models.

1. Introduction

The transport of cosmic rays (CRs) throughout the heliosphere is determined by the large-scale solar wind (SW) flow and the magnetized turbulence embedded in it. The CR diffusion tensor is the critical factor in simulating the observed temporal and spatial variation in the CR spectra and describes the scattering of CRs by fluctuations in the interplanetary magnetic field (IMF) [1]. Currently, there are several numerical models that couple CR propagation with magnetic turbulence transport and the background IMF evolution [e.g., 2-4]. It is well established that the background IMF periodically changes from solar minimum to maximum ($\sim$ 11 years). However, a detailed understanding of the dependence of turbulence quantities on solar activity, which is another important component of modeling the time-dependent CR modulation, is not yet well understood.

The evolution of low-frequency turbulent fluctuations in the large-scale radially expanding SW plasma can be interpreted in terms of magnetohydrodynamic (MHD) theory [5]. Zank et al. [6] develop a turbulence transport model to describe the evolution of the total energy density in forward- and backward-propagating modes, the residual energy, the corresponding correlation lengths and the cross-helicity provided the plasma beta is large (> 1). Adhikari et al. [7] study the effect of solar cycle on the evolution of various turbulence quantities based on the Zank et al. [6] model. The SW velocity, source terms, and inner boundary conditions are treated as time dependent. They find that the temporal SW introduces a periodic variability in the magnetic energy fluctuations and correlation length beyond 1 au. Shiota et al. [8] incorporate the Zank et al. [6] model in a three-dimensional (3D) inhomogeneous SW and find reasonable agreement with *Helios* 2 and *Ulysses* measurements. Recently, Zank et al. [9] have derived a coupled 2D and slab turbulence model based on the nearly incompressible (NI) MHD theory, which includes all basic turbulence variables and shows extensive agreement with *Voyager* 2 observations [10]. It is thought that SW fluctuations are highly anisotropic with respect to the direction of the large-scale mean magnetic field. Adhikari et al. [11] investigated anisotropy in magnetic field fluctuations by assuming a 80:20, 70:30, 60:40, and 55:45 initial energy ratio between 2D and slab turbulence at 1 au. They found that the evolving anisotropy ratios in both the energy-containing and inertial range are ordered by the inner boundary ratios, and exhibit similar trends for each initial anisotropy ratio. For slow and intermediate SW in the ecliptic plane at 1 au as used in our study, 2D turbulence dominants (50% - 85%) in the inertial range [12, 13]. Since the 2D and slab energy ratio 80:20 is usually used in theoretical work based on the assumption of a plasma beta ~ 1 or $\ll 1$ in SW or solar corona [e.g., 14-17], we employ an 80:20 energy ratio between 2D and slab turbulence at 1 au. For the correlation scale variability in the SW fluctuation, Weygand et al. [18] suggest that the ratio of slab to 2D correlation scales is 2.55 ± 0.76 (with 0.76 the uncertainty) for the slow SW, 2.15 ± 0.18 for the intermediate SW, and 0.71 ± 0.29 in the fast stream. Following prior studies [e.g., 19-21], we assume the ratio of slab to 2D correlation as 2.0.

Once the turbulence model is determined, the CR diffusion coefficients can be derived from an appropriate energetic particle diffusion theory. Zhao et al. [22] present a detailed analysis of the radial and rigidity dependence of the CR diffusion tensor using the Zank et al. [9] two-component turbulence model. Here, we continue to use quasi-linear theory (QLT, [1]) and nonlinear guiding center (NLGC, [23]) theory, respectively, to evaluate the CR parallel mean free path (MFP) and perpendicular MFP at 1 au. More sophisticated CR diffusion theories have since been developed [e.g., 24-26], but QLT and NLGC provide simple reasonably tractable expressions that have been employed in many energetic particle transport studies. For our purpose, they are sufficiently accurate.

The organization of this paper is as follows: Section 2 presents the data analysis method we employed. Section 3 presents the temporal evolution of various turbulence quantities from 1995 to 2017. Section 4 illustrates the effect of solar activity on CR mean free paths. The last section provides a summary and conclusions.

2. Data Analysis Process

In this paper, we analyze *OMNI* 1-minute resolution data for the magnetic and bulk velocity fields, from 1995 to 2017, covering about two solar cycles. We first calculate the total fluctuation energy density in forward- and backward-propagating modes, the residual energy, and the corresponding correlation functions at 1 au in view of an Elssässer description. Based on these turbulence variables, we further obtain the total fluctuating magnetic field variance and the correlation length as defined in Zank et al. [6, 9], which are essential factors of CR diffusion coefficients calculation.

The method to find the above-mentioned turbulence quantities is similar to that of Zank et al. [27]. The various quantities are calculated for the inwardly and outwardly directed magnetic field separately, because the orientation of the IMF determines whether a mode is forward or backward propagating. The process of calculating various turbulence quantities is as follows: 2

hr interval data containing the X, Y, and Z components of the magnetic field and SW velocity, the proton density, and temperature are taken. The missing data within an interval are discarded, then we check whether the following criteria are satisfied: (i) the mean square fluctuations of the magnetic and bulk velocity field should be smaller than the square of the corresponding mean fields, and (ii) the intervals should contain at least N/2 good data points (N is the total number of data points in one interval). Criterion (i) is used for all three components of magnetic and velocity fields, since it excludes intervals with irregular magnetic and velocity fluctuations. Various features of the SW plasma are speed related. For example, Dasso et al. [12] found that the anisotropy of SW fluctuations differs in fast and slow SW. Since more than 92% of the SW velocity is less than 450 kms^{-1} after classification by criteria (i) and (ii) in our study, we do not consider SW speed bins here. The next step is to find the direction of the background IMF in each retained interval, which is neither along the radial nor the azimuthal direction in the ecliptic plane at 1 au. Here, we employ the minimum variance analysis (MVA) technique to determine the unit vector of the background IMF in each interval from the measured magnetic field vector data. The detailed theory and derivation of the MVA technique can be found in Sonnerup and Scheible [28]. The variance ellipsoid [28, 29], given by MVA technique, defines the variance along an arbitrary chosen direction in the GSE coordinate system. The vector with the smallest variance is close to the local mean field direction, and assumed to be the direction of the background IMF in each interval. Since we consider transverse fluctuations, the observed magnetic field and velocity field components in each interval are projected into the two directions that are perpendicular to the minimum variance direction. The following calculation is carried out in these two directions respectively. We first introduce the definition of the Elsässer variables:

$$\mathbf{z}^{\pm} = \mathbf{u} \pm \frac{\mathbf{b}}{\sqrt{\mu_0 \rho}}, \tag{1}$$

where μ_0 is the magnetic permeability, ρ is the mass density, $\mathbf{u}$ is the fluctuating velocity field, and $\mathbf{b}$ is the fluctuating magnetic field. The Elsässer variables $\mathbf{z}^+$ ($\mathbf{z}^-$) represent the forward (backward) propagating modes with respect to the IMF orientation. We further introduce the following moments of the Elsässer variables $\mathbf{z}^{\pm}$ and the lagged variables $\mathbf{z}^{\pm'}$ [6, 9]:

$$\langle z^{+2} \rangle \equiv \langle \mathbf{z}^+ \cdot \mathbf{z}^+ \rangle = \langle u^2 \rangle + \langle b^2/\mu_0\rho \rangle + 2\langle \mathbf{u} \cdot \mathbf{b}/\sqrt{\mu_0\rho} \rangle; \tag{2}$$

$$\langle z^{-2} \rangle \equiv \langle \mathbf{z}^- \cdot \mathbf{z}^- \rangle = \langle u^2 \rangle + \langle b^2/\mu_0\rho \rangle - 2\langle \mathbf{u} \cdot \mathbf{b}/\sqrt{\mu_0\rho} \rangle; \tag{3}$$

$$E_D \equiv \langle \mathbf{z}^+ \cdot \mathbf{z}^- \rangle = \langle u^2 \rangle - \langle b^2/\mu_0\rho \rangle; \tag{4}$$

$$L^+ \equiv \int \langle \mathbf{z}^+ \cdot \mathbf{z}^{+'} \rangle dr \equiv \langle \mathbf{z}^+ \cdot \mathbf{z}^+ \rangle \lambda^+ = \lambda^+ \langle z^{+2} \rangle; \tag{5}$$

$$L^- \equiv \int \langle \mathbf{z}^- \cdot \mathbf{z}^{-'} \rangle dr \equiv \langle \mathbf{z}^- \cdot \mathbf{z}^- \rangle \lambda^- = \lambda^- \langle z^{-2} \rangle; \tag{6}$$

$$L_D \equiv \int \langle \mathbf{z}^+ \cdot \mathbf{z}^{-'} + \mathbf{z}^{+'} \cdot \mathbf{z}^- \rangle dr \equiv \lambda_D E_D. \tag{7}$$

Here, $\langle z^{+2} \rangle$ and $\langle z^{-2} \rangle$ respectively represent the energy density in forward and backward propagating modes, E_D is the residual energy, L^+, L^- and L_D are the corresponding correlation functions, and λ^+, λ^- and λ_D are the corresponding correlation lengths.

According to CR diffusion theory, we need to calculate the 2D and slab turbulence magnetic energies as input for the CR diffusion coefficients. Here we first calculate the total fluctuating magnetic energy $\langle b^2 \rangle$ and correlation length λ using

$$\langle b^2 \rangle = \frac{\langle z^{+2} \rangle + \langle z^{-2} \rangle - 2E_D}{4} \mu_0 \rho; \tag{8}$$

$$\lambda = \frac{L^+ + L^- - L_D}{\langle z^{+2} \rangle + \langle z^{-2} \rangle - 2E_D}. \quad (9)$$

Then we assume that the total energy in fluctuations can be divided into majority 2D and minority slab energies with a fixed ratio 80:20, and the slab correlation length λ_s is assumed to be twice the 2D correlation scale λ_{2D}, thus $\lambda_s = 2\lambda_{2D} = 2\lambda$.

3. Results

3.1. Solar cycle dependence of turbulence quantities

Figures 1 and 2 show the temporal evolution of the energy density in forward- and backward-propagating modes $\langle z^{\pm 2} \rangle$ and the corresponding correlation functions $L^{\pm}$, the residual energy E_D, and the corresponding correlation functions L_D. Figure 1 shows the data sets for two solar cycles for the outwardly directed IMF, and Figure 2 is for the inwardly directed IMF. All of these turbulence quantities are derived from Equations (2)-(7). From Figure 1 and 2, we can see that all turbulence quantities vary with solar activity and show solar cycle dependence. It is

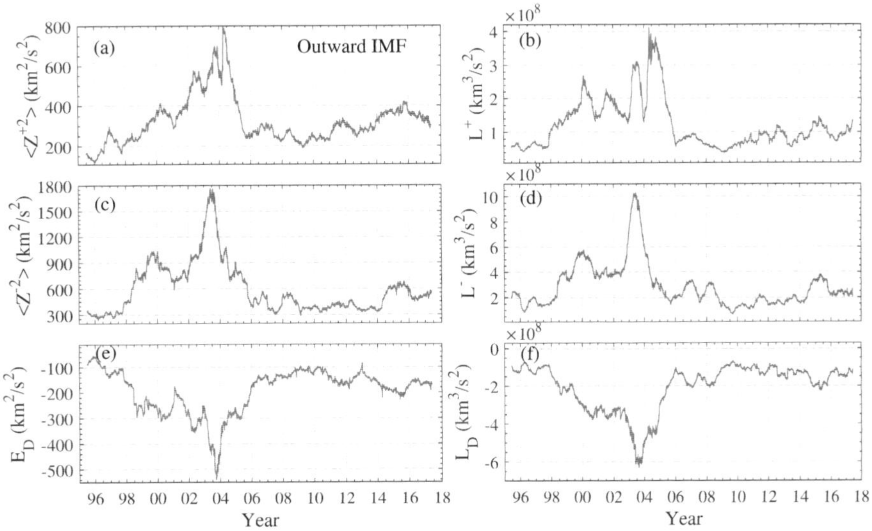

Figure 1. The temporal evolution of (a) the energy density in forward propagating modes $\langle z^{+2} \rangle$ and (b) the corresponding correlation function L^+; (c) the energy density in backward propagating modes $\langle z^{-2} \rangle$ and (d) the corresponding correlation function L^-; and (e) the residual energy E_D and (f) the corresponding correlation function L_D for the outwardly directed IMF.

interesting to notice that the values of the residual energy E_D and the corresponding correlation function L_D are always negative, which illustrates that the energy in magnetic fluctuations is larger than the energy in kinetic fluctuations, especially at solar maximum. The energy densities $\langle z^{\pm 2} \rangle$, residual energy $|E_D|$, and the associated correlation functions $L^{\pm}$ and $|L_D|$ are all positively correlated with solar cycle. For example, the energy density in forward-propagating mode is about 1500 km^2s^{-2} during solar maximum while it is about 360 km^2s^{-2} during solar minimum. The corresponding correlation function is about $9.0 \times 10^8 km^3s^{-2}$ for solar maximum and is about $1.2 \times 10^8 km^3s^{-2}$ during solar minimum. When calculating correlation functions in Equation (5)-(7), we assume the maximum time lag t is 1 hour. Most previous studies use the spatial lag r, where the auto-correlation function of the Elsässer variable becomes $1/e$ of the maximum values, to determine the correlation length. Since the auto-correlation function

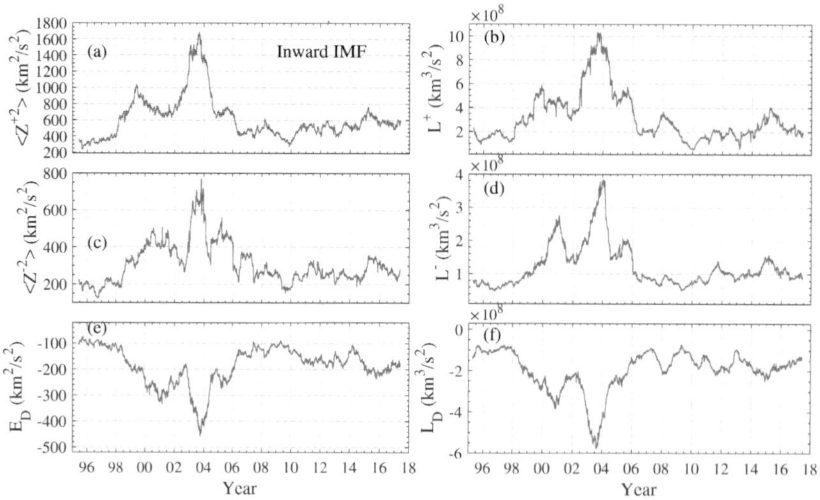

Figure 2. Same as Figure 1, but now for the inwardly directed IMF.

of Elsässer variable drops rapidly, a spatial lag at $1/e$ of the maximum value is closer to zero lag. Therefore, limiting our analysis to a specified maximum time lag seems reasonable.

Figure 3 depicts the influence of solar activity on a critical term in the shear source expression for 2D and slab turbulence transport equations. The monthly mean sunspot number is also

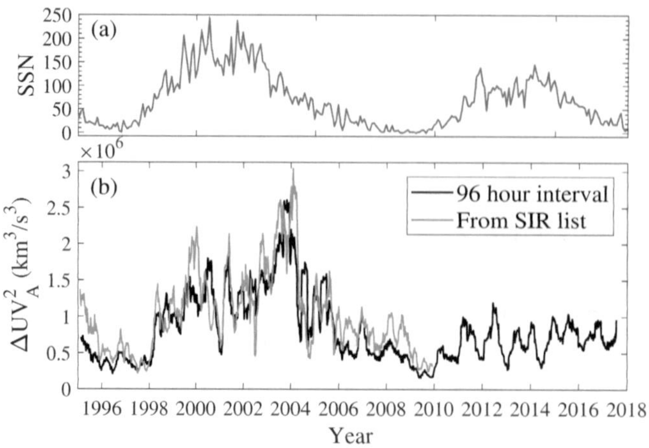

Figure 3. The top panel (a) shows the monthly averaged sunsport number (SSN). Panel (b) shows the temporal evolution of the critical term $\Delta U V_A^2$ for the shear source. The black line represents the result calculated in each 96 hour interval. The green line represents the result calculated in the time interval determined by start time and end time of each shear interaction region (SIR) event during for 1995-2009.

plotted as a reference. The complete form of shear source terms that we consider is given by Zank et al. [9] as

$$S_{\langle z^{\pm 2}\rangle} = C_{sh}^{\pm} \frac{r_0 |\Delta U| V_A^2}{r^2}; \qquad S_{E_D} = C_{sh}^{E_D} \frac{r_0 |\Delta U| V_A^2}{r^2}, \tag{10}$$

IOP Conf. Series: Journal of Physics: Conf. Series **1100** (2018) 012029 doi:10.1088/1742-6596/1100/1/012029

where the parameters $C_{sh}^{\pm}$ and $C_{sh}^{E_D}$ are parameterized strengths of the shear source of energy in forward and backward modes and the residual energy, respectively. $C_{sh}^{+} \neq C_{sh}^{-} \geq 0$, and $C_{sh}^{E_D}$ can be 0, positive, or negative. r_0 is a reference location. ΔU is the difference between fast and slow SW speeds, and V_A is the background Alfv́en speed. In Figure 3, we calculate the product $\Delta U V_A^2$ in each 96 hr interval to include the structure of streams. The green line is calculated in the time interval determined by start time and end time of each stream interaction region (SIR) event during 1995-2009, which were compiled by Jian et al. [30, 31]. Our result agrees well with the SIR list for 1995-2009, and shows a rather clear solar cycle dependence, which indicates that the shear source term of turbulence is also correlated with solar activity.

Figure 4 shows the monthly averaged 2D and slab turbulence magnetic energies and the corresponding correlation lengths respectively calculated from Equation 8 and 9. The red dashed lines are for inwardly directed IMF, and the blue lines are for outwardly directed IMF. We can see that there is little difference between inward and outward IMF modes. We have assumed $\langle b_{2D}^2 \rangle : \langle b_s^2 \rangle = 80\% : 20\%$ and $\lambda_s = 2\lambda_{2D} = 2\lambda$. We can see a clear solar cycle dependence in the turbulence energy. Solar maximum carries more turbulent energy than solar minimum, which is caused by the frequent generation of shocks that generate turbulence [27]. In contrast, the correlation lengths show no discernible solar cycle variation. Wicks et al. [32] studied the spatial correlation properties of solar wind using *ACE* and *Wind* measurements, and found that the correlation length for components of the magnetic field does not show an obvious solar cycle dependence, which is consistent with our results. From Figure 4, we notice that the slab turbulence energy $\langle b_s^2 \rangle$ in year 2003 is four times larger than that in year 2009, and the correlation length λ_s in 2003 is two times larger than in 2009. The averaged value of λ_s is

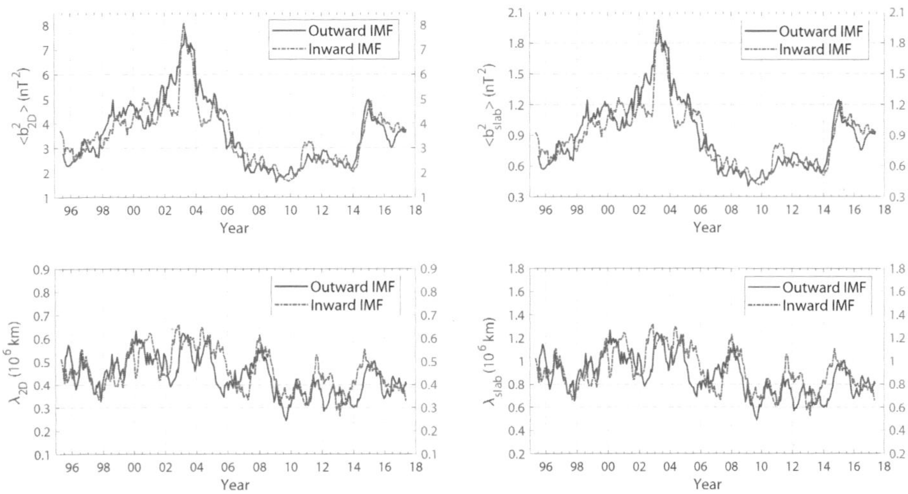

Figure 4. The temporal evolution of monthly averaged 2D turbulence magnetic energy $\langle b_{2D}^2 \rangle$ (left top) and the corresponding correlation length λ_{2D} (left bottom), slab turbulence magnetic energy $\langle b_s^2 \rangle$ (right top) and the corresponding correlation length λ_s (right bottom). A ratio of 2D and slab energy of 80:20 and $\lambda_s = 2\lambda_{2D} = 2\lambda$ has been assumed.

about 0.88×10^6 km. This value is a little smaller than previous studies of correlation lengths for magnetic field fluctuations in the solar wind. For example, Tu and Marsch [33] gave a value of the correlation length for the Elsässer variables as $\lambda \approx 265 R_E \approx 1.7 \times 10^6$ km using single-spacecraft measurements. Matthaeus et al. [34] investigated the value of λ using multiple-spacecraft measurements, and suggested a value of $\lambda \approx 1.3 \times 10^6$ km. However, these previous results didn't consider the effect of solar activity. One can get larger correlation lengths by increasing the length of interval. Shiota et al. [8] also have analyzed extensively the effect on correlation lengths when using longer intervals. They found that a longer interval data set may not be helpful in determining correlations within the inertial range of SW turbulence. This is because a longer interval may introduce some structures associated with large scale flows into the analysis, which then leads to an increase in the correlation functions and correlation lengths. For our analysis, 2 hours is the most suitable length of interval that yields reasonable correlation lengths when compared with *Voyager* 2 observations [10].

3.2. The effects of solar activity on CR diffusion

Figures 5 shows the effects of solar activity on CR parallel $\lambda_\parallel$ (blue line), perpendicular $\lambda_\perp$ (red line), and radial mean free path λ_{rr} (grey line) for a proton with rigidity 445 MV (corresponding to a 100 MeV proton) for the outward IMF mode. As described in Zank et al. [19], the parallel mean free path (mfp) based on standard QLT and assuming magnetostatic turbulence is approximated by

$$\lambda_\parallel = 6.2742\frac{B_0^{5/3}}{\langle b_s^2\rangle}\left(\frac{P}{c}\right)^{1/3}\lambda_s^{2/3}\left[1+\frac{7A/9}{(q+1/3)(q+7/3)}\right], \tag{11}$$

where $q = \frac{5s^2/3}{1+s^2-(1+s^2)^{1/6}}$, $A = (1+s^2)^{5/6}-1$, and $s = 0.746834R_L/\lambda_s$. R_L is the particle Larmor radius; P is the particle rigidity; B_0 is the mean magnetic field strength. The analytic form of the perpendicular mfp based on NLGC theory is given by Shalchi et al. [25]

$$\lambda_\perp = \left[\sqrt{3}\pi a^2 C\frac{\langle b_{2D}^2\rangle}{B_0^2}\lambda_{2D}\right]^{2/3}\lambda_\parallel^{1/3}, \tag{12}$$

where $a^2 = 1/3$ is a factor related to the gyrocenter velocity. $C = \frac{\Gamma(\nu)}{2\sqrt{\pi}\Gamma(\nu-1/2)}$ is a constant such that $\nu = 5/6$ yields a Kolmogorov [36] spectrum. Note that Equation 12 was derived under the assumption of a specific form of 2D wave spectrum, which is a constant at large turbulence scales. It means that the 2D turbulence spectrum is independent of wavenumber in the energy range in Equation 12.

The radial mfp, λ_{rr}, can be expressed as a function of the heliospheric magnetic field winding angle ψ,

$$\lambda_{rr} = \lambda_\parallel\cos^2\psi + \lambda_\perp\sin^2\psi, \tag{13}$$

where $\tan\psi = \frac{B_Y}{B_X}$, and B_X (B_Y) is the averaged X (Y) component of the magnetic field in each interval. The winding angle is not therefore a constant in our calculation, but varies with time. In Figure 5, we employ the monthly averaged 2D and slab turbulence energy and corresponding correlation lengths shown in Figure 4 to calculate CR mfps in the ecliptic plane at 1 au. In this figure, we use a mean value of $B_0 = 5.5$ nT at 1 au to focus on the effects of the small-scale turbulence quantities on CR diffusion coefficients. From Figure 5, we can clearly see the solar cycle dependence of $\lambda_\parallel$, $\lambda_\perp$, and λ_{rr}. The parallel mfp $\lambda_\parallel$ during solar maximum (i.e., 2003) is lower than that during solar minimum (i.e., 2009). In contrast, the perpendicular mfp $\lambda_\perp$ in the solar maximum is higher than that in the solar minimum. $\lambda_\parallel$ is nearly three orders of magnitude larger than the perpendicular mfp $\lambda_\perp$ at 1 au. The radial mfp λ_{rr} is dominated by parallel diffusion $\lambda_\parallel$ in the inner heliosphere [22], and also has a positive correlation with solar activity. Chhiber et al. [21] studied the effect of solar activity on CR parallel and perpendicular diffusion by varying the turbulence energy amplitude only (normal, half, and double turbulence energy) at the inner boundary, and found that increasing the turbulence magnitude leads to a decrease in parallel mfp and an increase in perpendicular mfp, which is in agreement with Figure 5 where we assumed a fixed value of B_0.

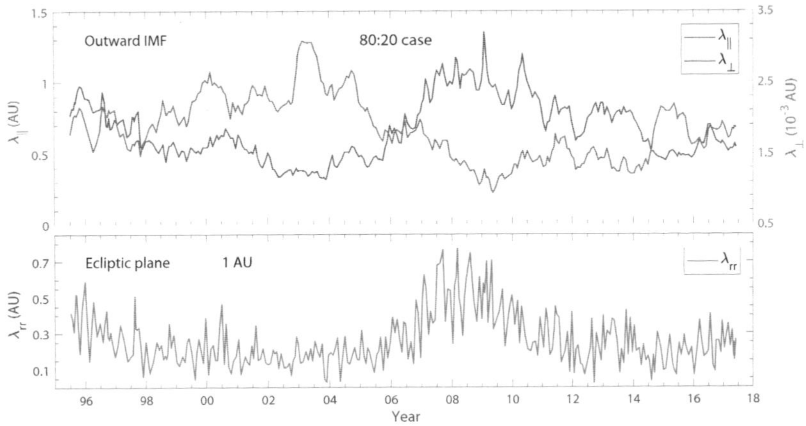

Figure 5. The temporal evolution of CR parallel $\lambda_{\parallel}$ (blue line), perpendicular $\lambda_{\perp}$ (red line), and radial λ_{rr} (grey line) mean free paths for a proton with rigidity 445 MV (corresponding to a 100 MeV proton) in the ecliptic plane at 1 au for outwardly directed IMF. For this example, we have assumed that the background IMF magnitude is constant, $B_0 = 5.5$ nT.

In Figure 6, we use the solar cycle dependent background IMF magnitude obtained from *OMNI* observations to calculate the parallel $\lambda_{\parallel}$ (blue lines), perpendicular $\lambda_{\perp}$ (red lines), and radial λ_{rr} (cyan lines) mfps for the outwardly directed IMF. 2D and slab fluctuation magnetic energies and the corresponding correlation lengths are the same as in Figure 5. By considering the variation of both turbulence quantities and the background IMF magnitude, we find that the solar cycle dependence of CR mean free paths is reduced, since the background IMF magnitude B_0 and turbulence energy $\langle b_s^2 \rangle$ (or $\langle b_{2D}^2 \rangle$) each have the opposite effect on $\lambda_{\parallel}$ (or $\lambda_{\perp}$), as shown by Equation 11 and 12. For example, for a fixed value of background IMF, an increase in $\langle b_s^2 \rangle$ leads to a decrease in $\lambda_{\parallel}$ as shown in Figure 5, while for a fixed $\langle b_s^2 \rangle$, an increase in B_0 leads to an increase in $\lambda_{\parallel}$. Therefore, when we take into account both effects, the solar cycle dependence of $\lambda_{\parallel}$ is somewhat masked but still identifiable. There is an anti-correlation between the parallel mfp $\lambda_{\parallel}$ and perpendicular mfp $\lambda_{\perp}$. In this figure, the averaged value of $\lambda_{\parallel}$ is about 0.9 AU, and the averaged value of $\lambda_{\perp}$ is about 1.6×10^{-3} AU.

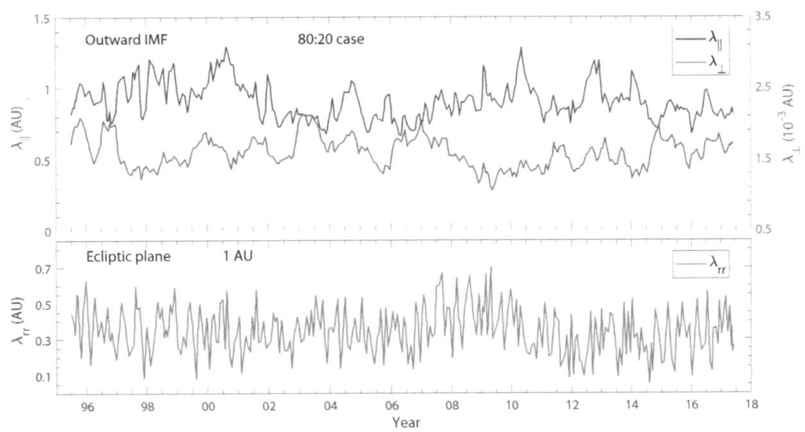

Figure 6. The same as Figure 5, but now for solar cycle dependent background IMF magnitude obtained from the observation.

4. Summary

In this article, we present a detailed analysis of various turbulence quantities and CR mean free paths from 1995-2017 to investigate the effects of solar cycle. Compared to previous empirical time-dependent solutions [7], we use *OMNI* 1-minute resolution data over the last two solar cycles to study the temporal evolution of various turbulence quantities at 1 au for inward and outward IMF modes respectively. The fluctuating magnetic energy densities $\langle z^{\pm 2}\rangle$ and the residual energy E_D all exhibit an obvious solar cycle dependence. This is consistent with previous studies [37, 38] that show a solar cycle dependence for magnetic energy density fluctuations within the inertial range, using *Wind* and *ACE* measurements. The value of $\langle z^{+2}\rangle$ for the inward IMF mode is approximately equal to $\langle z^{-2}\rangle$ for the outward IMF mode, and vice versa, which can be explained from their definitions (Equation 2 and 3). The residual energy E_D, which represents the energy difference between velocity fluctuations and magnetic field fluctuations (Equation 4), is always negative for both inward IMF and outward IMF modes. This illustrates that the energy in magnetic fluctuations is larger than the energy in kinetic fluctuations, especially at solar maximum (Figures 1(e) and 2(e)).

We introduce the correlation function L^+, L^-, and L_D for the Elsässer variables $\mathbf{z}^{\pm}$ and the lagged variables $\mathbf{z}^{\pm'}$ to calculate the correlation length λ of magnetic field fluctuations through Equation 9. We find that the correlation functions evolve similarly to the corresponding energy densities $\langle z^{\pm 2}\rangle$ or the residual energy E_D, and are also correlated with solar cycle. It indicates that the correlation length λ for magnetic fluctuations may not be sensitive to solar cycle. As shown in Figure 4, the correlation length for slab turbulence λ_s (assuming $\lambda_s = 2\lambda_{2D} = 2\lambda$), of which the averaged value is about 0.88×10^6 km, does not show significant solar cycle variation. One can increase the value of the correlation length by using a longer interval length, but it may introduce some structures associated with large-scale flows [8]. Magnetic turbulence energy in 2D $\langle b_{2D}^2\rangle$ and slab $\langle b_s^2\rangle$ fluctuations is positively correlated with solar cycle. The magnetic turbulence energy $\langle b^2\rangle$ in year 2003 (solar maximum) is almost 4 times larger than in 2009 (solar minimum), and also 1.5 times larger than in 2015 (weak solar maximum). There is little difference between inward and outward IMF modes in the evolution of these turbulence quantities, which illustrates that the small-scale SW fluctuations may not depend on the direction of the background magnetic field. Finally, the temporally varying shear source of turbulence, which is most important in the inner heliosphere, also depends on solar cycle.

The above solar cycle dependent turbulence quantities at 1 au can be used as input for both turbulence transport models, which in turn describe the radial evolution of such quantities, and CR diffusion coefficient models. To investigate the solar cycle variation of CR parallel and perpendicular diffusion, one should consider the variation of both turbulence quantities and background large-scale IMF magnitude. We employ quasi-linear theory and nonlinear guiding center theory to model the CR parallel mean free path (mfp) $\lambda_\parallel$ and the perpendicular mfp $\lambda_\perp$, respectively. The temporal evolution of $\lambda_\parallel$, $\lambda_\perp$ and λ_{rr} in the ecliptic plane at 1 au over the recent two solar cycles is calculated. The parallel mfp $\lambda_\parallel$ is about three orders of magnitude larger than the perpendicular mfp $\lambda_\perp$, and dominates the radial mfp λ_{rr} in the inner heliosphere. When we use a fixed value of the IMF, the CR mfps all have a clear solar cycle dependence. The parallel mfp $\lambda_\parallel$ and radial mfp λ_{rr} are anti-correlated with solar cycle while the perpendicular mfp $\lambda_\perp$ is positively correlated with solar cycle. This illustrates that high levels of solar wind fluctuations will increase CR perpendicular diffusion and decrease parallel diffusion if the background IMF strength remains unchanged. However, if the background IMF changes in concert with turbulence levels in response to solar activity, this trend can be masked.

In this study, we present the effect of solar activity on various turbulence quantities and the related cosmic ray diffusion coefficients by analyzing magnetic field and bulk velocity field observed by *OMNI* from 1995-2017. These results provide quantitative input for both cosmic ray propagation models and turbulence transport models, and also provide useful insight for time-dependent turbulence modeling to understand the long-term cosmic ray modulation which requires turbulence quantities over at least one solar cycle. The solar cycle influence on the global properties of turbulence beyond 1 au still needs further investigation.

Acknowledgments
We acknowledge the partial support of the NSF EPSCoR RII-Track-1 Cooperative Agreement OIA-1655280, NSF-DOE award 1707247, NASA grants NNX08AJ33G, Subaward 37102-2, NNX14AC08G, NNX14AJ53G, A99132BT, RR185-447/4944336 and NNX12AB30G. G.P.Z. is partly supported by the International Space Science Institute (ISSI), both through the award of the 2017 Johannes Geiss Fellowship and in the framework of an International Team Project 504 entitled "Current Sheets, Turbulence, Structures and Particle Acceleration in the Heliosphere". Q.H. acknowledge partial support from SAO subcontract SV4-84017 and NASA grant NNX17AB85G.

References
[1] Jokipii J R 1966 *ApJ* **146** 480
[2] Florinski V, Zank G P and Pogorelov N V 2003 *J. Geophys. Res.* **108** 1228
[3] Engelbrecht N E and Burger R A 2013 *ApJ* **772** 46
[4] Guo X and Florinski V 2016 *ApJ* **826** 65
[5] Zhou Y and Matthaeus W H 1990 *J. Geophys. Res.* **95** 10291
[6] Zank G P, Dosch A, Hunana P, Florinski V, Matthaeus W H and Webb G M 2012 *ApJ* **745** 35
[7] Adhikari L, Zank G P, Hu Q and Dosch A 2014 *ApJ* **793** 52
[8] Shiota D, Zank G P, Adhikari L, Hunana P, Telloni D and Bruno R 2017 *ApJ* **837** 75
[9] Zank G P, Adhikari L, Hunana P, Shiota D, Bruno R and Telloni D 2017 *ApJ* **835** 147
[10] Adhikari L, Zank G P, Hunana P, Shiota D, Bruno R, Hu Q and Telloni D 2017 *ApJ* **841** 85
[11] Adhikari L, Zank G P, Telloni D, Hunana P, Bruno R and Shiota D 2017 *ApJ* **851** 117
[12] Dasso S, Milano L J, Matthaeus W H and Smith C W 2005 *ApJL* **635** L181
[13] Oughton S, Matthaeus W H, Wan M and Osman K T 2015 *Phil. Trans. R. Soc. A* **373** 20410152
[14] Zank G P and Matthaeus W H 1992 *J. Geophys. Res.* **97** 17189
[15] Bieber J W, Matthaeus W H, Shalchi A and Qin G 2004 *GeoRL* **31** L10805
[16] Hunana P and Zank G P 2010 *ApJ* **718** 148
[17] Zank G P, Adhikari L, Hunana P, Tiwari S K, Moore R, Shiota D, Bruno R and Telloni D 2018 *ApJ* **854** 32
[18] Weygand J M, Matthaeus W H, Dasso S and Kivelson M G 2011 *J. Geophys. Res.* **116** A08102
[19] Zank G P, Matthaeus W H and Bieber J W 1998 *J. Geophys. Res.* **103** 2085
[20] Pei C, Bieber J W, Breech B, Burger R A, Clem J and Matthaeus W H 2010 *J. Geophys. Res.* **115** A03103
[21] Chhiber R, Subedi P, Usmanov A V, Matthaeus W H, Ruffolo D, Goldstein M L and Parashar T N 2017 *ApJ* **230** 21
[22] Zhao L L, Adhikari L, Zank G P, Hu Q and Feng X S 2017 *ApJ* **849** 88
[23] Matthaeus W H, Qin G, Bieber J W and Zank G P 2003 *ApJL* **590** L53
[24] Shalchi A 2005 *Physics of Plasmas* **12** 052324
[25] Shalchi A, Li G and Zank G P 2010 *Astrophysics and Space Science* **325** 99
[26] Ruffolo D, Painpanit T, Matthaeus W H and Chuychai P 2012 *ApJ* **747** L34
[27] Zank G P, Matthaeus W H and Smith C W 1996 *J. Geophys. Res.* **101** 17093
[28] Sonnerup B U Ö and Scheible M 1998 *ISSI Scientific Reports Series* **1** 185
[29] Zhao L L, Adhikari L, Zank G P, Hu Q and Feng X S 2018 *ApJ* **856** 94
[30] Jian L, Russell C T, Luhmann J G and Skoug R M 2006 *Solar Physics* **239** 337
[31] Jian L, Russell C T and Luhmann J G 2011 *Solar Physics* **274** 321
[32] Wicks R T, Chapman S C and Dendy R O 2009 *ApJ*, 690, 734
[33] Tu C Y and Marsch E 1995 *Space Science Reviews* **73** 1
[34] Matthaeus W H, Dasso S, Weygand J M, Milano L J, Smith C W and Kivelson M G 2005 *Physical Review Letters* **95** 231101
[35] Shalchi A, Li G and Zank G P 2010 *Astrophysics and Space Science* **325** 99
[36] Kolmogorov A 1941 *Akademiia Nauk SSSR Doklady* **30** 301
[37] Kiyani K, Chapman S C, Hnat B and Nicol R M 2007 *Physical Review Letters* **98** 211101
[38] Hnat B, Chapman S C, Kiyani K, Rowlands G and Watkins N W 2007 *Geophysical Research Letters* **34** L15108